A
FIRST
COURSE
IN
ABSTRACT
ALGEBRA

third edition

John B. Fraleigh

Department of Mathematics
University of Rhode Island

ADDISON-WESLEY
PUBLISHING COMPANY
Reading, Massachusetts
Menlo Park, California • London
Amsterdam • Don Mills, Ontario
Sydney

Library of Congress Cataloging in Publication Data

Fraleigh, John B.
  A first course in abstract algebra.

  Bibliography: p.
  Includes index.
  1. Algebra, Abstract.   I. Title.
QA162.F7   1982   512′.02   81-14938
ISBN 0-201-10405-9   AACR2

Reprinted with corrections, November 1982

ISBN 0-201-10405-9
ABCDEFGHIJK-HA-898765432

To the memory of my father
PERCY A. FRALEIGH

Preface
to the
Third
Edition

As in the previous editions, my aim remains to teach students as much about groups, rings, and fields as is possible n a first course.

The four starred chapters dealing with algebraic topology in the previous editions have been eliminated. I feel those sections were rarely covered in class. Copies of earlier editions are available in libraries and on many bookshelves. Anyone who now wishes to read the brief, intuitive introduction to algebraic topology may hunt them up.

Some instructors objected to the omission of proofs in previous editions, where important theorems of group theory were simply stated in unstarred sections. Accordingly, I have inserted starred sections that prove those theorems. I have also included chapters on group action on a set, followed by applications to Burnside counting and the Sylow theorems, with complete proofs. An appendix on mathematical induction has been included.

I have added some exercises. In response to comments, I have omitted answers to even-numbered exercises, as well as to any exercises that ask for proofs. Exercises on proofs are rather pointless if the proofs are only thirty seconds away.

I am gratified at the response to the first and second editions, not only from undergraduates and faculty but from graduate students preparing for their comprehensive examinations. I hope the third edition continues to be helpful.

Over the years, I have received many suggestions and corrections of errors, for which I am very grateful. I especially want to thank George Bergman, who sent me twelve closely typed pages of comments and suggestions and also some supplementary material, based on his classroom experience with the text. His opinions greatly influenced this revision.

J.B.F.

Kingston, Rhode Island
December 1981

Preface
to the
First
Edition

The primary objective of this work is to provide a text from which *average* students of mathematics can acquire as much depth and comprehension in their study of abstract algebra, exclusive of linear algebra, as is possible in a first course. Since algebra frequently provides a student's first encounter with an abstract mathematical discipline, a secondary objective is to sow the seeds from which a modern mathematical attitude may grow. Mastery of this text should constitute a firm foundation for more specialized work in algebra and also should be of great help in any further axiomatic study of mathematics.

In line with the stated secondary objective, the text begins with an introductory section dealing with the role of definitions in mathematics. This role is too seldom mentioned. To emphasize the importance of definitions, throughout the text each term appears in boldface type in its definition.

Part I is concerned with groups. The study of groups, and indeed all the work of the text, is motivated as much as possible in terms of the student's past experience with algebra. The important concept of a factor group is often a difficult one for a student to grasp. Accordingly, the treatment of factor groups and homomorphisms is delayed until the student has had time to acquire a little feeling for the group concept, and then the discussion is gradual and detailed.

In unstarred portions of Part I, several important results are presented with considerable discussion and many examples, but without proof. It is my feeling that in view of the vast body of mathematics it is important to train students to understand and use established results without feeling that they must first carefully check every detail of the proofs. Of course, professional mathematicians have been doing this for years. Such a policy is also in line with my objective of achieving a little depth in algebra, particularly since in many schools only a single semester is allotted to the study of the subject matter of this text.

Part II deals with rings and fields. Every effort is made to point out analogies with the previous study of groups. The main emphasis of Part II is on field theory, leading up to and including Galois theory. The treatment of vector spaces is very brief, designed only to develop the concepts of linear independence and dimension, which are needed for field theory. Since students often find field theory quite difficult, I have attempted to make the treatment gradual and always to make it clear what we are trying to accomplish and how we intend to do it.

Properties of the rational integers with which students are familiar, even though they may never have seen rigorous justifications of them, are used throughout the text without comment or apology. I have found that average students have difficulty understanding the reason for undertaking a formal study of results they have known for years. After acquiring an overall picture of the nature of algebraic structures, students will come to view these properties in a new light. Also, this treatment is consistent with my primary objective of attempting to achieve some depth in a first course.

In view of my desire to have students learn as much *algebra* as possible, I have chosen to treat set-theoretic material very intuitively and only as needed. There are two ways to acquire a working knowledge of set-theoretic material: to study it *per se,* and to plunge in and use it as it is needed. It is my experience that students find a study of "set-theoretic prerequisites" at the start of an algebra course the most discouraging part of the whole course. My approach in this matter is characteristic of my willingness throughout the text to sacrifice elegance of mathematical presentation, and sometimes even of language, for comprehension in this first course.

The text contains sufficient material for a two-semester course with an average class. *However, the sections not marked with a star were specifically designed to constitute a one-semester course. These sections are self-contained; no use is made in them of starred material. They represent my attempt to present material of some depth in algebra, including Galois theory, to an average class in a single semester.* Of course, a variety of other one-semester courses can be formed from the subject matter at hand. Some starred chapters are quite suitable for assignment as outside reading, especially Chapters 10, 37, 39, and 48. If there is not time to finish the field theory in the text, either Chapter 35, containing Kronecker's theorem, which is placed in the limelight, or Chapter 39 is a satisfying terminal section. In my opinion, there is little to be gained in starting Chapter 40 if there is not time to finish the unstarred material.

The exercises at the end of a chapter are usually divided into two groups by a horizontal rule. Those above the rule are recommended for the average class and would probably be assigned by the author to his classes at U.R.I. In an attempt to make the students' transition to abstract mathematics as easy as possible, these exercises in the first group have been made largely computational. Average students are completely lost when faced with a set of exercises all starting with the words *prove* or *show*. Of course, training in proofs is important. The first group of exercises usually contains one marked with a dagger, that asks for a proof. It is the author's policy to collect these marked exercises, to read them, and to make the students rewrite them, if necessary, to try to train them to write mathematics and not nonsense. The exercises in the second group usually include many asking for proofs, as well as some additional ones of a computational nature. *A star on an exercise does not denote difficulty, but rather indicates that the exercise depends on some starred material of the text.* In view of my desire to promote a good mathematical attitude, some exercises, especially near the beginning of the text, are of a somewhat metamathematical nature. Answers for, or comments concerning, almost all exercises that do not ask for proofs are given at the back of the text. Proofs requested in exercises are not given in the answers; I do not think that it is sound pedagogy to have such proofs so readily available.

A mimeographed first draft of this text was used in the spring semester, 1966, at the University of Rhode Island. I wish to express here my debt to George E. Martin, who taught one of the sections of the course. His comments and suggestions were of great value to me in the preparation of this version for publication.

Kingston, Rhode Island   J.B.F.
September 1966

Contents

A Very Few Preliminaries

0.1 THE ROLE OF DEFINITIONS

Many students do not realize the great importance of definitions to mathematics. This importance stems partly from the need for mathematicians to communicate with each other regarding their work. If two people who are trying to communicate on some subject have different ideas as to what is meant by certain technical terms, there can be misunderstanding, friction, and perhaps even bloodshed. Imagine the predicament of a butcher facing an irate customer who is trying to buy what most people call a rib roast of beef, but who insists on calling it a bottom round roast. The utopian ideal of complete standardization of terminology unfortunately does not seem to be achievable, even among such precise beings as mathematicians. For example, when speaking of functions in mathematics, various mathematicians have given the term *range* two distinct meanings. Consequently, there is a tendency now to avoid using this ambiguous term and to use either *image* or *codomain* in its place. *In mathematics, one should strive to avoid ambiguity.*

A very important ingredient of mathematical creativity is the ability to formulate useful definitions, ones that will lead to interesting results. A

1

mathematics student commencing graduate study may find that he or she spends a great deal of time discussing definitions with other graduate students. When the author was in graduate school, a physics graduate student once complained to him that at the evening meal the mathematics students always sat together and argued, and that the subject of their argument was always a definition. Graduate students are usually asked to give several definitions on oral examinations. If they cannot explain the meaning of a term, they probably cannot give sensible answers to questions involving that concept.

Every definition is understood to be an *if and only if* type of statement, even though it is customary to suppress the *only if*. Thus one may define: "A triangle is **isosceles** if it has two sides of equal length," really meaning that a triangle is isosceles if and only if it has two sides of equal length. Now you must not feel that you should memorize a definition word for word. The important thing is to *understand* the concept so that you can define precisely the same concept in your own words. Thus the definition "An **isosceles** triangle is one having two equal sides" is perfectly correct. The definition "An **isosceles** triangle is one having two equal angles" is also correct, for exactly the same triangles are called isosceles in all these definitions.

It is very important for you to note that if some concept has just been defined and you are asked to prove something concerning the concept, you *must* use the definition as an integral part of the proof. *Immediately after a concept is defined, the definition is the only information one has available regarding the concept.*

Throughout the text, a term that appears in **boldface** type is being *defined* at that point. Specifically labeled definitions are used for the main algebraic concepts with which we are concerned. Many other terms are defined, using the boldface convention, outside a labeled definition. You will find ideas defined in this fashion in text paragraphs, theorems, and exercises.

0.2 SETS

The basic importance of definitions to mathematics is also a structural weakness for the reason that not every concept used can be defined. Suppose, for example, one defines the term *set* by "A **set** is a well-defined collection of objects." One naturally asks what is meant by a *collection*. Perhaps then one defines: "A **collection** is an aggregate of things." What then is an *aggregate*? Now our language is finite, so after some time we will run out of new words to use and have to repeat some words already questioned. The definition is then circular and obviously worthless. Mathematicians realize that there must be some undefined or primitive concept. At the moment they have agreed that *set* shall be such a primitive concept. We shall not define *set*, but shall just hope that when such expressions as "the set of all real numbers" or "the set of all members of the United States Senate" are used, people's various ideas of what is meant are sufficiently similar to make communication feasible.

We summarize briefly some of the things we shall simply assume about sets.

1. A set S is made up of **elements**, and if a is one of these elements, we shall denote this fact by $a \in S$.
2. There is exactly one set with no elements. It is the **empty set** and is denoted by $\varnothing$.
3. We may describe a set either by giving a characterizing property of the elements, such as "the set of all members of the United States Senate," or by listing the elements. The standard way to describe a set by listing elements is to enclose the designations of the elements, separated by commas, in braces, for example, $\{1, 2, 15\}$. If a set is described by a characterizing property $P(x)$ of its elements x, the brace notation $\{x \mid P(x)\}$ is also often used, and is read "the set of all x such that the statement $P(x)$ about x is true." Thus

$$\{2, 4, 6, 8\} = \{x \mid x \text{ is an even whole positive number} \leq 8\}$$
$$= \{2x \mid x = 1, 2, 3, 4\}.$$

4. A set is **well defined,** meaning that if S is a set and a is some object, then either a is definitely in S, denoted by $a \in S$, or a is definitely not in S, denoted by $a \notin S$. Thus one should never say, "Consider the set S of some positive numbers," for it is not definite whether $2 \in S$ or $2 \notin S$. On the other hand, one can consider the set T of all prime positive integers. Every positive integer is definitely either prime or not prime. Thus $5 \in T$ and $14 \notin T$. It may be hard to actually determine whether an object is in a set. For example, as this book goes to press it is unknown whether $2^{(2^{17})} + 1$ is in T. However, $2^{(2^{17})} + 1$ is certainly either prime or not prime.

It will not be feasible for the student for whom this text is intended to push every definition back to the concept of a set. The author is well aware that he is building on some very naive definitions, especially at the beginning of the text. The first definition we will meet in Chapter 1 says, "A **binary operation on a set** is a rule . . . set." What on earth is a rule?

Throughout this text, much work will be done involving familiar sets of numbers. Let us take care of notation for these sets once and for all:

Z is the set of all integers (that is, whole numbers: positive, negative, and zero).

$\mathbf{Z}^+$ is the set of all positive integers. (Zero is excluded.)

Q is the set of all rational numbers (that is, numbers that can be expressed as quotients m/n of integers, where $n \neq 0$).

$\mathbf{Q}^+$ is the set of all positive rational numbers.

R is the set of all real numbers.

$\mathbf{R}^+$ is the set of all positive real numbers.

C is the set of all complex numbers.

0.3 PARTITIONS AND EQUIVALENCE RELATIONS

We just described **Q** as the set of all numbers that can be expressed as quotients m/n of integers, where $n \neq 0$. It would be incorrect to describe **Q** as the set S of all "quotient expressions" m/n for m and n in **Z** and $n \neq 0$. For surely $\frac{2}{3}$ and $\frac{4}{6}$ are distinguishable quotient expressions, but we know they represent the *same* rational number. In fact, each element of **Q** is represented by an infinite number of different elements of S. When doing arithmetic, we *identify* in our mind elements of S that represent the same rational number in **Q**.

The illustration in the preceding paragraph is typical of several situations in which we will consider different elements of a set to be arithmetically or algebraically equivalent, so that our set becomes *partitioned* into cells, each of which we may consider to be a single arithmetic or algebraic entity. If b is an element of such a set, we usually let $\overline{b}$ represent the cell of all elements being identified with b. For example, in our previous set S of formal quotients, we have

$$\overline{2/3} = \left\{ \frac{2}{3}, \frac{-2}{-3}, \frac{4}{6}, \frac{-4}{-6}, \frac{6}{9}, \frac{-6}{-9}, \cdots \right\}$$

$$= \left\{ \frac{2n}{3n} \middle| n \in \mathbf{Z} \text{ and } n \neq 0 \right\}.$$

Let us give a precise definition of such a partition.

Definition   A *partition of a set* is a decomposition of the set into cells such that every element of the set is in *exactly one* of the cells.

Two cells (or sets) having no elements in common are **disjoint.** Thus the cells in a partition of a set are disjoint.

How do we know whether two quotient expressions m/n and r/s in our set S above are in the same cell, that is, represent the same rational number? One way to decide is to reduce both fractions to lowest terms. This may not be easy to do; for example, 1909/4897 and 1403/3599 represent the same rational number since

$$\frac{1909}{4897} = \frac{23 \cdot 83}{59 \cdot 83} \quad \text{and} \quad \frac{1403}{3599} = \frac{23 \cdot 61}{59 \cdot 61}.$$

However, finding these factorizations, even with a hand calculator, is a somewhat tedious trial-and-error task. But as you know, in fraction arithmetic we have $m/n = r/s$ if and only if $ms = nr$. This gives us a more efficient criterion for our problem, namely

$$(1909)(3599) = (4897)(1403) = 6870491.$$

Let $a \sim b$ denote that a is in the same cell as b for a given partition of a set containing both a and b. Clearly the following properties are always satisfied:

$a \sim a$. The element a is in the same cell as itself.

If $a \sim b$ *then* $b \sim a$. If a is in the same cell as b, then b is in the same cell as a.

If $a \sim b$ *and* $b \sim c$, *then* $a \sim c$. If a is in the same cell as b and b is in the same cell as c, then a is in the same cell as c.

The theorem that follows is fundamental. It asserts that a relation $\sim$ between elements of a set that satisfies the three properties just described yields a natural partition of the set. Exhibiting a relation with these properties is frequently the most economical way of describing a partition of a set, and it is for this reason we are discussing this material now.

Theorem 0.1 *Let S be a nonempty set and let $\sim$ be a relation between elements of S that satisfies the following properties.*

1. *(Reflexive)* $a \sim a$ *for all* $a \in S$.
2. *(Symmetric) If* $a \sim b$, *then* $b \sim a$.
3. *(Transitive) If* $a \sim b$ *and* $b \sim c$, *then* $a \sim c$.

Then $\sim$ *yields a natural partition of S, where*

$$\bar{a} = \{x \in S \,|\, x \sim a\}$$

is the cell containing a for all $a \in S$. Conversely, each partition of S gives rise to a natural relation $\sim$ satisfying the reflexive, symmetric, and transitive properties if $a \sim b$ is defined to mean that $a \in \bar{b}$.

Proof. We have already proved the "converse" part of the theorem.

For the direct statement, it only remains to show that the cells defined by $\bar{a} = \{x \in S \,|\, x \sim a\}$ do constitute a partition of S, that is, that every element of S is in *exactly one* cell. Let $a \in S$. Then $a \in \bar{a}$ by condition 1, so a is in *at least one* cell.

Suppose now that a were in a cell $\bar{b}$ also. We need to show that $\bar{a} = \bar{b}$ as sets; this would show that a can't be in more than one cell. We do this by showing that each element of $\bar{a}$ is in $\bar{b}$ and each element of $\bar{b}$ is in $\bar{a}$. Let $x \in \bar{a}$. Then $x \sim a$. But $a \in \bar{b}$, so $a \sim b$. Then, by the transitive condition (3), $x \sim b$ so $x \in \bar{b}$. Thus $\bar{a}$ is part of $\bar{b}$. Now let $y \in \bar{b}$. Then $y \sim b$. But $a \in \bar{b}$, so $a \sim b$ and, by symmetry (2), $b \sim a$. Then by transitivity, $y \sim a$ so $y \in \bar{a}$. Hence $\bar{b}$ is part of $\bar{a}$ also, so $\bar{b} = \bar{a}$ and our proof is complete. ∎

> **Definition**   A relation $\sim$ on a set S satisfying the reflexive, symmetric, and transitive properties described in Theorem 0.1 is an ***equivalence relation on*** S. Each cell $\bar{a}$ in the natural parition given by an equivalence relation is an ***equivalence class.***

The symbol $\sim$ is usually reserved for an equivalence relation. We will use $\mathscr{R}$ for a relation between elements of a set S which is not necessarily an equivalence relation on S.

The term *natural*, appearing twice in Theorem 0.1, has the following significance. If you start with an equivalence relation, form the partition of equivalence classes, and then consider the relation given by this partition, it is your original equivalence relation. Similarly, starting with a partition, going to the equivalence relation, and then forming the equivalence classes yields the original partition.

Example 0.1   Let us verify directly that

$$m/n \sim r/s \text{ if and only if } ms = nr$$

is an equivalence relation on the set S of formal quotient expressions we considered earlier.

Reflexive. $m/n \sim m/n$ since $mn = nm$.

Symmetric. If $m/n \sim r/s$, then $ms = nr$. Consequently, $rn = sm$ so $r/s \sim m/n$.

Transitive. If $m/n \sim r/s$ and $r/s \sim u/v$, then $ms = nr$ and $rv = su$. Reordering terms and substituting, we obtain $mvs = vms = vnr = nrv = nsu = nus$. Since $s \neq 0$, we deduce that $mv = nu$, so $m/n \sim u/v$.

Each equivalence class of S is considered to *be* a rational number. ∎

Our discussion of the set S of formal quotient expressions, culminating in Example 0.1, is a special case of work we will do in Chapter 26.

Example 0.2 Let us define a relation $\mathcal{R}$ on the set $\mathbf{Z}$ by $n\,\mathcal{R}\,m$ if and only if $nm \geq 0$, and determine whether $\mathcal{R}$ is an equivalence relation.

Reflexive. $a\,\mathcal{R}\,a$, since $a^2 \geq 0$ for all $a \in \mathbf{Z}$.

Symmetric. If $a\,\mathcal{R}\,b$, then $ab \geq 0$, so $ba \geq 0$ and $b\,\mathcal{R}\,a$.

Transitive. If $a\,\mathcal{R}\,b$ and $b\,\mathcal{R}\,c$, then $ab \geq 0$ and $bc \geq 0$. Thus $ab^2c = acb^2 \geq 0$. If we knew $b^2 > 0$, we could deduce $ac \geq 0$ whence $a\,\mathcal{R}\,c$. We have to examine the case $b = 0$ separately. A moment of thought shows $-3\,\mathcal{R}\,0$ and $0\,\mathcal{R}\,5$ but $-3\,\not\mathcal{R}\,5$, so the relation $\mathcal{R}$ is not transitive, and hence not an equivalence relation. ∎

For each $n \in \mathbf{Z}^+$ we have a very important equivalence relation on $\mathbf{Z}$, **congruence modulo** n. For $h,\ k \in \mathbf{Z}$, we define h **congruent to** k **modulo** n, written $h \equiv k\ (\mathrm{mod}\ n)$, if $h - k$ is evenly divisible by n, so that $h - k = ns$ for some $s \in \mathbf{Z}$. For example, $17 \equiv 33\ (\mathrm{mod}\ 8)$, since $17 - 33 = 8(-2)$. Equivalence classes for congruence modulo n are **residue classes modulo** n. Each of these residue classes contains an infinite number of elements. For example, with our congruence modulo 8, you can easily convince yourself that the residue class containing both 17 and 33 is

$$\{\cdots, -47, -39, -31, -23, -15, -7, 1, 9, 17, 25, 33, 41, 49, \cdots\}.$$

This residue class contains every eighth number, starting with 1. There are actually seven more residue classes in the partition given by congruence modulo 8.

We ask you to show that congruence modulo n is indeed an equivalence relation and to look at some more residue classes in Exercise 0.18.

Exercises

In Exercises 1 through 4, describe the set by listing its elements.

0.1 $\{x \in \mathbf{R} \mid x^2 = 3\}$ 　　　　　　　　 **0.2** $\{m \in \mathbf{Z} \mid m^2 = 3\}$

0.3 $\{m \in \mathbf{Z} \mid mn = 60$ for some $n \in \mathbf{Z}\}$ 　　 **0.4** $\{m \in \mathbf{Z} \mid m^2 - m < 115\}$

In Exercises 5 through 10, decide whether the object described is indeed a set (is well defined). Give an alternate description of each set.

0.5 $\{n \in \mathbf{Z}^+ \mid n$ is a large number$\}$

0.6 $\{n \in \mathbf{Z} \mid n^2 < 0\}$

0.7 $\{n \in \mathbf{Z} \mid 39 < n^3 < 57\}$

0.8 $\{x \in \mathbf{Q} \mid$ the denominator of x is greater than 100$\}$

0.9 $\{x \in \mathbf{Q} \mid x$ may be written with denominator greater than 100$\}$

0.10 $\{x \in \mathbf{Q} \mid x$ may be written with denominator less than 3$\}$

In Exercises 11 through 19, determine whether the given relation is an equivalence relation on the set. Describe the partition arising from each equivalence relation.

0.11 $n \mathcal{R} m$ in $\mathbf{Z}$ if $nm > 0$ **0.12** $x \mathcal{R} y$ in $\mathbf{R}$ if $x \geq y$

0.13 $x \mathcal{R} y$ in $\mathbf{R}$ if $|x| = |y|$ **0.14** $x \mathcal{R} y$ in $\mathbf{R}$ if $|x - y| \leq 3$

0.15 $n \mathcal{R} m$ in $\mathbf{Z}^+$ if n and m have the same number of digits in the usual base ten notation.

0.16 $n \mathcal{R} m$ in $\mathbf{Z}^+$ if n and m have the same final digit in the usual base ten notation.

0.17 $n \mathcal{R} m$ in $\mathbf{Z}^+$ if $n - m$ is evenly divisible by 2.

0.18 Let n be a particular integer in $\mathbf{Z}^+$. Show that congruence modulo n is an equivalence relation on $\mathbf{Z}$. Describe the residue classes for $n = 1, 2, 3$.

0.19 The following is a famous incorrect argument. Find the error. "The reflexive criterion is redundant in the conditions for an equivalence relation, for from $a \sim b$ and $b \sim a$ (symmetry) we deduce $a \sim a$ by transitivity."

In Exercises 20 through 24, find the number of different equivalence relations on a set S having the given number of elements. (Make use of Theorem 0.1.)

0.20 1 element   **0.21** 2 elements   **0.22** 3 elements

0.23 4 elements   **0.24** 5 elements

GROUPS

Manyun

427-1598

Cp acc 12JKH

login lucky

{ mkdir ⊔ directory
name.

cd yoke dir

D acct

lucky   56 AAAD

↑ cd

one

Binary Operations

1.1 MOTIVATION

What constitutes the basic ingredient of algebra? The first contact of children with algebra comes when they are taught to add and multiply numbers. Let us try to analyze what really happens here.

Suppose that you are a visitor to a strange civilization in a strange world and you are observing one of the creatures of this world drilling a class of fellow creatures in the addition of numbers. Suppose also that you have not been told that the class is learning to add, but that you were just placed as an observer in the room where this was going on. You are asked to give a report on exactly what happens. The teacher makes noises that sound to you approximately like *gloop, poyt*. The class responds with *bimt*. The teacher then gives *ompt, gaft*, and the class responds with *poyt*. What are they doing? You cannot report that they are adding numbers, for you do not even know that the sounds are representing numbers. Of course, you do realize that there is communication going on. All you can say with any certainty is that these creatures know some rule, so that when certain pairs of things are designated in their language, one after another, like *gloop*, *poyt*, they are able to agree on a response, *bimt*. This same procedure goes on in addition drill in our first grade classes where a teacher may say *four*, *seven*, and the class responds with *eleven*.

In our attempt to analyze addition and multiplication of numbers, we are thus led to the idea that addition is basically just a rule that people learn, enabling them to associate, with two numbers in a given order, some number as answer. Multiplication is also such a rule, but a different rule. Note finally that in playing this game with students, teachers have to be a little careful of what two things they give to the class. If a first grade teacher suddenly inserts *ten*, *sky*, the poor class will be very confused. The rule is only defined for pairs of things from some specified set.

1.2 DEFINITION AND PROPERTIES

As mathematicians, let us attempt to collect the core of these basic ideas in a useful definition. As we remarked in the introductory section, we do not attempt to define a set.

Definition A *binary operation* $*$ *on a set* is a rule that assigns to each ordered pair of elements of the set some element of the set.

The word *ordered* in this definition is very important, for it allows the possibility that the element assigned to the pair (a, b) may be different from the element assigned to the pair (b, a). Also, we were careful not to say that to each ordered pair of elements is assigned *another* or a *third* element, for we wish to permit cases such as occur in addition of numbers where $(0, 2)$ has assigned to it the number 2.

For the first few sections we shall let $a * b$ be the element assigned to the pair (a, b) by $*$. If we have several different binary operations under simultaneous discussion, we shall use subscripts or superscripts on the $*$ to distinguish them. The most important method of describing a particular binary operation $*$ on a given set is to characterize the element $a * b$ assigned to each pair (a, b) by some property defined in terms of a and b.

Example 1.1 On $\mathbf{Z}^+$, define a binary operation $*$ by $a * b$ equals the smaller of a and b or the common value if $a = b$. Thus $2 * 11 = 2$; $15 * 10 = 10$; and $3 * 3 = 3$. ∎

Example 1.2 On $\mathbf{Z}^+$, define a binary operation $*'$ by $a *' b = a$. Thus $2 *' 3 = 2$; $25 *' 10 = 25$; and $5 *' 5 = 5$. ∎

Example 1.3 On $\mathbf{Z}^+$, define a binary operation $*''$ by $a *'' b = (a * b) + 2$, where $*$ is defined in Example 1.1. Thus $4 *'' 7 = 6$; $25 *'' 9 = 11$; and $6 *'' 6 = 8$. ∎

It probably seems to you that these examples are of no importance, but consider for a moment. Suppose you go into a store to buy a nice, large, delicious chocolate bar. Suppose you see two identical bars side by side, the wrapper of one stamped 99¢ and the wrapper of the other stamped 94¢. Of course you pick up the one stamped 94¢. Your knowledge of which one you want depends on the fact that sometime in your life you learned the binary operation $*$ of Example 1.1. *It is a very important operation.* Likewise the binary operation $*'$ of Example 1.2 is clearly dependent on the ability to distinguish order. An often cited illustration of the importance of order is the mess that would result if you tried to put on your shoes first, and then your socks! Thus you should not be hasty about dismissing some binary operation as being of little significance. Of course, our usual operations of addition and multiplication of numbers have a practical importance well known to you.

Examples 1.1 and 1.2 were chosen to demonstrate that a binary operation may or may not depend on the order of the given pair. Thus in Example 1.1, $a * b = b * a$ for all $a, b \in \mathbf{Z}^+$, and in Example 1.2 this is not the case, for $5 *' 7 = 5$ but $7 *' 5 = 7$.

Now suppose one wishes to consider an expression of the form $a * b * c$. A binary operation $*$ enables you to combine only two elements, and here we have three. The obvious attempts to combine the three elements are to form either $(a * b) * c$ or $a * (b * c)$. With $*$ defined as in Example 1.1, $(2 * 5) * 9$ is computed by $2 * 5 = 2$ and then $2 * 9 = 2$. Likewise $2 * (5 * 9)$ is computed by $5 * 9 = 5$ and then $2 * 5 = 2$. Hence $(2 * 5) * 9 = 2 * (5 * 9)$, and it is easily seen that for this $*$,

$$(a * b) * c = a * (b * c),$$

so there is no ambiguity in writing $a * b * c$. But for $*''$ of Example 1.3,

$$(2 *'' 5) *'' 9 = 4 *'' 9 = 6,$$

while

$$2 *'' (5 *'' 9) = 2 *'' 7 = 4.$$

Thus $(a *'' b) *'' c$ need not equal $a *'' (b *'' c)$ and an expression $a *'' b *'' c$ may be ambiguous.

Definition   A binary operation $*$ on a set S is **commutative** if (and only if) $a * b = b * a$ for all $a, b \in S$. The operation $*$ is **associative** if (and only if) $(a * b) * c = a * (b * c)$ for all $a, b, c \in S$.

As was pointed out in the introductory section, it is customary in mathematics to omit the words *and only if* from a definition. Definitions are always understood to be if and only if statements. *Theorems are not always if and only if statements, and no such convention is ever used for theorems.*

It is not difficult to show that if ∗ is associative, then longer expressions such as $a * b * c * d$ are not ambiguous. Parentheses may be inserted in any fashion for purposes of computation; the final results of two such computations will be the same.

1.3 TABLES

For a finite set, a binary operation on the set can also be defined by means of a table. The next example shows how this will be done in this text.

Example 1.4 Table 1.1 defines the binary operation ∗ on $S = \{a, b, c\}$ by the rule:

    (ith *entry on the left*) ∗ (jth *entry on the top*)

             = (*entry in the* ith *row and* jth *column of the table body*).

Thus $a * b = c$ and $b * a = a$, so ∗ is not commutative. ∎

Table 1.1

∗	a	b	c
a	b	c	b
b	a	c	b
c	c	b	a

The student can easily see that *a binary operation defined by a table is commutative if and only if the entries in the table are symmetric with respect to the diagonal that starts at the upper left corner of the table and terminates at the lower right corner.* We always assume that the elements of the set are listed across the top of a table in the same order as they are listed at the left.

Except for Example 1.4, our examples of binary operations have been defined on sets of numbers. It is important to realize that binary operations may be defined on any sets. Indeed, we shall be studying many important binary operations on sets whose elements are not numbers. Some of the examples to be given in a moment involve sets with *functions* as elements. It is assumed that the student has some familiarity with certain functions from calculus or other courses. We realize that you may not understand the concept of a function at the moment and we shall say more about it later. However, we are anxious to tie in the notions being introduced with the mathematics that you have already had.

1.4 SOME WORDS OF WARNING

The author knows from his own experience the chaos that may result if a
student is given a set and asked to define some binary operation on it.
Observe that in an attempt to define a binary operation $*$ on a set S you
must be sure that

1. *exactly one element is assigned to each possible ordered pair of elements
   of S,*
2. *for each ordered pair of elements of S, the element assigned to it is again
   in S.*

Regarding condition 1, a student will often give a rule that assigns an element of S to "most" ordered pairs, but for a few pairs the rule determines
no element. In this event, $*$ has **not been defined.** It may also happen that
for some pairs, the rule could assign any of several elements of S, that is,
there is ambiguity. In any case of ambiguity, $*$ is **not well defined.** If condition 2 is violated, then S is **not closed under** $*$.

We now give several illustrations of attempts to define binary operations
on sets. Some of these attempts are worthless, as we point out. Since no
comparison between operations will be made, we shall denote them all by $*$.

Example 1.5 On **Q**, "define" $*$ by $a * b = a/b$. Here $*$ is *not defined*, for
no rational number is assigned by this rule to the pair $(2, 0)$. ∎

Example 1.6 On **Q**$^+$, define $*$ by $a * b = a/b$. Here both conditions 1
and 2 are satisfied and $*$ is a binary operation on **Q**$^+$. ∎

Example 1.7 On **Z**$^+$, "define" $*$ by $a * b = a/b$. Here condition 2 is violated,
for $1 * 3$ is not in **Z**$^+$. Thus $*$ is not a binary operation on **Z**$^+$, since **Z**$^+$
is *not closed under* $*$. ∎

Example 1.8 Let S be the set of all real-valued functions defined for all real
numbers. Define $*$ to give the usual sum of two functions, that is, $f * g = h$,
where $h(x) = f(x) + g(x)$ for $f, g \in S$ and $x \in \mathbf{R}$. This definition of $*$ satisfies
conditions 1 and 2 and gives a binary operation on S. ∎

Example 1.9 Let S be as in Example 1.8 and define $*$ to give the usual
product of two functions, that is, $f * g = h$, where $h(x) = f(x)g(x)$. Again this
definition is a good one and gives a binary operation on S. ∎

Example 1.10 Let S be as in Example 1.8 and "define" $*$ to give the usual
quotient of f by g, that is, $f * g = h$, where $h(x) = f(x)/g(x)$. Here condition
2 is violated, for the functions in S were to be defined for *all* real numbers,
and for some $g \in S$, $g(x)$ will be zero for some values of x in **R** and $h(x)$ would
not be defined at those numbers in **R**. For example, if $f(x) = \cos x$ and
$g(x) = x^2$, then $h(0)$ is undefined, so $h \notin S$. ∎

Example 1.11 Let S be as in Example 1.8 and "define" $*$ by $f * g = h$, where h is the function greater than both f and g. This "definition" is completely worthless. In the first place, we have not defined what it means for one function to be greater than another. Even if we had, any sensible definition would result in there being many functions greater than both f and g, and $*$ would still be *not well defined*. ∎

Example 1.12 Let S be a set consisting of twenty people, no two of whom are of the same height. Define $*$ by $a * b = c$, where c is the tallest person among the twenty in S. This is a perfectly good binary operation on the set, although not a particularly interesting one. ∎

Example 1.13 Let S be as in Example 1.12 and "define" $*$ by $a * b = c$, where c is the shortest person in S who is taller than both a and b. This $*$ is *not defined*, since if either a or b is the tallest person in the set, $a * b$ is not determined. ∎

Exercises

1.1 Let the binary operation $*$ be defined on $S = \{a, b, c, d, e\}$ by means of Table 1.2.

a) Compute $b * d$, $c * c$ and $[(a * c) * e] * a$ from the table.

b) Compute $(a * b) * c$ and $a * (b * c)$ from the table. Can you say on the basis of this computation whether $*$ is associative?

c) Compute $(b * d) * c$ and $b * (d * c)$ from the table. Can you say on the basis of this computation whether $*$ is associative?

d) Is $*$ commutative? Why?

Table 1.2

$*$	a	b	c	d	e
a	a	b	c	b	d
b	b	c	a	e	c
c	c	a	b	b	a
d	b	e	b	e	d
e	d	b	a	d	c

1.2 Complete Table 1.3 so as to define a commutative binary operation $*$ on $S = \{a, b, c, d\}$.

Table 1.3

$*$	a	b	c	d
a	a	b	c	
b	b	d		c
c	c	a	d	b
d	d			a

1.3 Table 1.4 may be completed to define an associative binary operation $*$ on $S = \{a, b, c, d\}$. Assume this is possible and compute the missing entries.

Table 1.4

$*$	a	b	c	d
a	a	b	c	d
b	b	a	c	d
c	c	d	c	d
d				

1.4 Determine whether each of the definitions of $*$ given below does give a binary operation on the given set. In the event that $*$ is not a binary operation, state whether condition 1, condition 2, or both of these conditions from Section 1.4 are violated.

a) On $\mathbf{Z}^+$, define $*$ by $a * b = a - b$.

b) On $\mathbf{Z}^+$, define $*$ by $a * b = a^b$.

c) On $\mathbf{R}$, define $*$ by $a * b = a - b$.

d) On $\mathbf{Z}^+$, define $*$ by $a * b = c$, where c is the smallest integer greater than both a and b.

e) On $\mathbf{Z}^+$, define $*$ by $a * b = c$, where c is at least 5 more than $a + b$.

f) On $\mathbf{Z}^+$, define $*$ by $a * b = c$, where c is the largest integer less than the product of a and b.

†**1.5** Prove that if $*$ is an associative and commutative binary operation on a set S, then

$$(a * b) * (c * d) = [(d * c) * a] * b$$

for all $a, b, c, d \in S$. Assume the associative law only for triples as in the definition, that is, assume only

$$(x * y) * z = x * (y * z)$$

for all $x, y, z \in S$.

1.6 For each binary operation $*$ defined, determine whether $*$ is commutative and whether $*$ is associative.

a) On $\mathbf{Z}$, define $*$ by $a * b = a - b$.

b) On $\mathbf{Q}$, define $*$ by $a * b = ab + 1$.

c) On $\mathbf{Q}$, define $*$ by $a * b = ab/2$.

d) On $\mathbf{Z}^+$, define $*$ by $a * b = 2^{ab}$.

e) On $\mathbf{Z}^+$, define $*$ by $a * b = a^b$.

1.7 Mark each of the following true or false.

___ a) If $*$ is any binary operation on any set S, then $a * a = a$ for all $a \in S$.

___ b) If $*$ is any commutative binary operation on any set S, then $a * (b * c) = (b * c) * a$ for all $a, b, c \in S$.

___ c) If $*$ is any associative binary operation on any set S, then $a * (b * c) = (b * c) * a$ for all $a, b, c \in S$.

— d) The only binary operations of any importance are those defined on sets of numbers.

— e) A binary operation $*$ on a set S is commutative if there exist $a, b \in S$ such that $a * b = b * a$.

— f) Every binary operation defined on a set having exactly one element is both commutative and associative.

— g) A binary operation on a set S assigns at least one element of S to each ordered pair of elements of S.

— h) A binary operation on a set S assigns at most one element of S to each ordered pair of elements of S.

— i) A binary operation on a set S assigns exactly one element of S to each ordered pair of elements of S.

— j) A binary operation on a set S may assign more than one element of S to some ordered pair of elements of S.

1.8 Give a set different from any of those described in the examples of the text and not a set of numbers. Define two different binary operations $*$ and $*'$ on this set. Be sure that your set is *well defined*.

1.9 Let S be a set having exactly one element. How many different binary operations can be defined on S? Answer the question if S has exactly 2 elements; exactly 3 elements; exactly n elements.

1.10 How many different commutative binary operations can be defined on a set of 2 elements? on a set of 3 elements? on a set of n elements?

1.11 Observe that the binary operations $*$ and $*'$ on the set $\{a, b\}$ given by the tables

$*$	a	b
a	a	a
b	a	b

and

$*'$	a	b
a	a	b
b	b	b

provide the *same type of algebraic structure* on $\{a, b\}$, in the sense that if the table for $*'$ is rewritten

$*'$	b	a
b	b	b
a	b	a

this table for $*'$ looks just like that for $*$ with the roles of a and b interchanged.

a) Try to give a natural definition of a concept of two binary operations $*$ and $*'$ on the same set giving *algebraic structures of the same type*, which generalizes this observation.

b) How many different types of algebraic structures are given by the 16 possible different binary operations on a set of 2 elements?

Groups

2.1 MOTIVATION

Let us continue the analysis of our past experience with algebra. Once we had mastered the computational problems of addition and multiplication of numbers, we were ready to apply these binary operations to the solution of problems. Often problems lead to equations involving some unknown number x, which is to be determined. The simplest equations are the linear ones of the forms $a + x = b$ for the operation of addition, and $ax = b$ for multiplication. The additive linear equation always has a numerical solution, and so has the multiplicative one, provided $a \neq 0$. Indeed, the need for solutions of additive linear equations such as $5 + x = 2$ is a very good motivation for the negative numbers. Similarly, the need for rational numbers is shown by equations such as $2x = 3$, and the need for the complex number i is shown by the equation $x^2 = -1$.

It is desirable for us to be able to solve linear equations involving our binary operations. This is not possible for every binary operation, however. For example, the equation $a * x = a$ has no solution in $S = \{a, b, c\}$ for the operation $*$ of Example 1.4. Let us see just what properties of the operation of addition on the integers **Z** enable us to solve the equation $5 + x = 2$ in **Z**. We must not refer to subtraction, for we are concerned with the solution phrased in terms of a single binary operation, in this case addition. The steps in the solution are as follows:

$$5 + x = 2, \qquad \text{given,}$$
$$-5 + (5 + x) = -5 + 2, \qquad \text{adding } -5,$$
$$(-5 + 5) + x = -5 + 2, \qquad \text{associative law,}$$
$$0 + x = -5 + 2, \qquad \text{computing } -5 + 5,$$
$$x = -5 + 2, \qquad \text{property of } 0,$$
$$x = -3, \qquad \text{computing } -5 + 2.$$

Strictly speaking, we have not shown here that -3 is a solution, but rather that it is the only possibility for a solution. To show that -3 is a solution, one merely computes $5 + (-3)$. A similar analysis could be made for the equation $2x = 3$ in the rational numbers:

$$2x = 3, \qquad \text{given,}$$
$$\tfrac{1}{2}(2x) = \tfrac{1}{2}(3), \qquad \text{multiplying by } \tfrac{1}{2},$$
$$(\tfrac{1}{2} \cdot 2)x = \tfrac{1}{2}3, \qquad \text{associative law,}$$
$$1 \cdot x = \tfrac{1}{2}3, \qquad \text{computing } \tfrac{1}{2}2,$$
$$x = \tfrac{1}{2}3, \qquad \text{property of } 1,$$
$$x = \tfrac{3}{2}, \qquad \text{compu.ing } \tfrac{1}{2}3.$$

Let us see what properties a set S and a binary operation $*$ on S would have to have to permit imitation of this procedure for an equation $a * x = b$ for $a, b \in S$. Basic to the procedure is the existence of an element e in S with the property that $e * x = x$ for all $x \in S$. For our additive example, 0 played the role of e, and 1 played the role for our multiplicative example. Then we need an element a' in S that has the property that $a' * a = e$. For our additive example, -5 played the role of a', and $\tfrac{1}{2}$ played the role for our multiplicative example. Finally we need the associative law. The remainder is just computation. The student will easily see that in order to solve the equation $x * a = b$ (remember that $a * x$ need not equal $x * a$) one would like to have an element e in S such that $x * e = x$ for all $x \in S$ and an a' in S such that $a * a' = e$. With all of these properties of $*$ on S, we could be sure of being able to solve linear equations. These are precisely the properties of a *group*.

2.2 DEFINITION AND ELEMENTARY PROPERTIES

Definition   A *group* $\langle G, * \rangle$ is a set G, together with a binary operation $*$ on G, such that the following axioms are satisfied:

$\mathscr{G}_1$. The binary operation $*$ is associative.

$\mathscr{G}_2$. There is an element e in G such that $e * x = x * e = x$ for all $x \in G$. (This element e is an ***identity element*** for $*$ on G.)[†]

$\mathscr{G}_3$. For each a in G, there is an element a' in G with the property that $a' * a = a * a' = e$. (The element a' is an ***inverse of a with respect to*** $*$.)

Many books have another axiom for a group, namely that G is **closed under the operation** $*$, that is $(a * b) \in G$ for all $a, b \in G$. For us, this is a consequence of our *definition* of a binary operation on G.

We should point out right now that we are going to be sloppy in notation. Observe that a group is not just a set G. Rather, a group $\langle G, * \rangle$ is made up of two entities, the set G and the binary operation $*$ on G. There are *two* ingredients involved. Denoting the group by the single set symbol G is logically incorrect. Nevertheless, as you get further into the theory, the logical extensions of the notation $\langle G, * \rangle$ become so unwieldy as to actually make the exposition hard to read. At some point, all authors give up and become sloppy, denoting the group by the single letter G. We choose to recognize this and be sloppy from the start. We emphasize, however, that when you are speaking of a specific group G, you must make it clear what the group operation on G is to be, since a set could conceivably have a variety of binary operations defined on it, all giving different groups. We shall sometimes resort to the notation $\langle G, * \rangle$ for reasons of clarity in our discussions.

Theorem 2.1   *If G is a group with binary operation $*$, then the **left and right cancellation laws** hold in G, that is, $a * b = a * c$ implies $b = c$, and $b * a = c * a$ implies $b = c$ for $a, b, c \in G$.*

Proof.   Suppose $a * b = a * c$. Then by $\mathscr{G}_3$, there exists a', and

$$a' * (a * b) = a' * (a * c).$$

By the associative law,

$$(a' * a) * b = (a' * a) * c.$$

By the definition of a' in $\mathscr{G}_3$, $a' * a = e$, so

$$e * b = e * c.$$

[†] Remember that boldface type indicates that a term is being defined. See the last paragraph of Section 0.1. Thus an **identity element** for a binary operation $*$ on a set S is any element e satisfying $e * x = x * e = x$ for all $x \in S$.

By the definition of e in $\mathscr{G}_2$,

$$b = c.$$

Similarly, from $b * a = c * a$ one can deduce that $b = c$ upon multiplication on the right by a' and use of the axioms for a group. ∎

Note how we had to use the definition of a group to prove this theorem.

Theorem 2.2   *If G is a group with binary operation $*$, and if a and b are any elements of G, then the linear equations $a * x = b$ and $y * a = b$ have unique solutions in G.*

Proof.   Note that

$$a * (a' * b) = (a * a') * b, \qquad \text{associative law,}$$
$$= e * b, \qquad \text{definition of } a',$$
$$= b, \qquad \text{property of } e.$$

Thus $x = a' * b$ is a solution of $a * x = b$. In a similar fashion, $y = b * a'$ is a solution of $y * a = b$.

To show that y is unique, suppose that $y * a = b$ and $y_1 * a = b$. Then $y * a = y_1 * a$ and, by Theorem 2.1, $y = y_1$. The uniqueness of x follows similarly. ∎

Of course, to prove the uniqueness in the last theorem we could have followed the procedure we used in motivating the definition of a group, showing that if $a * x = b$, then $x = a' * b$. However, we chose to illustrate the standard way to prove an object is unique. Suppose you have two such objects, and then prove they must be the same. Note that the solutions $x = a' * b$ and $y = b * a'$ need not be the same unless $*$ is commutative.

Definition   A group G is ***abelian*** if its binary operation $*$ is commutative.

Let us give some examples of some sets with binary operations that give groups and also of some that do not give groups.

Example 2.1   The set Z^+ with operation $+$ is *not* a group. There is no identity element for $+$ in Z^+. ∎

Example 2.2   The set of all nonnegative integers (including 0) with operation $+$ is still *not* a group. There is an identity element 0, but no inverse for 2. ∎

Example 2.3   The set Z with operation $+$ *is* a group. All conditions of the definition are satisfied. The group is abelian. ∎

Example 2.4   The set Z^+ with operation multiplication is *not* a group. There is an identity 1, but no inverse of 3. ∎

Example 2.5 The set $\mathbf{Q}^+$ with operation multiplication *is* a group. All conditions of the definition are satisfied. The group is abelian. ∎

Example 2.6 Define $*$ on $\mathbf{Q}^+$ by $a * b = ab/2$. Then

$$(a * b) * c = \frac{ab}{2} * c = \frac{abc}{4},$$

and likewise

$$a * (b * c) = a * \frac{bc}{2} = \frac{abc}{4}.$$

Thus $*$ is associative. Clearly,

$$2 * a = a * 2 = a$$

for all $a \in \mathbf{Q}^+$, so 2 is an identity element for $*$. Finally,

$$a * \frac{4}{a} = \frac{4}{a} * a = 2,$$

so $a' = 4/a$ is an inverse for a. Hence $\mathbf{Q}^+$ with the operation $*$ is a group. ∎

There is one other result about groups we would like to prove in this section.

Theorem 2.3 *In a group G with operation $*$, there is only one identity e such that*

$$e * x = x * e = x$$

for all $x \in G$. Likewise for each $a \in G$, there is only one element a' such that

$$a' * a = a * a' = e.$$

In summary, the identity and inverses are unique in a group.

Proof. Suppose $e * x = x * e = x$ and also $e_1 * x = x * e_1 = x$ for all $x \in G$. We let e and e_1 compete. Now regarding e as identity, $e * e_1 = e_1$. But regarding e_1 as identify, $e * e_1 = e$. Thus

$$e_1 = e * e_1 = e,$$

and the identify of a group is unique.

Now suppose $a' * a = a * a' = e$ and $a'' * a = a * a'' = e$. Then

$$a * a'' = a * a' = e$$

and, by Theorem 2.1,

$$a'' = a',$$

so the inverse of a in a group is unique. ∎

For the student's information, we remark that algebraic structures consisting of sets with binary operations for which not all of the group axioms hold have also been studied quite extensively. Of these weaker structures, the **semigroup,** a set with an associative binary operation, has perhaps had the most attention. Recently, nonassociative structures have also been studied.

Finally, it is possible to give formally weaker axioms for a group $\langle G, * \rangle$, namely:

1. The binary operation $*$ on G is associative.
2. There exists a **left identity** e in G such that $e * x = x$ for all $x \in G$.
3. For each $a \in G$, there exists a **left inverse** a' in G such that $a' * a = e$.

From this *one-sided definition*, one can prove that the left identity is also a right identity and a left inverse is also a right inverse for the same element. Thus these axioms should not be called weaker, since they result in exactly the same structures being called groups. It is conceivable that it might be easier in some cases to check these *left axioms* than to check our *two-sided axioms*. Of course, by symmetry it is clear that there are also *right axioms* for a group.

2.3 FINITE GROUPS AND GROUP TABLES

Thus far all our examples have been of infinite groups, that is, groups where the set G has an infinite number of elements. The student may wonder whether there can be a group structure on some finite set. The answer is yes, and indeed such structures are very important.

Since a group has to have at least one element, namely the identity, a smallest set that might give rise to a group is a one-element set $\{e\}$. The only possible binary operation $*$ on $\{e\}$ is defined by $e * e = e$. The student can check at once that the three group axioms hold. The identity element is always its own inverse in every group.

Let us try to put a group structure on a set of two elements. Since one of the elements must play the role of identity element, we may as well let the set be $\{e, a\}$. Let us attempt to find a table for a binary operation $*$ on $\{e, a\}$ that gives a group structure on $\{e, a\}$. When giving a table for a group operation, we shall always list the elements in the same order across the top as down the left side, with the identity listed first, as in the following table.

$*$	e	a
e		
a		

Since e is to be the identity, so

$$e * x = x * e = x$$

for all $x \in \{e, a\}$, we are forced to fill in the table as shown below, if $*$ is to give a group.

$*$	e	a
e	e	a
a	a	

Also, a must have an inverse a' such that

$$a * a' = a' * a = e.$$

In our case, a' must be either e or a. Since $a' = e$ obviously does not work, we must have $a' = a$, so we have to complete the table as below.

$*$	e	a
e	e	a
a	a	e

All the group axioms are now satisfied except possibly the associative law. We will see later in a more general situation that this operation $*$ is associative. You are asked either to accept it here or to go through the tedious chore of checking various cases.

With these examples as background, we should be able to list some necessary conditions that a table giving a binary operation on a finite set must satisfy for the operation to give a group structure on the set. There must be one element of the set, which we may as well denote by e, that acts as identity. The condition $e * x = x$ means that the row of the table opposite e at the extreme left must contain exactly the elements appearing across the very top of the table in the same order. Similarly, the condition $x * e = x$ means that the column of the table under e at the very top must contain exactly the elements appearing at the extreme left in the same order. The fact that every element a has a right and a left inverse means that in the row opposite a at the extreme left, the element e must appear, and in the column under a at the very top, the e must appear. Thus e must appear in each row and in each column. We can do even better than this, however. By Theorem 2.2, not only the equations $a * x = e$ and $y * a = e$ have unique solutions, but also the equations $a * x = b$ and $y * a = b$. By a similar argument, this means that *each element b of the group must appear once and only once in each row and column of the table.*

Suppose conversely that a table for a binary operation on a finite set is such that there is an element acting as identity and that in each row and each column each element of the set appears exactly once. Then it can be seen that the structure is a group structure if and only if the associative law holds. If a binary operation $*$ is given by a table, the associative law is usually messy to check. If the operation $*$ is defined by some characterizing property of $a * b$, the associative law is often easy to check. Fortunately this second case turns out to be the one usually encountered.

We saw that there was essentially only one group of two elements in the sense that if the elements are denoted by e and a with the identity e appearing first, the table must be as follows.

$*$	e	a
e	e	a
a	a	e

Suppose that a set has three elements. As before, we may as well let the set be $\{e, a, b\}$. For e to be an identity, a binary operation $*$ on this set has to have a table of the form shown in Table 2.1. This leaves four places to be filled in. The student can quickly see that Table 2.1 must be completed as shown in Table 2.2 if each row and each column are to contain each element exactly once. Again you are asked to accept without proof the fact that this operation is associative, so that $*$ does give a group structure on $G = \{e, a, b\}$.

Table 2.1

$*$	e	a	b
e	e	a	b
a	a		
b	b		

Table 2.2

$*$	e	a	b
e	e	a	b
a	a	b	e
b	b	e	a

Now suppose that G' is any other group of three elements and imagine a table for G' with identity element appearing first. Since our filling out of the table for $G = \{e, a, b\}$ could be done in only one way, we see that if we take the table for G' and rename the identity e, the next element listed a, and the last element b, the resulting table for G' must be the same as the one we had for G. In other words, the *structural* features are the same for the two groups, and one group can be made to look exactly like the other by a renaming of the elements. *Thus any two groups of three elements are structurally the same.* This is our introduction to the concept of *isomorphism*. The groups G and G' are *isomorphic*. This concept is sometimes a bit sticky for the student. We say no more about it now, but we shall make it more precise later.

Exercises

2.1 For each binary operation ∗ defined on a set below, say whether ∗ gives a group structure on the set. If no group results, give the first axiom in the order $\mathscr{G}_1, \mathscr{G}_2, \mathscr{G}_3$ from Section 2.2 that does not hold.

a) Define ∗ on **Z** by $a * b = ab$.

b) Define ∗ on **Z** by $a * b = a - b$.

c) Define ∗ on $\mathbf{R}^+$ by $a * b = ab$.

d) Define ∗ on **Q** by $a * b = ab$.

e) Define ∗ on the set of all nonzero real numbers by $a * b = ab$.

f) Define ∗ on **C** by $a * b = a + b$.

2.2 Consider our axioms $\mathscr{G}_1, \mathscr{G}_2$, and $\mathscr{G}_3$ for a group. We gave them in the order $\mathscr{G}_1\mathscr{G}_2\mathscr{G}_3$. Conceivable other orders to state the axioms are $\mathscr{G}_1\mathscr{G}_3\mathscr{G}_2$, $\mathscr{G}_2\mathscr{G}_1\mathscr{G}_3$, $\mathscr{G}_2\mathscr{G}_3\mathscr{G}_1$, $\mathscr{G}_3\mathscr{G}_1\mathscr{G}_2$, and $\mathscr{G}_3\mathscr{G}_2\mathscr{G}_1$. Of these six possible orders, exactly three are acceptable for a definition. Which orders aren't acceptable, and why? (Remember this. Most instructors ask the student to define a group on at least one test.)

2.3 Show by computation and by Theorem 2.3 that if G is a group with binary operation ∗, then for all $a, b \in G$, we have $(a * b)' = b' * a'$. What is a similar expression for $(a * b' * c)'$?

2.4 Proceed as follows to show that there are two possible different types of group structures on a set of four elements. Let the set be $\{e, a, b, c\}$, with e the identity element for the group operation. A group table would then have to start in the manner shown in Table 2.3. The square indicated by the question mark can't be filled in with a. It must be filled in either with the identity e or with an element different from both e and a. In this latter case, it is no loss of generality to assume that this element is b. If this square is filled in with e, the table can then be completed in two ways

Table 2.3

∗	e	a	b	c
e	e	a	b	c
a	a	?		
b	b			
c	c			

to give a group. Find these two tables. (You need not check the associative law.) If this square is filled in with b, then the table can only be completed in one way to give a group. Find this table. (Again you need not check the associative law.) Of the three tables you now have, two give the same type of group structure. Determine which two tables these are, and show how the elements in one table would have to be renamed for these two tables to be the same. Are all groups of 4 elements commutative?

†2.5 Show that if G is a finite group with identity e and with an even number of elements, then there is $a \neq e$ in G such that $a * a = e$.

2.6 Mark each of the following true or false.

— a) A group may have more than one identity element.

— b) Any two groups of three elements are isomorphic.

— c) In a group, each linear equation has a solution.

  — d) The proper attitude toward a definition is to memorize it so that you can reproduce it word for word as in the text.

  — e) Any definition a person gives for a group is correct provided that everything that is a group by that person's definition is also a group by the definition in the text.

  — f) Any definition a person gives for a group is correct provided he or she can show that everything that satisfies the definition satisfies the one in the text and conversely.

  — g) Every finite group of at most three elements is abelian.

  — h) An equation of the form $a * x * b = c$ always has a unique solution in a group.

  — i) The empty set can be considered a group.

  — j) The text has as yet given no examples of groups that are not abelian.

2.7 Give a table for a binary operation on the set $\{e, a, b\}$ of three elements satisfying axioms $\mathcal{G}_2$ and $\mathcal{G}_3$ for a group but not axiom $\mathcal{G}_1$.

2.8 According to Exercise 1.9, there are 16 possible binary operations on a set of 2 elements. How many of these give a structure of a group? How many of the 19,683 possible binary operations on a set of 3 elements give a group structure?

2.9 Let S be the set of all real numbers except -1. Define $*$ on S by

$$a * b = a + b + ab.$$

a) Show that $*$ gives a binary operation on S.

b) Show that $\langle S, * \rangle$ is a group.

c) Find the solution of the equation $2 * x * 3 = 7$ in S.

2.10 Let $\mathbf{R}^*$ be the set of all real numbers except 0. Define $*$ on $\mathbf{R}^*$ by $a * b = |a|b$.

a) Show that $*$ gives an associative binary operation on $\mathbf{R}^*$.

b) Show that there is a left identity for $*$ and a right inverse for each element in $\mathbf{R}^*$.

c) Is $\mathbf{R}^*$ with this binary operation a group?

d) Explain the significance of this exercise.

2.11 If $*$ is a binary operation on a set S, an element x of S is an **idempotent for** $*$ if $x * x = x$. Prove that a group has exactly one idempotent element. (You may use any theorems proved so far in the text.)

2.12 Show that every group G with identity e and such that $x * x = e$ for all $x \in G$ is abelian. [*Hint:* Consider $(ab)^2$.]

2.13 Prove that a set G, together with a binary operation $*$ on G satisfying the left axioms 1, 2, and 3 given at the end of Section 2.2, is a group.

2.14 Prove that a nonempty set G, together with an associative binary operation $*$ on G such that equations

$$a * x = b \text{ and } y * a = b \text{ have solutions in } G \text{ for all } a, b \in G,$$

is a group [*Hint:* Use Exercise 2.13.]

2.15 The following "definitions" of a group are taken verbatim, including spelling and punctuation, from students' examination papers. Criticize them.

a) A group G is a set of elements together with a binary operation $*$ such that the following conditions are satisfied

$*$ is associative

There exists $e \in G$ such that

$$e * x = x * e = x = \text{identity.}$$

For every $a \in G$ there exists an a' (inverse) such that

$$a \cdot a' = a' \cdot a = e$$

b) A group is a set G such that

The operation on G is associative.

there is an identity element (e) in G.

for every $a \in G$, there is an a' (inverse for each element)

c) A group is a set with a binary operation such

the binary operation is defined

an inverse exists

an identity element exists

d) A set G is called a group over the binery operation $*$ such that for all $a, b \in G$

Binary operation $*$ is associative under addition

there exist an element $\{e\}$ such that

$$a * e = e * a = e$$

Fore every element a there exists an element a' such that

$$a * a' = a' * a = e$$

three

Subgroups

3.1 NOTATION AND TERMINOLOGY

It is time to explain some conventional notation and terminology used in group theory. Algebraists as a rule do not use a special symbol $*$ to denote a binary operation different from the usual addition and multiplication. They stick with the conventional additive or multiplicative notation and even call the operation *addition* or *multiplication*, depending on the symbol used. The symbol for addition is of course $+$, and usually multiplication is denoted by juxtaposition without a dot, if no confusion results. Thus in place of the notation $a * b$, we shall be using either $a + b$ to be read "the *sum* of a and b," or ab to be read "the *product* of a and b." There is a sort of gentlemen's agreement that the symbol $+$ should be used only to designate commutative operations. Algebraists feel very uncomfortable when they see $a + b \neq b + a$. For this reason when developing our group theory in a general situation where the operation may or may not be commutative, we shall always use multiplicative notation.

Mathematicians frequently use the symbol 0 to denote an additive identity and the symbol 1 to denote a multiplicative identity, even though they

29

may not be actually denoting the integers 0 and 1. Of course, if a person is also talking about numbers at the same time, so that confusion would result, symbols such as e or u are used as identity elements. Thus a table for a group of three elements might be one like Table 3.1 or, since such a group is commutative, the table might look like Table 3.2. In general situations we shall continue to use e to denote the identity element of a group.

Table 3.1

	1	a	b
1	1	a	b
a	a	b	1
b	b	1	a

Table 3.2

+	0	a	b
0	0	a	b
a	a	b	0
b	b	0	a

It is customary to denote the inverse of an element a in a group by a^{-1} in multiplicative notation and by $-a$ in additive notation. From now on we shall be using these notations in place of the symbol a'.

Let us explain one more term that is used so often it merits a special definition.

Definition   If G is a finite group, then the ***order*** $|G|$ ***of*** G is the number of elements in G. In general, for any finite set S, $|S|$ is the number of elements in S.

Finally, in place of the phrase *with the binary operation of* we shall use the word *under*, so that "the group **R** with the binary operation of addition" becomes "the group **R** under addition."

3.2 SUBSETS AND SUBGROUPS

You may have noticed that we sometimes have had groups contained within larger groups. For example, the group **Z** under addition is contained within the group **Q** under addition, which in turn is contained within the group **R** under addition. When we view the group $\langle \mathbf{Z}, + \rangle$ as contained in the group $\langle \mathbf{R}, + \rangle$, it is very important to notice that the operation $+$ on integers n and m as elements of $\langle \mathbf{Z}, + \rangle$ produces the same element $n + m$ as would result if you were to think of n and m as elements in $\langle \mathbf{R}, + \rangle$. Thus we should *not* regard the group $\langle \mathbf{Q}^+, \cdot \rangle$ as contained in $\langle \mathbf{R}, + \rangle$, even though $\mathbf{Q}^+$ is contained in **R** as a set. In this instance, $2 \cdot 3 = 6$ in $\langle \mathbf{Q}^+, \cdot \rangle$, while $2 + 3 = 5$

in $\langle \mathbf{R}, + \rangle$. We are requiring not only that the set of one group be contained in the set of the other, but also that the group operation on the smaller set assign the same element to each ordered pair from this smaller set as is assigned by the group operation of the larger set. Let us give a sequence of definitions to make these ideas really precise.

> **Definition** A set B is a *subset of a set* A, denoted by $B \subseteq A$ or $A \supseteq B$, if every element of B is in A. The notations $B \subset A$ or $A \supset B$ will be used for $B \subseteq A$ but $B \neq A$.

Note that according to this definition, for any set A, A itself and $\varnothing$ are both subsets of A.

> **Definition** If A is any set, then A is the *improper subset of A*. Any other subset of A is a *proper subset of A*.

> **Definition** Let G be a group and let S be a subset of G. If for every $a, b \in S$ it is true that the product ab computed in G is also in S, then S is *closed under the group operation* of G. The binary operation on S thus defined is the *induced operation on S from G*.

We can now make the concept of one group being contained in another precise.

> **Definition** If H is a subset of a group G closed under the group operation of G, and if H is itself a group under this induced operation, then H is a *subgroup of G*. We shall let $H \leq G$ or $G \geq H$ denote that H is a subgroup of G, and $H < G$ or $G > H$ shall mean $H \leq G$ but $H \neq G$.

Thus $\langle \mathbf{Z}, + \rangle < \langle \mathbf{R}, + \rangle$ but $\langle \mathbf{Q}^+, \cdot \rangle$ is *not* a subgroup of $\langle \mathbf{R}, + \rangle$, even though as sets, $\mathbf{Q}^+ \subset \mathbf{R}$. Every group G has as subgroups G itself and $\{e\}$, where e is the identity element of G.

> **Definition** If G is a group, then G is the *improper subgroup of G*. All other subgroups are *proper subgroups*. Also, $\{e\}$ is the *trivial subgroup* of G. All other subgroups are *nontrivial*.

We turn to some illustrations.

Example 3.1 $\mathbf{Q}^+$ under multiplication is a proper subgroup of $\mathbf{R}^+$ under multiplication. ∎

Example 3.2 There are two different types of group structures of order 4 (see Exercise 2.4). We describe them by their group tables (Tables 3.3 and 3.4). The group V is the **Klein 4-group,** and the notation V comes from the German word *Viergruppe*.

Table 3.3

$\mathbf{Z}_4$:	+	0	1	2	3
	0	0	1	2	3
	1	1	2	3	0
	2	2	3	0	1
	3	3	0	1	2

Table 3.4

V:		e	a	b	c
	e	e	a	b	c
	a	a	e	c	b
	b	b	c	e	a
	c	c	b	a	e

The only nontrivial proper subgroup of $\mathbf{Z}_4$ is $\{0, 2\}$. Note that $\{0, 3\}$ is *not* a subgroup of $\mathbf{Z}_4$, since $\{0, 3\}$ is *not closed* under $+$. For example, $3 + 3 = 2$, and $2 \notin \{0, 3\}$. However, the group V has three nontrivial proper subgroups, $\{e, a\}, \{e, b\}$, and $\{e, c\}$. Here $\{e, a, b\}$ is *not* a subgroup, since $\{e, a, b\}$ is not closed under the operation of V. For example, $ab = c$, and $c \notin \{e, a, b\}$. ∎

It is often useful to draw a *lattice diagram* of the subgroups of a group. In such a diagram, a line running downward from a group G to a group H means that H is a subgroup of G. Thus the larger group is placed nearer the top of the diagram. Figure 3.1 contains the lattice diagrams for the groups $\mathbf{Z}_4$ and V of Example 3.2.

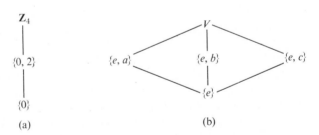

Fig. 3.1 (a) Lattice diagram for $\mathbf{Z}_4$. (b) Lattice diagram for V.

Note that if $H \leq G$ and $a \in H$, then by Theorem 2.2 the equation $ax = a$ must have a unique solution, namely the identity element of H. But this equation can also be viewed as one in G, and we see that this unique solution must also be the identity e of G. A similar argument then applied to the equation $ax = e$, viewed in both H and G, shows that the inverse a^{-1} of a in G is also the inverse of a in the subgroup H.

It is convenient to have a routine criterion for determining whether a subset of a group G is a subgroup of G. The next theorem gives such a criterion. While more compact criteria are available, involving only one condition, we prefer this more transparent one for a first course.

Theorem 3.1 *A subset H of a group G is a subgroup of G if and only if*

1. *H is closed under the binary operation of G,*
2. *the identity e of G is in H,*
3. *for all $a \in H$ it is true that $a^{-1} \in H$ also.*

Proof. The fact that if $H \leq G$ then conditions 1, 2, and 3 must hold follows at once from the definition of a subgroup and from the remarks preceding the statement of the theorem.

Conversely, suppose H is a subset of a group G such that conditions 1, 2, and 3 hold. By 2 we have at once that $\mathscr{G}_2$ is satisfied. Also $\mathscr{G}_3$ is satisfied by 3. It remains to check the associative axiom, $\mathscr{G}_1$. But surely for all $a, b, c \in H$ it is true that $(ab)c = a(bc)$ in H, for we may actually view this as an equation in G, where the associative law holds. Hence $H \leq G$. ∎

3.3 CYCLIC SUBGROUPS

We remarked in Example 3.2 that $\{0, 3\}$ is not a subgroup of $\mathbf{Z}_4$. Let us see how big a subgroup H of $\mathbf{Z}_4$ would have to be if it contained 3. It would have to contain the identity 0 and the inverse of 3, which is 1. Also H would have to contain $3 + 3$, which is 2. Hence the only subgroup of $\mathbf{Z}_4$ containing 3 is $\mathbf{Z}_4$ itself.

Let us imitate this reasoning in a general situation. As we remarked before, for a general argument we always use multiplicative notation. Let G be a group and let $a \in G$. A subgroup of G containing a must, by Theorem 3.1, contain aa, which we denote by a^2. Then it must contain a^2a, which we denote by a^3. In general, it must contain a^n, the result of computing products of a and itself for n factors for every positive integer n. (In additive notation we would denote this by na.) These positive integral powers of a do give a set closed under multiplication. It is possible, however, that the inverse of a is not in this set. Of course, a subgroup containing a must also contain a^{-1}, and then $a^{-1}a^{-1}$, which we denote by a^{-2}, and, in general, it must contain a^{-m} for all $m \in \mathbf{Z}^+$. It must contain the identity $e = aa^{-1}$. For obvious symbolic reasons, we agree to let a^0 be e. Summarizing, we have shown that *a subgroup of G containing a must contain all elements a^n (or na for additive groups) for all $n \in \mathbf{Z}$.* That is, a subgroup containing a must contain $\{a^n \mid n \in \mathbf{Z}\}$. Observe that these powers a^n of a need not be distinct. For example, in the group V of Example 3.2,

$$a^2 = e, \qquad a^3 = a, \qquad a^4 = e, \qquad a^{-1} = a, \qquad \text{and so on.}$$

It is easy to see that our usual law of exponents, $a^m a^n = a^{m+n}$ for $m, n \in \mathbf{Z}$, holds. For $m, n \in \mathbf{Z}^+$, it is clear. We illustrate another type of case by an example:

$$a^{-2}a^5 = a^{-1}a^{-1}aaaaa = a^{-1}(a^{-1}a)aaaa = a^{-1}eaaaa = a^{-1}(ea)aaa$$
$$= a^{-1}aaaa = (a^{-1}a)aaa = eaaa = (ea)aa = aaa = a^3.$$

The boring details of a careful proof for the general case are left to the student who does not mind the tedium. We have almost proved the next theorem.

Theorem 3.2   *Let G be a group and let $a \in G$. Then*

$$H = \{a^n \mid n \in \mathbf{Z}\}$$

is a subgroup of G and is the smallest subgroup of G that contains a, that is, every subgroup containing a contains H.[†]

Proof.   We check the three conditions given in Theorem 3.1 for a subset of a group to give a subgroup. Since $a^r a^s = a^{r+s}$ for r, $s \in \mathbf{Z}$, we see that the product in G of two elements of H is again in H. Thus H is closed under the group operation of G. Also $a^0 = e$, so $e \in H$, and for $a^r \in H$, $a^{-r} \in H$ and $a^{-r}a^r = e$. Hence all the conditions are satisfied, and $H \leq G$.

Our arguments prior to the statement of the theorem showed that any subgroup of G containing a must contain H, so H is the smallest subgroup of G containing a. ∎

Definition   The group H of Theorem 3.2 is the **cyclic subgroup of G generated by** a, and will be denoted by $\langle a \rangle$.

Definition   An element a of a group G **generates** G and is a **generator for** G if $\langle a \rangle = G$. A group G is **cyclic** if there is some element a in G that generates G.

Example 3.3   Let $\mathbf{Z}_4$ and V be the groups of Example 3.2. Then $\mathbf{Z}_4$ is cyclic and both 1 and 3 are generators, that is,

$$\langle 1 \rangle = \langle 3 \rangle = \mathbf{Z}_4.$$

However, V is *not* cyclic, for $\langle a \rangle$, $\langle b \rangle$, and $\langle c \rangle$ are proper subgroups of 2 elements. Of course $\langle e \rangle$ is the trivial subgroup of one element. ∎

Example 3.4   The group **Z** under addition is a cyclic group. Both 1 and -1 are generators for the group. ∎

[†] You may find occasion to distinguish between the terms *minimal* and *smallest* as applied to subsets of a set S that have some property. A subset H of S is minimal with respect to the property if H has the property, and no subset $K \subset H$, $K \neq H$, has the property. If H has the property and $H \subseteq K$ for every subset K with the property, then H is the smallest subset with the property. There may be many minimal subsets, but there can only be one smallest subset. To illustrate, $\{e, a\}$, $\{e, b\}$, and $\{e, c\}$ are all minimal nontrivial subgroups of the group V. (See Fig. 3.1.) However, V contains no smallest nontrivial subgroup.

Example 3.5   Consider the group **Z** under addition. Let us find $\langle 3 \rangle$. Here the notation is additive, and $\langle 3 \rangle$ must contain

$$3, \qquad 3 + 3 = 6, \qquad 3 + 3 + 3 = 9, \qquad \text{and so on,}$$
$$0, \qquad -3, \qquad -3 + -3 = -6, \qquad -3 + -3 + -3 = -9, \qquad \text{and so on.}$$

In other words, the cyclic subgroup generated by 3 consists of all multiples of 3, positive, negative, and zero. We denote this subgroup by 3**Z** as well as $\langle 3 \rangle$. In a similar way, we shall let $n\mathbf{Z}$ be the cyclic subgroup $\langle n \rangle$ of **Z**. Note that 6**Z** < 3**Z**. ∎

Exercises

3.1   Determine which of the following subsets of the complex numbers are subgroups under addition of the group **C** of complex numbers under addition.

a) **R**   b) $\mathbf{Q}^+$ c) 7**Z**

d) The set $i\mathbf{R}$ of pure imaginary numbers including 0

e) The set $\pi\mathbf{Q}$ of rational multiples of π f) The set $\{\pi^n \mid n \in \mathbf{Z}\}$

3.2 A number of groups are given below. Give a *complete* list of all relations of one group being a subgroup of one of the groups listed.

   $G_1 = \mathbf{Z}$ under addition

   $G_2 = 12\mathbf{Z}$ under addition

   $G_3 = \mathbf{Q}^+$ under multiplication

   $G_4 = \mathbf{R}$ under addition

   $G_5 = \mathbf{R}^+$ under multiplication

   $G_6 = \{\pi^n \mid n \in \mathbf{Z}\}$ under multiplication

   $G_7 = 3\mathbf{Z}$ under addition

   $G_8 =$ the set of all integral multiples of 6 under addition

   $G_9 = \{6^n \mid n \in \mathbf{Z}\}$ under multiplication

3.3 Write at least 5 elements of each of the following cyclic groups.

a) 25**Z** under addition

b) $\{(\tfrac{1}{2})^n \mid n \in \mathbf{Z}\}$ under multiplication

c) $\{\pi^n \mid n \in \mathbf{Z}\}$ under multiplication

3.4 Which of the following groups are cyclic? For each cyclic group, give all the generators of the group.

   $G_1 = \langle \mathbf{Z}, + \rangle$ $G_2 = \langle \mathbf{Q}, + \rangle$ $G_3 = \langle \mathbf{Q}^+, \cdot \rangle$ $G_4 = \langle 6\mathbf{Z}, + \rangle$

   $G_5 = \{6^n \mid n \in \mathbf{Z}\}$ under multiplication

   $G_6 = \{a + b\sqrt{2} \mid a, b \in \mathbf{Z}\}$ under addition

3.5 Study the structure of the table for the group Z_4 of Example 3.2.

a) By analogy, complete Table 3.5 to give a cyclic group Z_6 of 6 elements. (You need not prove the associative law.)

Table 3.5

Z_6:	+	0	1	2	3	4	5
	0	0	1	2	3	4	5
	1	1	2	3	4	5	0
	2	2					
	3	3					
	4	4					
	5	5					

b) Compute the subgroups $\langle 1 \rangle$, $\langle 2 \rangle$, $\langle 3 \rangle$, $\langle 4 \rangle$, and $\langle 5 \rangle$ of the group Z_6 given in part (a).

c) Which elements are generators for the group Z_6 of part (a)?

†**3.6** Show that if H and K are subgroups of an abelian group G, then $\{hk \mid h \in H$ and $k \in K\}$ is a subgroup of G.

3.7 Mark each of the following true or false.

— a) The associative law holds in every group.

— b) There may be a group in which the cancellation law fails.

— c) Every group is a subgroup of itself.

— d) Every group has exactly two improper subgroups.

— e) In every cyclic group, every element is a generator.

— f) This text has still given no example of a group that is not abelian.

— g) Every set of numbers that is a group under addition is also a group under multiplication.

— h) A subgroup may be defined as a subset of a group.

— i) Z_4 is a cyclic group.

— j) Every subset of every group is a subgroup under the induced operation.

3.8 Find the flaw in the following argument: "Condition 2 of Theorem 3.1 is redundant, since it can be derived from 1 and 3, for let $a \in H$. Then $a^{-1} \in H$ by 3, and by 1 $aa^{-1} = e$ is an element of H, proving 2.

3.9 Show that a nonempty subset H of a group G is a subgroup of G if and only if $ab^{-1} \in H$ for all $a, b \in H$. (This is one of the *more compact criteria* referred to prior to Theorem 3.1.)

3.10 Prove that a cyclic group with only one generator can have at most 2 elements.

3.11 Prove that if G is an abelian group with identity e, then all elements x of G satisfying the equation $x^2 = e$ form a subgroup H of G.

3.12 Repeat Exercise 3.11 for the general situation of the set H of all solutions x of the equation $x^n = e$ for a fixed integer $n \geq 1$ in an abelian group G with identity e.

3.13 Show that if $a \in G$, where G is a finite group with identity e, then there exists $n \in \mathbf{Z}^+$ such that $a^n = e$.

3.14 Let the binary operation of a group G be closed on a nonempty finite subset H of G. Show that H is a subgroup of G.

3.15 Let G be a group and let a be a fixed element of G. Show that

$$H_a = \{x \in G \mid xa = ax\}$$

is a subgroup of G.

3.16 Generalizing Exercise 3.15, let S be any subset of a group G.

a) Show that $H_S = \{x \in G \mid xs = sx \text{ for all } s \in S\}$ is a subgroup of G.

b) In reference to part (a), the subgroup H_G is the **center of** G. Show that H_G is an abelian group.

3.17 Let H be a subgroup of a group G. For $a, b \in G$, let $a \sim b$ if and only if $ab^{-1} \in H$. Show that $\sim$ is an equivalence relation on G.

3.18 For sets H and K, we define the **intersection** $H \cap K$ by

$$H \cap K = \{x \mid x \in H \text{ and } x \in K\}.$$

Show that if $H \leq G$ and $K \leq G$, then $H \cap K \leq G$.

3.19 Show by means of an example that it is possible for the quadratic equation $x^2 = e$ to have more than two solutions in some group G with identity e.

four

Permutations I

4.1 FUNCTIONS AND PERMUTATIONS

In this chapter and the one that follows, we consider groups whose elements are entities called *permutations*. These groups will provide us with our first examples of groups that are not abelian. Indeed, we shall show in a later chapter that any group is structurally the same as some group of permutations. Unfortunately, this result, which sounds very powerful, does not turn out to be particularly useful to us.

You are probably familiar with the idea of a permutation of a set as a rearrangement of the elements of the set. Thus for the set {1, 2, 3, 4, 5}, a rearrangement of the elements could be given schematically as in Fig. 4.1, resulting in the new arrangement {4, 2, 5, 3, 1}. Let us think of this schematic diagram in Fig. 4.1 as a carrying or a *mapping* of each element listed in the left column into a single (not necessarily different) element from the same set listed at the right. Thus 1 is carried into 4, 2 is mapped into 2, and so on. Furthermore, to be a permutation of the set, this mapping must be such that each element appears in the right column once and only once. For example,

$$1 \rightarrow 4 \qquad\qquad 1 \rightarrow 3$$

$$2 \rightarrow 2 \qquad\qquad 2 \rightarrow 2$$

$$3 \rightarrow 5 \qquad\qquad 3 \rightarrow 4$$

$$4 \rightarrow 3 \qquad\qquad 4 \rightarrow 5$$

$$5 \rightarrow 1 \qquad\qquad 5 \rightarrow 3$$

Fig. 4.1   **Fig. 4.2**

the diagram in Fig. 4.2 does *not* give a permutation, for 3 appears twice while 1 does not appear at all in the right column. We shall be defining a permutation as such a mapping. However, the general idea of assigning to each element of some set an element from the same or possibly from a different set will arise so often in our work that we first give a separate definition for this concept. The concept is that of a *function*, a term that you have already encountered.

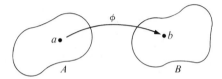

Fig. 4.3

Definition A *function* or *mapping* ϕ *from a set A into a set B* is a rule that assigns to each element a of A exactly one element b of B. We say that ϕ *maps a into b*, and that ϕ *maps A into B*.

The classic notation to denote that ϕ maps a into b is

$$\phi(a) = b.$$

However, we shall usually use the notation

$$a\phi = b.$$

The notation $a^\phi = b$ is also found in the literature. The element b is the **image of a under** ϕ. The fact that ϕ maps A into B will be symbolically expressed by

$$\phi: A \rightarrow B.$$

It may help the student to visualize a function in terms of Fig. 4.3. As to the three possible notations for a being mapped into b by ϕ given after the definition, the student is acquainted with the notation $\phi(a) = b$ from previous

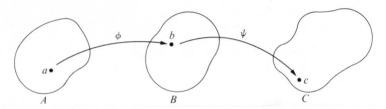

Fig. 4.4

courses. Many algebraists are using the notations $a\phi = b$ and $a^\phi = b$ for the following reason. If ϕ and ψ are functions with $\phi : A \to B$ and $\psi : B \to C$, then there is a natural function mapping A into C, as illustrated in Fig. 4.4. That is, you can get from A to C via B, using the functions ϕ and ψ. This function mapping A into C is the **composite function** consisting of ϕ followed by ψ. In classic notation $\phi(a) = b$ and $\psi(b) = c$, so

$$\psi(\phi(a)) = c,$$

and one denotes the composite function by $\psi\phi$. The symbol $\psi\phi$ for ϕ followed by ψ then has to be read from right to left. In the more recent notations we have $a\phi = b$ and $b\psi = c$ with

$$a(\phi\psi) = (a\phi)\psi = c,$$

or $a^\phi = b$ and $b^\psi = c$ with

$$a^{(\phi\psi)} = (a^\phi)^\psi = c.$$

Thus the composite function in these notations is $\phi\psi$ and can be read from left to right. We suggest that the student read the notations $a\phi = b$ and $a^\phi = b$ as "the image of a under ϕ is b. You should realize that all this talk is not about the concept involved, but about notation. However, a bad choice of notation may greatly hinder the development of a mathematical theory.

Returning to permutations, according to our definition, we see that the assignment of Fig. 4.2 is a function from $\{1, 2, 3, 4, 5\}$ into itself. We do not wish to call this a permutation, however, We need to pick out those functions ϕ such that *each* element of the set has *exactly one* element mapped into it. There is again a terminology for a more general situation.

Definition A function from a set A into a set B is **one to one** if each element of B has at most one element of A mapped into it, and is **onto B** if each element of B has at least one element of A mapped into it.

In terms of Fig. 4.3, a function $\phi : A \to B$ is one to one if each $b \in B$ has *at most one* arrow coming into it. To say that ϕ is onto B is to say that *every* $b \in B$ has *at least one* arrow coming into it. Since we will often be proving that certain functions are one to one or onto or both, it is worth outlining the technique always used.

1. To show that ϕ is one to one, you show that $a_1\phi = a_2\phi$ implies $a_1 = a_2$.
2. To show that ϕ is onto B, you show that for each $b \in B$, there exists $a \in A$ such that $a\phi = b$.

Finally, we remark that for $\phi : A \to B$, the set A is the **domain of** ϕ, the set B is the **codomain of** ϕ, and the set $A\phi = \{a\phi \mid a \in A\}$ is the **image of** A **under** ϕ.

Since for a permutation of a set A we want each element of A to have both at most one element of A and at least one element of A mapped into it, we arrive at the following definition.

> **Definition** A *permutation of a set* A is a function from A into A that is both one to one and onto. In other words, a permutation of A is a one-to-one function from A onto A.

One sometime writes

$$\phi : A \xrightarrow[\text{onto}]{1\text{-}1} B$$

for a one-to-one function ϕ mapping A onto B.

You are urged to take a little time to study and to try to understand these ideas. It will make things easier for you throughout the course. The terminology is still standard, although another terminology is coming more and more into vogue, having been spread by the disciples of N. Bourbaki. While we shall not use this terminology, we give it here so that you may realize its meaning if you run across it elsewhere. In the new terminology, a one-to-one map is an **injection,** an onto map is a **surjection,** and a map that is both one to one and onto is a **bijection.**

4.2 GROUPS OF PERMUTATIONS

A natural binary operation, *permutation multiplication*, is defined on the permutations of a set. Let A be a set, and let σ and τ be permutations of A so that σ and τ are both one-to-one functions mapping A onto A. The composite function $\sigma\tau$ as illustrated in Fig. 4.4, with $B = C = A$, $\phi = \sigma$, and $\psi = \tau$, gives a mapping of A into A. Now $\sigma\tau$ will be a permutation if it is one to one and onto A. We are using the notation of writing functions on the right so that $\sigma\tau$ can be read from left to right. Let us show that $\sigma\tau$ is one to one. If

$$a_1(\sigma\tau) = a_2(\sigma\tau),$$

then

$$(a_1\sigma)\tau = (a_2\sigma)\tau,$$

and since τ is given to be one to one, we know that $a_1\sigma = a_2\sigma$. But then, since σ is one to one, this gives $a_1 = a_2$. Hence $\sigma\tau$ is one to one. To show that $\sigma\tau$ is onto A, let $a \in A$. Since τ is onto A, there exists $a' \in A$ such that $a'\tau = a$. Since σ is onto A, there exists $a'' \in A$ such that $a' = a''\sigma$. Then

$$a = a'\tau = (a''\sigma)\tau = a''(\sigma\tau),$$

so $\sigma\tau$ is onto A.

To illustrate, suppose that

$$A = \{1, 2, 3, 4, 5\}$$

and that σ is the permutation given by Fig. 4.1. We write σ in a more standard notation as

$$\sigma = \begin{pmatrix} 1 & 2 & 3 & 4 & 5 \\ 4 & 2 & 5 & 3 & 1 \end{pmatrix},$$

so that $1\sigma = 4$, $2\sigma = 2$, and so on. Let

$$\tau = \begin{pmatrix} 1 & 2 & 3 & 4 & 5 \\ 3 & 5 & 4 & 2 & 1 \end{pmatrix}.$$

Then

$$\sigma\tau = \begin{pmatrix} 1 & 2 & 3 & 4 & 5 \\ 4 & 2 & 5 & 3 & 1 \end{pmatrix}\begin{pmatrix} 1 & 2 & 3 & 4 & 5 \\ 3 & 5 & 4 & 2 & 1 \end{pmatrix} = \begin{pmatrix} 1 & 2 & 3 & 4 & 5 \\ 2 & 5 & 1 & 4 & 3 \end{pmatrix}.$$

For example,

$$1(\sigma\tau) = (1\sigma)\tau = 4\tau = 2.$$

We now show that the collection of all permutations of a nonempty set A forms a group under this permutation multiplication.

Theorem 4.1 *Let A be a nonempty set, and let S_A be the collection of all permutations of A. Then S_A is a group under permutation multiplication.*

Proof. We have three axioms to check. Since permutations are functions, in order to show for permutations σ, τ, and μ that

$$(\sigma\tau)\mu = \sigma(\tau\mu),$$

we have to show that each composite function maps each $a \in A$ into the same image in A. That is, we must show that

$$a[(\sigma\tau)\mu] = a[\sigma(\tau\mu)]$$

for all $a \in A$. We have

$$a[(\sigma\tau)\mu] = [a(\sigma\tau)]\mu = [(a\sigma)\tau]\mu = (a\sigma)(\tau\mu) = a[\sigma(\tau\mu)].$$

Thus $(\sigma\tau)\mu$ and $\sigma(\tau\mu)$ map each $a \in A$ into the same element $[(a\sigma)\tau]\mu$ and hence are the same permutation. Since we made no use of the fact that σ, τ, and μ are one to one and onto, we have actually proved that *function composition is associative.* Hence $\mathscr{G}_1$ is satisfied.

The permutation $\imath$ such that $a\imath = a$ for all $a \in A$ obviously acts as identity. Therefore $\mathscr{G}_2$ is satisfied.

For a permutation σ, define σ^{-1} to be the permutation that reverses the direction of the mapping σ, that is, $a\sigma^{-1}$ is to be the element a' of A such that $a = a'\sigma$. The existence of exactly one such element a' is a consequence of the fact that, as a function, σ is both one to one and onto. (See Exercise 4.18.) It is clear that

$$a\imath = a = a'\sigma = (a\sigma^{-1})\sigma = a(\sigma^{-1}\sigma)$$

and also that

$$a'\imath = a' = a\sigma^{-1} = (a'\sigma)\sigma^{-1} = a'(\sigma\sigma^{-1}),$$

so that $\sigma^{-1}\sigma$ and $\sigma\sigma^{-1}$ are both the permutation $\imath$. Thus $\mathscr{G}_3$ is satisfied. ∎

There was nothing in our definition of a permutation to require that the set A be finite. However, most of our examples of permutation groups will be concerned with permutations of finite sets. Clearly, if A and B both have the same number of elements, then the group of all permutations of A has the same structure as the group of all permutations of B. One group can be obtained from the other by just renaming elements. This is again the concept of *isomorphic groups* mentioned in Chapter 2 and about which more will be said later.

Definition If A is the finite set $\{1, 2, \ldots, n\}$, then the group of all permutations of A is the **symmetric group on n letters,** and is denoted by S_n.

Note that S_n has $n!$ elements, where

$$n! = n(n - 1)(n - 2) \ldots (3)(2)(1).$$

4.3 TWO IMPORTANT EXAMPLES

Example 4.1 An interesting example for us is the group S_3 of $3! = 6$ elements. Let the set A be $\{1, 2, 3\}$. We list the permutations of A and assign to each a subscripted Greek letter for a name. The reasons for the choice of names and for the shading in the table will be clear later. Let

$$\rho_0 = \begin{pmatrix} 1 & 2 & 3 \\ 1 & 2 & 3 \end{pmatrix}, \qquad \mu_1 = \begin{pmatrix} 1 & 2 & 3 \\ 1 & 3 & 2 \end{pmatrix},$$

$$\rho_1 = \begin{pmatrix} 1 & 2 & 3 \\ 2 & 3 & 1 \end{pmatrix}, \qquad \mu_2 = \begin{pmatrix} 1 & 2 & 3 \\ 3 & 2 & 1 \end{pmatrix},$$

$$\rho_2 = \begin{pmatrix} 1 & 2 & 3 \\ 3 & 1 & 2 \end{pmatrix}, \qquad \mu_3 = \begin{pmatrix} 1 & 2 & 3 \\ 2 & 1 & 3 \end{pmatrix}.$$

You can check that the multiplication table given in Table 4.1 is correct. Note that this group is not abelian! It is our first such example. We have seen that any group of at most 4 elements is abelian. Later we will see that a group of 5 elements is also abelian. Thus S_3 has the smallest order of any nonabelian group. ∎

Table 4.1

	ρ_0	ρ_1	ρ_2	μ_1	μ_2	μ_3
ρ_0	ρ_0	ρ_1	ρ_2	μ_1	μ_2	μ_3
ρ_1	ρ_1	ρ_2	ρ_0	μ_2	μ_3	μ_1
ρ_2	ρ_2	ρ_0	ρ_1	μ_3	μ_1	μ_2
μ_1	μ_1	μ_3	μ_2	ρ_0	ρ_2	ρ_1
μ_2	μ_2	μ_1	μ_3	ρ_1	ρ_0	ρ_2
μ_3	μ_3	μ_2	μ_1	ρ_2	ρ_1	ρ_0

There is a natural correspondence between the elements of S_3 in Example 4.1 and the ways in which two copies of an equilateral triangle with vertices 1, 2, and 3 (see Fig. 4.5) can be placed, one covering the other. For this reason, S_3 is also the **group D_3 of symmetries of an equilateral triangle.** Naively, we used ρ_i for *rotations* and μ_i for *mirror images* in bisectors of angles. The

Fig. 4.5

Fig. 4.6

notation D_3 stands for the third dihedral group. The *n*th **dihedral group** D_n is the group of symmetries of the regular *n*-gon.

Example 4.2 Let us form the dihedral group D_4 of permutations corresponding to the ways that two copies of a square with vertices 1, 2, 3, and 4 can be placed, one covering the other (see Fig. 4.6). D_4 will then be the **group of symmetries of the square.** It is also called the **octic group.** Again we choose seemingly arbitrary notation and use shading that we shall explain later. Naively, we are using ρ_i for *rotations*, μ_i for *mirror images* in perpendicular bisectors of sides, and δ_i for *diagonal flips*. There are eight permutations involved here. Let

$$\rho_0 = \begin{pmatrix} 1 & 2 & 3 & 4 \\ 1 & 2 & 3 & 4 \end{pmatrix}, \qquad \mu_1 = \begin{pmatrix} 1 & 2 & 3 & 4 \\ 2 & 1 & 4 & 3 \end{pmatrix},$$

$$\rho_1 = \begin{pmatrix} 1 & 2 & 3 & 4 \\ 2 & 3 & 4 & 1 \end{pmatrix}, \qquad \mu_2 = \begin{pmatrix} 1 & 2 & 3 & 4 \\ 4 & 3 & 2 & 1 \end{pmatrix},$$

$$\rho_2 = \begin{pmatrix} 1 & 2 & 3 & 4 \\ 3 & 4 & 1 & 2 \end{pmatrix}, \qquad \delta_1 = \begin{pmatrix} 1 & 2 & 3 & 4 \\ 3 & 2 & 1 & 4 \end{pmatrix},$$

$$\rho_3 = \begin{pmatrix} 1 & 2 & 3 & 4 \\ 4 & 1 & 2 & 3 \end{pmatrix}, \qquad \delta_2 = \begin{pmatrix} 1 & 2 & 3 & 4 \\ 1 & 4 & 3 & 2 \end{pmatrix}.$$

You can check that the table for D_4 given in Table 4.2 is correct. Note that D_4 is again nonabelian. This group is simply beautiful. It will provide us with nice examples for almost all concepts we will introduce in group theory. Look at the lovely symmetries in that table!

Table 4.2

	ρ_0	ρ_1	ρ_2	ρ_3	μ_1	μ_2	δ_1	δ_2
ρ_0	ρ_0	ρ_1	ρ_2	ρ_3	μ_1	μ_2	δ_1	δ_2
ρ_1	ρ_1	ρ_2	ρ_3	ρ_0	δ_2	δ_1	μ_1	μ_2
ρ_2	ρ_2	ρ_3	ρ_0	ρ_1	μ_2	μ_1	δ_2	δ_1
ρ_3	ρ_3	ρ_0	ρ_1	ρ_2	δ_1	δ_2	μ_2	μ_1
μ_1	μ_1	δ_1	μ_2	δ_2	ρ_0	ρ_2	ρ_1	ρ_3
μ_2	μ_2	δ_2	μ_1	δ_1	ρ_2	ρ_0	ρ_3	ρ_1
δ_1	δ_1	μ_2	δ_2	μ_1	ρ_3	ρ_1	ρ_0	ρ_2
δ_2	δ_2	μ_1	δ_1	μ_2	ρ_1	ρ_3	ρ_2	ρ_0

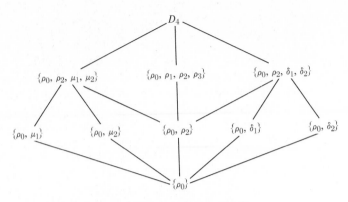

Fig. 4.7 Lattice diagram for D_4.

Finally, we give in Fig. 4.7 the lattice diagram for the subgroups of D_4. You can verify that it is correct. ∎

Exercises

4.1 Consider the three permutations in S_6:

$$\sigma = \begin{pmatrix} 1 & 2 & 3 & 4 & 5 & 6 \\ 3 & 1 & 4 & 5 & 6 & 2 \end{pmatrix},$$

$$\tau = \begin{pmatrix} 1 & 2 & 3 & 4 & 5 & 6 \\ 2 & 4 & 1 & 3 & 6 & 5 \end{pmatrix},$$

$$\mu = \begin{pmatrix} 1 & 2 & 3 & 4 & 5 & 6 \\ 5 & 2 & 4 & 3 & 1 & 6 \end{pmatrix},$$

Compute

a) $\sigma\tau$,   b) $\sigma\tau^2$,   c) $\sigma^2\mu$,   d) $\tau\sigma^{-2}$,   e) $\sigma\tau\sigma^{-1}$.

4.2 Which of the following functions from $\mathbf{R}$ into $\mathbf{R}$ are permutations of $\mathbf{R}$?

a) $f_1: \mathbf{R} \to \mathbf{R}$ defined by $f_1(x) = x + 1$

b) $f_2: \mathbf{R} \to \mathbf{R}$ defined by $f_2(x) = x^2$

c) $f_3: \mathbf{R} \to \mathbf{R}$ defined by $f_3(x) = -x^3$

d) $f_4: \mathbf{R} \to \mathbf{R}$ defined by $f_4(x) = e^x$

e) $f_5: \mathbf{R} \to \mathbf{R}$ defined by $f_5(x) = x^3 - x^2 - 2x$

4.3 Consider the group S_3 of Example 4.1.

a) Find the cyclic subgroups $\langle \rho_1 \rangle$, $\langle \rho_2 \rangle$, and $\langle \mu_1 \rangle$ of S_3.

b) Find *all* subgroups, proper and improper, of S_3 and give the lattice diagram for them.

4.4 Give the multiplication table for the cyclic subgroup of S_5 generated by

$$\rho = \begin{pmatrix} 1 & 2 & 3 & 4 & 5 \\ 2 & 4 & 5 & 1 & 3 \end{pmatrix}.$$

There will be 6 elements. Let them be ρ, ρ^2, ρ^3, ρ^4, ρ^5, and $\rho^0 = \rho^6$. Is this group isomorphic to S_3?

†**4.5** Let A be a set and let a be one particular element of A. Let T_a be the set of all permutations of A having the property that $a\sigma = a$. Show that T_a is a subgroup of the group S_A of all permutations of A that is given by Theorem 4.1.

4.6 Mark each of the following true or false.

— a) Every permutation is a one-to-one function.

— b) Every function is a permutation if and only if it is one to one.

— c) Every function from a finite set onto itself must be one-to-one.

— d) The text still has given no example of a group which is nonabelian.

— e) Every subgroup of an abelian group is abelian.

— f) Every element of a group generates a cyclic subgroup of the group.

— g) The symmetric group S_{10} has 10 elements.

— h) The symmetric group S_3 is cyclic.

— i) S_n is not cyclic for any n.

— j) Every group is isomorphic to some group of permutations.

4.7 Show by an example that every proper subgroup of a nonabelian group may be abelian.

4.8 For the permutations σ, τ, and μ in Exercise 4.1, find the following.

a) $|\langle\sigma\rangle|$ b) $|\langle\tau^2\rangle|$ c) σ^{100} d) μ^{100}

4.9 In analogy with Examples 4.1 and 4.2, consider a regular plane n-gon for $n \geq 3$. Each way that two copies of such an n-gon can be placed, with one covering the other, corresponds to a certain permutation of the vertices. The set of these permutations is a group, the n**th dihedral group** D_n, under permutation multiplication. Find the order of this group D_n. Argue *geometrically* that this group has a subgroup having just half as many elements as the whole group has.

4.10 Consider a cube that exactly fills a certain cubical box. As in Examples 4.1 and 4.2, the ways in which the cube can be placed into the box correspond to a certain group of permutations of the vertices of the cube. This group is the **group of rigid motions of the cube.** (It should not be confused with the *group of symmetries of the cube*, which will be discussed in the exercises of Chapter 10.) How many elements does this group have? Argue *geometrically* that this group has at least three different subgroups of order 4 and at least four different subgroups of order 3.

4.11 Show that S_n is a nonabelian group for $n \geq 3$.

4.12 Strengthening Exercise 4.11, show that if $n \geq 3$, then the only element σ of S_n satisfying $\sigma\gamma = \gamma\sigma$ for all $\gamma \in S_n$ is $\sigma = \iota$, the identity permutation.

4.13 Let A be a set, B a subset of A, and b one particular element of B. Which of the following must be a subgroup of S_A?

a) $\{\sigma \in S_A \mid b\sigma = b\}$

b) $\{\sigma \in S_A \mid b\sigma \in B\}$

c) $\{\sigma \in S_A \mid B\sigma \subseteq B\}$

d) $\{\sigma \in S_A \mid B\sigma = B\}$

4.14 Let A be a set and let $\sigma \in S_A$. For a fixed $a \in A$, the set

$$\mathcal{O}_{a,\sigma} = \{a\sigma^n \mid n \in \mathbf{Z}\}$$

is the **orbit of** a **under** σ. Find the orbits of 1 under each of the permutations in Exercise 4.1.

4.15 Referring to the concept defined in Exercise 4.14, show that if for $a, b \in A$, $\mathcal{O}_{a,\sigma}$ and $\mathcal{O}_{b,\sigma}$ have an element in common, then $\mathcal{O}_{a,\sigma} = \mathcal{O}_{b,\sigma}$.

4.16 If A is a set, then a subgroup H of S_A is **transitive on** A if for each $a, b \in A$ there exists $\sigma \in H$ such that $a\sigma = b$. Show that if A is a nonempty finite set, then there exists a finite cyclic subgroup H of S_A with $|H| = |A|$ which is transitive on A.

4.17 Referring to Exercises 4.14 and 4.16, show that for $\sigma \in S_A$, $\langle \sigma \rangle$ is transitive on A if and only if $\mathcal{O}_{a,\sigma} = A$ for some $a \in A$.

The following exercise is of a set-theoretic nature. You are asked to prove something we glossed over as intuitive in the text.

4.18 Let $\phi : A \to B$. A map $\phi^{-1} : B \to A$ is called an **inverse of** ϕ if $(x\phi)\phi^{-1} = x$ for all $x \in A$ and $(y\phi^{-1})\phi = y$ for all $y \in B$.

a) Show that ϕ is a bijection if and only if it has an inverse.

b) Show that the inverse of a bijection ϕ is unique.

Permutations II

5.1 CYCLES AND CYCLIC NOTATION

There is another notation for permutations that is often used. Suppose the five numbers 2, 4, 3, 6, 8 are distributed evenly on the circumference of a circle, as shown in Fig. 5.1. Suppose the circle is then rotated $2\pi/5$ radians counterclockwise so that 2 is carried into the position formerly occupied by 4, 4 into that occupied by 3, and so on. Let σ be the permutation in S_8 that leaves 1, 5, and 7 fixed and acts on the remaining elements by rotating the circle as just described. Then

$$\sigma = \begin{pmatrix} 1 & 2 & 3 & 4 & 5 & 6 & 7 & 8 \\ 1 & 4 & 6 & 3 & 5 & 8 & 7 & 2 \end{pmatrix}.$$

This permutation σ is a *cycle of length* 5 and we introduce a new, more compact notation

$$\sigma = (2, 4, 3, 6, 8)$$

for it. The new notation is *cyclic notation*. Each element appearing in (2, 4,

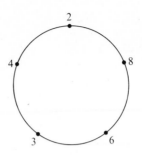

Fig. 5.1

3, 6, 8) is mapped onto the element following it except the last one, which is mapped onto the first. Elements not appearing in the notation are considered fixed by the permutation.

Definition A permutation σ of a set A is a *cycle of length* n if there exist $a_1, a_2, \ldots, a_n \in A$ such that

$$a_1\sigma = a_2, \qquad a_2\sigma = a_3, \qquad \ldots, \qquad a_{n-1}\sigma = a_n, \qquad a_n\sigma = a_1$$

and $x\sigma = x$ for $x \in A$ but $x \notin \{a_1, a_2, \ldots, a_n\}$. We write $\sigma = (a_1, a_2, \ldots, a_n)$.

When using cyclic notation, the set A must be clear from the context.

Example 5.1 If $A = \{1, 2, 3, 4, 5\}$, then

$$(1, 3, 5, 4) = \begin{pmatrix} 1 & 2 & 3 & 4 & 5 \\ 3 & 2 & 5 & 1 & 4 \end{pmatrix}.$$

Observe that

$$(1, 3, 5, 4) = (3, 5, 4, 1) = (5, 4, 1, 3) = (4, 1, 3, 5). \ \blacksquare$$

Of course since cycles are special types of permutations, they can be multiplied just as any two permutations. The product of two cycles need not again be a cycle, however.

Example 5.2 Let $(1, 4, 5, 6)$ and $(2, 1, 5)$ be cycles in the group S_6 of all permutations of $\{1, 2, 3, 4, 5, 6\}$. Then

$$(1, 4, 5, 6)(2, 1, 5) = \begin{pmatrix} 1 & 2 & 3 & 4 & 5 & 6 \\ 4 & 1 & 3 & 2 & 6 & 5 \end{pmatrix}$$

and

$$(2, 1, 5)(1, 4, 5, 6) = \begin{pmatrix} 1 & 2 & 3 & 4 & 5 & 6 \\ 6 & 4 & 3 & 5 & 2 & 1 \end{pmatrix}.$$

Neither of these permutations is a cycle. $\blacksquare$

The cycles in a collection of cycles are **disjoint** if there is no element of A that appears in the notations for two different cycles of the collection, that is, if no element of A is moved by two different cycles of the collection. In terms of mappings, every element of A must be mapped onto itself by all but at most one of the cycles in the collection, if the cycles are to be disjoint.

Let us agree that any cycle of length 1 represents the identity permutation.

We are going to show that any permutation of a finite set is a product of disjoint cycles. The proof we shall give of this is constructive, that is, the steps of the proof can be used to actually find a representation of a given permutation as a product of disjoint cycles. We believe you will understand more from this proof than from a more formal and elegant induction argument. We first illustrate the technique with an example.

Example 5.3 Consider the permutation

$$\begin{pmatrix} 1 & 2 & 3 & 4 & 5 & 6 \\ 6 & 5 & 2 & 4 & 3 & 1 \end{pmatrix}.$$

Let us write it as a product of disjoint cycles. First, 1 is moved to 6 and then 6 to 1, giving the cycle $(1, 6)$. Then 2 is moved to 5, which is moved to 3, which is moved to 2, or $(2, 5, 3)$. This takes care of all elements but 4, which is left fixed. Thus

$$\begin{pmatrix} 1 & 2 & 3 & 4 & 5 & 6 \\ 6 & 5 & 2 & 4 & 3 & 1 \end{pmatrix} = (1, 6)(2, 5, 3).$$

Multiplication of *disjoint* cycles is clearly commutative, so the order of the factors $(1, 6)$ and $(2, 5, 3)$ is not important. ∎

Theorem 5.1 *Every permutation σ of a finite set A is a product of disjoint cycles.*

Proof. It is no loss of generality to assume that $A = \{1, 2, 3, \ldots, n\}$. Consider the elements

$$1, 1\sigma, 1\sigma^2, 1\sigma^3, \ldots$$

Since A is finite, these elements cannot all be distinct. Let $1\sigma^r$ be the first term in the sequence that has appeared previously. Then $1\sigma^r = 1$, for if $1\sigma^r = 1\sigma^s$, with $0 < s < r$, we would have $1\sigma^{r-s} = 1$, with $r - s < r$, contradicting our choice of r. Let

$$\tau_1 = (1, 1\sigma, 1\sigma^2, \ldots, 1\sigma^{r-1}).$$

We see that τ_1 has the same effect as σ on all elements of A appearing in this cyclic notation for τ_1.

Let i be the first element of A not appearing in this cyclic notation for τ_1. Repeating the above argument with the sequence

$$i, i\sigma, i\sigma^2, \ldots,$$

we arrive at a cycle τ_2. Now τ_2 and τ_1 are disjoint, for if they had any element j of A in common, they would be identical, since each cycle could be constructed by repeated application of the permutation σ starting at j.

Continuing, we pick the first element in A not appearing in the cyclic notations of either τ_1 or τ_2 and construct τ_3, and so on. Since A is finite, this process must terminate with some τ_m. The product

$$\tau_1 \tau_2 \cdots \tau_m$$

then clearly has the same effect on each element of A as σ does, so

$$\sigma = \tau_1 \tau_2 \cdots \tau_m. \quad \blacksquare$$

You can easily convince yourself that the representation of a permutation as a product of disjoint cycles, none of which is the identity permutation, is unique up to the order of the factors.

5.2 EVEN AND ODD PERMUTATIONS

Definition   A cycle of length 2 is a ***transposition.***

Thus a transposition leaves all elements but two fixed, and maps each of these onto the other. A computation shows that

$$(a_1, a_2, \ldots, a_n) = (a_1, a_2)(a_1, a_3) \cdots (a_1, a_n).$$

Therefore any cycle is a product of transpositions. We then have the following as a corollary to Theorem 5.1.

Corollary   *Any permutation of a finite set of at least two elements is a product of transpositions.*

Naively, this corollary just states that any rearrangement of n objects can be achieved by successively interchanging pairs of them.

Example 5.4   Following the remarks prior to the corollary, we see that $(1, 6,)(2, 5, 3)$ is the product $(1, 6)(2, 5)(2, 3)$ of transpositions. $\blacksquare$

We have seen that every permutation of a finite set with at least 2 elements is a product of transpositions. The transpositions may not be disjoint, and a representation of the permutation in this way is not unique. For example, one can always insert at the beginning the transposition (a, b) twice, since $(a, b)(a, b)$ is the identity permutation. What is true is that the number of transpositions used to represent a given permutation must either always be even or always be odd. This is an important fact, and the usual proof of it, which may be found in the first edition of this text, involves a rather artificial construction. In 1971 William I. Miller published a proof that we like much better and that we give here.[†]

Theorem 5.2 *No permutation of a finite set can be expressed both as a product of an even number of transpositions and as a product of an odd number of transpositions.*

Proof. It is no loss of generality to let the set be $A = \{1, 2, \ldots, n\}$, and we are assuming that $n \geq 2$ so that transpositions exist.

We first study the special case of the identity permutation ι. Of course ι can be expressed as a product of an even number of transpositions, for $\iota = (1, 2)(1, 2)$. We must show that if

$$\iota = \tau_1 \tau_2 \cdots \tau_k, \tag{5.1}$$

where each τ_i is a transposition, then k must be even. Choose any integer m that appears in one of the transpositions in Eq. (5.1) and let τ_j be the first transposition, counting from left to right, in which m occurs. We can't have $j = k$, or ι would not leave m fixed. Now $\tau_j \tau_{j+1}$ must have the form on the left-hand side of one of the following easily verified identities.

$$
\begin{aligned}
(m, x)(m, x) &= \iota \\
(m, x)(m, y) &= (x, y)(m, x) \\
(m, x)(y, z) &= (y, z)(m, x) \\
(m, x)(x, y) &= (x, y)(m, y)
\end{aligned}
\tag{5.2}
$$

If we substitute the correct identity in Eq. (5.2) for $\tau_j \tau_{j+1}$ in Eq. (5.1), we either reduce the number k of transpositions in Eq. (5.1) by 2 or shift the first occurrence of m one step to the right. We repeat this process until m is eliminated from the expression of Eq. (5.1); recall that m can't appear for the first time in the final transposition, so eventually the situation in the first identity Eq. (5.2) must occur to eliminate m completely. We then choose another integer in A appearing in our reduced Eq. (5.1), and eliminate it from Eq. (5.1) by a similar process, and continue until the right-hand side of Eq. (5.1) is reduced to a sequence $\iota\iota \cdots \iota$. Since the number k was either left unchanged or reduced by 2 at each substitution of an identity from Eq. (5.2), we see that k must have been even.

It is now easy to prove the theorem from our special case for ι. Suppose that

$$\sigma = \tau_1 \tau_2 \cdots \tau_r = \tau'_1 \tau'_2 \cdots \tau'_s.$$

[†] William I. Miller, "Even and Odd Permutations," *Mathematics Associations of Two-Year Colleges, Journal* 5(1971):32.

Since every transposition is its own inverse, we obtain

$$\iota = \sigma\sigma^{-1} = \tau_1\tau_2 \cdots \tau_r(\tau_1'\tau_2' \cdots \tau_s')^{-1} = \tau_1\tau_2 \cdots \tau_r\tau_s' \cdots \tau_2'\tau_1'.$$

Our special case now shows that $r + s$ is an even number, so r and s are either both odd numbers or both even numbers. ∎

Definition   A permutation of a finite set is **even** or **odd** according to whether it can be expressed as the product of an even number of transpositions or the product of an odd number of transpositions, respectively.

5.3 THE ALTERNATING GROUPS

We claim that for $n \geq 2$, the number of even permutations in S_n is the same as the number of odd permutations, that is, S_n is split equally and both numbers are $(n!)/2$. To show this, let A_n be the set of even permutations in S_n and let B_n be the set of odd permutations for $n \geq 2$. We proceed to define a one-to-one function from A_n onto B_n. This is exactly what is needed to show that A_n and B_n have the same number of elements.

Let τ be any fixed transposition in S_n; it exists since $n \geq 2$. We may as well suppose that $\tau = (1, 2)$. We define a function

$$\lambda_\tau : A_n \to B_n$$

by

$$\sigma\lambda_\tau = \tau\sigma,$$

that is, $\sigma \in A_n$ is mapped into $(1, 2)\sigma$ by λ_τ. Observe that since σ is even, the permutation $(1, 2)\sigma$ appears as a product of a $(1 + \text{even number})$, or odd number, of transpositions, so $(1, 2)\sigma$ is indeed in B_n. If for σ and $\mu \in A_n$ it is true that $\sigma\lambda_\tau = \mu\lambda_\tau$, then

$$(1, 2)\sigma = (1, 2)\mu,$$

and since S_n is a group, we have $\sigma = \mu$. Thus λ_τ is a one-to-one function. Finally,

$$\tau = (1, 2) = \tau^{-1},$$

so if $\rho \in B_n$, then

$$\tau^{-1}\rho \in A_n,$$

and

$$(\tau^{-1}\rho)\lambda_\tau = \tau(\tau^{-1}\rho) = \rho.$$

Thus λ_τ is onto B_n. Hence the number of elements in A_n is the same as the number in B_n since there is a one-to-one correspondence between the elements of the sets.

Note that the product of two even permutations is again even. Also since $n \geq 2$, A has two elements a and b, and $\iota = (a, b)(a, b)$ is an even permutation. Finally, note that if σ is expressed as a product of transpositions, the product of the same transpositions taken in just the opposite order is σ^{-1}. Thus if σ is an even permutation, σ^{-1} must also be even. Referring to Theorem 3.1, we see that we have proved:

> **Theorem 5.3** *If $n \geq 2$, the collection of all even permutations of $\{1, 2, 3, \ldots, n\}$ forms a subgroup of order $n!/2$ of the symmetric group S_n.*

> **Definition** The subgroup of S_n consisting of the even permutations of n letters is the **alternating group A_n on n letters.**

Both S_n and A_n are very important groups. We have already mentioned without proof that every finite group is structurally identical to some subgroup of S_n for some n. The importance of A_n will appear later.

Exercises

5.1 Compute the indicated product of cycles that are permutations of $\{1,2,3,4,5,6,7,8\}$.

a) $(1, 4, 5)(7, 8)(2, 5, 7)$ b) $(1, 3, 2, 7)(4, 8, 6)$

c) $(1, 2)(4, 7, 8)(2, 1)(7, 2, 8, 1, 5)$

5.2 Express each of the following permutations of $\{1, 2, 3, 4, 5, 6, 7, 8\}$ as a product of disjoint cycles, and then as a product of transpositions.

a) $\begin{pmatrix} 1 & 2 & 3 & 4 & 5 & 6 & 7 & 8 \\ 8 & 2 & 6 & 3 & 7 & 4 & 5 & 1 \end{pmatrix}$ b) $\begin{pmatrix} 1 & 2 & 3 & 4 & 5 & 6 & 7 & 8 \\ 3 & 6 & 4 & 1 & 8 & 2 & 5 & 7 \end{pmatrix}$

c) $\begin{pmatrix} 1 & 2 & 3 & 4 & 5 & 6 & 7 & 8 \\ 3 & 1 & 4 & 7 & 2 & 5 & 8 & 6 \end{pmatrix}$

†**5.3** Prove the following about S_n if $n \geq 2$.

a) Every permutation in S_n can be written as a product of at most $n - 1$ transpositions.

b) Every permutation in S_n that is not a cycle can be written as a product of at most $n - 2$ transpositions.

c) Every odd permutation in S_n can be written as a product of $2n + 3$ transpositions, and every even permutation as a product of $2n + 8$ transpositions.

5.4 Which of the permutations in S_3 of Example 4.1 are even permutations? Give the table for the alternating group A_3.

5.5 An element a of a group G with identity e has **order** $r > 0$ if $a^r = e$ and no smaller positive power of a is the identity. Consider the group S_8 as in Exercise 5.1.

a) What is the order of the cycle $(1, 4, 5, 7)$?

b) State a theorem suggested by part (a).

c) What is the order of $\sigma = (4, 5)(2, 3, 7)$? of $\tau = (1, 4)(3, 5, 7, 8)$?

d) Find the order of each of the permutations given in Exercise 5.2 by looking at its decomposition into a product of disjoint cycles.

e) State a theorem suggested by parts (c) and (d). [*Hint:* The important words you are looking for are *least common multiple.*]

5.6 Mark each of the following true or false.

— a) Every permutation is a cycle.

— b) Every cycle is a permutation.

— c) The definition of even and odd permutations could have been given equally well before Theorem 5.2.

— d) Every nontrivial subgroup of S_9 containing some odd permutations contains a transposition.

— e) A_5 has 120 elements.

— f) S_n is not cyclic for any $n \geq 1$.

— g) A_3 is a commutative group.

— h) S_7 is isomorphic to the subgroup of all those elements of S_8 that leave the number 8 fixed.

— i) S_7 is isomorphic to the subgroup of all those elements of S_8 that leave the number 5 fixed.

— j) The odd permutations in S_8 form a subgroup of S_8.

5.7 Show that for every subgroup H of S_n for $n \geq 2$, either all the permutations in H are even or exactly half of them are even.

5.8 Let σ be a permutation of a set A. We shall say "σ **moves** $a \in A$" if $a\sigma \neq a$. If A is a finite set, how many elements are moved by a cycle $\sigma \in S_A$ of length n?

5.9 Let A be an infinite set. Let H be the set of all $\sigma \in S_A$ that move (see Exercise 5.8) only a finite number of elements of A. Show that H is a subgroup of S_A.

5.10 Let A be an infinite set. Let K be the set of all $\sigma \in S_A$ that move (see Exercise 5.8) at most 50 elements of A. Is K a subgroup of S_A? Why?

5.11 Give a more elegant proof of Theorem 5.1 than the one given in the text, by making an induction argument on the number of elements moved (see Exercise 5.8) by σ.

5.12 Consider S_n for a fixed $n \geq 2$ and let σ be a fixed odd permutation. Show that every odd permutation in S_n is a product of σ and some permutation in A_n.

5.13 Show that if σ is a cycle, then σ^2 is a cycle, provided that the length of σ is an odd integer.

5.14 Following the line of thought opened by Exercise 5.13, complete the following with a condition involving n and r so that the resulting statement is a theorem:

If σ is a cycle of length n, then σ^r is also a cycle if and only if . . .

5.15 Let G be a group and let a be a fixed element of G. Show that the map $\lambda_a : G \to G$, given by $g\lambda_a = ag$ for $g \in G$, is a permutation of the set G.

5.16 Referring to Exercise 5.15 show that $H = \{\lambda_a \mid a \in G\}$ is a subgroup of S_G, the group of all permutations of G.

5.17 Referring to Exercise 4.16, show that H of Exercise 5.16 is transitive on the set G. [*Hint:* This is an immediate corollary of one of the theorems in Chapter 2.]

six

Cyclic Groups

6.1 ELEMENTARY PROPERTIES

Recall the following facts from Chapter 3:

If G is a group and $a \in G$, then

$$H = \{a^n \mid n \in \mathbf{Z}\}$$

is a subgroup of G (Theorem 3.2). This group is the **cyclic subgroup of G generated by** a. Also, given a group G and an element $a \in G$, if

$$G = \{a^n \mid n \in \mathbf{Z}\},$$

then a is a **generator of G** and the group $G = \langle a \rangle$ is **cyclic.**

It is the purpose of this section to classify all cyclic groups and all subgroups of cyclic groups.

Theorem 6.1   *Every cyclic group is abelian.*

Proof.   Let G be a cyclic group and let a be a generator of G so that

$$G = \langle a \rangle = \{a^n \mid n \in \mathbf{Z}\}.$$

If g_1 and g_2 are any two elements of G, there exist integers r and s such that $g_1 = a^r$ and $g_2 = a^s$. Then

$$g_1 g_2 = a^r a^s = a^{r+s} = a^{s+r} = a^s a^r = g_2 g_1,$$

so G is abelian. ∎

   We shall continue to use multiplicative notation for our general work on cyclic groups, even though they are abelian.

   There is a weak but very important converse of Theorem 6.1 that we will discuss in greater detail later. Namely, it can be shown that every "sufficiently small" abelian group can be built up from cyclic groups in a certain fashion. Thus cyclic groups are very fundamental in the study of abelian groups. Cyclic groups are sort of elementary types of abelian groups. One would hope that a portion of an elementary type is again an elementary type. The next theorem shows that this is indeed so. We first give a seemingly trivial but very fundamental number-theoretic lemma.

   Lemma 6.1 (Division Algorithm for Z)   *If m is a positive integer and n is any integer, then there exist unique integers q and r such that*

$$n = mq + r \qquad and \qquad 0 \leq r < m.$$

Proof.   We give an intuitive diagrammatic explanation, using Fig. 6.1. On the real x-axis of analytic geometry, mark off the multiples of m and the position of n. Now n falls either on a multiple qm of m and r can be taken as 0, or n falls between two multiples of m. If the latter is the case, let qm be the first multiple of m to the left of n. Then r is as shown in Fig. 6.1. Note that $0 \leq r < m$. After a little thought, you will see that the uniqueness of q and r is clear from the diagrams. ∎

Fig. 6.1

   Theorem 6.2   *A subgroup of a cyclic group is cyclic.*

Proof.   Let G be a cyclic group generated by a and let H be a subgroup of G. If $H = \{e\}$, then $H = \langle e \rangle$ is cyclic. If $H \neq \{e\}$, then $a^n \in H$ for some $n \in \mathbf{Z}^+$. Let $m \in \mathbf{Z}^+$ be minimal such that $a^m \in H$.

We claim that $c = a^m$ generates H, that is,

$$H = \langle a^m \rangle = \langle c \rangle.$$

We must show that every $b \in H$ is a power of c. Since $b \in H$ and $H \leq G$, we have $b = a^n$ for some n. Find q and r such that

$$n = mq + r \qquad \text{for} \quad 0 \leq r < m$$

by Lemma 6.1. Then

$$a^n = a^{mq+r} = (a^m)^q a^r,$$

so

$$a^r = (a^m)^{-q} a^n.$$

Now since $a^n \in H$, $a^m \in H$, and H is a group, both $(a^m)^{-q}$ and a^n are in H. Thus

$$(a^m)^{-q} a^n \in H, \qquad \text{that is,} \quad a^r \in H.$$

Since m was the smallest positive integer such that $a^m \in H$ and $0 \leq r < m$, we must have $r = 0$. Thus $n = qm$ and

$$b = a^n = (a^m)^q = c^q,$$

so b is a power of c. ∎

Theorem 6.2 is for some reason one that the student is often asked to prove on a test in this course, in an oral examination for a master's degree, and so forth. It is easy, but will test whether the student has any ability in understanding and devising proofs. Note first that since the theorem concerns cyclic groups and we have proved practically nothing about them yet, the *definition* of a cyclic group *must* be used. That is, you want to show that a subgroup H of a cyclic group G is cyclic. You *must* realize that G cyclic means that there exists $a \in G$ such that every element of G is of the form a^n for $n \in \mathbf{Z}$, and that you have to pull out of thin air one element c of H that will do the same for H. *This all comes from just the definition of a cyclic group.* The ingenuity comes in defining c. But it is a natural choice, being about the only element of H you can get hold of in terms of a. Then you have to show that c works, that is, that c generates H.

As noted in Examples 3.4 and 3.5, $\mathbf{Z}$ under addition is cyclic, and for a positive integer n the set $n\mathbf{Z}$ of all multiples of n is a subgroup of $\mathbf{Z}$ under addition, the cyclic subgroup generated by n. Theorem 6.2 shows that these cyclic subgroups are the only subgroups of $\mathbf{Z}$ under addition. We state this as a corollary.

Corollary   *The subgroups of $\mathbf{Z}$ under addition are precisely the groups $n\mathbf{Z}$ under addition for $n \in \mathbf{Z}$.*

6.2 THE CLASSIFICATION OF CYCLIC GROUPS

Let G be a cyclic group with generator a. We consider two cases.

CASE I.   *G has an infinite number of elements, that is, the order of G is infinite.* In this case we claim that no two distinct exponents h and k can give equal elements a^h and a^k of G. Suppose $a^h = a^k$ and say $h > k$. Then

$$a^h a^{-k} = a^{h-k} = e,$$

the identity, and $h - k > 0$. Let m be the smallest positive integer such that $a^m = e$ (note the similarity with the construction in the proof of Theorem 6.2). We claim that G would then only have the distinct elements $e, a, a^2, \ldots, a^{m-1}$. Let $a^n \in G$, and find q and r such that

$$n = mq + r \qquad \text{for} \quad 0 \le r < m$$

by Lemma 6.1. Then

$$a^n = a^{mq+r} = (a^m)^q a^r = e^q a^r = a^r$$

for $0 \le r < m$. This would mean that G would be finite, contradicting our assumption for Case I. *Thus all powers of a are distinct.*

Suppose that G' is another infinite cyclic group with generator b. Clearly, if b^n is renamed a^n, G' can be made to look exactly like G, that is, the groups are isomorphic. This will be done again carefully in the next chapter. *Thus all infinite cyclic groups are just alike except for the names of the elements and operations.* We will take $\mathbf{Z}$ with the operation of addition as a prototype of any infinite cyclic group. From now on in Part I, "the group $\mathbf{Z}$" will always mean "the group $\mathbf{Z}$ under addition."

Example 6.1   It may seem quite odd to you that $\mathbf{Z}$ and $3\mathbf{Z}$, both being infinite cyclic groups under addition, are structurally identical although $3\mathbf{Z} < \mathbf{Z}$. You may say that $1 \in \mathbf{Z}$ but $1 \notin 3\mathbf{Z}$, so how can they be structurally the same? But names don't matter, and if 1 is renamed 3, 2 is renamed 6, and in general n is renamed $3n$, you have converted $\mathbf{Z}$ into $3\mathbf{Z}$ as an additive group. ∎

CASE II.   *G has finite order.* In this case, not all positive powers of a generator a of G can be distinct, so for some h and k we must have $a^h = a^k$. Following the argument of Case I, there exists an integer m such that $a^m = e$ and no smaller positive power of a is e. The group G then consists of the distinct elements $e, a, a^2, \ldots, a^{m-1}$.

Since it is usual to use n for the order of a general finite cyclic group, we shall change notation, setting $m = n$ for what follows.

Example 6.2   It is nice to visualize the elements $e = a^0, a^1, a^2, \ldots, a^{n-1}$ of a cyclic group of order n as being distributed evenly on the circumference of a circle (see Fig. 6.2). The element a^h is located h of these equal units counterclockwise along the circumference, measured from the bottom where $e = a^0$ is located. To multiply a^h and a^k diagrammatically, you start from a^h and go k additional units around counterclockwise. To see arithmetically where you end up, find q and r such that

$$h + k = nq + r \qquad \text{for} \qquad 0 \le r < n.$$

The nq takes you all the way around the circle q times, and you then wind up at a^r. ∎

Definition   Let n be a fixed positive integer and let h and k be any integers. The number r such that

$$h + k = nq + r \qquad \text{for} \quad 0 \le r < n$$

is the ***sum of h and k modulo n***.

Theorem 6.3   *The set $\{0, 1, 2, \ldots, n-1\}$ is a cyclic group $\mathbf{Z}_n$ of elements under addition modulo n.*

Congruence modulo n was discussed in Chapter 0; we see that if $h + k = r$ in $\mathbf{Z}_n$, then for addition in $\mathbf{Z}$ we have $h + k \equiv r \pmod{n}$.

The proof of Theorem 6.3 is easy and will give you practice using the division algorithm. (See Exercise 6.8.) You have to check $\mathcal{G}_1$, $\mathcal{G}_2$, and $\mathcal{G}_3$. Check them mentally. Think of the diagram in Fig. 6.3 as explained in Example 6.2. This really just amounts to letting the element a^h of Example 6.2 be renamed h.

Thus there is a cyclic group of order n for every positive integer n. Throughout Part I, we will let $\mathbf{Z}_n$ be the group given by Theorem 6.3. Just as in the infinite case, it is clear that if G and G' are two cyclic groups of n elements each and generators a and b respectively, then if b^r is renamed a^r, G' can be made to look just like G, that is, *any two cyclic groups of the same finite order are isomorphic.*

Fig. 6.2

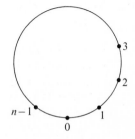

Fig. 6.3

6.3 SUBGROUPS OF FINITE CYCLIC GROUPS

We have completed our classification of cyclic groups and turn now to their subgroups. The corollary of Theorem 6.2 gives us complete information about subgroups of infinite cyclic groups. Let us give the basic theorem regarding generators of subgroups for the finite cyclic groups.

> **Theorem 6.4** Let G be a cyclic group with n elements and generated by a. Let $b \in G$ and let $b = a^s$. Then b generates a cyclic subgroup H of G containing n/d elements, where d is the greatest common divisor (abbreviated gcd) of n and s.

Proof. That b generates a cyclic subgroup H of G is known from Theorem 3.2. We need show only that H has n/d elements. Following the argument of the preceding Case I, you can see that H has as many elements as the smallest power of b that gives the identity. Now $b = a^s$ and $b^m = e$ if and only if $(a^s)^m = a^{ms} = e$, or if and only if n divides ms. What is the smallest value of m such that n divides ms? If d is the largest number dividing both n and s, then in the expression $n = d(n/d)$, the d factor of n will divide the s factor of ms. No prime factors of n/d can be absorbed by s in addition to the factor d, since we chose d as the largest integer dividing both n and s. Thus all of n/d must be absorbed in m, so the smallest such m is $m = (n/d)$. ∎

Example 6.3 Consider $\mathbf{Z}_{12}$, with a generator $a = 1$. Since the greatest common divisor (gcd) of 3 and 12 is 3, $3 = 3 \cdot 1$ generates a subgroup of $\frac{12}{3} = 4$ elements, namely

$$\langle 3 \rangle = \{0, 3, 6, 9\}.$$

Since the gcd of 8 and 12 is 4, 8 generates a subgroup of $\frac{12}{4} = 3$ elements, namely

$$\langle 8 \rangle = \{0, 4, 8\}.$$

Since the gcd of 12 and 5 is 1, 5 generates a subgroup of $\frac{12}{1} = 12$ elements, that is, 5 is a generator of the whole group $\mathbf{Z}_{12}$. ∎

The following corollary follows immediately from the theorem.

> **Corollary** If a is a generator of a finite cyclic group G of order n, then the other generators of G are the elements of the form a^r, where r is **relatively prime to** n, that is, where the greatest common divisor of r and n is 1.

Example 6.4 Let us find all subgroups of $\mathbf{Z}_{18}$ and give their lattice diagram. All subgroups are cyclic. By the corollary of Theorem 6.4, the elements 1, 5, 7, 11, 13, and 17 are all generators of $\mathbf{Z}_{18}$. Starting with 2,

$$\langle 2 \rangle = \{0, 2, 4, 6, 8, 10, 12, 14, 16\}$$

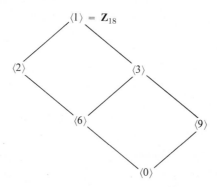

Fig. 6.4 Lattice diagram for $\mathbf{Z}_{18}$.

is of order 9 and has as generators elements of the form $h2$, where h is relatively prime to 9, namely $h = 1, 2, 4, 5, 7$, and 8, so $h2 = 2, 4, 8, 10, 14$, and 16. The element 6 of $\langle 2 \rangle$ generates $\{0, 6, 12\}$ and 12 also is a generator of this subgroup.

We have thus far found all subgroups generated by 0, 1, 2, 4, 5, 6, 7, 8, 10, 11, 12, 13, 14, and 16. This leaves just 3, 9, and 15 to consider.

$$\langle 3 \rangle = \{0, 3, 6, 9, 12, 15\},$$

and 15 also generates this group of order 6, since $15 = 5 \cdot 3$ and the gcd of 5 and 6 is 1. Finally,

$$\langle 9 \rangle = \{0, 9\}.$$

The lattice diagram for these subgroups of $\mathbf{Z}_{18}$ is given in Fig. 6.4.

This example is very easy; we are afraid we wrote it out in such hideous detail that it may look hard to you. The exercises will give you some practice along these lines. ∎

Exercises

6.1 Find the number of generators of cyclic groups of orders 6, 8, 12, and 60.

†**6.2** Show that a group that has only a finite number of subgroups must be a finite group.

6.3 Find the number of elements in each of the indicated cyclic groups.

a) The cyclic subgroup of $\mathbf{Z}_{30}$ generated by 25

b) The cyclic subgroup of $\mathbf{Z}_{42}$ generated by 30

c) The cyclic subgroup $\langle i \rangle$ of the group C^* of nonzero complex numbers under multiplication

d) The cyclic subgroup of the group C^* of part (c) generated by $(1 + i)/\sqrt{2}$

e) The cyclic subgroup of the group C^* of part (c) generated by $1 + i$

6.4 For each of the following groups, find all the subgroups and give the lattice diagram for these subgroups.

a) $\mathbf{Z}_{12}$ b) $\mathbf{Z}_{36}$ c) $\mathbf{Z}_8$

6.5 Find all orders of subgroups of $\mathbf{Z}_6, \mathbf{Z}_8, \mathbf{Z}_{12}, \mathbf{Z}_{60}$, and $\mathbf{Z}_{17}$.

6.6 Mark each of the following true or false.

___ a) Every cyclic group is abelian.

___ b) Every abelian group is cyclic.

___ c) $\mathbf{Q}$ under addition is a cyclic group.

___ d) Every element of every cyclic group generates the group.

___ e) There is at least one abelian group of every finite order > 0.

___ f) Every group of order ≤ 4 is cyclic.

___ g) All generators of $\mathbf{Z}_{20}$ are prime numbers.

___ h) S_3 is a cyclic group.

___ i) A_3 is a cyclic group.

___ j) Every cyclic group of order > 2 has at least two distinct generators.

6.7 Show by a counterexample that the following "converse" of Theorem 6.2 is not a theorem: "If a group G is such that every proper subgroup is cyclic, then G is cyclic."

6.8 Let $+_n$ be addition modulo n in $\mathbf{Z}_n = \{1, 2, 3, \ldots, n\}$. Prove that $(\mathbf{Z}_n, +_n)$ is a group. [*Hint:* Associativity is the only nontrivial axiom. Use the division algorithm, and show that both $r +_n (s +_n t)$ and $(r +_n s) +_n t$ are just the remainder of $r + s + t$ when divided by n.]

6.9 Let G be a group and suppose $a \in G$ generates a cyclic subgroup of order 2 and is the *unique* such element. Show that $ax = xa$ for all $x \in G$. (*Comment:* You have probably noticed by now that it may be hard to find a proof in algebra, even when easy proofs exist. There are usually no "pictures" you can draw that would help you devise a proof. Often you have to invent the right "trick." Coming up with the right trick takes experience, insight, and sometimes just luck. One of the foremost algebraists of this century once remarked that the way to do research in algebra is to think of a trick and then find some problem the trick will solve, rather than try to find a way to solve a specific problem. Well, try this exercise, and if you have trouble, we indicate the "trick" in the answer section.)

6.10 Let $G = \langle a \rangle$ be a finite cyclic group of order n.

a) Show that every subgroup $H \le G$ has the form $\langle a^m \rangle$ where $m > 0$ is some divisor of n, and show that for positive integers m and m' dividing n, you have $\langle a^m \rangle = \langle a^{m'} \rangle$ if and only if $m = m'$.

b) Let $D(n)$ be the set of positive integer divisors of n, and let $S(G)$ be the set of subgroups of G. Give a translation of the result of part (a) as a statement that a certain map from $D(n)$ to $S(G)$ is one to one and onto.

c) Show that if two subgroups of the finite cyclic group G have the same order, then they are the same. What integers are orders of subgroups of G?

d) Give an example to show that for *noncyclic* finite groups G, the conclusion of part (c) need not be true.

6.11 Let p and q be prime numbers. Find the number of generators of the cyclic group Z_{pq}.

6.12 Let p be a prime number. Find the number of generators of the cyclic group Z_{p^r}, where r is an integer ≥ 1.

6.13 Show that in a finite cyclic group G of order n, the equation $x^m = e$ has exactly m solutions x in G for each positive integer m that divides n.

6.14 With reference to Exercise 6.13, what is the situation if $1 < m < n$ and m does not divide n?

6.15 Show that Z_p has no proper subgroups if p is a prime number.

6.16 Let G be an abelian group and let H and K be finite cyclic subgroups with $|H| = r$ and $|K| = s$.

a) Show that if r and s are relatively prime, then G contains a cyclic subgroup of order rs.

b) Generalizing part (a), show that G contains a cyclic subgroup of order the least common multiple of r and s.

Isomorphism

7.1 DEFINITION AND ELEMENTARY PROPERTIES

We now come to the business of making more mathematically precise the idea that two groups, G and G', are structurally the same or *isomorphic*. We have tried to give you the feeling that G and G' are isomorphic if the groups are identical except for the names of the elements and operations. Thus we should be able to obtain G' from G by renaming an element x in G with the name of a certain element x' in G'. That is, to each $x \in G$ is assigned its counterpart $x' \in G'$. This is really nothing but a *function* ϕ with domain G. Clearly two different elements x and y in G should have two different counterparts $x' = x\phi$ and $y' = y\phi$ in G', that is, the function ϕ must be one to one. Also every element of G' must be the counterpart of some element of G, that is, the function ϕ must be onto G'. This renames the elements. Finally, if the groups are to be structurally the same and if for the moment we denote the group operation of G by $*$ and that of G' by $*'$, then the counterpart of $x * y$

should be $x' *' y'$, or $(x * y)\phi$ should be $(x\phi) *' (y\phi)$. Usually we drop the notations $*$ and $*'$ for the operations and use multiplicative notations, that is,

$$(xy)\phi = (x\phi)(y\phi).$$

Note that the multiplication xy on the left side in $(xy)\phi = (x\phi)(y\phi)$ is the multiplication in G, whereas the multiplication $(x\phi)(y\phi)$ on the right is in G'. We collect these ideas in a definition.

Definition An *isomorphism of a group G with a group G'* is a one-to-one function ϕ mapping G onto G' such that for all x and y in G,

$$(xy)\phi = (x\phi)(y\phi).$$

The groups G and G' are then *isomorphic.* The usual notation is $G \simeq G'$.

Let us prove a theorem that is very obvious if we consider an isomorphism to be a renaming of one group so that it is just like another. We shall prove it from our definition of an isomorphism, of course.

Theorem 7.1 *If $\phi:G \to G'$ is an isomorphism of G with G', and e is the identity of G, then $e\phi$ is the identity in G'. Also,*

$$a^{-1}\phi = (a\phi)^{-1} \qquad \text{for all} \quad a \in G.$$

Loosely, an isomorphism maps the identity onto the identity and inverses onto inverses.

Proof. Let $x' \in G'$. Since ϕ is onto, there is $x \in G$ such that $x\phi = x'$. Then

$$x' = x\phi = (ex)\phi = (e\phi)(x\phi) = (e\phi)x'.$$

Likewise

$$x' = x\phi = (xe)\phi = (x\phi)(e\phi) = x'(e\phi).$$

Thus for every $x' \in G'$, we have

$$(e\phi)x' = x' = x'(e\phi),$$

so $e\phi$ is the identity of G'.

Also for $a \in G$, we have

$$e\phi = (a^{-1}a)\phi = (a^{-1}\phi)(a\phi).$$

Likewise

$$e\phi = (aa^{-1})\phi = (a\phi)(a^{-1}\phi).$$

Thus $a^{-1}\phi = (a\phi)^{-1}$. ∎

7.2 HOW TO SHOW THAT GROUPS
ARE ISOMORPHIC

In the past, some of the author's students have had a hard time understanding and using the concept of isomorphism. We introduced it several sections before we made it precise in the hope that you would really comprehend the importance and meaning of the concept. Regarding its use, we now give an outline showing how the mathematician would proceed from the definition to show that two groups, G and G', are isomorphic.

STEP 1. *Define the function ϕ that gives the isomorphism of G with G'.* Now this means that you have to describe, in some fashion, what $x\phi$ is to be in G' for every $x \in G$.

STEP 2. *Show that ϕ is a one-to-one function.*

STEP 3. *Show that ϕ is onto G'.*

STEP 4. *Show that $(xy)\phi = (x\phi)(y\phi)$ for all $x, y \in G$.* This is just a question of computation. One computes both sides of the equation and sees whether they are the same.

We illustrate this technique with an example.

Example 7.1   Let us show that $\mathbf{R}$ under addition is isomorphic to $\mathbf{R}^+$ under multiplication.

STEP 1.   For $x \in \mathbf{R}$, define $x\phi = e^x$. This gives a mapping $\phi: \mathbf{R} \to \mathbf{R}^+$.

STEP 2.   If $x\phi = y\phi$, then $e^x = e^y$, so $x = y$. Thus ϕ is one to one.

STEP 3.   If $r \in \mathbf{R}^+$, then

$$(\ln r)\phi = e^{\ln r} = r,$$

where $(\ln r) \in \mathbf{R}$. Thus ϕ is onto $\mathbf{R}^+$.

STEP 4.   For $x, y \in \mathbf{R}$, we have

$$(x + y)\phi = e^{x+y} = e^x e^y = (x\phi)(y\phi). \quad \blacksquare$$

We illustrate this technique again in a theorem.

Theorem 7.2   *Any infinite cyclic group G is isomorphic to the group $\mathbf{Z}$ of integers under addition.*

Proof.   We suppose that G has a generator a and use multiplicative notation for the operation in G. Thus

$$G = \{a^n \mid n \in \mathbf{Z}\}.$$

Our argument in Case 1 in Section 6.2 for infinite cyclic groups showed that the elements a^n of G are all distinct, that is, $a^n \neq a^m$ if $n \neq m$.

STEP 1.   Define $\phi : G \to \mathbf{Z}$ by $a^n \phi = n$ for all $a^n \in G$.

STEP 2.   If $a^n \phi = a^m \phi$, then $n = m$ and $a^n = a^m$. Thus ϕ is one to one.

STEP 3.   For any $n \in \mathbf{Z}$, the element $a^n \in G$ is mapped onto n by ϕ. Thus ϕ is onto $\mathbf{Z}$.   proof ?

STEP 4.   Now $(a^n a^m)\phi = a^{n+m}\phi = n + m$. (Note that the binary operation here was in the group G.) It remains to compute $(a^n \phi) + (a^m \phi)$, the $+$ appearing since the operation in $\mathbf{Z}$ is addition. But $(a^n \phi) + (a^m \phi)$ is again $n + m$. Thus $(a^n a^m)\phi = (a^n \phi) + (a^m \phi)$. ∎

The preceding proof was certainly very easy; be sure you understand the steps involved.

It is immediate that every group G is isomorphic to itself; the identity function ι defined by $g\iota = g$ for all $g \in G$ shows this. If G is isomorphic to G', then G' is isomorphic to G; the function $\phi^{-1}: G' \to G$ for an isomorphism $\phi : G \to G'$ shows this (see Exercise 7.6). Finally, if G is isomorphic to G' and G' is isomorphic to G'', then G is isomorphic to G''; if $\phi : G \to G'$ and $\psi : G' \to G''$ are isomorphisms, then the composite function $\phi\psi$ shows this (see Exercise 7.7). You should recognize that we have shown that the property of isomorphism is an equivalence relation on a collection of groups. In view of Theorem 0.1, this means that *given a nonempty collection of groups, you can always partition the collection into cells (equivalence classes) such that any two groups in the same cell are isomorphic and no two groups in different cells are isomorphic.*

We have seen that any two groups of order 3 are isomorphic. We express this by saying that *there is only one group of order 3 up to isomorphism.*

Example 7.2   There is only one group of order 1, one of order 2, and one of order 3 up to isomorphism. We saw in Example 3.2 that there are exactly two different groups of order 4 up to isomorphism, the group $\mathbf{Z}_4$ and the Klein 4-group V. There are at least two different groups of order 6 up to isomorphism, namely $\mathbf{Z}_6$ and S_3. ∎

7.3 HOW TO SHOW THAT GROUPS ARE NOT ISOMORPHIC

We turn now to a topic discussed in few algebra texts, namely:

How do you demonstrate that two groups G and G' are not isomorphic, if this is the case?

This would mean that there is no one-to-one function ϕ from G onto G' with the property $(xy)\phi = (x\phi)(y\phi)$. In general, it is clearly not feasible to try every possible one-to-one function to find out whether it has the above property, except in the case where there are *no* one-to-one functions. This is the case, for example, if G and G' are of finite order and have different numbers of elements.

Example 7.3 $\mathbf{Z}_4$ and S_6 are *not* isomorphic. There is no one-to-one function from $\mathbf{Z}_4$ onto S_6. ■

In the infinite case, it is not always clear whether there are any one-to-one onto functions. For example, you may think that $\mathbf{Q}$ has "more" elements than $\mathbf{Z}$, but your instructor can show you in five minutes (ask him or her to!) that *there are lots of one-to-one functions from $\mathbf{Z}$ onto $\mathbf{Q}$*. However, it is true that $\mathbf{R}$ *has too many elements to be put into a one-to-one correspondence with* $\mathbf{Z}$. This will take your instructor only another five minutes to show you.

Example 7.4 $\mathbf{Z}$ under addition is *not* isomorphic with $\mathbf{R}$ under addition, for there is no one-to-one function from $\mathbf{Z}$ onto $\mathbf{R}$. ■

In the event that there are one-to-one mappings of G onto G', one usually shows that the groups are not isomorphic (if this is the case) by showing that one group has some structural property that the other does not possess. A **structural property of a group** *is one that must be shared by any isomorphic group. It must not depend on the names or some other nonstructural characteristics of the elements. The following are examples of some structural properties and some nonstructural properties of groups.*

Possible Structural Properties	*Possible Nonstructural Properties*
1. The group is cyclic.	1′. The group contains 5.
2. The group is abelian.	2′. All elements of the group are numbers.
3. The group has order 8.	3′. The group operation is called "composition."
4. The group is finite.	4′. The elements of the group are permutations.
5. The group has exactly two elements of order 6.	5′. The group operation is denoted by juxtaposition.
6. The equation $x^2 = a$ has a solution for each element a in the group.	6′. The group is a subgroup of $\langle \mathbf{R}, + \rangle$.

Of course there are many other possible structural properties we could have listed. The fact that each of the properties 1 through 6 is indeed structural is a little theorem about isomorphic groups. We ask you to prove a couple of such theorems in the exercises. (See Exercises 7.8 and 7.9.) We shall regard them as obviously structural in our work in the text.

Example 7.5   You can't say that $\mathbf{Z}$ and $3\mathbf{Z}$ under addition are not isomorphic because $17 \in \mathbf{Z}$ and $17 \notin 3\mathbf{Z}$. These are *not* structural properties but rather have to do with just the names of the elements. Actually $\mathbf{Z}$ and $3\mathbf{Z}$ are isomorphic under the map $\phi : \mathbf{Z} \to 3\mathbf{Z}$, where $n\phi = 3n$. ∎

Example 7.6   You can't say that $\mathbf{Z}$ and $\mathbf{Q}$, both under addition, are not isomorphic because $\frac{1}{2} \in \mathbf{Q}$ and $\frac{1}{2} \notin \mathbf{Z}$. But you can say that they are not isomorphic because $\mathbf{Z}$ is cyclic and $\mathbf{Q}$ is not. ∎

Example 7.7   The group $\mathbf{Q}^*$ of nonzero elements of $\mathbf{Q}$ under multiplication is not isomorphic to the group $\mathbf{R}^*$ of nonzero elements of $\mathbf{R}$ under multiplication. One argument is that there is no one-to-one correspondence between them. Another is that every element in $\mathbf{R}^*$ is the cube of some element of $\mathbf{R}^*$, that is, for $a \in \mathbf{R}^*$, the equation $x^3 = a$ has a solution in $\mathbf{R}^*$. This is not true for $\mathbf{Q}^*$; for example, the equation $x^3 = 2$ has no solution in $\mathbf{Q}^*$. ∎

Example 7.8   The group $\mathbf{R}^*$ of nonzero real numbers under multiplication is not isomorphic to the group $\mathbf{C}^*$ of nonzero complex numbers under multiplication. Every element of $\mathbf{R}^*$ generates an infinite cyclic subgroup except for 1 and -1, which generate subgroups of orders 1 and 2, respectively. But in $\mathbf{C}^*$, i generates the cyclic subgroup $\{i, -1, -i, 1\}$ of order 4. Or, for another argument, the equation $x^2 = a$ has a solution x in $\mathbf{C}^*$ for every $a \in \mathbf{C}^*$, but $x^2 = -1$ has no solution in $\mathbf{R}^*$. ∎

Example 7.9   The group $\mathbf{R}^*$ of nonzero real numbers under multiplication is not isomorphic to the group $\mathbf{R}$ of real numbers under addition. An equation $x + x = a$ always has a solution in $\langle \mathbf{R}, + \rangle$ for every $a \in \mathbf{R}$, but the corresponding equation $x \cdot x = a$ does not always have a solution in $\langle \mathbf{R}^*, \cdot \rangle$, for example if $a = -1$. ∎

7.4 CAYLEY'S THEOREM

Look at any group table in the text. Note how each row of the table gives a permutation of the set of elements of the group, as listed at the top of the table. Similarly, each column of the table gives a permutation of the group set, as listed at the left of the table. In view of these observations, it is not surprising that at least every finite group G is isomorphic to a subgroup of the group S_G of all permutations of G. The same is true for infinite groups; Cayley's Theorem states that *every* group is isomorphic to some group consisting of permutations under permutation multiplication. This is a nice and intriguing result, although we shall have no important use for it. However, the theorem is a classic of group theory and appears in almost all algebra texts. Moreover, it is the first theorem we come to of any complexity; it brings together various ideas and techniques to which the student has been exposed separately before. Every student should know the statement of Cayley's Theorem. We mark the proof with a star, indicating that we do not consider this result to be of basic importance for this text.

To make it easier for you to follow the proof, we outline the steps to be taken. Starting with any given group G, we shall proceed as follows:

STEP 1.   Find a set G' of permutations that is a candidate for forming a group under permutation multiplication isomorphic to G.

STEP 2.   Prove that G' is a group under permutation multiplication.

STEP 3.   Define a mapping $\phi: G \rightarrow G'$, and show that ϕ is an isomorphism of G with G'.

Theorem 7.3 (Cayley)   *Every group is isomorphic to a group of permutations.*

Proof.   Let G be a given group.

STEP 1.   Our first task is to find a set G' of permutations that is a candidate to form a group isomorphic to G. Think of G as just a set, and let S_G be the group of all permutations of G given by Theorem 4.1. (Note that in the finite case if G has n elements, S_G has $n!$ elements. Thus, in general, S_G is clearly too big to be isomorphic to G.) We define a certain subset of S_G. For $a \in G$, let ρ_a be the mapping of G into G given by

$$x\rho_a = xa$$

for $x \in G$. (We can think of ρ_a as meaning *right multiplication by a*.) If $x\rho_a = y\rho_a$, then $xa = ya$, so $x = y$ by Theorem 2.1. Thus ρ_a is a one-to-one function. Also, if $y \in G$, then

$$(ya^{-1})\rho_a = (ya^{-1})a = y,$$

so ρ_a maps G onto G. Since $\rho_a: G \rightarrow G$ is both one to one and onto G, ρ_a is a permutation of G, that is, $\rho_a \in S_G$. Let

$$G' = \{\rho_a \mid a \in G\}.$$

STEP 2.   We claim that G' is a subgroup of S_G. We must show that G' is closed under permutation multiplication, contains the identity permutation, and contains an inverse for each of its elements. First, we claim that

$$\rho_a \rho_b = \rho_{ab}.$$

To show that these functions are the same, we must show that they have the same action on each $x \in G$. Now

$$x(\rho_a \rho_b) = (x \rho_a) \rho_b = (xa) \rho_b = (xa)b = x(ab) = x \rho_{ab}.$$

Thus $\rho_a \rho_b = \rho_{ab}$, so G' is closed under multiplication. Clearly for all $x \in G$,

$$x \rho_e = xe = x,$$

where e is the identity element of G, so ρ_e is the identity permutation ι of S_G and is in G'. Since $\rho_a \rho_b = \rho_{ab}$, we have

$$\rho_a \rho_{a^{-1}} = \rho_{aa^{-1}} = \rho_e,$$

and also

$$\rho_{a^{-1}} \rho_a = \rho_e.$$

Hence

$$(\rho_a)^{-1} = \rho_{a^{-1}},$$

so $(\rho_a)^{-1} \in G'$. Thus G' is a subgroup of S_G.

STEP 3.   It remains for us now to prove that G is isomorphic to this group G' that we have described. Define $\phi : G \rightarrow G'$ by

$$a\phi = \rho_a$$

for $a \in G$. If $a\phi = b\phi$, then ρ_a and ρ_b must be the same permutation of G. In particular,

$$e\rho_a = e\rho_b,$$

so $ea = eb$, and $a = b$. Thus ϕ is one to one. It is immediate that ϕ is onto G' by the definition of G'. Finally, $(ab)\phi = \rho_{ab}$, while

$$(a\phi)(b\phi) = \rho_a \rho_b.$$

But we saw above that ρ_{ab} and $\rho_a \rho_b$ are the same permutation of G. Thus

$$(ab)\phi = (a\phi)(b\phi). \quad \blacksquare$$

For the proof of the theorem, we could have considered equally well the permutations λ_a of G defined by

$$x\lambda_a = ax$$

for $x \in G$. (We can think of λ_a as meaning *left multiplication by a*.) These permutations would have formed a subgroup G'' of S_G, again isomorphic to G but under the map $\psi : G \rightarrow G''$ defined by

$$a\psi = \lambda_{a^{-1}}.$$

Definition   The group G' in the proof of Theorem 7.3 is the **right regular representation of** G, and the group G'' in the preceding comment is the **left regular representation of** G.

Table 7.1			
	e	a	b
e	e	a	b
a	a	b	e
b	b	e	a

Table 7.2			
	ρ_e	ρ_a	ρ_b
ρ_e	ρ_e	ρ_a	ρ_b
ρ_a	ρ_a	ρ_b	ρ_e
ρ_b	ρ_b	ρ_e	ρ_a

Example 7.10 Let us compute the right regular representation of the group given by the group table Table 7.1. By "compute" we mean give the elements of the right regular representation and the group table. Here the elements are

$$\rho_e = \begin{pmatrix} e & a & b \\ e & a & b \end{pmatrix}, \qquad \rho_a = \begin{pmatrix} e & a & b \\ a & b & e \end{pmatrix}, \qquad \text{and} \qquad \rho_b = \begin{pmatrix} e & a & b \\ b & e & a \end{pmatrix}.$$

The table for this representation is just like the original table with x renamed ρ_x, as seen in Table 7.2. *This "renaming" is the basic idea of an isomorphism.* For example,

$$\rho_a \rho_b = \begin{pmatrix} e & a & b \\ a & b & e \end{pmatrix} \begin{pmatrix} e & a & b \\ b & e & a \end{pmatrix} = \begin{pmatrix} e & a & b \\ e & a & b \end{pmatrix} = \rho_e. \quad \blacksquare$$

For a finite group given by a group table, ρ_a is the permutation of the elements corresponding to their order in the column under a at the very top, and λ_a is the permutation corresponding to the order of the elements in the row opposite a at the extreme left. The notations ρ_a and λ_a were chosen to suggest right and left multiplication by a, respectively.

Exercises

7.1 Give two arguments showing that $\mathbf{Z}_4$ is not isomorphic to the Klein 4-group V of Example 3.2.

7.2 Partition the following collection of groups into subcollections of isomorphic groups, as discussed following Theorem 7.2. Here a * superscript means all nonzero elements of the set.

Z under addition

$\mathbf{Z}_6$

$\mathbf{Z}_2$

S_6

17Z under addition

Q under addition

3Z under addition

R under addition

S_2

R* under multiplication

R$^+$ under multiplication

Q* under multiplication

C* under multiplication

The subgroup $\langle \pi \rangle$ of R* under multiplication.

The subgroup G of S_8 generated by (1, 3, 4)(2, 6)

7.3 Give a formal proof (for example, as was done in Theorem 7.1) of the statement that if ϕ is an isomorphism of a group G with a group G', and H is a subgroup of G, then

$$H\phi = \{h\phi \mid h \in H\}$$

is a subgroup of G'. (This is obvious from the motivation of our definition of an isomorphism, but it will do you good to try to write down a formal proof based on just the *definition* of an isomorphism.)

7.4 Let G be a cyclic group with generator a, and let G' be a group isomorphic to G. If $\phi: G \to G'$ is an isomorphism, show that, for every $x \in G$, $x\phi$ is completely determined by the value $a\phi$.

7.5 Mark each of the following true or false.

___ a) Any two groups of order 3 are isomorphic.

___ b) There is, up to isomorphism, only one cyclic group of a given finite order.

___ c) Any two finite groups with the same number of elements are isomorphic.

___ d) Every isomorphism is a one-to-one function.

___ e) Every one-to-one function between groups is an isomorphism.

___ f) The property of being cyclic (or not being cyclic, as the case may be) is a structural property of a group.

___ g) A structural property of a group must be shared by every isomorphic group.

___ h) An abelian group can't be isomorphic to a nonabelian group.

___ i) An additive group can't be isomorphic to a multiplicative group.

___ j) **R** under addition is isomorphic to a group of permutations.

7.6 Let $\phi: G \to G'$ be an isomorphism of a group G with a group G'. Show that the map $\phi^{-1}: G' \to G$, defined by $x'\phi^{-1} = x$ such that $x\phi = x'$ for $x' \in G'$, is a well-defined function and is an isomorphism of G' with G.

7.7 Let $\phi: G \to G'$ be an isomorphism of a group G with a group G', and $\psi: G' \to G''$ an isomorphism of G' with a group G''. Show that $\phi\psi: G \to G''$ is an isomorphism of G with G''.

7.8 Let G be an abelian group. Prove that being abelian is a structural property of G by showing that if G' is isomorphic to G, then G' is abelian also.

7.9 Let G be a cyclic group. Prove that the property of being cyclic is a structural property of G. (See Exercise 7.8.)

7.10 An isomorphism of a group with itself is an **automorphism of the group.** How many automorphisms are there of $\mathbf{Z}_2$? of $\mathbf{Z}_6$? of $\mathbf{Z}_8$? of $\mathbf{Z}$? of $\mathbf{Z}_{17}$? [*Hint:* Use Exercise 7.4.]

7.11 Let $\langle G, \cdot \rangle$ be a group. Consider the binary operation $*$ on the set G defined by

$$a * b = b \cdot a$$

for $a, b \in G$. Show that $\langle G, * \rangle$ is a group and that $\langle G, * \rangle$ is actually isomorphic to $\langle G, \cdot \rangle$. [*Hint:* Consider the map ϕ with $a\phi = a^{-1}$ for $a \in G$.]

    Comments: This is one instance where the notations $\langle G, \cdot \rangle$ and $\langle G, * \rangle$ are *very* handy. See the discussion following the definition of a group. Note that if G is finite, then we obtain the group table for $\langle G, * \rangle$ from the group table for $\langle G, \cdot \rangle$ by reading this table from the top to the left instead of from the left to the top.

7.12 Prove in a manner similar to that used in Theorem 7.2 that every finite cyclic group of order n is isomorphic to $\mathbf{Z}_n$.

7.13 Let G be a group and let g be a fixed element of G. Show that the map i_g, such that $xi_g = gxg^{-1}$ for $x \in G$, is an isomorphism of G with itself, that is, an automorphism of G (see Exercise 7.10).

***7.14** Compute the left regular representation of $\mathbf{Z}_4$. Compute the right regular representation of S_3 using the notation of Example 4.1.

7.15 Let $\langle S, * \rangle$ be the group of all real numbers except -1 under the operation $*$ defined by $a * b = a + b + ab$ (see Exercise 2.9). Show that $\langle S, * \rangle$ is isomorphic to the group $\mathbf{R}^*$ of nonzero real numbers under multiplication. Actually define an isomorphism $\psi : \mathbf{R}^* \to S$.

7.16 Let ϕ be an isomorphism of a group G with a group G'. If for $x \in G$ we view $x\phi$ as a new name for x or consider $x\phi$ as x renamed, then the condition $(xy)\phi = (x\phi)(y\phi)$ corresponds to the statement that the diagram in Fig. 7.1 is commutative. By "the diagram is commutative" we mean that starting in the upper left corner and following the path to the lower right corner, given by (arrow down) (arrow across), gives the same result as following the path (arrow across) (arrow down). Illustrating with the isomorphism ψ of our answer to Exercise 7.15, if we consider $x\psi$ as x renamed for $x \in \mathbf{R}^*$, we get the diagram in Fig. 7.2 for $x = 2$ and $y = 5$.

Fig. 7.1

Fig. 7.2

a) Starting with the group $\mathbf{R}^*$ under multiplication, suppose x is renamed $x - 4$ for $x \in \mathbf{R}^*$. Let S_1 be the set of all real numbers except -4. Define $*_1$ on S_1 such that $\langle S_1, *_1 \rangle$ is isomorphic to $\mathbf{R}^*$ under multiplication by means of this renaming.

b) Repeat part (a) if $x \in \mathbf{R}^*$ is renamed $x - t$ for fixed $t \in \mathbf{R}$. First determine the required set S_2.

c) Repeat part (a) if $x \in \mathbf{R}^*$ is renamed $x^3 + 1$. First determine the required set S_3.

eight

Direct Products

8.1 EXTERNAL DIRECT PRODUCTS

Let us take a moment to review our present stockpile of groups. Starting with finite groups, we have the cyclic group $\mathbf{Z}_n$, the symmetric group S_n, and the alternating group A_n for each positive integer n. We also have the octic group D_4 of Example 4.2 and the Klein 4-group V. Of course we know that subgroups of these groups exist, and indeed Cayley's theorem applied to finite groups shows that every finite group is isomorphic to a subgroup of some S_n. But we have no easy way of computing all subgroups of a given group. Turning to infinite groups, we have groups consisting of sets of numbers under the usual addition or multiplication, as for example $\mathbf{Z}$ and $\mathbf{R}$ under addition.

One purpose of this chapter is to show a constructive method for using known groups as building blocks to form more groups. The Klein 4-group will be recovered from the cyclic groups. As we shall describe in the next chapter, employing this procedure with the cyclic groups will give us a large class of abelian groups that includes all abelian groups of finite order. Let us start with a set-theoretic definition.

Definition The *Cartesian product of sets* $S_1, S_2, \ldots, S_n$ is the set of all ordered n-tuples $(a_1, a_2, \ldots, a_n)$, where $a_i \in S_i$. The Cartesian product is denoted by either

$$S_1 \times S_2 \times \cdots \times S_n$$

or by

$$\prod_{i=1}^{n} S_i.$$

One can also define the Cartesian product of an infinite number of sets, but the definition is considerably more sophisticated and we shall not need it.

Now let $G_1, G_2, \ldots, G_n$ be groups, and let us use multiplicative notation for all the group operations. Regarding the G_i as sets, we can form $\prod_{i=1}^{n} G_i$. Let us show that one can make $\prod_{i=1}^{n} G_i$ into a group by means of a binary operation of *multiplication by components*. We point out again that we are being sloppy in our use of the same notation for a group as for the set of elements of the group.

Theorem 8.1 *Let $G_1, G_2, \ldots, G_n$ be groups. For $(a_1, a_2, \ldots, a_n)$ and $(b_1, b_2, \ldots, b_n)$ in $\prod_{i=1}^{n} G_i$, define $(a_1, a_2, \ldots, a_n)(b_1, b_2, \ldots, b_n)$ to be $(a_1 b_1, a_2 b_2, \ldots, a_n b_n)$. Then $\prod_{i=1}^{n} G_i$ is a group, the **external direct product of the groups** G_i, under this binary operation.*

Proof. Note that since $a_i \in G_i$, $b_i \in G_i$, and G_i is a group, we have $a_i b_i \in G_i$. Thus the definition of the binary operation on $\prod_{i=1}^{n} G_i$ given in the statement of the theorem makes sense, that is, $\prod_{i=1}^{n} G_i$ is closed under the binary operation.

The associative law in $\prod_{i=1}^{n} G_i$ is thrown back onto the associative law in each component as follows:

$$
\begin{aligned}
(a_1, a_2, \ldots, & a_n)[(b_1, b_2, \ldots, b_n)(c_1, c_2, \ldots, c_n)] \\
&= (a_1, a_2, \ldots, a_n)(b_1 c_1, b_2 c_2, \ldots, b_n c_n) \\
&= (a_1(b_1 c_1), a_2(b_2 c_2), \ldots, a_n(b_n c_n)) \\
&= ((a_1 b_1)c_1, (a_2 b_2)c_2, \ldots, (a_n b_n)c_n) \\
&= (a_1 b_1, a_2 b_2, \ldots, a_n b_n)(c_1, c_2, \ldots, c_n) \\
&= [(a_1, a_2, \ldots, a_n)(b_1, b_2, \ldots, b_n)](c_1, c_2, \ldots, c_n).
\end{aligned}
$$

If e_i is the identity element in G_i, then clearly, with multiplication by components, $(e_1, e_2, \ldots, e_n)$ is an identity in $\prod_{i=1}^{n} G_i$. An inverse of $(a_1, a_2, \ldots, a_n)$ is $(a_1^{-1}, a_2^{-1}, \ldots, a_n^{-1})$; just compute the product by components. Hence $\prod_{i=1}^{n} G_i$ is a group. ∎

In the event that the operation in each G_i is commutative, we sometimes use additive notation in $\prod_{i=1}^{n} G_i$ and refer to $\prod_{i=1}^{n} G_i$ as the **external direct sum of the groups** G_i. The notation $\bigoplus_{i=1}^{n} G_i$ is sometimes used in this case in place of $\prod_{i=1}^{n} G_i$, especially with abelian groups with operation $+$. The direct sum of abelian groups $G_1, G_2, \ldots, G_n$ may be written $G_1 \oplus G_2 \oplus \cdots \oplus G_n$. We leave as an exercise the trivial proof that an external direct product of abelian groups is again abelian.

It is easily seen that if the set S_i has r_i elements for $i = 1, \ldots, n$, then $\prod_{i=1}^{n} S_i$ has $r_1 r_2 \cdots r_n$ elements, for in an n-tuple, there are r_1 choices for the first component from S_1, and for each of these there are r_2 choices for the next component from S_2, and so on.

Example 8.1 Consider the group $\mathbf{Z}_2 \times \mathbf{Z}_3$ which has $2 \cdot 3 = 6$ elements, namely $(0, 0)$, $(0, 1)$, $(0, 2)$, $(1, 0)$, $(1, 1)$, and $(1, 2)$. We claim that $\mathbf{Z}_2 \times \mathbf{Z}_3$ is cyclic. It is only necessary to find a generator. Let us try $(1, 1)$. Here the operations in $\mathbf{Z}_2$ and $\mathbf{Z}_3$ are written additively, so we do the same in the external direct product $\mathbf{Z}_2 \times \mathbf{Z}_3$.

$$(1, 1) = (1, 1)$$
$$2(1, 1) = (1, 1) + (1, 1) = (0, 2)$$
$$3(1, 1) = (1, 1) + (1, 1) + (1, 1) = (1, 0)$$
$$4(1, 1) = 3(1, 1) + (1, 1) = (1, 0) + (1, 1) = (0, 1)$$
$$5(1, 1) = 4(1, 1) + (1, 1) = (0, 1) + (1, 1) = (1, 2)$$
$$6(1, 1) = 5(1, 1) + (1, 1) = (1, 2) + (1, 1) = (0, 0)$$

Thus $(1, 1)$ generates all of $\mathbf{Z}_2 \times \mathbf{Z}_3$. Since there is only one cyclic group of a given order up to isomorphism, we have $(\mathbf{Z}_2 \times \mathbf{Z}_3) \simeq \mathbf{Z}_6$. ∎

Example 8.2 Consider $\mathbf{Z}_3 \times \mathbf{Z}_3$. This is a group of nine elements. We claim that $\mathbf{Z}_3 \times \mathbf{Z}_3$ is *not* isomorphic to $\mathbf{Z}_9$. We need only show that $\mathbf{Z}_3 \times \mathbf{Z}_3$ is not cyclic. Since the addition is by components, and since in $\mathbf{Z}_3$ every element added to itself three times gives the identity, the same is true in $\mathbf{Z}_3 \times \mathbf{Z}_3$. Thus no element can generate the group, for a generator added to itself successively could only give the identity after nine summands. We have found another group of order 9. A similar argument shows that $\mathbf{Z}_2 \times \mathbf{Z}_2$ is not cyclic. Thus $\mathbf{Z}_2 \times \mathbf{Z}_2$ must be isomorphic to the Klein 4-group. ∎

The preceding examples illustrate the following theorem:

Theorem 8.2 *The group $\mathbf{Z}_m \times \mathbf{Z}_n$ is isomorphic to $\mathbf{Z}_{mn}$ if and only if m and n are relatively prime, that is, the gcd of m and n is 1.*

Proof. Consider the cyclic subgroup of $\mathbf{Z}_m \times \mathbf{Z}_n$ generated by $(1, 1)$ as described by Theorem 3.2. As our previous work has shown, the order of this cyclic subgroup is the smallest power of $(1, 1)$ that gives the identity $(0, 0)$. Here taking a power of $(1, 1)$ in our additive notation will involve adding $(1, 1)$ to itself repeatedly. Under addition by components, the first component $1 \in \mathbf{Z}_m$ yields 0 only after m summands, $2m$ summands and so on, and the second component $1 \in \mathbf{Z}_n$ yields 0 only after n summands, $2n$ summands, and so on. For them to yield 0 simultaneously, the number of summands must be a multiple of both m and n. The smallest number that is a multiple of both m and n will be mn if and only if the gcd of m and n is 1; in this case $(1, 1)$ generates a cyclic subgroup of order mn, which is the order of the whole group. This shows that $\mathbf{Z}_m \times \mathbf{Z}_n$ is isomorphic to $\mathbf{Z}_{mn}$ if m and n are relatively prime.

If the gcd of m and n is $d > 1$, then mn/d is divisible by both m and n. Consequently, for any (r, s) in $\mathbf{Z}_m \times \mathbf{Z}_n$, we have

$$\underbrace{(r, s) + (r, s) + \cdots + (r, s)}_{mn/d \text{ summands}} = (0, 0).$$

Hence no element (r, s) in $\mathbf{Z}_m \times \mathbf{Z}_n$ can generate the entire group, so $\mathbf{Z}_m \times \mathbf{Z}_n$ is not cyclic and therefore not isomorphic to $\mathbf{Z}_{mn}$. ■

It is clear that this theorem can be extended to a product of more than two factors by an induction argument. We state this as a corollary without going through the details of the proof.

Corollary   *The group $\prod_{i=1}^{n} \mathbf{Z}_{m_i}$ is cyclic and isomorphic to $\mathbf{Z}_{m_1 m_2 \cdots m_n}$ if and only if the numbers m_i for $i = 1, \ldots, n$ are such that the gcd of any two of them is 1.*

Example 8.3   The preceding corollary shows that if n is written as a product of powers of distinct prime numbers, as in

$$n = (p_1)^{n_1} (p_2)^{n_2} \cdots (p_r)^{n_r},$$

then $\mathbf{Z}_n$ is isomorphic to

$$\mathbf{Z}_{(p_1)^{n_1}} \times \mathbf{Z}_{(p_2)^{n_2}} \times \cdots \times \mathbf{Z}_{(p_r)^{n_r}}.$$

In particular, $\mathbf{Z}_{72}$ is isomorphic to $\mathbf{Z}_8 \times \mathbf{Z}_9$. ■

We have had several occasions to use the concept of the smallest positive power of an element of a group that gives the identity. It is time we introduced the usual terminology for this.

Definition   Let G be a group and $a \in G$. If there is a positive integer n such that $a^n = e$, the least such positive integer n is the **order of** a. If no such n exists, then a is of **infinite order.**

It is immediate that *if a is an element of a group G, the order of a is the same as the order of the cyclic subgroup generated by a.* This is a very useful fact to remember.

Theorem 8.3 *Let* $(a_1, a_2, \ldots, a_n) \in \prod_{i=1}^{n} G_i$. *If* a_i *is of finite order* r_i *in* G_i, *then the order of* $(a_1, a_2, \ldots, a_n)$ *in* $\prod_{i=1}^{n} G_i$ *is equal to the least common multiple of all the* r_i.

Proof. This follows by a repetition of the argument used in the proof of Theorem 8.2. For a power of $(a_1, a_2, \ldots, a_n)$ to give $(e_1, e_2, \ldots, e_n)$, the power must simultaneously be a multiple of r_1 so that this power of the first component a_1 will yield e_1, a multiple of r_2 so that this power of the second component a_2 will yield e_2, and so on. ∎

It is obvious that if $\prod_{i=1}^{n} G_i$ is an external direct product of groups G_i, the subset

$$\bar{G}_i = \{(e_1, e_2, \ldots, e_{i-1}, a_i, e_{i+1}, \ldots, e_n) \mid a_i \in G_i\},$$

that is, the set of all n-tuples with the identity elements in all places but the ith, is a subgroup of $\prod_{i=1}^{n} G_i$. It is also clear that this subgroup $\bar{G}_i$ is naturally isomorphic to G_i under the correspondence given by the projection mapping π_i, where

$$(e_1, e_2, \ldots, e_{i-1}, a_i, e_{i+1}, \ldots, e_n)\pi_i = a_i.$$

The group G_i is mirrored in the ith component of the elements of $\bar{G}_i$, and the e_j in the other components just ride along. We consider $\prod_{i=1}^{n} G_i$ to be the *internal direct product* of these subgroups $\bar{G}_i$. The terms *internal* and *external*, as applied to a direct product of groups, just reflect whether or not (respectively) you are regarding the component groups as subgroups of the product group. After this section, we shall usually omit the words *external* and *internal* and just say *direct product*. Which term we mean will be clear from the context. Section 8.2 (starred) gives a careful treatment of the internal direct product for those who wish to see it. A basic set-theoretic definition and an equally basic group-theoretic theorem will be needed. We give these here, since we shall have other use for them later.

Definition Let $\{S_i \mid i \in I\}$ be a collection of sets. Here I may be any set of indices. The *intersection* $\bigcap_{i \in I} S_i$ *of the sets* S_i is the set of all elements that are in all the sets S_i, that is,

$$\bigcap_{i \in I} S_i = \{x \mid x \in S_i \text{ for all } i \in I\}.$$

If I is finite, $I = \{1, 2, \ldots, n\}$, we may denote $\bigcap_{i \in I} S_i$ by

$$S_1 \cap S_2 \cap \cdots \cap S_n.$$

Theorem 8.4 *The intersection of subgroups* H_i *of a group* G *for* $i \in I$ *is again a subgroup of* G.

Proof. Let us show closure. Let $a \in \bigcap_{i \in I} H_i$ and $b \in \bigcap_{i \in I} H_i$, so that $a \in H_i$ for all $i \in I$ and $b \in H_i$ for all $i \in I$. Then $ab \in H_i$ for all $i \in I$, since H_i is a group. Thus $ab \in \bigcap_{i \in I} H_i$.

Since H_i is a subgroup for all $i \in I$, we have $e \in H_i$ for all $i \in I$, and hence $e \in \bigcap_{i \in I} H_i$.

Finally, for $a \in \bigcap_{i \in I} H_i$, we have $a \in H_i$ for all $i \in I$, so $a^{-1} \in H_i$ for all $i \in I$, which implies that $a^{-1} \in \bigcap_{i \in I} H_i$. ∎

*8.2 INTERNAL DIRECT PRODUCTS

Definition   Let a group G have subgroups H_i for $i = 1, \ldots, n$. G is the *internal direct product of the subgroups* H_i if the map $\phi: \prod_{i=1}^{n} H_i \to G$ given by

$$(h_1, h_2, \ldots, h_n)\phi = h_1 h_2 \cdots h_n$$

is an isomorphism.

Note that under this isomorphism ϕ, the subgroup $\bar{H}_i$ of $\prod_{i=1}^{n} H_i$ is mapped in a natural way onto H_i. In view of the isomorphism appearing in this definition, everything we observe for either an external direct product or an internal direct product has an immediate interpretation for the other.

Theorem 8.5   *If G is the internal direct product of subgroups H_1, $H_2, \ldots, H_n$, then each $g \in G$ can be uniquely written as $g = h_1 h_2 \cdots h_n$, where $h_i \in H_i$.*

Proof.   Using the isomorphism of our definition, we need only show that the corresponding statement is true for the external direct product $\prod_{i=1}^{n} H_i$ with respect to its subgroups $\bar{H}_i$, which are naturally isomorphic to H_i. We have to show that every element $(h_1, h_2, \ldots, h_n)$ of $\prod_{i=1}^{n} H_i$ can be uniquely written as a product

$$(a_1, e_2, \ldots, e_n)(e_1, a_2, \ldots, e_n) \cdots (e_1, e_2, \ldots, a_n),$$

where $a_i \in H_i$. This is obvious. We must have $a_i = h_i$. ∎

The preceding definition and theorem suggest that it would be interesting to examine products of elements from various subgroups of a group. For the rest of this section we work with only two subgroups of a group, although the definitions and theorems can be generalized to more than two subgroups.

Let H and K be subgroups of a group G. We want to look at $\{hk \,|\, h \in H, k \in K\}$, which we denote by HK. Unfortunately, *this set HK need not be a subgroup of G*, since $h_1 k_1 h_2 k_2$ need not be of the form hk. Of course, if G is abelian, or even if each element h of H **commutes** with each element k of K, that is, $hk = kh$, then

$$h_1 k_1 h_2 k_2 = h_1 h_2 k_1 k_2 = h_3 k_3,$$

where $h_3 = h_1 h_2$ and $k_3 = k_1 k_2$ are elements of H and K, respectively. It is easily checked that in this case we have a subgroup, for $ee = e$ and $(hk)^{-1} = k^{-1}h^{-1} = h^{-1}k^{-1}$.

Let us try to recover a subgroup in the noncommutative case. Note that there is always at least one subgroup of G containing HK, namely G itself.

Definition Let H and K be subgroups of a group G. The *join* $H \vee K$ *of* H *and* K is the intersection of all subgroups of G containing $HK = \{hk \mid h \in H, k \in K\}$.

Clearly, this intersection will be the smallest possible subgroup of G containing HK, and if elements in H and K commute, in particular, if G is abelian, we have $H \vee K = HK$. Note that since $h = he$ and $k = ek$, $H \subseteq HK$ and $K \subseteq HK$, so $H \le H \vee K$ and $K \le H \vee K$. But clearly, $H \vee K$ would be contained in any subgroup containing both H and K. Thus we see that $H \vee K$ *is the smallest subgroup of* G *containing both* H *and* K.

We conclude with a standard theorem for which we will have use in later starred sections.

Theorem 8.6 *A group G is the internal direct product of subgroups H and K if and only if*

1. $G = H \vee K$,
2. $hk = kh$ for all $h \in H$ and $k \in K$,
3. $H \cap K = \{e\}$.

Proof. Let G be the internal direct product of H and K. We claim that conditions 1, 2, and 3 are obvious if one will regard G as isomorphic to the external direct product of H and K under the map ϕ, with $(h, k)\phi = hk$. Under this map,
$$\bar{H} = \{(h, e) \mid h \in H\}$$
corresponds to H, and
$$\bar{K} = \{(e, k) \mid k \in K\}$$
corresponds to K. Then conditions 1, 2, and 3 follow immediately from the corresponding assertions regarding $\bar{H}$ and $\bar{K}$ in $H \times K$, which are obvious.

Conversely, let conditions 1, 2, and 3 hold. We must show that the map ϕ of the external direct product $H \times K$ into G, given by $(h, k)\phi = hk$, is an isomorphism. The map ϕ has already been defined.

Suppose
$$(h_1, k_1)\phi = (h_2, k_2)\phi.$$

Then $h_1 k_1 = h_2 k_2$; consequently $h_2^{-1} h_1 = k_2 k_1^{-1}$. But $h_2^{-1} h_1 \in H$ and $k_2 k_1^{-1} \in K$, and they are the same element and thus in $H \cap K = \{e\}$ by condition 3. Therefore $h_2^{-1} h_1 = e$ and $h_1 = h_2$. Likewise $k_1 = k_2$, so $(h_1, k_1) = (h_2, k_2)$. This shows that ϕ is one to one.

The fact that $hk = kh$ by condition 2 for all $h \in H$ and $k \in K$ means that

$$HK = \{hk \,|\, h \in H, k \in K\}$$

is a group, for we have seen that this is the case if elements of H commute with those of K. Thus by condition 1, $HK = H \vee K = G$, so ϕ is onto G.

Finally,

$$[(h_1, k_1)(h_2, k_2)]\phi = (h_1 h_2, k_1 k_2)\phi = h_1 h_2 k_1 k_2,$$

while

$$[(h_1, k_1)\phi][(h_2, k_2)\phi] = h_1 k_1 h_2 k_2.$$

But by condition 2 we have $k_1 h_2 = h_2 k_1$. Thus

$$[(h_1, k_1)(h_2, k_2)]\phi = [(h_1, k_1)\phi][(h_2, k_2)\phi]. \quad \blacksquare$$

Exercises

8.1 List the eight elements of $Z_2 \times Z_4$. Find the order of each of the elements. Is this group cyclic?

8.2 What is the largest order among the orders of all the cyclic subgroups of $Z_6 \times Z_8$? of $Z_{12} \times Z_{15}$?

†**8.3** Prove that an external direct product of abelian groups is abelian.

8.4 Find all proper nontrivial subgroups of $Z_2 \times Z_2$.

8.5 Disregarding the order of the factors, write direct products of two or more groups of the form Z_n so that the resulting product is isomorphic to Z_{60} in as many ways as possible.

8.6 Fill in the blanks.

a) The cyclic subgroup of Z_{24} generated by 18 has order _____.

b) $Z_3 \times Z_4$ is of order _____.

c) The element (4, 2) of $Z_{12} \times Z_8$ has order _____.

d) The Klein 4-group is isomorphic to Z__ $\times$ Z__.

e) $Z_2 \times Z \times Z_4$ has _____ elements of finite order.

8.7 Mark each of the following true or false.

___ a) If G_1 and G_2 are any groups, then $G_1 \times G_2$ is always isomorphic to $G_2 \times G_1$.

___ b) Computation in an external direct product of groups is easy if you know how to compute in each component group.

___ c) Groups of finite order must be used when forming an external direct product.

___ d) A group of prime order could not be the internal direct product of two proper nontrivial subgroups.

___ e) $Z_2 \times Z_4$ is isomorphic to Z_8.

___ f) $Z_2 \times Z_4$ is isomorphic to S_8.

— g) $\mathbf{Z}_3 \times \mathbf{Z}_8$ is isomorphic to S_4.

— h) Every element in $\mathbf{Z}_4 \times \mathbf{Z}_8$ has order 8.

— i) The order of $\mathbf{Z}_{12} \times \mathbf{Z}_{15}$ is 60.

— j) $\mathbf{Z}_m \times \mathbf{Z}_n$ has mn elements whether m and n are relatively prime or not.

8.8 Find all proper nontrivial subgroups of $\mathbf{Z}_2 \times \mathbf{Z}_2 \times \mathbf{Z}_2$.

8.9 Find all subgroups of $\mathbf{Z}_2 \times \mathbf{Z}_2 \times \mathbf{Z}_4$ which are isomorphic to the Klein 4-group.

8.10 Give an example illustrating that not every abelian group is the internal direct product of two proper nontrivial subgroups. (See Exercise 8.16 for a corresponding example for nonabelian groups.)

8.11 Let G be an abelian group. Let H be the subset of G consisting of the identity e together with all elements of G of order 2. Show that H is a subgroup of G.

8.12 Following up the idea of Exercise 8.11, determine whether H will always be a subgroup for every abelian group G if H consists of the identity e together with all elements of G of order 3; of order 4. For what positive integers n will H always be a subgroup for every abelian group G, if H consists of the identity e together with all elements of G of order n? Compare with Exercise 3.12.

8.13 Find a counterexample for Exercise 8.11 with the hypothesis that G is abelian omitted.

8.14 You sometimes hear people carelessly say, "An intersection of groups is a group." Why is it incorrect to say this?

*8.15 State a theorem similar to the theorem of Exercise 8.3, but for the case of a group that is the internal direct product of subgroups.

*8.16 Show that S_3 of Example 4.1 is not the internal direct product of the subgroups

$$H = \{\rho_0, \rho_1, \rho_2\} \qquad \text{and} \qquad K = \{\rho_0, \mu_1\}.$$

*8.17 Let $n = rs$, where r and s are relatively prime integers, that is, the gcd of r and s is 1. Show that $\mathbf{Z}_n$ is the internal direct product of its cyclic subgroups $\langle r \rangle$ and $\langle s \rangle$.

*8.18 Consider the subgroups $H = \langle 2 \rangle$ and $K = \langle 6 \rangle$ of $\mathbf{Z}_{12}$. Find HK and $H \vee K$.

*8.19 Consider the subgroups

$$H = \{\rho_0, \mu_1\} \qquad \text{and} \qquad K = \{\rho_0, \mu_2\}$$

of S_3 of Example 4.1. Find HK and $H \vee K$.

*8.20 Consider the subgroups

$$H = \{\rho_0, \delta_1\} \qquad \text{and} \qquad K = \{\rho_0, \delta_2\}$$

of the group D_4 of symmetries of the square in Example 4.2. Find HK and $H \vee K$.

*8.21 Let G be a group. Let h and k be elements of G that commute and are of relatively prime orders r and s. Apply Theorem 8.6 to the subgroups $\langle h \rangle$ and $\langle k \rangle$ of $\langle h \rangle \vee \langle k \rangle$ to show that hk is of order rs.

nine

Finitely Generated Abelian Groups

Some theorems of abstract algebra are easy to understand and use, although their proofs may be quite technical and time-consuming to present. This is the first of several sections of the text in which we explain the meaning and significance of some theorems and ask you to use them without proof. Proofs are generally given in later starred sections. The body of mathematical literature is so great now that mathematicians who insist on checking every last detail of the proof of every result they use can't get to important frontiers of the subject. Theorems that we present without proof in unstarred sections are well within your understanding, and we feel you should be acquainted with them. It would be impossible to get to all these fascinating facts in a one-semester undergraduate course if we were to insist you see complete proofs of them.

9.1 GENERATORS AND TORSION

The first concept we define is of great importance to us. Contrary to our previous practice, we shall first give an elegant definition and then explain it naively in a theorem. Recall that by Theorem 8.4 an intersection of sub-

86

groups of a group is again a group. Let G be a group, and let a_i be elements of G for $i \in I$, where I is some indexing set. There is at least one subgroup of G containing all the a_i, namely G itself. Obviously, *the intersection of all subgroups of G containing all the a_i is the smallest subgroup of G containing all the a_i for $i \in I$.*

Definition Let G be a group and let $a_i \in G$ for $i \in I$. The smallest subgroup of G containing $\{a_i | i \in I\}$ is the **subgroup generated by** $\{a_i | i \in I\}$. If this subgroup is all of G, then $\{a_i | i \in I\}$ **generates** G and the a_i are **generators of** G. If there is a finite set $\{a_i | i \in I\}$ that generates G, then G is **finitely generated.**

Note that this definition is consistent with our previous definition of a generator for a cyclic group. Be sure to note that the next theorem is stated and proved for *any* group G, rather than only for abelian groups.

Theorem 9.1 *If G is a group and $a_i \in G$ for $i \in I$, then the subgroup H of G generated by $\{a_i | i \in I\}$ has as elements precisely those elements of G that are finite products of integral powers of the a_i, where powers of a fixed a_i may occur several times in the product.*

Proof. The reason we may have to have powers of a fixed a_i occur several times in a product is that G is not assumed to be abelian. If G is abelian, then $(a_1)^{-3}(a_2)^5(a_1)^7$ could be simplified to $(a_1)^4(a_2)^5$, but this may not be true in the nonabelian case.

Let K denote the set of all finite products of integral powers of the a_i. Clearly, $K \subseteq H$. We need only observe that K is a subgroup and then, since H is the smallest subgroup containing a_i for $i \in I$, we will be done. It is obvious that a product of elements in K is again in K. Since $(a_1)^0 = e$, we have $e \in K$. For every element k in K, if you form from the product giving k a new product with the order of the a_i reversed and the opposite sign on all exponents, you have k^{-1}, which is thus in K. For example,

$$[(a_1)^3(a_2)^2(a_1)^{-7}]^{-1} = (a_1)^7(a_2)^{-2}(a_1)^{-3},$$

which is again in K. ∎

Example 9.1 $\mathbf{Z} \times \mathbf{Z}_2$ is generated by $\{(1, 0), (0, 1)\}$. ∎

The student who has studied Section *8.2 should note that if A and B are subgroups of G, then the subgroup join $A \vee B$ is precisely the subgroup generated by $A \cup B$.

Although we are concerned chiefly with abelian groups in this section, we shall consistently use multiplicative notation in our general discussions. It is our experience that the student understands the notation a^n more readily than na. Given the latter notation, a student tends to make the error of thinking of n as an element of the group.

Definition   A group G is a ***torsion group*** if every element in G is of finite order. G is ***torsion free*** if no element other than the identity is of finite order.

Theorem 9.2   *In an abelian group G, the set T of all elements of G of finite order is a subgroup of G, the **torsion subgroup of** G.*

Proof.   We use multiplicative notation. Let a and b be elements of T. Then there are positive integers m and n such that $a^m = b^n = e$. Now since G is abelian,

$$(ab)^{mn} = a^{mn}b^{mn},$$

so

$$(ab)^{mn} = a^{mn}b^{mn} = (a^m)^n(b^n)^m = e^ne^m = e.$$

Thus ab is of finite order and hence in T. This shows that T is closed under group multiplication.

Of course, e is of finite order and consequently in T.

Finally, if $a \in T$ and $a^m = e$, then

$$e = e^m = (aa^{-1})^m = a^m(a^{-1})^m = e(a^{-1})^m = (a^{-1})^m,$$

so a^{-1} is of finite order and thus in T. ∎

Example 9.2   Every finite group is a torsion group, while $\mathbf{Z}$ under addition is torsion free. If we consider $\mathbf{Z} \times \mathbf{Z}_2$, the element $(1, 0)$ is not of finite order but the element $(0, 1)$ is of order 2. It is clear that $T = \{(0, 0), (0, 1)\}$ is the torsion subgroup of $\mathbf{Z} \times \mathbf{Z}_2$. ∎

9.2 THE FUNDAMENTAL THEOREM

We now state some lemmas leading up to the main theorem of the section, Theorem 9.3. The lemmas are not proved. Theorem 9.3 is proved in starred Chapter 20. The proof there is not constructed via the sequence of lemmas given here; we selected these introductory lemmas to lead you gradually to an understanding of the group structure described in the theorem.

Lemma 9.1   *If G is a finitely generated abelian group with a torsion subgroup T, then G is an (internal) direct product $T \times F$ for some subgroup F of G that is torsion free.*

Proof.   The proof will follow from Theorem 9.3, which is proved in starred Chapter 20. ∎

Lemma 9.2   *A finitely generated torsion free abelian group F is isomorphic to* $\mathbf{Z} \times \mathbf{Z} \times \cdots \times \mathbf{Z}$ *for some number m of factors. The number m, the **betti number of** F, is unique.*

Proof.   The proof will follow from Theorem 9.3, which is proved in starred Chapter 20. ■

Lemma 9.3   *A finite abelian group T is isomorphic to two different types of direct products of cyclic groups as follows:*

1. *T is isomorphic to a product*

$$\mathbf{Z}_{(p_1)^{r_1}} \times \mathbf{Z}_{(p_2)^{r_2}} \times \cdots \times \mathbf{Z}_{(p_n)^{r_n}},$$

where the p_i are primes and are not necessarily distinct. This direct product of cyclic groups of prime power order isomorphic to T is unique except for a rearrangement of the factors.

2. *T is isomorphic to a direct product*

$$\mathbf{Z}_{m_1} \times \mathbf{Z}_{m_2} \times \cdots \times \mathbf{Z}_{m_r},$$

*where m_i divides m_{i+1}. The numbers m_i, the **torsion coefficients of** T, are unique.*

Proof.   The proof will follow from Theorem 9.3, which is proved in starred Chapter 20. ■

The terms *betti number* and *torsion coefficients* come from algebraic topology, where they play an important role.

It is unavoidable that the primes p_i appearing in form 1 of Lemma 9.3 may not be distinct. We have seen, for example, that $\mathbf{Z}_5 \times \mathbf{Z}_5 \times \mathbf{Z}_9$ is not isomorphic to $\mathbf{Z}_{25} \times \mathbf{Z}_9$, for the former group has elements at most of order 45, while the latter is cyclic of order 225.

Just think for a moment of the significance and tremendous power of Lemma 9.3. *It gives us a description, up to isomorphism, of all finite abelian groups.*

Let us describe a method of finding a group, expressed as in form 2 of Lemma 9.3, which is isomorphic to a given direct product of cyclic groups of prime power order. For each prime appearing in the order of the group, write the subscripts in the direct product involving that prime in a row in order of increasing magnitude. Keep the right-hand ends of the rows aligned. Thus starting with $\mathbf{Z}_2 \times \mathbf{Z}_4 \times \mathbf{Z}_3 \times \mathbf{Z}_3 \times \mathbf{Z}_5$, we form the array

$$
\begin{array}{cc}
2 & 4 \\
3 & 3 \\
  & 5
\end{array}
$$

Then take the product of the numbers in each column, getting in this case the numbers 6 and 60. Then $\mathbf{Z}_2 \times \mathbf{Z}_4 \times \mathbf{Z}_3 \times \mathbf{Z}_3 \times \mathbf{Z}_5$ is isomorphic to

$Z_6 \times Z_{60}$. Likewise, $Z_2 \times Z_2 \times Z_2 \times Z_3 \times Z_3 \times Z_5$ gives rise to the array

$$
\begin{array}{ccc}
2 & 2 & 2 \\
  & 3 & 3 \\
  &   & 5 \\
\hline
2 & 6 & 30
\end{array}
$$

so it is isomorphic to $Z_2 \times Z_6 \times Z_{30}$. We give no formal proof of the validity of this algorithm. From the theory we developed in Chapter 8, especially Theorem 8.2, you can easily see why it works.

Example 9.3 Let us find all abelian groups (up to isomorphism) of order 360. First we express 360 as a product of prime powers $2^3 3^2 5$. Then using form 1 of Lemma 9.3, we get as possibilities

1. $Z_2 \times Z_2 \times Z_2 \times Z_3 \times Z_3 \times Z_5$
2. $Z_2 \times Z_4 \times Z_3 \times Z_3 \times Z_5$
3. $Z_2 \times Z_2 \times Z_2 \times Z_9 \times Z_5$
4. $Z_2 \times Z_4 \times Z_9 \times Z_5$
5. $Z_8 \times Z_3 \times Z_3 \times Z_5$
6. $Z_8 \times Z_9 \times Z_5$

Thus there are six different abelian groups (up to isomorphism) of order 360. We write these six cases below as in form 2 of Lemma 9.3, keeping the groups (up to isomorphism) in the same order. That is, the first group listed above is isomorphic to the first listed below, and so on. This is easy to verify from the remarks preceding this example.

1. $Z_2 \times Z_6 \times Z_{30}$ 2. $Z_6 \times Z_{60}$
3. $Z_2 \times Z_2 \times Z_{90}$ 4. $Z_2 \times Z_{180}$
5. $Z_3 \times Z_{120}$ 6. Z_{360} ∎

Except for the knowledge that a subgroup of a finitely generated abelian group is again finitely generated, a proof of which is contained in Chapter 20, our main theorem follows immediately from the preceding lemmas.

Theorem 9.3 (Fundamental Theorem of Finitely Generated Abelian Groups) *Every finitely generated abelian group G is isomorphic to a direct product of cyclic groups in the form*

1. $\qquad Z_{(p_1)^{r_1}} \times Z_{(p_2)^{r_2}} \times \cdots \times Z_{(p_n)^{r_n}} \times Z \times Z \times \cdots \times Z,$

where the p_i are primes, not necessarily distinct, and also in the form

2. $\qquad\qquad Z_{m_1} \times Z_{m_2} \times \cdots \times Z_{m_r} \times Z \times Z \times \cdots \times Z,$

where m_i divides m_{i+1}.

> *In both cases the direct product is unique except for possible rearrangement of the factors, that is, the number (**betti number** of G) of factors of **Z** is unique, the **torsion coefficients** m_i of G are unique, and prime powers $(p_i)^{r_i}$ are unique.*

Proof. A complete proof can be found in starred Chapter 20. ∎

> *Do you really comprehend the significance of these results? Among other things, they give us complete information about all finite abelian groups.*

*9.3 APPLICATIONS

We conclude this section with a sampling of the many theorems we could now prove regarding abelian groups. Some follow from work in Chapter 8; for others we will need our powerful Theorem 9.3.

> **Definition** A group G is **decomposable** if it is isomorphic to a direct product of two proper nontrivial subgroups. Otherwise G is **indecomposable.**

> **Theorem 9.4** *The finite indecomposable abelian groups are exactly the cyclic groups with order a power of a prime.*

Proof. Let G be a finite indecomposable abelian group. Then by Theorem 9.3 (or Lemma 9.3), G is isomorphic to a direct product of cyclic groups of prime power order. Since G is indecomposable, this direct product must consist of just one cyclic group of prime power order.

Conversely, let p be a prime. Our work in Chapter 8 shows that $\mathbf{Z}_{p^r}$ is indecomposable, for if $\mathbf{Z}_{p^r}$ were isomorphic to $\mathbf{Z}_{p^i} \times \mathbf{Z}_{p^j}$, where $i + j = r$, then every element would have an order at most $p^{\max(i,j)} < p^r$. ∎

> **Theorem 9.5** *If m divides the order of a finite abelian group G, then G has a subgroup of order m.*

Proof. By Theorem 9.3 (or Lemma 9.3), we can think of G as being

$$\mathbf{Z}_{(p_1)^{r_1}} \times \mathbf{Z}_{(p_2)^{r_2}} \times \cdots \times \mathbf{Z}_{(p_n)^{r_n}},$$

where not all primes p_i need be distinct. Since $(p_1)^{r_1}(p_2)^{r_2} \cdots (p_n)^{r_n}$ is the order of G, then m must be of the form $(p_1)^{s_1}(p_2)^{s_2} \cdots (p_n)^{s_n}$, where $0 \leq s_i \leq r_i$. By Theorem 6.4, $(p_i)^{r_i - s_i}$ generates a cyclic subgroup of $\mathbf{Z}_{(p_i)^{r_i}}$ of order equal to the quotient of $(p_i)^{r_i}$ by the gcd of $(p_i)^{r_i}$ and $(p_i)^{r_i - s_i}$. But the gcd of $(p_i)^{r_i}$ and $(p_i)^{r_i - s_i}$ is $(p_i)^{r_i - s_i}$. Thus $(p_i)^{r_i - s_i}$ generates a cyclic subgroup $\mathbf{Z}_{(p_i)^{r_i}}$ of order

$$[(p_i)^{r_i}]/[(p_i)^{r_i - s_i}] = (p_i)^{s_i}.$$

Recalling that $\langle a \rangle$ denotes the cyclic subgroup generated by a, we see that

$$\langle (p_1)^{r_1 - s_1} \rangle \times \langle (p_2)^{r_2 - s_2} \rangle \times \cdots \times \langle (p_n)^{r_n - s_n} \rangle$$

is the required subgroup of order m. ∎

Theorem 9.6   *If m is a square free integer, that is, m is not divisible by the square of any prime, then every abelian group of order m is cyclic.*

Proof.   Let G be an abelian group of square free order m. Then by Theorem 9.3 (or Lemma 9.3), G is isomorphic to

$$Z_{(p_1)^{r_1}} \times Z_{(p_2)^{r_2}} \times \cdots \times Z_{(p_n)^{r_n}},$$

where $m = (p_1)^{r_1}(p_2)^{r_2} \cdots (p_n)^{r_n}$. Since m is square free, we must have all $r_i = 1$ and all p_i distinct primes. The corollary of Theorem 8.2 then shows that G is isomorphic to $Z_{p_1 p_2 \cdots p_n}$, so G is cyclic. ∎

Exercises

9.1 Find all abelian groups (up to isomorphism) of order 720; of order 1089. Express them in both forms 1 and 2 of Lemma 9.3, and pair up isomorphic groups of forms 1 and 2.

9.2 Find the torsion coefficients and the betti number of the group

$$Z \times Z_6 \times Z \times Z \times Z_{12} \times Z_{10}.$$

9.3 Find the subgroup of Z_{12} generated by $\{2, 3\}$; generated by $\{4, 6\}$; generated by $\{8, 6, 10\}$.

9.4 Find the order of the torsion subgroup of $Z_4 \times Z \times Z_3$; of $Z_{12} \times Z \times Z_{12}$.

†**9.5** Show that a finite abelian group is not cyclic if and only if it contains a subgroup isomorphic to $Z_p \times Z_p$ for some prime p.

9.6 Mark each of the following true or false.

___ a) Every abelian group of prime order is cyclic.

___ b) Every abelian group of prime power order is cyclic.

___ c) Z_8 is generated by $\{4, 6\}$.

___ d) Z_8 is generated by $\{4, 5, 6\}$.

___ e) All finite abelian groups are classified up to isomorphism by Lemma 9.3.

___ f) Any two finitely generated abelian groups with the same betti number are isomorphic.

___ g) Every abelian group of order divisible by 5 contains a cyclic subgroup of order 5.

___ h) Every abelian group of order divisible by 4 contains a cyclic subgroup of order 4.

___ i) Every abelian group of order divisible by 6 contains a cyclic subgroup of order 6.

___ j) Every finite abelian group has a betti number of 0.

9.7 How many abelian groups (up to isomorphism) are there of order 24? of order 25? of order (24)(25)?

9.8 Following the idea suggested in Exercise 9.7, let m and n be relatively prime positive integers. Show that if there are (up to isomorphism) r abelian groups of order m and s of order n, then there are (up to isomorphism) rs abelian groups of order mn.

9.9 Use Exercise 9.8 to determine the number of abelian groups (up to isomorphism) of order $(10)^5$.

9.10 Let G be an abelian group of order 72.

a) Can you say how many subgroups of order 8 G has? Why?

b) Can you say how many subgroups of order 4 G has? Why?

9.11 Prove that if a finite abelian group has order a power of a prime p, then the order of every element in the group is a power of p.

9.12 For what positive integers n is it true that the only abelian groups of order n are cyclic?

9.13 Let p and q be distinct prime numbers. How does the number (up to isomorphism) of abelian groups of order p^r compare with the number (up to isomorphism) of abelian groups of order q^r?

9.14 (For students knowing a bit about complex numbers, especially DeMoivre's theorem.) Find the torsion subgroup T of the multiplicative group $\mathbf{C}^*$ of nonzero complex numbers.

9.15 Show that S_n is generated by $\{(1, 2), (1, 2, 3, \ldots, n)\}$. [*Hint:* Show that as r varies, $(1, 2, 3, \ldots, n)^{n-r}(1, 2)(1, 2, 3, \ldots, n)^r$ gives all the transpositions $(1, 2), (2, 3), (3, 4), \ldots$, $(n - 1, n), (n, 1)$. Then show that any transposition is a product of some of these transpositions and use the corollary of Theorem 5.1.]

9.16 What is the least number of elements that can be used to generate S_3 of Example 4.1? the group D_4 of symmetries of the square in Example 4.2? the group $\mathbf{Z}_2 \times \mathbf{Z}_2 \times \mathbf{Z}_2$?

9.17 What is wrong with the following argument?

"By Exercise 9.15, S_n can be generated by two elements. By Cayley's theorem, every finite group is isomorphic to some subgroup of some S_n. Therefore, every finite group can be generated by two elements."

Note that the third part of Exercise 9.16 shows that this conclusion is false.

9.18 Let G, H, and K be finitely generated abelian groups. Show that if $G \times K \simeq H \times K$, then $G \simeq H$.

9.19 In $\mathbf{Z} \times \mathbf{Z}$ let $G = \{(a, b) \mid a \equiv b \,(\mathrm{mod}\ 10)\}$.

a) Prove that G is a subgroup of $\mathbf{Z} \times \mathbf{Z}$ and is torsion free.

b) Show that G is finitely generated.

c) Find an isomorphism $\phi : G \to \mathbf{Z} \times \cdots \times \mathbf{Z}$ for some number of factors.

ten

Groups in Geometry and Analysis

We interrupt our purely algebraic study to give a rather vague indication of the significance of the group concept in geometry and analysis. Since we are not trying to teach either geometry or analysis, our discussion as it pertains to these fields will not be very precise.

*10.1 GROUPS IN GEOMETRY

Definition By a *transformation of a set* A, a geometer means a permutation of the set, that is, a one-to-one function of A onto itself.

According to Theorem 4.1, the transformations of a set form a group under transformation (permutation) multiplication. Remember that this multiplication is nothing but function composition. That is, if ϕ and ψ are transformations of A, then the product $\mu = \phi\psi$ is defined by $a\mu = (a\phi)\psi$ for $a \in A$.

Felix Klein (1849–1925) gave a famous definition of a *geometry* in an address (the *Erlanger Programm*) in 1872 when he accepted a chair at the

94

University of Erlangen. From the present-day geometer's point of view, Klein's definition is not inclusive enough, but it will serve for our purpose.

Definition (*Klein*) A *geometry* is the study of those properties of a space (set) that remain invariant under some fixed subgroup of the full transformation group.

Let us illustrate this definition as it applies to the classical Euclidean geometry of the Euclidean line, plane, 3-space, and so on. Here we have sets for which there is an idea of *distance between elements* defined. If we let $d(x, y)$ be the distance between two elements x and y, then we can talk about transformations that preserve distance.

Definition If A is a set on which an idea of distance is defined, a transformation ϕ of A is an *isometry* if $d(x, y) = d(x\phi, y\phi)$, that is, if ϕ preserves distance.

It is clear that the subset of the full transformation group consisting of all isometries of the set is a subgroup. The *Euclidean geometry* of the line, plane, 3-space, and so on is exactly the study of those properties left invariant under the group of isometries. Thus in Euclidean geometry one can talk about the concepts of the length of a line segment, the size of an angle, and the number of sides of a polygon, for these are all invariant under an isometry.

Let us describe some of the isometries of the Euclidean plane. A *translation* of the plane is a transformation that moves each point a fixed distance in a fixed direction. In terms of coordinates, a translation $\tau_{(a,b)}$ moves a point (x, y) to $(x + a, y + b)$. Clearly,

$$\tau_{(a,b)}\tau_{(c,d)} = \tau_{(a+c, b+d)}.$$

It is immediate that the translations form a subgroup of the group of isometries isomorphic to $\mathbf{R} \times \mathbf{R}$ under addition. A **rotation** $\rho_{(P,\theta)}$ is a transformation that rotates the plane about the point P counterclockwise through the angle θ, where $0 \leq \theta < 2\pi$. The rotations do not form a subgroup of the isometries, for $\rho_{(P,\theta_1)}\rho_{(Q,\theta_2)}$ is not a rotation if $P \neq Q$ and $\theta_1 + \theta_2 = 2\pi$. But clearly,

$$\rho_{(P,\theta_1)}\rho_{(P,\theta_2)} = \rho_{(P, \theta_1 + \theta_2 \bmod 2\pi)},$$

where

$$(\theta_1 + \theta_2 \bmod 2\pi) = \begin{cases} \theta_1 + \theta_2 & \text{if } (\theta_1 + \theta_2) < 2\pi, \\ \theta_1 + \theta_2 - 2\pi & \text{if } (\theta_1 + \theta_2) \geq 2\pi. \end{cases}$$

The rotations about a fixed point P do form a subgroup of the isometries. This group is not isomorphic to $\mathbf{R}$ under addition, for it has cyclic subgroups of finite order. For example, $\rho_{(P,\pi/2)}$ generates a cyclic subgroup of order 4. All rotations about a fixed point actually form a group isomorphic to the multiplicative group of complex numbers of absolute value 1. Finally, a **reflection** in the plane is a function μ that carries each point of a fixed line l into itself and every point not on the line into the mirror image point straight across l and the same distance from l, as indicated in Fig. 10.1. It can be shown that the translations, rotations, and reflections generate (in the sense of Chapter 9) the whole isometry group of the plane. Actually this set of generators is much larger than is necessary. *The reflections alone can be shown to generate the isometry group, every plane isometry being expressible as a product of at most three reflections.* You will have no difficulty in convincing yourself, for example, that a translation can be written as the product of two reflections in lines that are perpendicular to the direction of the translation and of distance apart equal to half the length of the translation. We refer the interested student to Coxeter [44].

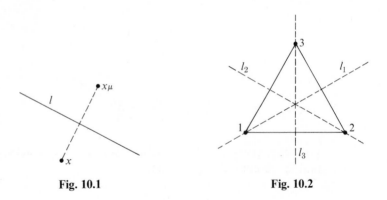

Fig. 10.1　　　　　　　　　　　　　Fig. 10.2

　　The group S_3 given in Example 4.1 has a nice geometric interpretation. Consider an equilateral triangle as shown in Fig. 10.2. We let

$$\rho_1 = \begin{pmatrix} 1 & 2 & 3 \\ 2 & 3 & 1 \end{pmatrix},$$

where ρ denotes a rotation, and we can think of this as a rotation counterclockwise through $2\pi/3$ radians. Similarly,

$$\rho_2 = \begin{pmatrix} 1 & 2 & 3 \\ 3 & 1 & 2 \end{pmatrix}$$

is a rotation counterclockwise through $4\pi/3$ radians, and

$$\rho_0 = \begin{pmatrix} 1 & 2 & 3 \\ 1 & 2 & 3 \end{pmatrix}$$

is a rotation through 0 radians. Also,

$$\mu_1 = \begin{pmatrix} 1 & 2 & 3 \\ 1 & 3 & 2 \end{pmatrix},$$

where μ denotes a mirror image, corresponds to a reflection of the triangle in the line l_1, and in general our μ_i of Example 4.1 was a reflection in the line l_i. A reflection μ_i corresponds to a turning over of the triangle about an axis l_i. The reason for our choice of the notation in Example 4.1 is clear. Look again at Table 4.1 of Example 4.1. Note that it divides itself into four blocks, which we brought out by shading. This block arrangement is shown again in Table 10.1. Algebraically, Table 10.1 gives a group of order 2. Geometrically, this table means that a product of two rotations is a rotation, a product of a rotation and a reflection is a reflection, and a product of two reflections is a rotation. This splitting up of a group into blocks that themselves form a group will be the next topic in our algebraic study.

Table 10.1

	ρ terms	μ terms
ρ terms	ρ terms	μ terms
μ terms	μ terms	ρ terms

Your first guess might be that the symmetries of a square should give S_4, but be careful. The permutation

$$\begin{pmatrix} 1 & 2 & 3 & 4 \\ 1 & 3 & 4 & 2 \end{pmatrix}$$

is not an isometry of the square in Fig. 10.3, for the distance from vertex 1 to vertex 2 would become the longer distance from vertex 1 to vertex 3. The group of symmetries of the square, or the octic group, was computed in Example 4.2. The computation of the dihedral group of symmetries of the regular n-gon in the plane for $n \geq 3$ is left for the exercises at the end of this

Fig. 10.3

section. Note that the group of symmetries of the regular n-gon for $n \geq 3$ is S_n only for $n = 3$.

One can recover S_4 geometrically as the group of symmetries of the regular tetrahedron, each face of which is an equilateral triangle, as shown in Fig. 10.4. You may say that you can't achieve the permutation

$$\begin{pmatrix} 1 & 2 & 3 & 4 \\ 1 & 3 & 2 & 4 \end{pmatrix}$$

by a rigid motion, but just as you have to go out of the plane to turn the equilateral triangle over and to obtain all of S_3, you have to go out of 3-space to 4-space to "turn over" the tetrahedron and to get all of S_4. This permutation amounts to a reflection in the plane containing the line through vertices 1 and 4 perpendicular to the line through vertices 2 and 3.

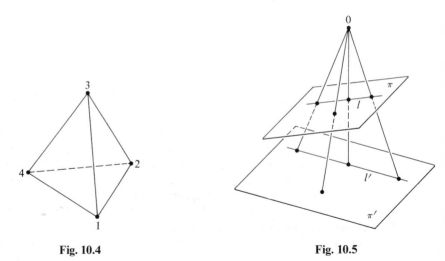

Fig. 10.4   Fig. 10.5

We give two more illustrations of Klein's definition. If we augment the usual plane $\mathbf{R} \times \mathbf{R}$ by adding a *line at infinity* containing one point for each direction in $\mathbf{R} \times \mathbf{R}$, we obtain the **projective plane.** In the projective plane, there are no such things as parallel lines, for two lines that are parallel in $\mathbf{R} \times \mathbf{R}$ are defined as intersecting at the point on the line at infinity corresponding to their common direction in the projective plane. Now let π and π' be two such projective planes with their usual $\mathbf{R} \times \mathbf{R}$ portions viewed as part of 3-space and not necessarily parallel as 3-space planes. A **projection of π onto π'** is a mapping of π onto π' by means of projection from a point not in either plane, or projection by parallel rays. Projection from a point is illustrated in Fig. 10.5. By means of two projections, we can project π

onto π' and then π' back onto π, the composition giving a transformation of π onto itself. The **group of projective transformations of** π is the subgroup of all transformations of π generated by the type of mappings of π onto itself just described. Clearly, lines are carried into lines and quadrilaterals into quadrilaterals, that is, these are concepts of *projective geometry*. However, distance is not preserved, so distance is not a projective geometry concept. One can show, however, that the so called *cross ratio*

$$\frac{(CA/CB)}{(DA/DB)}$$

of four points on a line, as shown in Fig. 10.6, is an invariant of the projective group. This cross ratio is about the only numerical quantity one can get hold of, and thus it plays a very important role in projective geometry.

$$A \quad B \quad\quad C \quad D$$

Fig. 10.6

Finally, if for a given space we have enough of an idea of when points are near each other so that we can talk about *continuous transformations* (such a set is a *topological space*), then we can define **topology** as the study of those properties of such spaces that are invariant under the group of all transformations that are continuous and whose inverses also are continuous. The Euclidean line, plane, 3-space, and so on are all topological spaces. Vaguely, a continuous transformation with a continuous inverse is one that we can achieve by bending, stretching, and twisting the space without tearing or cutting it. For Euclidean spaces these transformations include all isometries, and for projective planes they include all the projective transformations. Topologically, one can't tell the difference between a football and a basketball, for one can be deformed, without being torn, to look just like the other. Similarly, a square and a circle are topologically the same. However, you can't make a solid chocolate mint disk look like a lifesaver candy without tearing a hole in it. Thus these are topologically different. Topology is by far the most active field of all the geometries at the present time.

*10.2 GROUPS IN ANALYSIS

Turning to analysis, we describe very briefly a few situations in which groups arise naturally. Of course analysis works with subgroups primarily of the complex numbers, so they appear in that obvious way. In addition, consider the function f of one real variable given by $f(x) = \sin x$. It has a well-known graph, shown in Fig. 10.7. Also

$$\sin x = \sin (x + 2\pi n)$$

for every integer n, that is, the function *sine* of one real variable is invariant under a transformation of its domain by an element of the infinite cyclic subgroup $\langle 2\pi \rangle$ of the translation group of **R**. Such a function of one real variable that is invariant under a transformation of its domain by an element of an infinite cyclic group is a **periodic function.**

Fig. 10.7

Recall that the complex numbers can be viewed as filling up the Euclidean plane. A function on the plane that is invariant under an element of a group of transformations generated by two translations in different (not opposite) directions, as in Fig. 10.8, is a **doubly periodic function.** The group here is isomorphic to **Z** × **Z**. An **elliptic function** is defined as a doubly periodic meromorphic function of a complex variable. The term *meromorphic function* is explained in a first undergraduate course in functions of a complex

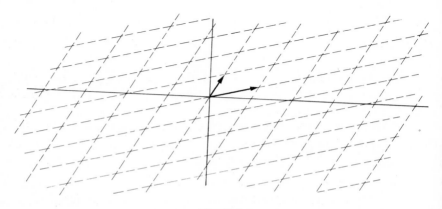

Fig. 10.8

variable. An elliptic function is thus known everywhere if it is known in one fundamental region, that is, one of the diamond-shaped regions in Fig. 10.8. Both the trigonometric and elliptic functions are special cases of **automorphic functions,** which are functions invariant under a discrete group of transformations. The term *discrete* means roughly that no two group elements are close together.

Finally, in *measure theory* one likes to assign a numerical size to certain "nice" subsets of a set. If the set has a group structure $\langle G, \cdot \rangle$ and a is an element of the group, it is often convenient to have a measure such that the size of a subset S of G is the same as the size of the subset

$$aS = \{a \cdot s \mid s \in S\}.$$

By analogy with **R** and the operation of addition, one often calls aS a **left translation of S by** a; similarly, Sa is a **right translation of S by** a. A **left invariant measure** is a measure such that the size of aS is the same as the size of S for every $a \in G$ and every nice set S. A **right invariant measure** is similarly defined. Thus for **R** under addition, our ordinary idea of size (length) for nice sets (intervals) is a left and also a right invariant measure. Similarly, our usual idea of area in the plane **C** (or $\mathbf{R} \times \mathbf{R}$) under addition is a left and also a right invariant measure. However, our idea of area in **C***, the group of nonzero complex numbers under multiplication, that is, the plane minus the origin, is neither a left nor a right invariant measure. For example, if S is the inside of the circle of radius 1 about the origin minus the origin, it has area π, while $2S$, which has radius 2, has area 4π. However, Haar has shown that there is a left invariant measure (and also a right invariant measure), **Haar measure,** on every locally compact topological group.[†] The classification *locally compact topological group* covers many natural groups composed of complex numbers including **C***.

Exercises

*10.1 Review Exercise 4.9, or do it now if you did not work it out before.

√ *10.2 Show that the nth dihedral group D_n of Exercise 4.9 can be generated by two elements. Argue geometrically.

*10.3 Review Exercise 4.10, or do it now if you did not work it out before.

*10.4 Show that the group of rigid motions of the cube given in Exercise 4.10 can be generated by two elements. Argue geometrically.

[†] A. Haar, "Der Massbegriff in der Theorie der kontinuierlichen Gruppen," *Ann. of Math.* (2), **34**, 147–169 (1933).

√*10.5 Consider the group of *all* symmetries (isometries) of the cube. This group includes all rigid motions and also all reflections of the cube. What is the order of the group? Show that this group can be generated by three elements.

*10.6 Consider the *finite affine geometry* of four points A, B, C, D and six lines AB, AC, AD, BC, BD, CD, as indicated schematically in Fig. 10.9. Here each line contains just two points. A **collineation of an affine geometry** is a one-to-one map of the set of points onto itself that carries lines onto lines. (It is not necessary to know what a point or a line really is to do this exercise. It is based on your naive idea that a line is composed of points, and so on.)

a) Show that *every* one-to-one map of the set of points of this four-point affine geometry onto itself is a collineation.

b) Show that for any affine geometry, the collineations form a group, the **affine group,** under function composition.

c) To what group that we have seen before is the affine group for the geometry of Fig. 10.9 isomorphic?

Fig. 10.9

*10.7 Following the ideas of Exercise 10.6, consider the affine geometry of nine points and twelve lines, each containing three points, given schematically in Fig. 10.10.

a) Show that not every one-to-one map of the set of points onto itself is a collineation for this nine-point geometry.

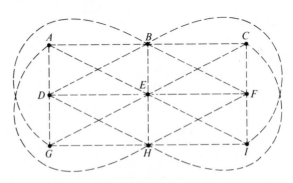

Fig. 10.10

b) Show that a collineation in this nine-point geometry is completely determined by its values on any three points not all on the same line.

c) What is the order of the affine group, that is, the group of all collineations, for this nine-point geometry?

d) Consider the subgroup H of the affine group of part (c) that consists of those collineations leaving each point on the line ABC fixed. What is the order of this subgroup? To what group that we have defined before is this subgroup H isomorphic?

eleven

Groups of Cosets

11.1 INTRODUCTION

You probably noticed that the thirty-six entries in the table for S_3 in Example 4.1 divide themselves naturally into four blocks, each block consisting just of terms ρ_i or of terms μ_i. These blocks were brought out in Table 4.1 by shading. Thus the group S_3 itself was partitioned into two cells B_ρ and B_μ of equal size, and the set $\{B_\rho, B_\mu\}$ forms a group with a table derived from that in Table 4.1 and shown here in Table 11.1. This partitioning of a group into cells such that the set of cells itself forms a group is a notion of basic importance in algebra. Let us call an element of a cell a **representative of the cell.** The equation

$$B_\rho B_\mu = B_\mu$$

means that *any* representative of B_ρ multiplied by *any* representative of B_μ gives some representative of B_μ.

Passing to the general case, we would like to determine precise conditions under which a group G can be partitioned into cells B_i such that *any* repre-

104

Table 11.1

	B_ρ	B_μ
B_ρ	B_ρ	B_μ
B_μ	B_μ	B_ρ

sentative of a fixed cell B_r multiplied by *any* representative of another fixed cell B_s always produces a representative of one and the same cell B_t, which we would then consider as the product $B_r B_s$. The cell product $B_r B_s$ is *defined* to be the cell B_t obtained by multiplying representatives of B_r and B_s, and in order to have the binary operation of cell multiplication on $\{B_i\}$ *well defined,* as explained in Chapter 1, the final cell B_t containing the product of the representatives must be the same no matter which representatives one chooses from B_r and from B_s. This binary operation of cell multiplication on the set $\{B_i\}$ is the **operation induced on** $\{B_i\}$ **by the operation of** G. *Only if this operation is well defined does it make sense to ask whether the set $\{B_i\}$ is a group under the operation.*

Theorem 11.1 *If a group G can be partitioned into cells with the induced operation described above well defined, and if the cells form a group under this induced operation, then the cell containing the identity e of G must be a subgroup of G.*

Proof. Let G be partitioned into cells with the induced operation well defined and giving a group, and let B_e be the cell containing the identity. In computing $B_e B_r$ we can take *any* representatives from B_e and from B_r and compute their product in G. Let us choose $e \in B_e$ and say $r \in B_r$. Then $er = r$ and $r \in B_r$, so $B_e B_r = B_r$. Similarly $B_r B_e = B_r$. Thus B_e must act as the identity cell in our group of cells. Therefore,

$$B_e B_e = B_e,$$

which shows, if we choose all possible representatives, that B_e is closed under the multiplication of the group G.

By its definition, B_e contains e.

Let $a \in B_e$. Now a^{-1} is in some cell B_k. Since B_e is the identity cell, we know $B_e B_k = B_k$. Choosing representatives $a \in B_e$ and $a^{-1} \in B_k$ and using them to compute $B_e B_k$, we see that we must have $B_e B_k = B_e$. Thus $B_k = B_e$, and $a^{-1} \in B_e$.

Hence B_e is a subgroup of G. ■

11.2 COSETS

Suppose a group G can be partitioned into cells with the induced operation well defined and giving a group. Let B_e be the cell containing the identity. The preceding theorem shows that B_e is a subgroup of G. Let B_a be the cell containing $a \in G$. The equation $B_a B_e = B_a$ shows, if we choose the representative $a \in B_a$ and all representatives of B_e, that the set

$$aB_e = \{ax \,|\, x \in B_e\}$$

must be contained in B_a. This suggests that these *translations* or *cosets* aB_e of a subgroup B_e might be important.

Definition   Let H be a subgroup of a group G, and let $a \in G$. The **left coset aH of H** is the set $\{ah \,|\, h \in H\}$. The **right coset Ha** is similarly defined.

We have seen that if G can be partitioned into cells with the induced operation well defined and giving a group, then

$$aB_e \subseteq B_a.$$

Let $a^{-1} \in B_k$. Then $B_k B_a = B_e$, so choosing representatives $a^{-1} \in B_k$ and any $x \in B_a$, we have $a^{-1}x \in B_e$. Thus $a^{-1}x = b$ and $x = ab$, where $b \in B_e$. This shows that

$$B_a \subseteq aB_e,$$

so

$$B_a = aB_e.$$

Of course, we also have $B_a = B_e a$ by a similar argument. We summarize these results in a theorem.

Theorem 11.2   *If a group G can be partitioned into cells with the induced operation well defined and giving a group, then the cells must be precisely the left (and also the right) cosets of a subgroup of G. In particular, every left coset must be a right coset.*

Example 11.1   Let us determine how the left cosets of $3\mathbf{Z}$ as a subgroup of $\mathbf{Z}$ under addition look. The notation here is additive. Of course $3\mathbf{Z} = 0 + 3\mathbf{Z}$ is itself a left coset. Another left coset is $1 + 3\mathbf{Z}$. Upon a moment's reflection, we can see that $1 + 3\mathbf{Z}$ consists of all integers that leave a remainder of 1 when divided by 3 in the sense of Lemma 6.1. Likewise the left coset $2 + 3\mathbf{Z}$ consists of all integers leaving a remainder of 2 when divided by 3. Since Lemma 6.1 shows that the remainder of any integer divided by 3 is an integer r, where $0 \le r < 3$, the only possibilities are 0, 1, and 2. Thus these are all the left cosets. ■

We might ask whether, given a subgroup H of a group G, the left (or right) cosets of H always do give a partition of G into distinct cells. Of course, any such partition corresponds to an equivalence relation on G by Theorem 0.1. Note that $b \in aH$ if and only if $b = ah$ for some $h \in H$, or if and only if $a^{-1}b \in H$. This suggests that we examine the relation $a \sim b$ if and only if $a^{-1}b \in H$ to see if it is an equivalence relation. As indicated in the statement of the theorem that follows, this relation is given a special notation.

Theorem 11.3 *Let H be a subgroup of a group G. The relations*

$$a \equiv_\ell b \pmod{H} \qquad \text{if and only if} \quad a^{-1}b \in H$$

and

$$a \equiv_r b \pmod{H} \qquad \text{if and only if} \quad ab^{-1} \in H$$

*are equivalence relations on G, **left congruence modulo** H and **right congruence modulo** H respectively. The equivalence classes of left (right) congruence modulo H are the left (right) cosets of H. All cosets of H have the same number of elements.*

Proof.   We prove the statement for left congruence and left cosets. Similar proofs hold for right congruence and right cosets. Let us first show that left congruence modulo H is an equivalence relation.

Reflexive. $a \equiv_\ell a \pmod{H}$ since $a^{-1}a = e$ is in H.

Symmetric. If $a \equiv_\ell b \pmod{H}$, then $a^{-1}b \in H$, and H is a subgroup so $(a^{-1}b)^{-1} = b^{-1}a$ is in H also; consequently $b \equiv_\ell a \pmod{H}$.

Transitive. If $a \equiv_\ell b \pmod{H}$ and $b \equiv_\ell c \pmod{H}$, then $a^{-1}b \in H$ and $b^{-1}c \in H$. Since H is a subgroup, $(a^{-1}b)(b^{-1}c) = a^{-1}c$ is in H, so $a \equiv_\ell c \pmod{H}$.

Thus left congruence modulo H is an equivalence relation.

The equivalence class $\bar{a}$ containing a is easily computed as

$$\bar{a} = \{x \in G \mid x \equiv_\ell a \pmod{H}\} = \{x \in G \mid a^{-1}x \in H\}$$
$$= \{x \in G \mid a^{-1}x = h \in H\} = \{x \in G \mid x = ah \text{ for some } h \in H\}$$
$$= aH.$$

Thus the equivalence classes of left congruence modulo H are precisely the left cosets of H.

To show that any two left cosets have the same number of elements, consider the map $\lambda_a : H \to aH$ given by $h\lambda_a = ah$. This map λ_a is clearly onto aH. If h_1 and h_2 are in H, and $ah_1 = ah_2$, then $h_1 = h_2$ by the group cancellation law. Thus λ_a maps H one to one and onto aH. Hence every left coset has the same number of elements as H, and therefore all have the same number of elements. ∎

We now have a pretty good idea of how a partitioning of a group into disjoint cells looks if the induced operation is well defined and gives a group. The cells are always cosets (left and right cosets are the same in this case by Theorem 11.2) of some subgroup. We shall drop the term *cell* from now on and always use the term *coset* instead.

Unfortunately, the left cosets of a subgroup of a group do not always form a group under the induced operation. *The difficulty lies in the fact that the induced operation may not be well defined.* We give an example of this.

Example 11.2 Consider the group S_3 of Example 4.1 and the subgroup $H = \{\rho_0, \mu_1\}$. To find the left cosets of a subgroup H of a finite group G, one hunts for an element a in G not in H and finds the left coset aH. Then hunting for $b \in G$ with b not in either H or aH, one finds a new left coset bH. Continuing this process, one finds all left cosets of H in G. For our example, it is easy to see that the left cosets of $H = \{\rho_0, \mu_1\}$ in S_3 are

$$H = \{\rho_0, \mu_1\},$$
$$\rho_1 H = \{\rho_1, \mu_2\},$$
$$\rho_2 H = \{\rho_2, \mu_3\}.$$

Let us write the group table for S_3 again but with the elements in the order

$$\rho_0, \mu_1 | \rho_1, \mu_2 | \rho_2, \mu_3.$$

In Table 11.2 we shade lightly squares containing elements of the coset $\{\rho_1, \mu_2\}$ and heavily those containing elements of $\{\rho_2, \mu_3\}$. One glance at this table will show whether the induced operation is well defined and these left cosets form a group. We see that it doesn't work. The product of two cells does not always produce elements of just one shade, that is, elements in just one coset. By multiplying elements in the left coset $\{\rho_0, \mu_1\}$ by elements in the left coset $\{\rho_1, \mu_2\}$, one can get all the elements in $\{\rho_1, \rho_2, \mu_2, \mu_3\}$,

Table 11.2

	ρ_0	μ_1	ρ_1	μ_2	ρ_2	μ_3
ρ_0	ρ_0	μ_1	ρ_1	μ_2	ρ_2	μ_3
μ_1	μ_1	ρ_0	μ_3	ρ_2	μ_2	ρ_1
ρ_1	ρ_1	μ_2	ρ_2	μ_3	ρ_0	μ_1
μ_2	μ_2	ρ_1	μ_1	ρ_0	μ_3	ρ_2
ρ_2	ρ_2	μ_3	ρ_0	μ_1	ρ_1	μ_2
μ_3	μ_3	ρ_2	μ_2	ρ_1	μ_1	ρ_0

Table 11.3

	0	3	1	4	2	5
0	0	3	1	4	2	5
3	3	0	4	1	5	2
1	1	4	2	5	3	0
4	4	1	5	2	0	3
2	2	5	3	0	4	1
5	5	2	0	3	1	4

which is not a left coset. *The induced operation is not well defined on these left cosets.* ∎

Example 11.3 Let us see whether the group $\mathbf{Z}_6$ can be partitioned into a group of left cosets of the subgroup $H = \{0, 3\}$. Here, using additive notation, our left cosets are

$$H = \{0, 3\},$$
$$1 + H = \{1, 4\},$$
$$2 + H = \{2, 5\}.$$

The table for $\mathbf{Z}_6$ with elements in the order

$$0, 3 \,|\, 1, 4 \,|\, 2, 5,$$

again with shaded squares according to the coset containing the element, is given in Table 11.3. It works! Isn't that lovely? Such a delightful display of symmetry should make shivers of joy run up and down the spine of anyone with any mathematical sensitivity. ∎

A group of left cosets formed from a group G gives some information about G. You don't know precisely what the product of any two elements of G is if you only know the left coset group, but you do know the *type* of element resulting from the product of two *types* of elements. This is the power of the concept. We have illustrated it with S_3, where elements are of two types, type ρ (rotations) and type μ (reflections). Another analysis is that those of type ρ are even permutations and those of type μ are odd permutations. The group given by the two left cosets of A_3 in S_3 then has a simple interpretation in terms of classifying products of permutations as *even* or *odd* as shown in Table 11.4. It is not surprising that subgroups whose left cosets form a group play a fundamental role in the theory of groups.

Table 11.4

	Even	Odd
Even	Even	Odd
Odd	Odd	Even

To complete this discussion, it would only remain to characterize exactly the types of subgroups of a group G whose cosets do form a group under the induced operation. We delay this until the next section, mainly because this is very important material, and we want you to have lots of time to assimilate it.

11.3 APPLICATIONS

Instead of putting the finishing touches on this theory now, we are going to prove several lovely results about finite groups that follow very easily from the work we have done thus far.

> **Theorem 11.4 (Lagrange)** Let G be a group of finite order n and H a subgroup of G. The order of H divides the order of G.

Proof. Let H have m elements. Consider the collection of left cosets of H. By Theorem 11.3, these left cosets are disjoint, have the same number m of elements as H has, and every element of G is in some left coset. Thus, if there are r left cosets, we must have $n = rm$, so m divides n. ∎

Note that this elegant and important theorem comes from the simple counting done in Theorem 11.3, which showed that all cosets have the same number of elements. *Never underestimate a theorem that counts something.*

Our Theorem 9.3 shows at once that any *abelian* group of prime order is cyclic. But as a corollary to Theorem 11.4, we have

> **Corollary** Every group of prime order is cyclic.

Proof. Let G be of prime order p, and let a be an element of G different from the identity. Then the cyclic subgroup $\langle a \rangle$ of G generated by a has at least two elements, a and e. But by Theorem 11.4, the order $m \geq 2$ of $\langle a \rangle$ must divide the prime p. Thus we must have $m = p$ and $\langle a \rangle = G$, so G is cyclic. ∎

Thus there is only one group (up to isomorphism) of a given prime order. Now doesn't this elegant result follow easily from the theorem of Lagrange, a *counting* theorem? *Never underestimate a theorem that counts something.* The preceding corollary is a favorite examination question.

> **Theorem 11.5** The order of an element of a finite group divides the order of the group.

Proof. Remembering that the order of an element is the same as the order of the cyclic subgroup generated by the element, we see that this theorem follows directly from Theorem 11.4. ∎

> **Definition** Let H be a subgroup of a group G. The number of left cosets of H in G is the *index* $(G:H)$ *of H in* G.

The index $(G:H)$ just defined may be finite or infinite. If G is finite, then obviously $(G:H)$ is finite and $(G:H) = |G|/|H|$, since every coset of H contains $|H|$ elements. Exercise 11.5 shows the index $(G:H)$ could be equally well defined as the number of right cosets of H in G. We state a basic theorem concerning indices of subgroups, and leave the proof to the exercises (see Exercise 11.9).

> **Theorem 11.6** Suppose H and K are subgroups of a group G such that $K \leq H \leq G$, and suppose $(H : K)$ and $(G : H)$ are both finite. Then $(G : K)$ is finite, and $(G : K) = (G : H)(H : K)$.

Theorem 11.4 shows that if there is a subgroup H of a finite group G, then the order of H divides the order of G. The student may wonder whether the converse is true. That is, if G is a group of order n, and m divides n, is there always a subgroup of order m? The answer is no, although from Theorem 9.3 it follows easily that the converse is true for abelian groups (see Theorem 9.5). However, A_4 can be shown to have no subgroup of order 6, which gives a counterexample for nonabelian groups.

Exercises

11.1 Find the number of left cosets of each of the following subgroups:

a) The subgroup $\langle 18 \rangle$ of $\mathbf{Z}_{36}$.

b) The subgroup $\langle 1 \rangle \times \langle 0 \rangle \times \langle 0 \rangle$ of $\mathbf{Z}_3 \times \mathbf{Z}_2 \times \mathbf{Z}_4$.

c) The subgroup $\langle 0 \rangle \times \langle 1 \rangle \times \langle 2 \rangle$ of $\mathbf{Z}_3 \times \mathbf{Z}_2 \times \mathbf{Z}_4$.

11.2 Three subgroups of the group of symmetries of the square of Example 4.2 are $H_1 = \{\rho_0, \rho_2\}$, $H_2 = \{\rho_0, \mu_1\}$, and $H_3 = \{\rho_0, \rho_1, \rho_2, \rho_3\}$. For each of these subgroups, write the table for the group of symmetries of the square in an order exhibiting the left cosets as we did in Examples 11.2 and 11.3, and determine whether the induced operation is well defined and whether the left cosets form a group. The use of a different color for each coset will make your tables very striking.

11.3 Write the left cosets of the cyclic subgroup $\langle (1, 2) \rangle$ of $\mathbf{Z}_2 \times \mathbf{Z}_4$, and determine whether the induced operation is well defined and whether the left cosets form a group. See Exercise 11.2 regarding the use of colors.

11.4 How many groups are there of order 17, up to isomorphism?

†**11.5** Show that there are the same number of left as right cosets of a subgroup H of a group G, that is, exhibit a one-to-one map of the collection of left cosets onto the collection of right cosets. (Note that this result is obvious by counting for finite groups. Your proof must hold for any group.)

11.6 Mark each of the following true or false.

  — a) Every subgroup of every group has left cosets.

  — b) The number of left cosets of a subgroup of a finite group divides the order of the group.

  — c) Every group of prime order is abelian.

  — d) One cannot have left cosets of a finite subgroup of an infinite group.

— e) A subgroup of a group is a left coset of itself.

— f) Only subgroups of finite groups can have left cosets.

— g) A_n is of index 2 in S_n for $n > 1$.

— h) The theorem of Lagrange is a nice result.

— i) Every finite group contains an element of every order that divides the order of the group.

— j) Every finite cyclic group contains an element of every order that divides the order of the group.

11.7 Exercise 2.5 showed that every finite group of even order $2n$ contains an element of order 2. Using the theorem of Lagrange but not Theorem 9.3, show that if n is odd, an abelian group of order $2n$ contains precisely one element of order 2.

11.8 Show that a group with at least two elements but with no proper nontrivial subgroups must be finite and of prime order.

11.9 Prove Theorem 11.6. [*Hint:* Let $\{a_i H \mid i = 1, \ldots, r\}$ be the collection of distinct left cosets of H in G and $\{b_j K \mid j = 1, \ldots, s\}$ be the collection of distinct left cosets of K in H. Show that

$$\{(a_i b_j) K \mid i = 1, \ldots, r; j = 1, \ldots, s\}$$

is the collection of distinct left cosets of K in G.]

11.10 Show that if H is a subgroup of an abelian group G, then every left coset of H is also a right coset of H.

11.11 Show that if H is a subgroup of index 2 in a finite group G, then every left coset of H is also a right coset of H.

11.12 Taking our notation for S_3 in Example 4.1, try to decide, without writing out the table, whether the induced operation is well defined on the left cosets of the subgroup

$$\{(\rho_0, 0), (\rho_0, 3), (\rho_1, 0), (\rho_1, 3), (\rho_2, 0), (\rho_2, 3)\}$$

of $S_3 \times \mathbf{Z}_6$ and whether they form a group. How about left cosets of

$$\{(\rho_0, 0), (\rho_0, 3), (\mu_1, 0), (\mu_1, 3)\}?$$

11.13 Show that the left cosets of the subgroup $\{0\} \times \mathbf{Z}_2$ of $\mathbf{Z} \times \mathbf{Z}_2$ form a group isomorphic to $\mathbf{Z}$ under the induced operation on left cosets.

11.14 Show that if a group G with identity e has finite order n, then $a^n = e$ for all $a \in G$.

11.15 Show that every left coset of the subgroup $\mathbf{Z}$ of the additive group of real numbers contains exactly one representative x in $\mathbf{R}$ such that $0 \leq x < 1$.

11.16 Show that the function *sine* assigns the same value to each representative of any fixed left coset of the subgroup $\langle 2\pi \rangle$ of the additive group $\mathbf{R}$ of real numbers. (Thus *sine* induces a well-defined function on the set of cosets; the value of the function on a coset is obtained when we choose a representative x of the coset and compute sin x.)

11.17 Show that a finite cyclic group of order n has exactly one subgroup of each order d dividing n, and that these are all the subgroups it has.

11.18 The **Euler phi-function** is defined for positive integers n by $\phi(n) = s$, where s is the number of positive integers less than or equal to n that are relatively prime to n. Use

Exercise 11.17 to show that

$$n = \sum_{d \mid n} \phi(d),$$

the sum being taken over all positive integers d dividing n. [*Hint*: Note that the number of generators of $\mathbf{Z}_d$ is $\phi(d)$ by the corollary of Theorem 6.4.]

11.19 Let G be a finite group. Show that if for each positive integer m the number of solutions x of the equation $x^m = e$ in G is at most m, then G is cyclic. [*Hint*: Use Theorem 11.5 and Exercise 11.18 to show that G must contain an element of order $n = |G|$.]

twelve

Normal Subgroups and Factor Groups

12.1 CRITERIA FOR THE EXISTENCE OF A COSET GROUP

We now return to the problem of deciding for what subgroups H of a group G the left (right) cosets will form a group under the induced operation. The crucial thing is whether the induced operation is well defined.

> **Lemma 12.1** *If H is a subgroup of G, and if the induced operation of coset multiplication on left (right) cosets of H is well defined, then the collection of left (right) cosets of H is a group under this induced coset multiplication.*

Proof. Recall that a coset product $(aH)(bH)$ of left cosets was defined as the coset containing the product of *any* two representatives, one from aH and one from bH. We are assuming that this coset product is well defined, that is, independent of the choice of representatives of aH and bH. We can thus take x as a representative of the coset xH for purposes of computation.

114

The group axioms for the collection of cosets are thrown back on the group axioms in G, since multiplication of cosets is defined in terms of multiplication of elements of G. To prove the associative law we must show that

$$aH[(bH)(cH)] = [(aH)(bH)]cH.$$

Computing, we have

$$aH[(bH)(cH)] = (aH)(bcH) = a(bc)H.$$

Also

$$[(aH)(bH)](cH) = (abH)(cH) = (ab)cH.$$

But by the associative law in G, we know that

$$a(bc) = (ab)c,$$

so coset multiplication is also associative. Likewise eH acts as identity, since the representative e acts as identity in G, and the inverse of aH is $a^{-1}H$. ∎

Theorem 12.1 *If H is a subgroup of a group G, then the operation of induced multiplication is well defined on the left (right) cosets of H if and only if every left coset is a right coset.*

Proof. Suppose the induced operation is well defined. The left (right) cosets then form a group by Lemma 12.1, and by Theorem 11.2 the left cosets are the same as the right cosets.

Conversely, let us assume every left coset aH is also a right coset. Since $g \in gH$, and the right coset containing g is Hg, where $g \in G$, we are assuming that $gH = Hg$ as sets for all $g \in G$. We wish to show that if one attempts to define $(aH)(bH)$ by multiplying representatives, the left coset in which the product of the representatives is found is the same for all choices of these representatives. To this end, suppose a_1 and a_2 are representatives of aH, and b_1 and b_2 are representatives of bH. We have to show that a_1b_1 and a_2b_2 are in the same left coset of H. Since $aH = a_1H = a_2H$ and $bH = b_1H = b_2H$, we can write $a_1 = a_2h_1$ and $b_1 = b_2h_2$ for some h_1 and h_2 in H. Then

$$a_1b_1 = a_2h_1b_2h_2.$$

Now since $b_2H = Hb_2$ by assumption, we have $h_1b_2 = b_2h_3$ for some $h_3 \in H$. Thus

$$a_1b_1 = a_2b_2h_3h_2, \qquad \text{so} \quad a_1b_1 \in a_2b_2H,$$

that is, a_1b_1 and a_2b_2 are in the same left coset. This shows that the induced operation is well defined. ∎

12.2 INNER AUTOMORPHISMS AND NORMAL SUBGROUPS

We now know that cosets of a subgroup H of a group G form a group under the induced operation of coset multiplication if and only if $gH = Hg$ for every $g \in G$. We can rewrite $gH = Hg$ as $H = g^{-1}Hg$ for all $g \in G$, where, of course,

$$g^{-1}Hg = \{g^{-1}hg \,|\, h \in H\}.$$

To get more insight into the meaning of $g^{-1}Hg$, we are led to the idea of studying for each $g \in G$ the mapping $i_g : G \to G$ given by

$$xi_g = g^{-1}xg.$$

Definition   An isomorphism of a group G with itself is an **automorphism** *of the group* G.

Theorem 12.2   *For each* $g \in G$, *the mapping* $i_g : G \to G$ *given by* $xi_g = g^{-1}xg$ *is an automorphism of* G, *the* **inner automorphism of** G **under** **conjugation by** g.

Proof.   We must show that i_g is an isomorphism of G with itself. The map is defined, so we show that it is one to one, onto, and that

$$(xy)i_g = (xi_g)(yi_g).$$

First, for one to one, if $xi_g = yi_g$, then $g^{-1}xg = g^{-1}yg$, so $x = y$ by the group cancellation law. For onto, if $x \in G$, then

$$(gxg^{-1})i_g = g^{-1}(gxg^{-1})g = x.$$

Finally,

$$(xy)i_g = g^{-1}xyg,$$

and also,

$$(xi_g)(yi_g) = (g^{-1}xg)(g^{-1}yg) = g^{-1}xyg. \quad \blacksquare$$

Since an automorphism of G is an isomorphism of G with itself, it just interchanges names of elements of G, preserving all algebraic structural features. That is, i_g will map a subgroup of G one to one onto a (possibly different) subgroup of G, an element of order n of G onto a (possibly different element of order n in G, and so on. Now saying that $gH = Hg$ is the same as saying that

$$H = g^{-1}Hg = \{g^{-1}hg \,|\, h \in H\}.$$

This means that i_g carries H onto itself. It does not mean that necessarily $hi_g = h$ for all $h \in H$, but rather that $hi_g \in H$ for all $h \in H$. That is, $g^{-1}Hg = H$ means that i_g gives a *permutation* of H.

Definition   A subgroup H of a group G is a ***normal*** (or ***invariant***) ***subgroup of*** G if $g^{-1}Hg = H$ for all $g \in G$, that is, if H is left invariant by every inner automorphism of G.

Thus normal subgroups are precisely those important subgroups of a group having the property that, for the cosets (left and right are the same), the induced operation is well defined, and the cosets form a group.

It is worth noting that if H is a subgroup of G such that $g^{-1}Hg \subseteq H$ for all $g \in G$, then $g^{-1}Hg = H$ for all $g \in G$, that is, H is then a normal subgroup of G. For suppose $g^{-1}Hg \subseteq H$ for all $g \in G$. We want to show that $H \subseteq g^{-1}Hg$. Let $h \in H$. Then $(g^{-1})^{-1}Hg^{-1} \subseteq H$ so $(g^{-1})^{-1}hg^{-1} = h_1$ and $h_1 \in H$. Thus $ghg^{-1} = h_1$ so $h = g^{-1}h_1g$ and $h \in g^{-1}Hg$. Thus to show that a subgroup H of a group G is a normal subgroup, one usually shows that $g^{-1}hg \in H$ for all $h \in H$ and all $g \in G$.

Theorem 12.3   *Every subgroup of an abelian group is a normal subgroup.*

Proof.   This is easy since if H is a subgroup of an abelian group G, then for all $g \in G$ and $h \in H$, we have

$$g^{-1}hg = g^{-1}gh = eh = h. \qquad \blacksquare$$

Example 12.1   Example 11.2 shows that $\{\rho_0, \mu_1\}$ is not a normal subgroup of S_3 of Example 4.1. Indeed,

$$(\rho_1)^{-1}\mu_1\rho_1 = \rho_2\mu_1\rho_1 = \mu_2 \notin \{\rho_0, \mu_1\}.$$

Here i_{ρ_1} maps $\{\rho_0, \mu_1\}$ onto the subgroup $\{\rho_0, \mu_2\}$. $\blacksquare$

Definition   Two subgroups H and K of a group G are ***conjugate*** if $H = a^{-1}Ka$ for some $a \in G$, that is, if one is mapped onto the other by some inner automorphism of G.

Thus Example 12.1 shows that $\{\rho_0, \mu_1\}$ and $\{\rho_0, \mu_2\}$ are conjugate subgroups of S_3 of Example 4.1.

12.3 FACTOR GROUPS

Definition   If N is a normal subgroup of a group G, the group of cosets of N under the induced operation is the ***factor group of*** G ***modulo*** N, and is denoted by G/N. The cosets are ***residue classes of*** G ***modulo*** N.

Example 12.2   Referring to Example 11.1, we see that since $\mathbf{Z}$ is abelian, $3\mathbf{Z}$ is a normal subgroup, and thus $\mathbf{Z}/3\mathbf{Z}$ is the factor group of the three residue classes

$$0 + 3\mathbf{Z}, \qquad 1 + 3\mathbf{Z}, \qquad 2 + 3\mathbf{Z}.$$

This is a group of order 3 and hence is cyclic and isomorphic to $\mathbf{Z}_3$. $\blacksquare$

Example 12.3 The subgroup $n\mathbf{Z}$ is normal in $\mathbf{Z}$ for all $n \in \mathbf{Z}^+$. There are n residue classes:

$$0 + n\mathbf{Z}, \quad 1 + n\mathbf{Z}, \quad \ldots, \quad (n - 1) + n\mathbf{Z}.$$

Since we perform the addition of residue classes by choosing representatives, it is immediate that the map $\phi_n : \mathbf{Z}/n\mathbf{Z} \to \mathbf{Z}_n$ given by

$$(m + n\mathbf{Z})\phi_n = m \qquad \text{for} \quad 0 \le m < n$$

is an isomorphism. ∎

By abuse of notation, we may sometimes write $\mathbf{Z}/n\mathbf{Z} = \mathbf{Z}_n$, and think of $\mathbf{Z}_n$ as the additive group of residue classes of $\mathbf{Z}$ modulo $\langle n \rangle$, or again by abuse of notation, the group of residue classes of $\mathbf{Z}$ modulo n. Recall from Chapter 0 that two integers a and b are congruent modulo n, denoted by $a \equiv b \pmod{n}$, if and only if n divides $a - b$. This is precisely the criterion that a and b be in the same coset of $\mathbf{Z}/n\mathbf{Z}$.

We point out here that the formation of $\mathbf{Z}/n\mathbf{Z}$ is the *elegant* approach to the demonstration of the existence of a cyclic group of order n, as opposed to our naive approach in Theorem 6.3.

Example 12.4 Let us compute the factor group $(\mathbf{Z}_4 \times \mathbf{Z}_6)/\langle (0, 1) \rangle$. Here $\langle (0, 1) \rangle$ is the cyclic subgroup H of $\mathbf{Z}_4 \times \mathbf{Z}_6$ generated by $(0, 1)$. Thus

$$H = \{(0, 0), (0, 1), (0, 2), (0, 3), (0, 4), (0, 5)\}.$$

Since $\mathbf{Z}_4 \times \mathbf{Z}_6$ has twenty-four elements and H has six elements, all cosets of H must have six elements and $(\mathbf{Z}_4 \times \mathbf{Z}_6)/H$ must have order 4. Since $\mathbf{Z}_4 \times \mathbf{Z}_6$ is abelian, so is $(\mathbf{Z}_4 \times \mathbf{Z}_6)/H$ (remember, you compute in a factor group by means of representatives from the original group). By "computing" $(\mathbf{Z}_4 \times \mathbf{Z}_6)/H$, we mean: determine this finite abelian group according to our classification in the Fundamental Theorem of Finitely Generated Abelian Groups (Theorem 9.3). In additive notation, the cosets are

$$H = (0, 0) + H, \qquad (1, 0) + H, \qquad (2, 0) + H, \qquad (3, 0) + H.$$

Since we can compute by choosing the representatives $(0, 0), (1, 0), (2, 0)$, and $(3, 0)$, it is clear that $(\mathbf{Z}_4 \times \mathbf{Z}_6)/H$ is isomorphic to $\mathbf{Z}_4$. Note that this is what you would expect, since in a factor group modulo H, everything in H becomes the identity, that is, you are essentially setting everything in H equal to zero. Thus the whole second factor $\mathbf{Z}_6$ of $\mathbf{Z}_4 \times \mathbf{Z}_6$ is collapsed, leaving just the first factor $\mathbf{Z}_4$. This is an example of a general situation given in Theorem 12.7. ∎

Example 12.5 Let us compute the factor group $(\mathbf{Z}_4 \times \mathbf{Z}_6)/\langle (0, 2) \rangle$. Now $(0, 2)$ generates the subgroup

$$H = \{(0, 0), (0, 2), (0, 4)\}$$

of $\mathbf{Z}_4 \times \mathbf{Z}_6$ of order 3. Here the first factor $\mathbf{Z}_4$ of $\mathbf{Z}_4 \times \mathbf{Z}_6$ is left alone. The $\mathbf{Z}_6$ factor, on the other hand, is essentially collapsed by a subgroup of order 3, giving a factor group in the second factor of order 2 that must be isomorphic to $\mathbf{Z}_2$. Thus $(\mathbf{Z}_4 \times \mathbf{Z}_6)/\langle(0, 2)\rangle$ is isomorphic to $\mathbf{Z}_4 \times \mathbf{Z}_2$. ∎

Example 12.6 Let us compute the factor group $(\mathbf{Z}_4 \times \mathbf{Z}_6)/\langle(2, 3)\rangle$. *Be careful!* There is a great temptation to say that you are setting the 2 of $\mathbf{Z}_4$ and the 3 of $\mathbf{Z}_6$ both equal to zero, so that $\mathbf{Z}_4$ is collapsed to a factor group isomorphic to $\mathbf{Z}_2$ and $\mathbf{Z}_6$ to one isomorphic to $\mathbf{Z}_3$, giving a total factor group isomorphic to $\mathbf{Z}_2 \times \mathbf{Z}_3$. *This is wrong!* Note that

$$H = \langle(2, 3)\rangle = \{(0, 0), (2, 3)\}$$

is of order 2, so $(\mathbf{Z}_4 \times \mathbf{Z}_6)/\langle(2, 3)\rangle$ has order 12, not 6. You see, setting $(2, 3)$ equal to zero does not make $(2, 0)$ and $(0, 3)$ equal to zero individually, so the factors don't collapse separately.

The possible abelian groups of order 12 are $\mathbf{Z}_4 \times \mathbf{Z}_3$ and $\mathbf{Z}_2 \times \mathbf{Z}_2 \times \mathbf{Z}_3$, and we must decide to which one our factor group is isomorphic. These two groups are most easily distinguished in that $\mathbf{Z}_4 \times \mathbf{Z}_3$ has an element of order 4, and $\mathbf{Z}_2 \times \mathbf{Z}_2 \times \mathbf{Z}_3$ does not. We claim that the coset $(1, 0) + H$ is of order 4 in the factor group $(\mathbf{Z}_4 \times \mathbf{Z}_6)/H$. To find the smallest power of a coset giving the identity in a factor group modulo H, we must, by choosing representatives, find the smallest power of a representative that is in the subgroup H. Clearly,

$$4(1, 0) = (1, 0) + (1, 0) + (1, 0) + (1, 0) = (0, 0)$$

is the first time that $(1, 0)$ added to itself gives an element of H. Thus $(\mathbf{Z}_4 \times \mathbf{Z}_6)/\langle(2, 3)\rangle$ has an element of order 4 and is isomorphic to $\mathbf{Z}_4 \times \mathbf{Z}_3$ or $\mathbf{Z}_{12}$. ∎

Example 12.7 Let us compute (that is, classify as in Theorem 9.3) the group $(\mathbf{Z} \times \mathbf{Z})/\langle(1, 1)\rangle$. We may visualize $\mathbf{Z} \times \mathbf{Z}$ as the points in the plane with both coordinates integers, as indicated by the dots in Fig. 12.1. The subgroup $\langle(1, 1)\rangle$ consists of those points that lie on the 45° line through the origin, indicated in the figure. The coset $(1, 0) + \langle(1, 1)\rangle$ consists of those dots on the 45° line through the point $(1, 0)$, also shown in the figure. Continuing, we easily see that each coset consists of those dots lying on one of the 45° lines in the figure. We may choose the representatives

$$\ldots, (-3, 0), (-2, 0), (-1, 0), (0, 0), (1, 0), (2, 0), (3, 0), \ldots$$

of these cosets to compute in the factor group. Since these representatives correspond precisely to the points of $\mathbf{Z}$ on the x-axis, we see at once that the factor group $(\mathbf{Z} \times \mathbf{Z})/\langle(1, 1)\rangle$ is isomorphic to $\mathbf{Z}$. ∎

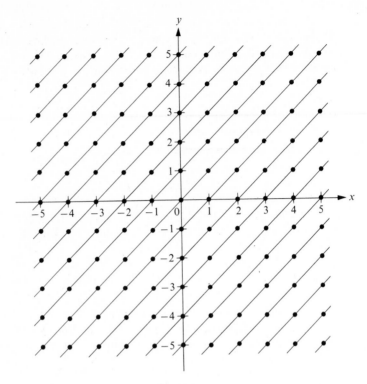

Fig. 12.1

12.4 SIMPLE GROUPS

As we mentioned in the preceding section, one feature of a factor group is that it gives crude information about the structure of the whole group. Of course, sometimes there may be no nontrivial proper normal subgroups. For example, Theorem 11.4 shows that a group of prime order can have no nontrivial proper subgroups of any sort.

Example 12.8   It is clear that both the improper subgroup G and the trivial subgroup $\{e\}$ of a group G are normal subgroups. Obviously G/G is the trivial group of one element while $G/\{e\}$ is isomorphic to G under the natural map carrying $g\{e\}$ into g for each $g \in G$. These factor groups are worthless so far as telling anything further about the structure of G is concerned. ■

Definition   A group is **simple** if it has no proper nontrivial normal subgroups.

Theorem 12.4   *The alternating group A_n is simple for $n \geq 5$.*

Proof.   See Exercise 22. ■

We will have an important use for Theorem 12.4 in the very last chapter of this text. There are many simple groups other than those given above. For example, A_5 is of order 60 and A_6 is of order 360, and there is a simple group of nonprime order, namely 168, between these orders.

The complete determination and classification of all finite simple groups were recently completed. Hundreds of mathematicians have worked on this task over the past three decades. Chapter 14 on series of groups will indicate that a finite group has a sort of factorization into simple groups, where the factors are unique up to order. The situation is similar to the factorization of positive integers into primes. The new knowledge of all finite simple groups can now be used to solve some problems of finite group theory.

We have seen in this text that a finite simple abelian group is isomorphic to $\mathbf{Z}_p$ for some prime p. In 1964, Thompson and Feit [21] proved a long-standing conjecture of Burnside by showing that every finite nonabelian simple group is of even order. Further great strides toward the complete classification were made by Aschbacher in the 1970s. Early in 1980, Griess announced that he had constructed a predicted "monster" simple group of order

$$808, 017, 424, 794, 512, 875, 886, 459, 904, 961,$$
$$710, 757, 005, 754, 368, 000, 000, 000.$$

Aschbacher added the final details of the classification in August 1980. The research papers contributing to the entire classification fill roughly five thousand journal pages.

*12.5 APPLICATIONS

For the remainder of this section we try to illustrate further various aspects of factor groups (proving subgroups normal, computation with factor groups, their significance, uses, and so on). To illustrate how easy it is to compute in a factor group if you can compute in the whole group, we prove the following theorem.

Theorem 12.5   *A factor group of a cyclic group is cyclic.*

Proof.   Let G be cyclic with generator a, and let N be a normal subgroup of G. We claim the coset aN generates G/N. We must compute all powers of aN. But this amounts to computing, in G, all powers of the representative a and all these powers give all elements in G. Hence the powers of aN certainly give all cosets of N and G/N is cyclic. ∎

Note that in forming a factor group of G modulo a subgroup N, you are essentially putting every element in G that is in N equal to e, for N forms your new identity in the factor group. This indicates another use for factor groups. Suppose, for example, that you are studying the structure of a non-abelian group G. Since Theorem 9.3 gives complete information about the structure of all sufficiently small abelian groups, it might be of interest to try to form an abelian group as much like G as possible, *an abelianized version of* G, by starting with G and then requiring that $ab = ba$ for all a and b in your new group structure. To require that $ab = ba$ is to say that $aba^{-1}b^{-1} = e$ in your new group. An element $aba^{-1}b^{-1}$ in a group is a **commutator of the group.** Thus we wish to attempt to form an abelianized version of G by replacing every commutator of G by e. By the first observation of this paragraph, we should then attempt to form the factor group of G modulo the smallest normal subgroup we can find that contains all commutators of G.

> **Theorem 12.6** *The set of all commutators* $aba^{-1}b^{-1}$ *of a group* G *generates a normal subgroup* G' (*the* **commutator subgroup**) *of* G, *and* G/G' *is abelian. Furthermore,* G/N *is abelian if and only if* $G' \leq N$.

Proof. The commutators certainly generate a subgroup G'; we must show that it is normal in G. Note that the inverse $(aba^{-1}b^{-1})^{-1}$ of a commutator is again a commutator, namely $bab^{-1}a^{-1}$. Also $e = eee^{-1}e^{-1}$ is a commutator. Theorem 9.1 then shows that G' consists precisely of all finite products of commutators. For $x \in G'$, we must show that $g^{-1}xg \in G'$ for all $g \in G$, or that if x is a product of commutators, so is $g^{-1}xg$ for all $g \in G$. By inserting $e = gg^{-1}$ between each product of commutators occurring in x, we see that it is sufficient to show for each commutator $cdc^{-1}d^{-1}$ that $g^{-1}(cdc^{-1}d^{-1})g$ is in G'. But

$$\begin{aligned}
g^{-1}(cdc^{-1}d^{-1})g &= (g^{-1}cdc^{-1})(e)(d^{-1}g)\\
&= (g^{-1}cdc^{-1})(gd^{-1}dg^{-1})(d^{-1}g)\\
&= [(g^{-1}c)d(g^{-1}c)^{-1}d^{-1}][dg^{-1}d^{-1}g],
\end{aligned}$$

which is in G'. Thus G' is normal in G.

The rest of the theorem is obvious if you have acquired the proper feeling for factor groups. One doesn't visualize it this way, but writing out that G/G' is abelian follows from

$$\begin{aligned}
(aG')(bG') = abG' &= ab(b^{-1}a^{-1}ba)G'\\
&= (abb^{-1}a^{-1})baG' = baG' = (bG')(aG').
\end{aligned}$$

Furthermore, if G/N is abelian, then $(a^{-1}N)(b^{-1}N) = (b^{-1}N)(a^{-1}N)$, that is, $aba^{-1}b^{-1}N = N$, so $aba^{-1}b^{-1} \in N$, and $G' \leq N$. Finally, if $G' \leq N$, then

$$\begin{aligned}
(aN)(bN) = abN &= ab(b^{-1}a^{-1}ba)N\\
&= (abb^{-1}a^{-1})baN = baN = (bN)(aN). \quad \blacksquare
\end{aligned}$$

Theorem 12.7 *If G is the internal direct product of subgroups H and K, then H and K are normal subgroups of G. Also G/H is isomorphic to K in a natural way.*

Proof. We can regard G as isomorphic to the external direct product $H \times K$, and we need to show that $\bar{H} = \{(h, e) | h \in H\}$ is normal in $H \times K$ and that $(H \times K)/\bar{H}$ is isomorphic to $\bar{K} = \{(e, k) | k \in K\}$.

For normality, we need to show that

$$(h, k)^{-1} \bar{H}(h, k) = \bar{H}$$

for all $(h, k) \in H \times K$. But

$$(h, k)^{-1}(h_1, e)(h, k) = (h^{-1}, k^{-1})(h_1, e)(h, k)$$
$$= (h^{-1}h_1 h, k^{-1}ek) = (h^{-1}h_1 h, e),$$

and $(h^{-1}h_1 h, e) \in \bar{H}$. Thus $\bar{H}$ is normal in $H \times K$. The cosets of $\bar{H}$ are clearly all of the form $(e, k)\bar{H}$ for $k \in K$. It is obvious that the map

$$\phi : \bar{K} \to (H \times K)/\bar{H} \quad \text{given by} \quad (e, k)\phi = (e, k)\bar{H}$$

is an isomorphism. ∎

Exercises

Students often write nonsense when first proving theorems about factor groups. The first two exercises are designed to call attention to one basic type of error.

12.1 A student is asked to show that if H is a normal subgroup of an abelian group G, then G/H is abelian. The student's proof starts as follows:

We must show that G/H is abelian. Let a and b be two elements of G/H.

a) Why does the instructor reading this proof expect to find nonsense from here on in the student's paper?

b) What should the student have written?

c) Complete the proof.

12.2 A student is asked to prove that if G is a torsion group, then so is G/H for every normal subgroup H of G. The student writes

We must show that each element of G/H is of finite order. Let $x \in G/H$.

Answer the same questions as in Exercise 12.1.

12.3 Fill in the blanks.

a) The factor group $\mathbf{Z}_6/\langle 3 \rangle$ is of order _____.

b) The factor group $(\mathbf{Z}_4 \times \mathbf{Z}_{12})/(\langle 2 \rangle \times \langle 2 \rangle)$ is of order _____.

c) The factor group $(\mathbf{Z}_4 \times \mathbf{Z}_{12})/\langle (2, 2) \rangle$ is of order _____.

d) The coset $5 + \langle 4 \rangle$ is of order _____ in the factor group $\mathbf{Z}_{12}/\langle 4 \rangle$.

e) The coset $26 + \langle 12 \rangle$ is of order _____ in the factor group $\mathbf{Z}_{60}/\langle 12 \rangle$.

12.4 Show that A_n is a normal subgroup of S_n and compute S_n/A_n, that is, find a known group to which S_n/A_n is isomorphic.

12.5 Compute (that is, classify as in Theorem 9.3) the following.

a) $(\mathbf{Z} \times \mathbf{Z})/\langle (0, 1) \rangle$.

b) $(\mathbf{Z} \times \mathbf{Z})/\langle (1, 2) \rangle$.

c) $(\mathbf{Z} \times \mathbf{Z} \times \mathbf{Z})/\langle (1, 1, 1) \rangle$.

12.6 This exercise illustrates the fact that if G contains two isomorphic normal subgroups H and K, then G/H need not be isomorphic to G/K. It depends on how H and K are *embedded in* G.

a) Compute $(\mathbf{Z}_2 \times \mathbf{Z}_4)/\langle (1, 0) \rangle$. Note that $\langle (1, 0) \rangle$ is cyclic of order 2. "Compute" means discover to which of the two (up to isomorphism) groups of order 4 this factor group is isomorphic.

b) Repeat part (a) with $(\mathbf{Z}_2 \times \mathbf{Z}_4)/\langle (0, 2) \rangle$.

c) Repeat part (a) with $(\mathbf{Z}_2 \times \mathbf{Z}_4)/\langle (1, 2) \rangle$.

12.7 Find all subgroups of S_3 of Example 4.1 which are conjugate to $\{p_0, \mu_1\}$.

†**12.8** Prove that the torsion subgroup T of an abelian group G is a normal subgroup of G, and that G/T is torsion free. (Note that Lemma 9.1 cannot be used since G may not be finitely generated. Proceed directly from the definitions of a torsion group and a normal subgroup. Do not make the mistake the students made in Exercises 12.1 and 12.2.)

12.9 Mark each of the following true or false.

___ a) It makes sense to speak of the factor group G/N if and only if N is a normal subgroup of the group G.

___ b) Every subgroup of an abelian group G is a normal subgroup of G.

___ c) An inner automorphism of an abelian group must be just the identity map.

___ d) Every factor group of a finite group is again of finite order.

___ e) Every factor group of a torsion group is a torsion group.

___ f) Every factor group of a torsion-free group is torsion free.

___ g) Every factor group of an abelian group is abelian.

___ h) Every factor group of a nonabelian group is nonabelian.

___ i) $\mathbf{Z}/n\mathbf{Z}$ is cyclic of order n.

___ j) $\mathbf{R}/n\mathbf{R}$ is cyclic of order n, where $n\mathbf{R} = \{nr \mid r \in \mathbf{R}\}$ and $\mathbf{R}$ is under addition.

12.10 Describe all subgroups of order ≤ 4 of $\mathbf{Z}_4 \times \mathbf{Z}_4$, and in each case classify the factor group of $\mathbf{Z}_4 \times \mathbf{Z}_4$ modulo the subgroup as in Theorem 9.3, part 1. That is, describe the subgroup and say that the factor group of $\mathbf{Z}_4 \times \mathbf{Z}_4$ modulo the subgroup is isomorphic to $\mathbf{Z}_2 \times \mathbf{Z}_4$, or whatever the case may be. [*Hint:* $\mathbf{Z}_4 \times \mathbf{Z}_4$ has six different cyclic subgroups of order 4. Describe them by giving a generator, such as the subgroup $\langle (1, 0) \rangle$. There is one subgroup of order 4 that is isomorphic to the Klein 4-group. There are three subgroups of order 2.]

12.11 Let H be a normal subgroup of a finite group G, and let $m = (G:H)$. Show that $a^m \in H$ for every $a \in G$.

12.12 Show that an intersection of normal subgroups of a group G is again a normal subgroup of G.

12.13 Show that it makes sense to speak of the smallest normal subgroup of a group G that contains a fixed subset S of G. [*Hint:* Use Exercise 12.12.]

12.14 Show that if a finite group G has exactly one subgroup H of a given order, then H is a normal subgroup of G.

12.15 Show that if a finite group G contains a proper subgroup of index 2 in G, then G is not simple.

12.16 Show that if H and N are subgroups of a group G, and N is normal in G, then $H \cap N$ is normal in H. Show by an example that $H \cap N$ need not be normal in G.

12.17 Let G be a group containing at least one subgroup of a fixed finite order s. Show that the intersection of all subgroups of G of order s is a normal subgroup of G. [*Hint:* Use the fact that if H has order s, then so does $x^{-1}Hx$ for all $x \in G$.]

12.18 a) Show that all automorphisms of a group G form a group under function composition.

b) Show that the inner automorphisms of a group G form a normal subgroup of the group of all automorphisms of G under function composition. [*Warning:* Be sure to show that the inner automorphisms do form a subgroup.]

12.19 Show that the set of all $g \in G$ such that $i_g : G \to G$ is the identity inner automorphism i_e is a normal subgroup of a group G.

12.20 Let G be a group. Show that the relation $a \sim b$ if and only if $a = g^{-1}bg$ for some $g \in G$ is an equivalence relation on G. Some equivalence classes contain only one element c. Characterize those elements c.

12.21 Let G be a group. Show that the relation of $A \sim B$ if and only if A and B are conjugate subgroups of G, so that $A = g^{-1}Bg$ for some $g \in G$, is an equivalence relation on the collection of all subgroups of G. Some equivalence classes may contain a single subgroup K. Characterize those subgroups K.

12.22 Prove that A_n is simple for $n \geq 5$, following the steps and hints given.

a) Show A_n contains every 3-cycle if $n \geq 3$.

b) Show A_n is generated by the 3-cycles for $n \geq 3$. [*Hint:* Note that $(a, b)(c, d) = (a, c, d)$ (a, c, b) and $(a, b)(a, c) = (a, b, c)$.]

c) Let r and s be fixed elements of $\{1, 2, \ldots, n\}$ for $n \geq 3$. Show that A_n is generated by the n "special" 3-cycles of the form (r, s, i) for $1 \leq i \leq n$. [*Hint:* Show every 3-cycle is the product of "special" 3-cycles by computing

$$(r, s, i)^2, \qquad (r, s, i)^2(r, s, j), \qquad (r, s, i)(r, s, j)^2,$$

and
$$(r, s, i)(r, s, j)^2(r, s, k)(r, s, i)^2.$$

Observe that these products give all possible types of 3-cycles.]

d) Let N be a normal subgroup of A_n for $n \geq 3$. Show that if N contains a 3-cycle, then $N = A_n$. [*Hint:* Show that $(r, s, i) \in N$ implies that $(r, s, j) \in N$ for $j = 1, 2, \ldots, n$ by computing $((i, j)(r, s))^{-1}(r, s, i)^2((i, j)(r, s))$.]

e) Let N be a normal subgroup of A_n for $n \geq 5$. Show that one of the following cases must hold, and conclude in each case that $N = A_n$.

Case 1. N contains a 3-cycle.

Case 2. N contains a product of disjoint cycles, at least one of which has length greater than 3. [*Hint:* Suppose N contains the disjoint product $\sigma = (a_1, a_2, \ldots, a_r)\mu$. Show $(a_1, a_2, a_3)^{-1}\sigma(a_1, a_2, a_3)\sigma^{-1}$ is in N, and compute it.]

Case 3. N contains a disjoint product of the form $\sigma = (a_1, a_2, a_3)(a_4, a_5, a_6)\mu$. [*Hint:* Show $(a_1, a_2, a_4)^{-1}\sigma(a_1, a_2, a_4)\sigma^{-1}$ is in N, and compute it.]

Case 4. N contains a disjoint product of the form $\sigma = (a_1, a_2, a_3)\mu$ where μ is a product of disjoint 2-cycles. [*Hint:* Show $\sigma^2 \in N$ and compute it.]

Case 5. N contains a disjoint product σ of the form $\sigma = (a_1, a_2)(a_3, a_4)\mu$ where μ is a product of an even number of disjoint 2-cycles. [*Hint:* Show $(a_1, a_2, a_3)^{-1}\sigma(a_1, a_2, a_3)\sigma^{-1}$ is in N, and compute it to deduce that $\alpha = (a_1, a_3)(a_2, a_4)$ is in N. Using $n \geq 5$ for the first time, find $i \in \{1, 2, \ldots, n\}$ where $i \neq a_1, a_2, a_3, a_4$. Let $\beta = (a_1, a_3, i)$. Show $\beta^{-1}\alpha\beta\alpha \in N$ and compute it.]

*12.23 Find the commutator subgroup G' of the group D_4 of symmetries of the square in Example 4.2.

*12.24 a) Show that if N is a normal subgroup of G, and H is any subgroup of G, then $HN = NH = N \vee H$.

b) Show that if N and M are normal subgroups of G, then NM is also a normal subgroup of G.

*12.25 Show that if H and K are normal subgroups of a group G such that $H \cap K = \{e\}$, then $hk = kh$ for all $h \in H$ and $k \in K$. [*Hint:* Consider the commutator $hkh^{-1}k^{-1} = (hkh^{-1})k^{-1} = h(kh^{-1}k^{-1})$.]

thirteen

Homomorphisms

13.1 DEFINITION AND ELEMENTARY PROPERTIES

An isomorphism of a group G with a group G' was defined as a one-to-one map ϕ of G onto G' such that for all a and b in G, $(ab)\phi = (a\phi)(b\phi)$. If we drop the condition that ϕ be one to one and onto, just retaining $(ab)\phi = (a\phi)(b\phi)$, the map ϕ is then a *homomorphism*. Homomorphisms are closely connected with factor groups, as we shall see.

Definition   A map ϕ of a group G into a group G' is a ***homomorphism*** if

$$(ab)\phi = (a\phi)(b\phi)$$

for all elements a and b in G.

Let us examine the idea behind the requirement $(ab)\phi = (a\phi)(b\phi)$ for a homomorphism $\phi : G \rightarrow G'$. This requirement is the only one that distinguishes a homomorphism from just any map of G into G'. It asserts that ϕ is a *structure-relating map*. The algebraic structure of G is completely determined by the binary operation on G, and that of G' is completely determined by

127

the binary operation on G'. In the condition $(ab)\phi = (a\phi)(b\phi)$, the operation ab on the left-hand side takes place in G, while the operation $(a\phi)(b\phi)$ on the right-hand side takes place in G'. The homomorphism condition thus relates the structure of G to that of G'.

Example 13.1 The natural map γ of $\mathbf{Z}$ into $\mathbf{Z}_n$ given by $m\gamma = r$, where r is the remainder (in the sense of Lemma 6.1) of m when divided by n, is a homomorphism. We need to observe that

$$(s + t)\gamma = s\gamma + t\gamma.$$

If

$$(1)\ \ s = q_1 n + r_1 \qquad \text{and} \qquad (2)\ \ t = q_2 n + r_2$$

for $0 \le r_i < n$, then $s\gamma = r_1$ and $t\gamma = r_2$. Thus

$$s\gamma + t\gamma = (r_1 + r_2) \text{ modulo } n.$$

That is, if $r_1 + r_2 = q_3 n + r_3$ for $0 \le r_3 < n$, then

$$s\gamma + t\gamma = r_3.$$

Adding equations (1) and (2) we get

$$s + t = (q_1 + q_2)n + r_1 + r_2 = (q_1 + q_2 + q_3)n + r_3$$

and $0 \le r_3 < n$. Thus

$$(s + t)\gamma = r_3$$

also.

Viewing $\mathbf{Z}_n$ as the group $\mathbf{Z}/n\mathbf{Z}$ of residue classes modulo n, we see that γ assigns to each element of $\mathbf{Z}$ the coset or residue class in which it appears modulo n. This is an example of a general situation described in the next theorem. ∎

Theorem 13.1 *If N is a normal subgroup of a group G, then the canonical (or natural) map $\gamma: G \to G/N$ given by $a\gamma = aN$ for $a \in G$ is a homomorphism.*

Proof. This is an immediate consequence of the definition of coset multiplication in terms of multiplication of representatives, for

$$(ab)\gamma = abN = (aN)(bN) = (a\gamma)(b\gamma). \quad \blacksquare$$

Definition The *kernel of a homomorphism* ϕ of a group G into a group G' is the set of all elements of G mapped onto the identity element of G' by ϕ.

Example 13.2 For the canonical map $\mathbf{Z} \to \mathbf{Z}_n$ given in Example 13.1, the kernel is $n\mathbf{Z}$. Note that $n\mathbf{Z}$ is a normal subgroup of $\mathbf{Z}$, and that $\mathbf{Z}/n\mathbf{Z}$ is isomorphic to $\mathbf{Z}_n$. ∎

The preceding example illustrates the general connection between homomorphisms and factor groups, which we shall state and prove in Theorem 13.3.

Definition Let ϕ be a mapping of a set X into a set Y, and let $A \subseteq X$ and $B \subseteq Y$. The *image $A\phi$ of A in Y under* ϕ is $\{a\phi \,|\, a \in A\}$. The *inverse image $B\phi^{-1}$ of B in X* is $\{x \in X \,|\, x\phi \in B\}$.

The next theorem gives some structural features preserved under a homomorphism.

Theorem 13.2 *Let ϕ be a homomorphism of a group G into a group G'. If e is the identity in G, then $e\phi$ is the identity in G', and if $a \in G$, then $a^{-1}\phi = (a\phi)^{-1}$. If H is a subgroup of G, then $H\phi$ is a subgroup of G', and H normal in G implies that $H\phi$ is normal in $G\phi$. Going the other way, if K' is a subgroup of G', then $K'\phi^{-1}$ is a subgroup of G, and K' normal in $G\phi$ implies that $K'\phi^{-1}$ is normal in G. Loosely, subgroups correspond to subgroups and normal subgroups to normal subgroups under a homomorphism.*

Proof. Let ϕ be a homomorphism of G into G'. Then

$$a\phi = (ae)\phi = (a\phi)(e\phi).$$

Thus $e\phi$ must be the identity e' in G'. The equation

$$e\phi = (aa^{-1})\phi = (a\phi)(a^{-1}\phi)$$

then shows that $a^{-1}\phi = (a\phi)^{-1}$.

Let H be a subgroup of G, and let $a\phi$ and $b\phi$ be any two elements in $H\phi$. Then $(ab)\phi = (a\phi)(b\phi)$, so $(a\phi)(b\phi) \in H\phi$, that is, $H\phi$ is closed under the operation of G'. The fact that $e' = e\phi$ and $a^{-1}\phi = (a\phi)^{-1}$ completes the proof that $H\phi$ is a subgroup of $G\phi$. Suppose that H is normal in G, and let $g\phi \in G\phi$. Now

$$(g\phi)^{-1}(h\phi)(g\phi) = (g^{-1}\phi)(h\phi)(g\phi) = (g^{-1}hg)\phi.$$

Since $g^{-1}hg \in H$, we have $(g^{-1}hg)\phi \in H\phi$. Thus $H\phi$ is normal in $G\phi$.

Going the other way, let K' be a subgroup of G'. Suppose a and b are in $K'\phi^{-1}$. Then $(ab)\phi = (a\phi)(b\phi)$ and $(a\phi)(b\phi) \in K'$, so $ab \in K'\phi^{-1}$. Also K' must contain the identity $e\phi$, so $e \in K'\phi^{-1}$. If $a \in K'\phi^{-1}$, then $a\phi \in K'$, so $(a\phi)^{-1} \in K'$. But $(a\phi)^{-1} = a^{-1}\phi$, so we must have $a^{-1} \in K'\phi^{-1}$. Hence $K'\phi^{-1}$ is a subgroup of G. If K' is a normal subgroup of $G\phi$, then for $b \in K'\phi^{-1}$ and $g \in G$, we have

$$(g^{-1}bg)\phi = (g\phi)^{-1}(b\phi)(g\phi),$$

and $(g\phi)^{-1}(b\phi)(g\phi)$ is in K', so $g^{-1}bg \in K'\phi^{-1}$. Therefore $K'\phi^{-1}$ is normal in G. ∎

Theorem 13.2 may seem complex to you, but it is very easy, and the proof is really mechanical. It would be an excellent exercise for you to write out the whole proof without referring to the text. This will test whether you really understand the definitions involved.

13.2 THE FUNDAMENTAL HOMOMORPHISM THEOREM

Theorem 13.2 shows, in particular, that *for a homomorphism $\phi:G \to G'$, the kernel $K = \{e'\}\phi^{-1}$ is a normal subgroup of G.* We are now in a position to prove our main theorem.

Theorem 13.3   (Fundamental Homomorphism Theorem)   *Let ϕ be a homomorphism of a group G into a group G' with kernel K. Then $G\phi$ is a group, and there is a canonical (natural) isomorphism of $G\phi$ with G/K.*

Proof.   We saw in Theorem 13.2 that $G\phi$ is a group, for G is a special case of a subgroup of G. Let $aK \in G/K$, and let us attempt to define a map $\psi:G/K \to G\phi$ by

$$(aK)\psi = a\phi.$$

We thus define the map ψ on a coset by choosing a representative a of the coset; our first job is to show that ψ is well defined, that is, independent of the choice of representative. To this end, let $b \in aK$. We must show that $a\phi = b\phi$. But $b \in aK$ means that $b = ak_1$ for $k_1 \in K$ so $a^{-1}b = k_1$. Then

$$e' = k_1\phi = (a^{-1}b)\phi = (a^{-1}\phi)(b\phi) = (a\phi)^{-1}(b\phi).$$

Hence

$$b\phi = (a\phi)e' = a\phi.$$

Thus ψ is well defined.

To show that ψ is one to one, suppose that $(aK)\psi = (bK)\psi$. Then $a\phi = b\phi$, so

$$e' = (a\phi)^{-1}(b\phi) = (a^{-1}\phi)(b\phi) = (a^{-1}b)\phi.$$

Thus $a^{-1}b \in K$, by definition of K. But $a^{-1}b \in K$ implies that $b \in aK$, so $bK = aK$. Thus ψ is one to one.

It is obvious that ψ is onto $G\phi$.

The equation

$$[(aK)(bK)]\psi = (abK)\psi = (ab)\phi = (a\phi)(b\phi) = [(aK)\psi][(bK)\psi]$$

completes the proof that ψ is an isomorphism.

The map ψ is a canonical (or natural) map in the sense that if γ is the canonical homomorphism $\gamma: G \to G/K$ of Theorem 13.1, then

$$\phi = \gamma\psi.$$

One says that the diagram in Fig. 13.1 is *commutative*. ∎

Fig. 13.1

Theorem 13.3 is one that students often have trouble understanding. It really says that for a homomorphism ϕ of a group G, the image, except for the names of the elements, is just G/K, where K is the kernel, and that the homomorphism ϕ is essentially the canonical map $\gamma: G \to G/K$. In other words, Theorem 13.1 describes, in a sense, all homomorphisms.

We tell you here once and for all that any time you have a homomorphism, two things are of paramount importance: the image and the kernel.

Theorem 13.1 and Theorem 13.3 show that homomorphisms correspond to factor groups in a natural way. Namely, for each factor group G/N, there is a homomorphism $\gamma: G \to G/N$ with kernel N. Conversely, for each homomorphism $\phi: G \to G'$, the image $G\phi$ is essentially G/K, where K is the kernel of ϕ. "Essentially" means up to a canonical isomorphism.

Example 13.3   The student having some knowledge of complex number theory will see that the map $\phi: \mathbf{R} \to \mathbf{C}^*$ given by

$$x\phi = \cos x + i \sin x$$

is a homomorphism, where $\mathbf{R}$ is under addition and $\mathbf{C}^*$ is the multiplicative group of nonzero complex numbers. Note that $\cos x + i \sin x = 1$ if and only if $x = 2\pi n$ for some integer n. Thus the kernel of the homomorphism is the cyclic subgroup $\langle 2\pi \rangle$ of $\mathbf{R}$.

Theorem 13.3 shows that $\mathbf{R}/\langle 2\pi \rangle$ is isomorphic to $\mathbf{R}\phi$, which is the multiplicative group of complex numbers of absolute value 1, that is, the complex numbers on the unit circle. This isomorphism can be visualized geometrically. Every coset of $\mathbf{R}/\langle 2\pi \rangle$ has exactly one representative ≥ 0 and $< 2\pi$. Thus the interval $0 \leq x < 2\pi$ can be visualized as $\mathbf{R}/\langle 2\pi \rangle$, and if bent around so that the *open end* of the interval at 2π joins onto the *closed end* at 0, it forms a circle. The addition in $\mathbf{R}/\langle 2\pi \rangle$, viewed in this way as a circle, is just arc length (or central angle) addition, which is exactly what happens if two complex numbers on the unit circle are multiplied. ∎

13.3 APPLICATIONS

Definition   A *maximal normal subgroup of a group* G is a normal subgroup M not equal to G such that there is no proper normal subgroup N of G properly containing M.

Theorem 13.4   *M is a maximal normal subgroup of G if and only if G/M is simple.*

Proof.   Let M be a maximal normal subgroup of G. Consider the canonical homomorphism $\gamma: G \to G/M$ given by Theorem 13.1. Now γ^{-1} of any proper normal subgroup of G/M would be a proper normal subgroup of G properly containing M. But M is maximal so this can't happen. Thus G/M must be simple.

Conversely, Theorem 13.2 shows that if N is a normal subgroup of G properly containing M, then $N\gamma$ is normal in G/M. If also $N \neq G$, then

$$N\gamma \neq G/M \qquad \text{and} \qquad N\gamma \neq \{M\}.$$

Thus, if G/M is simple so that no such $N\gamma$ can exist, no such N can exist, and M is maximal. ∎

Note that a homomorphism ϕ gives an isomorphism of the domain of ϕ with the image of ϕ if and only if ϕ is a one-to-one map.

Theorem 13.5   *A homomorphism ϕ of a group G is a one-to-one function if and only if the kernel of ϕ is $\{e\}$.*

Proof.   Of course, if the map ϕ is one to one, the kernel is just $\{e\}$, since we know that $e\phi$ is the identity e' of the image.

Conversely, suppose that the kernel is $\{e\}$. If $a\phi = b\phi$, then

$$e' = e\phi = (a\phi)^{-1}(b\phi) = (a^{-1}\phi)(b\phi) = (a^{-1}b)\phi,$$

so $a^{-1}b$ is in the kernel. Since the kernel is $\{e\}$, we must have $a^{-1}b = e$, so $a = b$. Thus ϕ is one to one. ∎

In view of Theorem 13.5, we revise the list of steps a mathematician usually goes through to exhibit an isomorphism.

STEP 1.   Define the map.

STEP 2.   Prove that the map is a homomorphism.

STEP 3. Prove that the kernel of the map is $\{e\}$. It is then known that the map gives an isomorphism of the domain with the image.

While we shall not use this terminology, we state for your information that a homomorphism $\phi: G \to G'$ that is a one-to-one map is a **monomorphism** and ϕ is an **epimorphism** if it is onto G'.

Exercises

13.1 Determine which of the following maps are homomorphisms. A superscript* denotes nonzero elements. If the map is a homomorphism, describe the image and the kernel.

a) $\phi: \mathbf{Z} \to \mathbf{R}$ under addition given by $n\phi = n$

b) $\phi: \mathbf{R} \to \mathbf{Z}$ under addition given by $x\phi = $ the greatest integer $\leq x$

c) $\phi: \mathbf{R}^* \to \mathbf{R}^*$ under multiplication given by $x\phi = |x|$

d) $\phi: \mathbf{Z}_6 \to \mathbf{Z}_2$ given by $x\phi = $ the remainder of x when divided by 2, as in Lemma 6.1

e) $\phi: \mathbf{Z}_9 \to \mathbf{Z}_2$ given by $x\phi = $ the remainder of x when divided by 2, in the sense of Lemma 6.1

†**13.2** Let a group G be generated by $\{a_i | i \in I\}$, where I is some indexing set and $a_i \in G$. Let $\phi: G \to G'$ be a homomorphism of G into a group G'. Show that the value of ϕ on every element of G is completely determined by the values $a_i\phi$. Thus, for example, a homomorphism of a cyclic group is completely determined by the value of the homomorphism on a generator of the group. [*Hint:* Use Theorem 9.1 and, of course, the definition of a homomorphism.]

13.3 How many homomorphisms are there of $\mathbf{Z}$ onto $\mathbf{Z}$? of $\mathbf{Z}$ into $\mathbf{Z}_2$? of $\mathbf{Z}$ onto $\mathbf{Z}_2$? [*Hint:* Use Exercise 13.2. See also Exercise 13.11.]

13.4 How many homomorphisms are there of $\mathbf{Z}$ into $\mathbf{Z}_8$? of $\mathbf{Z}$ onto $\mathbf{Z}_8$? [*Hint:* Use Exercise 13.2. See also Exercise 13.11.]

13.5 How many homomorphisms are there of $\mathbf{Z}_{12}$ onto $\mathbf{Z}_5$? of $\mathbf{Z}_{12}$ into $\mathbf{Z}_6$? of $\mathbf{Z}_{12}$ onto $\mathbf{Z}_6$? of $\mathbf{Z}_{12}$ into $\mathbf{Z}_{14}$? of $\mathbf{Z}_{12}$ into $\mathbf{Z}_{16}$? [*Hint:* Use Exercise 13.2.]

13.6 What can one say about homomorphisms of a simple group?

13.7 Mark each of the following true of false.

— a) A_n is a normal subgroup of S_n.

— b) Every isomorphism is also a homomorphism.

— c) Every homomorphism is an isomorphism.

— d) A homomorphism is an isomorphism of the domain with the image if and only if the kernel consists of the group of the identity element alone.

— e) The image of a group of six elements under some homomorphism may have four elements.

— f) The image of a group of six elements under a homomorphism may have twelve elements.

— g) There is a homomorphism of some group of six elements into some group of twelve elements.

— h) There is a homomorphism of some group of six elements into some group of ten elements.

— i) All homomorphisms of a group of prime order are in some sense trivial.

— j) It is not possible to have a homomorphism of some infinite group into some finite group.

13.8 How many homomorphisms are there of $\mathbf{Z}_2 \times \mathbf{Z}_2$ into $\mathbf{Z}_2$? of $\mathbf{Z}_2 \times \mathbf{Z}_2$ onto $\mathbf{Z}_2$? of $\mathbf{Z}_2 \times \mathbf{Z}_2$ into $\mathbf{Z}_6$? of $\mathbf{Z}_2 \times \mathbf{Z}_2$ into $\mathbf{Z}_2 \times \mathbf{Z}_2 \times \mathbf{Z}_2$? of $\mathbf{Z}_2 \times \mathbf{Z}_2$ into $\mathbf{Z}_2 \times \mathbf{Z}_2 \times \mathbf{Z}_4$? [*Hint*: Use Exercise 13.2.]

13.9 The **sign of an even permutation** is $+1$ and the **sign of an odd permutation** is -1. Observe that the map $\text{sgn}_n : S_n \to \{1, -1\}$ defined by

$$\text{sgn}_n(\sigma) = \text{sign of } \sigma$$

is a homomorphism of S_n onto the group $\{1, -1\}$ under multiplication. What is the kernel?

13.10 For groups G_1 and G_2, consider the map $\pi_1 : (G_1 \times G_2) \to G_1$ given by $(x, y)\pi_1 = x$. Show that π_1 is a homomorphism. What is the kernel? To what group is the kernel isomorphic?

13.11 Let G be any group, and let a be any element of G. Let $\phi : \mathbf{Z} \to G$ be defined by $n\phi = a^n$. Show that ϕ is a homomorphism. Describe the image and the possibilities for the kernel of ϕ. [*Comment*: Using this exercise and Theorem 13.3, we get the *elegant proof* that every cyclic group is isomorphic to $\mathbf{Z}/n\mathbf{Z}$ for some nonnegative integer n, and also the *elegant proof* that every element of a group generates a cyclic subgroup of the group.]

13.12 Let G be a group.

a) If $\phi : \mathbf{Z} \times \mathbf{Z} \to G$ is a homomorphism and $(1, 0)\phi = h$ while $(0, 1)\phi = k$, find $(m, n)\phi$.

b) Let $h, k \in G$ and let $\psi : \mathbf{Z} \times \mathbf{Z} \to G$ be defined by $(m, n)\psi = h^m k^n$. Give a necessary and sufficient condition, involving h and k, for ψ to be a homomorphism. Prove your condition.

c) Find a necessary and sufficient condition on G such that the map ψ described in part (b) is a homomorphism for *all* choices of $h, k \in G$.

13.13 Let G be a finite abelian group of order n, and let r be a positive integer relatively prime to n.

a) Show that the map $\phi_r : G \to G$ given by $a\phi_r = a^r$ is an isomorphism of G onto itself. (Follow the outline given after Theorem 13.5.)

b) Deduce that the equation $x^r = a$ always has a unique solution in a finite abelian group G if r is relatively prime to the order of G. What can happen if r is not relatively prime to the order of G?

13.14 Show that if G, G', and G'' are groups and if $\phi : G \to G'$ and $\psi : G' \to G''$ are homomorphisms, then the composite function $\phi\psi : G \to G''$ is a homomorphism.

13.15 Let G be a group, and let $\mathscr{I}_G$ be the group of inner automorphisms of G given by Exercise 12.18. Show that the map $\phi : G \to \mathscr{I}_G$ given by $g\phi = i_g$ is a homomorphism of G onto $\mathcal{I}_G$. Show that the kernel (the **center of** G) is

$$\{a \in G \mid ax = xa \text{ for all } x \in G\}.$$

Describe when ϕ is an isomorphism.

13.16 Let G_1 and G_2 be groups, and let $\phi_1 : G_1 \to G_2$ and $\phi_2 : G_2 \to G_1$ be homomorphisms such that $\phi_1 \phi_2 = \phi_2 \phi_1 = \iota$, where ι is the identity map, that is, $\phi_1 \phi_2 : G_1 \to G_1$ and $\phi_2 \phi_1 : G_2 \to G_2$ are both the identity map. Show that both ϕ_1 and ϕ_2 are isomorphisms of G_1 with G_2, and that $\phi_1 = (\phi_2)^{-1}$.

13.17 Let G and G' be groups, and let H and H' be normal subgroups of G and G', respectively. Let ϕ be a homomorphism of G into G'. Show that ϕ induces a natural homomorphism $\phi_* : (G/H) \to (G'/H')$ if $H\phi \subseteq H'$. (This fact is used constantly in algebraic topology.)

fourteen

Series
of
Groups

14.1 SUBNORMAL AND
NORMAL SERIES

This chapter is concerned with the notion of a *series* of a group G, which gives insight into the structure of G. The results, presented here without proof, hold for both abelian and nonabelian groups. They are not too important for finitely generated abelian groups because of our strong Theorem 9.3. Many of our illustrations will be taken from abelian groups, however, for ease of computation. The results are proved in the following chapter.

> **Definition** A *subnormal* (or *subinvariant*) *series of a group* G is a finite sequence $H_0, H_1, \ldots, H_n$ of subgroups of G such that $H_i < H_{i+1}$ and H_i is a normal subgroup of H_{i+1} with $H_0 = \{e\}$ and $H_n = G$. A *normal* (or *invariant*) *series of* G is a finite sequence $H_0, H_1, \ldots, H_n$ of normal subgroups of G such that $H_i < H_{i+1}$, $H_0 = \{e\}$, and $H_n = G$.

Note that for abelian groups the notions of subnormal and normal series coincide, since every subgroup is normal. A normal series is always subnormal, but the converse need not be true. We defined a subnormal series before a normal series, since the concept of a subnormal series is more important for our work.

Example 14.1 Two examples of normal series of $\mathbf{Z}$ under addition are

$$\{0\} < 8\mathbf{Z} < 4\mathbf{Z} < \mathbf{Z}$$

and

$$\{0\} < 9\mathbf{Z} < \mathbf{Z}. \quad \blacksquare$$

Example 14.2 Consider the group D_4 of symmetries of the square in Example 4.2. The series

$$\{\rho_0\} < \{\rho_0, \mu_1\} < \{\rho_0, \rho_2, \mu_1, \mu_2\} < D_4$$

is a subnormal series, as you can easily check. It is not a normal series, since $\{\rho_0, \mu_1\}$ is not normal in D_4. $\quad\blacksquare$

Definition A subnormal (normal) series $\{K_j\}$ is a **refinement of a subnormal (normal) series** $\{H_i\}$ of a group G if $\{H_i\} \subseteq \{K_j\}$, that is, if each H_i is one of the K_j.

Example 14.3 The series

$$\{0\} < 72\mathbf{Z} < 24\mathbf{Z} < 8\mathbf{Z} < 4\mathbf{Z} < \mathbf{Z}$$

is a refinement of the series

$$\{0\} < 72\mathbf{Z} < 8\mathbf{Z} < \mathbf{Z}.$$

Two new terms, $4\mathbf{Z}$ and $24\mathbf{Z}$, have been inserted. $\quad\blacksquare$

Of interest in studying the structure of G are the factor groups H_{i+1}/H_i. These are defined for both normal and subnormal series, since H_i is normal in H_{i+1} in either case.

Definition Two subnormal (normal) series $\{H_i\}$ and $\{K_j\}$ of the same group G are **isomorphic** if there is a one-to-one correspondence between the collections of factor groups $\{H_{i+1}/H_i\}$ and $\{K_{j+1}/K_j\}$ such that corresponding factor groups are isomorphic.

Clearly, two isomorphic subnormal (normal) series must have the same number of groups.

Example 14.4 The two series of $\mathbf{Z}_{15}$,

$$\{0\} < \langle 5 \rangle < \mathbf{Z}_{15}$$

and

$$\{0\} < \langle 3 \rangle < \mathbf{Z}_{15},$$

are isomorphic. Both $\mathbf{Z}_{15}/\langle 5 \rangle$ and $\langle 3 \rangle/\{0\}$ are isomorphic to $\mathbf{Z}_5$, and $\mathbf{Z}_{15}/\langle 3 \rangle$ is isomorphic to $\langle 5 \rangle/\{0\}$, or to $\mathbf{Z}_3$. $\quad\blacksquare$

14.2 THE JORDAN-HÖLDER THEOREM

The following theorem is fundamental to the theory.

Theorem 14.1 (Schreier)   *Two subnormal (normal) series of a group G have isomorphic refinements.*

Proof. See Chapter 15 for proof of this theorem.   ■

The proof of Theorem 14.1 is really not too difficult. However, we know from experience that some students get lost in the proof and then tend to feel that they can't understand the theorem. We don't prove it in the unstarred sections, even though many students could follow it. Let us, however, illustrate this theorem.

Example 14.5   Let us try to find isomorphic refinements of the series

$$\{0\} < 8\mathbf{Z} < 4\mathbf{Z} < \mathbf{Z}$$

and

$$\{0\} < 9\mathbf{Z} < \mathbf{Z}$$

given in Example 14.1. Consider the refinement

$$\{0\} < 72\mathbf{Z} < 8\mathbf{Z} < 4\mathbf{Z} < \mathbf{Z}$$

of $\{0\} < 8\mathbf{Z} < 4\mathbf{Z} < \mathbf{Z}$ and the refinement

$$\{0\} < 72\mathbf{Z} < 18\mathbf{Z} < 9\mathbf{Z} < \mathbf{Z}$$

of $\{0\} < 9\mathbf{Z} < \mathbf{Z}$. In both cases the refinements have four factor groups isomorphic to $\mathbf{Z}_4, \mathbf{Z}_2, \mathbf{Z}_9$, and $72\mathbf{Z}$ or $\mathbf{Z}$. The *order* in which the factor groups occur is different to be sure.   ■

We now come to the real meat of the theory.

Definition   A subnormal series $\{H_i\}$ of a group G is a **composition series** if all the factor groups H_{i+1}/H_i are simple. A normal series $\{H_i\}$ of G is a **principal** or **chief series** if all the factor groups H_{i+1}/H_i are simple.

Note that for abelian groups the concepts of composition and principal series coincide. Also, since every normal series is subnormal, every principal series is a composition series for any group, abelian or not.

Example 14.6   We claim that $\mathbf{Z}$ has no composition (and also no principal) series. For if

$$\{0\} = H_0 < H_1 < \cdots < H_{n-1} < H_n = \mathbf{Z}$$

is a subnormal series, H_1 must be of the form $r\mathbf{Z}$ for some $r \in \mathbf{Z}^+$. But then H_1/H_0 is isomorphic to $r\mathbf{Z}$, which is infinite cyclic with many nontrivial proper normal subgroups, for example $2r\mathbf{Z}$. Thus $\mathbf{Z}$ has no composition (and also no principal) series. ∎

Example 14.7   The series

$$\{e\} < A_n < S_n$$

for $n \geq 5$ is a composition series (and also a principal series) of S_n, since $A_n/\{e\}$ is isomorphic to A_n, which is simple for $n \geq 5$, and S_n/A_n is isomorphic to $\mathbf{Z}_2$, which is simple. Likewise the two series given in Example 14.4 are composition series (and also principal series) of $\mathbf{Z}_{15}$. They are isomorphic, as shown in that example. This illustrates our main theorem, which will be stated shortly. ∎

Observe that by Theorem 13.4 H_{i+1}/H_i is simple if and only if H_i is a maximal normal subgroup of H_{i+1}. Thus for a composition series, each H_i must be maximal normal subgroup of H_{i+1}. *To form a composition series of a group G, we just hunt for a maximal normal subgroup H_{n-1} of G, then for a maximal normal subgroup of H_{n-1}, and so on. If this process terminates in a finite number of steps, we have a composition series.* Note that by Theorem 13.4, a composition series cannot have any further refinement. *To form a principal series, we have to hunt for a maximal normal subgroup H_{n-1} of G, then for a maximal normal subgroup of H_{n-1} that is also normal in G, and so on.* The main theorem is as follows.

Theorem 14.2 (Jordan-Hölder)   *Any two composition (principal) series of a group G are isomorphic.*

Proof.   Let $\{H_i\}$ and $\{K_j\}$ be two composition (principal) series of G. By Theorem 14.1, they have isomorphic refinements. But since all factor groups are already simple, Theorem 13.4 shows that neither series has any further refinement. Thus $\{H_i\}$ and $\{K_j\}$ must already be isomorphic.

For a finite group, you should regard a composition series as a type of factorization of the group into simple factor groups, analogous to the factorization of a positive integer into primes. In both cases, the factorization is unique, up to the order of the factors.

Theorem 14.3   *If G has a composition (principal) series, and if N is a proper normal subgroup of G, then there exists a composition (principal) series containing N.*

Proof.   The series

$$\{e\} < N < G$$

is both a subnormal and a normal series. Since G has a composition series $\{H_i\}$, then by Theorem 14.1, there is a refinement of $\{e\} < N < G$ to a subnormal series isomorphic to a refinement of $\{H_i\}$. But as a composition series, $\{H_i\}$ can have no further refinement. Thus $\{e\} < N < G$ can be refined to a subnormal series all of whose factor groups are simple, that is, to a composition series. A similar argument holds if we start with a principal series $\{K_j\}$ of G. ∎

Example 14.8   A composition (and also a principal) series of $\mathbf{Z}_4 \times \mathbf{Z}_9$ containing $\langle (0, 1) \rangle$ is

$$\{(0, 0)\} < \langle (0, 3) \rangle < \langle (0, 1) \rangle < \langle 2 \rangle \times \langle 1 \rangle < \langle 1 \rangle \times \langle 1 \rangle = \mathbf{Z}_4 \times \mathbf{Z}_9. \quad \blacksquare$$

The next definition is basic to the last chapter of this text, which deals with solutions of polynomial equations in terms of radicals.

Definition   A group G is **solvable** if it has a composition series $\{H_i\}$ such that all factor groups H_{i+1}/H_i are abelian.

By the Jordan-Hölder theorem, we see that for a solvable group, *every* composition series $\{H_i\}$ must have abelian factor groups H_{i+1}/H_i.

Example 14.9   The group S_3 is solvable, for the composition series

$$\{e\} < A_3 < S_3$$

has factor groups isomorphic to $\mathbf{Z}_3$ and $\mathbf{Z}_2$, which are abelian. The group S_5 is not solvable, for since A_5 is simple, the series

$$\{e\} < A_5 < S_5$$

is a composition series, and $A_5/\{e\}$, which is isomorphic to A_5, is not abelian. *This group A_5 of order 60 can be shown to be the smallest group that is not solvable.* This fact is closely connected with the fact that a polynomial equation of degree 5 is not in general solvable by radicals, but a polynomial equation of degree ≤ 4 is. ∎

*14.3 THE CENTER AND THE ASCENDING CENTRAL SERIES

We give one other type of series for a group.

Definition   The *center of a group* G is the set of all $a \in G$ such that $ax = xa$ for all $x \in G$, that is, the set of all elements of G that commute with every element of G.

Theorem 14.4   *The center of a group is a normal subgroup of the group*

Proof.   The proof of this is so easy and instructive that we save it for an exercise in this chapter. ∎

If you have the table for a finite group G, it is easy to find the center. An element a will clearly be in the center of G if and only if the elements in the row opposite a at the extreme left are given in the same order as the elements in the column under a at the very top of the table.

Now let G be a group, and let $Z(G)$ be the center of G. Since by Theorem 14.4, $Z(G)$ is normal in G, we can form the factor group $G/Z(G)$ and find the center $Z(G/Z(G))$ of this factor group. Since $Z(G/Z(G))$ is normal in $G/Z(G)$, if $\gamma : G \to G/Z(G)$ is the canonical map, then by Theorem 13.2, $[Z(G/Z(G))]\gamma^{-1}$ is a normal subgroup $Z_1(G)$ of G. We can then form the factor group $G/Z_1(G)$ and find its center, take $(\gamma_1)^{-1}$ of it to get $Z_2(G)$, and so on.

Definition   The series

$$\{e\} \leq Z(G) \leq Z_1(G) \leq Z_2(G) \leq \cdots$$

described in the preceding discussion is the **ascending central series of the group** G.

Example 14.10   The center of S_3 is just the identity $\{\rho_0\}$. Thus the ascending central series of S_3 is

$$\{\rho_0\} \leq \{\rho_0\} \leq \{\rho_0\} \leq \cdots$$

The center of the group D_4 of symmetries of the square in Example 4.2 is $\{\rho_0, \rho_2\}$. (Do you remember that we said that this group would give us nice examples of almost anything we discussed?) Since $G/\{\rho_0, \rho_2\}$ is of order 4 and hence abelian, its center is all of $G/\{\rho_0, \rho_2\}$. Thus the ascending central series of D_4 is

$$\{\rho_0\} \leq \{\rho_0, \rho_2\} \leq D_4 \leq D_4 \leq D_4 \leq \cdots . \quad \blacksquare$$

Exercises

14.1 Give isomorphic refinements of the two normal series $\{0\} < 60\mathbf{Z} < 20\mathbf{Z} < \mathbf{Z}$ and $\{0\} < 245\mathbf{Z} < 49\mathbf{Z} < \mathbf{Z}$ of $\mathbf{Z}$ under addition.

14.2 Find all composition series of $\mathbf{Z}_{60}$ and show that they are indeed all isomorphic.

14.3 Find all composition series of $\mathbf{Z}_5 \times \mathbf{Z}_5$.

14.4 Find all composition series of $S_3 \times \mathbf{Z}_2$.

†**14.5** Show that if

$$H_0 = \{e\} < H_1 < H_2 < \cdots < H_n = G$$

is a subnormal (normal) series for a group G, and if H_{i+1}/H_i is of finite order S_{i+1}, then G is of finite order $s_1 s_2 \cdots s_n$.

14.6 Mark each of the following true or false.

___ a) Every normal series is also subnormal.

___ b) Every subnormal series is also normal.

___ c) Every principal series is a composition series.

___ d) Every composition series is a principal series.

___ e) Every abelian group has exactly one composition series.

___ f) Every finite group has a composition series.

___ g) A group is solvable if and only if it has a composition series with simple factor groups.

___ h) S_7 is a solvable group.

___ i) The Jordan-Hölder theorem has some similarity with the Fundamental Theorem of Arithmetic, which states that every positive integer greater than 1 can be factored into a product of primes uniquely up to order.

___ j) Every finite group of prime order is solvable.

14.7 Show that an infinite abelian group can have no composition series. [*Hint:* Use Exercise 14.5, together with the fact that an infinite abelian group always has a proper normal subgroup.]

14.8 Find a composition series of $S_3 \times S_3$. Is $S_3 \times S_3$ solvable?

14.9 Show that a finite direct product of solvable groups is solvable.

14.10 Is the group D_4 of symmetries of the square in Example 4.2 solvable?

14.11 Find the center of $S_3 \times \mathbf{Z}_4$.

14.12 Prove that the center of a group is a normal subgroup of the group. [*Warning:* Don't forget that you have to prove that it is a subgroup before you can prove that it is normal.]

14.13 Find the ascending central series of $S_3 \times \mathbf{Z}_4$.

Isomorphism Theorems; Proof of the Jordan-Hölder Theorem

*15.1 THE ISOMORPHISM THEOREMS

There are several theorems concerning isomorphic factor groups that are known as the *isomorphism theorems* of group theory. The first of these is Theorem 13.3, which we restate for easy reference. The theorem is diagrammed in Fig. 15.1.

Theorem 15.1 (*First Isomorphism Theorem*) *Let $\phi:G \to G'$ be a homomorphism with kernel K, and let $\gamma_K: G \to G/K$ be the canonical homomorphism. Then there exists a unique isomorphism $\psi:G/K \to G\phi$ such that $x\phi = x(\gamma_K\psi)$ for each $x \in G$.*

Recall that if H and N are subgroups of a group G, then $HN = \{hn \mid h \in H, n \in N\}$. In Chapter 8, we defined the *join* $H \vee N$ to be the smallest subgroup of G containing HN. Of course $H \vee N$ is also the smallest subgroup of G containing both H and N, since any such subgroup must contain HN. In general, HN need not be a subgroup of G. However, we have the following lemma.

143

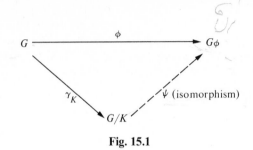

Fig. 15.1

Lemma 15.1 *If N is a normal subgroup of G and H is any subgroup of G, then $H \vee N = HN = NH$. If furthermore H is also normal in G, then HN is normal in G.*

Proof. We show HN is a subgroup of G, from which $H \vee N = HN$ follows at once. Let h_1, $h_2 \in H$ and n_1, $n_2 \in N$. Since N is a normal subgroup, we have $n_1 h_2 = h_2 n_3$ for some $n_3 \in N$. Then $(h_1 n_1)(h_2 n_2) = h_1(n_1 h_2)n_2 = h_1(h_2 n_3)n_2 = (h_1 h_2)(n_3 n_2) \in HN$, so HN is closed under the induced operation in G. Clearly $e = ee$ is in HN. For $h \in H$ and $n \in N$, we have $(hn)^{-1} = n^{-1}h^{-1} = h^{-1}n_4$ for some $n_4 \in N$, since N is a normal subgroup. Thus $(hn)^{-1} \in HN$, so $HN \leq G$. A similar argument shows NH is a subgroup, so $NH = H \vee N = HN$.

Now suppose H is also normal in G, and let $h \in H$, $n \in N$, and $g \in G$. Then $g^{-1}hng = (g^{-1}hg)(g^{-1}ng) \in HN$, so HN is indeed normal in G. ∎

We are now ready for the second isomorphism theorem.

Theorem 15.2 (Second Isomorphism Theorem) *Let H be a subgroup of G and let N be a normal subgroup of G. Then $(HN)/N \simeq H/(H \cap N)$.*

Proof. Since N is normal in G, you see at once that $H \cap N$ is normal in H (see Exercise 15.1). Let $h \in H$ and $n \in N$. We attempt to define $\phi : HN \to H/(H \cap N)$ by $(hn)\phi = h(H \cap N)$. We must show ϕ is well defined. Let $h_1 \in H$ and $n_1 \in N$, and suppose $h_1 n_1 = hn$. Then $h^{-1}h_1 = nn_1^{-1}$, so $h^{-1}h_1$ is in both H and N, and thus is in $H \cap N$. Consequently, $h(H \cap N) = h_1(H \cap N)$ in $H/(H \cap N)$. Thus $(h_1 n_1)\phi = (hn)\phi$, so ϕ is well defined.

We claim ϕ is a homomorphism onto $H/(H \cap N)$. Let n_1, $n_2 \in N$ and h_1, $h_2 \in H$. As in the preceding lemma, we can write $n_1 h_2 = h_2 n_3$ since N is normal in G. Then $[(h_1 n_1)(h_2 n_2)]\phi = [(h_1 h_2)(n_3 n_2)]\phi = h_1 h_2(H \cap N) = h_1(H \cap N) \cdot h_2(H \cap N) = (h_1 n_1)\phi \cdot (h_2 n_2)\phi$, so ϕ is a homomorphism. Since $(he)\phi = h(H \cap N)$ for all $h \in H$, we see ϕ is onto $H/(H \cap N)$.

The kernel of ϕ consists of all $hn \in HN$ such that $h \in H \cap N$; this kernel is thus $(H \cap N)N$. Clearly $(H \cap N)N = N$. Thus ϕ is a homomorphism onto $H/(H \cap N)$ with kernel N, so by Theorem 15.1, $HN/N \simeq H/(H \cap N)$. ∎

Example 15.1 Let $G = \mathbf{Z} \times \mathbf{Z} \times \mathbf{Z}$, $H = \mathbf{Z} \times \mathbf{Z} \times \{0\}$, and $N = \{0\} \times \mathbf{Z} \times \mathbf{Z}$. Then clearly $HN = \mathbf{Z} \times \mathbf{Z} \times \mathbf{Z}$ and $H \cap N = \{0\} \times \mathbf{Z} \times \{0\}$. We have $(HN)/N \simeq \mathbf{Z}$ and also $H/(H \cap N) \simeq \mathbf{Z}$. ∎

If H and K are two normal subgroups of G and $K \leq H$, then clearly H/K is a normal subgroup of G/K. The third isomorphism theorem concerns these groups.

> **Theorem 15.3 (*Third Isomorphism Theorem*)** *Let H and K be normal subgroups of a group G with $K \leq H$. Then $G/H \simeq (G/K)/(H/K)$.*

Proof. Let $\phi : G \to (G/K)/(H/K)$ be given by $a\phi = (aK)(H/K)$ for $a \in G$. Clearly ϕ is onto $(G/K)/(H/K)$, and for $a, b \in G$,

$$(ab)\phi = [(ab)K](H/K) = [(aK)(bK)](H/K) = [(aK)(H/K)][(bK)(H/K)]$$
$$= (a\phi)(b\phi),$$

so ϕ is a homomorphism. The kernel consists of those $x \in G$ such that $x\phi = H/K$. These x are just the elements of H. Then Theorem 15.1 shows that $G/H \simeq (G/K)/(H/K)$. ∎

A nice way of viewing Theorem 15.3 is to regard the canonical map $\gamma_H : G \to G/H$ as being factored via a normal subgroup K of G, $K \leq H \leq G$, to give

$$\gamma_H = \gamma_K \gamma_{H/K},$$

up to a natural isomorphism, as illustrated in Fig. 15.2. Another way of visualizing this theorem is to use the lattice diagram in Fig. 15.3, where each group is a normal subgroup of G and is contained in the one above it. *The larger the normal subgroup, the smaller the factor group.* Thus you can think of G collapsed by H, that is, G/H, as being smaller than G collapsed by K. Theorem 15.3 states that you can collapse G all the way down to G/H in two steps. First collapse to G/K, and then using H/K collapse this to $(G/K)/(H/K)$. The overall result is the same (up to isomorphism) as collapsing G by H.

Fig. 15.2   **Fig. 15.3**

Example 15.2   Consider $K = 6\mathbf{Z} < H = 2\mathbf{Z} < G = \mathbf{Z}$. Then $G/H = \mathbf{Z}/2\mathbf{Z} \simeq \mathbf{Z}_2$. Now $G/K = \mathbf{Z}/6\mathbf{Z}$ has as elements

$$6\mathbf{Z}, \quad 1 + 6\mathbf{Z}, \quad 2 + 6\mathbf{Z}, \quad 3 + 6\mathbf{Z}, \quad 4 + 6\mathbf{Z}, \quad \text{and} \quad 5 + 6\mathbf{Z}.$$

Of these six cosets, $6\mathbf{Z}$, $2 + 6\mathbf{Z}$, and $4 + 6\mathbf{Z}$ lie in $2\mathbf{Z}/6\mathbf{Z}$. Clearly $(\mathbf{Z}/6\mathbf{Z})/(2\mathbf{Z}/6\mathbf{Z})$ has two elements and is isomorphic to $\mathbf{Z}_2$ also. Alternatively, you see $\mathbf{Z}/6\mathbf{Z} \simeq \mathbf{Z}_6$, and $2\mathbf{Z}/6\mathbf{Z}$ corresponds *under this isomorphism* to the cyclic subgroup $\langle 2 \rangle$ of $\mathbf{Z}_6$. Thus $(\mathbf{Z}/6\mathbf{Z})/(2\mathbf{Z}/6\mathbf{Z}) \simeq \mathbf{Z}_6/\langle 2 \rangle \simeq \mathbf{Z}_2 \simeq \mathbf{Z}/2\mathbf{Z}$. ∎

*15.2 THE ZASSENHAUS (BUTTERFLY) LEMMA

The proof of the Jordan-Hölder theorem follows quite easily from a rather technical lemma developed by Zassenhaus. This lemma is also called the butterfly lemma, since Fig. 15.4, which accompanies the lemma, has a butterfly shape.

Let H and K be subgroups of a group G, and let H^* be a normal subgroup of H and K^* a normal subgroup of K. Applying the first statement in Lemma 15.1 to H^* and $H \cap K$ as subgroups of H, you see that $H^*(H \cap K)$ is a group. Similar arguments show that $H^*(H \cap K^*)$, $K^*(H \cap K)$, and $K^*(H^* \cap K)$ are also groups. It is very easy to show that $H^* \cap K$ is a normal subgroup of $H \cap K$ (see Exercise 15.2). The same argument using Lemma

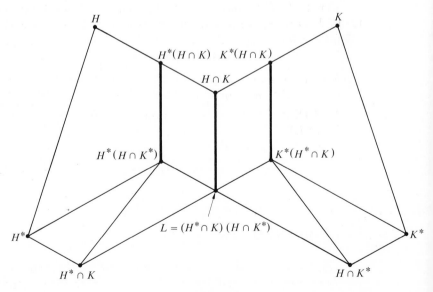

Fig. 15.4

15.1 applied to $H^* \cap K$ and $H \cap K^*$ as subgroups of $H \cap K$ shows that $L = (H^* \cap K)(H \cap K^*)$ is a group. Thus we have the lattice of subgroups shown in Fig. 15.4. You can easily verify the inclusion relations indicated by the diagram.

Since both $H \cap K^*$ and $H^* \cap K$ are normal subgroups of $H \cap K$, the second statement in Lemma 15.1 shows that $L = (H^* \cap K)(H \cap K^*)$ is a normal subgroup of $H \cap K$. We have denoted this particular normal subgroup relationship by the heavy middle line in Fig. 15.4. We claim the other two heavy lines also indicate normal subgroup relationships, and that the three factor groups given by the three normal subgroup relations are all isomorphic. To show this, we shall define a homomorphism $\phi : H^*(H \cap K) \to (H \cap K)/L$, and show ϕ is onto $(H \cap K)/L$ with kernel $H^*(H \cap K^*)$. It will then follow at once that $H^*(H \cap K^*)$ is normal in $H^*(H \cap K)$, and that $H^*(H \cap K)/H^*(H \cap K^*) \simeq (H \cap K)/L$. A similar result for the groups on the right-hand heavy line in Fig. 15.4 then follows by symmetry.

Let $\phi : H^*(H \cap K) \to (H \cap K)/L$ be defined as follows. For $h \in H^*$ and $x \in H \cap K$, let $(hx)\phi = xL$. We show ϕ is well defined and a homomorphism. Let $h_1, h_2 \in H^*$ and $x_1, x_2 \in H \cap K$. If $h_1 x_1 = h_2 x_2$, then $h_2^{-1} h_1 = x_2 x_1^{-1} \in H^* \cap (H \cap K) = H^* \cap K \subseteq L$, so $x_1 L = x_2 L$. Thus ϕ is well defined. Since H^* is normal in H, there is h_3 in H^* such that $x_1 h_2 = h_3 x_1$. Then

$$[(h_1 x_1)(h_2 x_2)]\phi = [(h_1 h_3)(x_1 x_2)]\phi = (x_1 x_2)L$$
$$= (x_1 L)(x_2 L) = (h_1 x_1)\phi \cdot (h_2 x_2)\phi.$$

Thus ϕ is a homomorphism.

Obviously ϕ is onto $(H \cap K)/L$. Finally if $h \in H^*$ and $x \in H \cap K$, then $(Hx)\phi = xL = L$ if and only if $x \in L$, or if and only if $hx \in H^* L = H^*(H^* \cap K)(H \cap K^*) = H^*(H \cap K^*)$. Thus $\mathrm{Ker}(\phi) = H^*(H \cap K^*)$.

We have proved the following lemma.

Lemma 15.2 (Zassenhaus) *Let H and K be subgroups of a group G and let H^* and K^* be normal subgroups of H and K respectively. Then*

1. $H^*(H \cap K^*)$ *is a normal subgroup of* $H^*(H \cap K)$.
2. $K^*(H^* \cap K)$ *is a normal subgroup of* $K^*(H \cap K)$.
3. $H^*(H \cap K)/H^*(H \cap K^*) \simeq K^*(H \cap K)/K^*(H^* \cap K)$

$$\simeq (H \cap K)/[(H^* \cap K)(H \cap K^*)].$$

*15.3 PROOF OF THE SCHREIER THEOREM

Chapter 14 showed that the Jordan-Hölder theorem follows at once from the Schreier theorem (Theorem 14.1). We restate the Schreier theorem here for easy reference, and give the proof.

Theorem 15.4 (Schreier)   *Two subnormal (normal) series of a group G have isomorphic refinements.*

Proof.   Let G be a group and let

$$\{e\} = H_0 < H_1 < H_2 < \cdots < H_n = G \tag{15.1}$$

and

$$\{e\} = K_0 < K_1 < K_2 < \cdots < K_m = G \tag{15.2}$$

be two subnormal series for G. For i where $0 \leq i \leq n - 1$, form the chain of groups

$$H_i = H_i(H_{i+1} \cap K_0) \leq H_i(H_{i+1} \cap K_1) \leq \cdots \leq H_i(H_{i+1} \cap K_m) = H_{i+1}.$$

This inserts $m - 1$ not necessarily distinct groups between H_i and H_{i+1}. If we do this for each i where $0 \leq i \leq n - 1$ and let $H_{i,j} = H_i(H_{i+1} \cap K_j)$, then we obtain the chain of groups

$$\{e\} = H_{0,0} \leq H_{0,1} \leq H_{0,2} \leq \cdots \leq H_{0,\, m-1} \leq H_{1,0}$$
$$\leq H_{1,1} \leq H_{1,2} \leq \cdots \leq H_{1,\, m-1} \leq H_{2,0}$$
$$\leq H_{2,1} \leq H_{2,2} \leq \cdots \leq H_{2,\, m-1} \leq H_{3,0}$$
$$\leq \cdots$$
$$\leq H_{n-1,\, 1} \leq H_{n-1,\, 2} \leq \cdots \leq H_{n-1,\, m-1} \leq H_{n-1,\, m} = G. \tag{15.3}$$

This chain (15.3) contains $nm + 1$ not necessarily distinct groups, and $H_{i,0} = H_i$ for each i. By the Zassenhaus lemma, (15.3) is a subnormal chain that is, each group is normal in the following group. This chain refines the series of (15.1).

In a symmetric fashion, we set $K_{j,i} = K_j(K_{j+1} \cap H_i)$ for $0 \leq j \leq m - 1$ and $0 \leq i \leq n$. This gives a subnormal chain

$$\{e\} = K_{0,0} \leq K_{0,1} \leq K_{0,2} \leq \cdots \leq K_{0,\, n-1} \leq K_{1,0}$$
$$\leq K_{1,1} \leq K_{1,2} \leq \cdots \leq K_{1,\, n-1} \leq K_{2,0}$$
$$\leq K_{2,1} \leq K_{2,2} \leq \cdots \leq K_{2,\, n-1} \leq K_{3,0}$$
$$\leq \cdots$$
$$\leq K_{m-1,\, 1} \leq K_{m-1,\, 2} \leq \cdots \leq K_{m-1,\, n-1} \leq K_{m-1,\, n} = G. \tag{15.4}$$

This chain (15.4) contains $mn + 1$ not necessarily distinct groups, and $K_{j,0} = K_j$ for each j. This chain refines the series of (15.2).

By the Zassenhaus lemma, we have

$$H_i(H_{i+1} \cap K_{j+1})/H_i(H_{i+1} \cap K_j) \simeq K_j(K_{j+1} \cap H_{i+1})/K_j(K_{j+1} \cap H_i),$$

or

$$H_{i,\, j+1}/H_{i,j} \simeq K_{j,\, i+1}/K_{j,i} \tag{15.5}$$

for $0 \leq i \leq n - 1$ and $0 \leq j \leq m - 1$. The isomorphisms of Eq. (15.5) give a one-to-one correspondence of isomorphic factor groups between the subnormal chains of Eq. (15.3) and Eq. (15.4). To verify this correspondence, note that $H_{i,0} = H_i$ and $H_{i,m} = H_{i+1}$, while $K_{j,0} = K_j$ and $K_{j,n} = K_{j+1}$. Each chain of Eq. (15.3) and Eq. (15.4) contains a rectangular array of mn symbols $\leq$. Each $\leq$ gives rise to a factor group. The factor groups arising from the rth row of $\leq$'s in Eq. (15.3) correspond to the factor groups arising from the rth column of $\leq$'s in Eq. (15.4). Deleting repeated groups from the chains in Eq. (15.3) and Eq. (15.4), we obtain subnormal series of distinct groups that are isomorphic refinements of Eq. (15.1) and Eq. (15.2). This establishes the theorem for subnormal series.

For normal series, where all H_i and K_j are normal in G, we merely observe that all the groups $H_{i,j}$ and $K_{j,i}$ formed above are also normal in G, so the same proof applies. This normality of $H_{i,j}$ and $K_{j,i}$ follows at once from the second assertion in Lemma 15.1 and from the fact that intersections of normal subgroups of a group yield normal subgroups. ■

Exercises

*15.1 Show that if H and N are subgroups of G, and N is normal in G, then $H \cap N$ is normal in H.

*15.2 Let H^*, H, and K be subgroups of G with H^* normal in H. Show that $H^* \cap K$ is normal in $H \cap K$.

*15.3 Let H, K, and L be normal subgroups of G with $H < K < L$. Let $A = G/H$, $B = K/H$, and $C = L/H$.

a) Show that B and C are normal subgroups of A, and $B < C$.

b) To what group is $(A/B)/(C/B)$ isomorphic?

†*15.4 Let K and L be normal subgroups of G with $K \vee L = G$, and $K \cap L = \{e\}$. Show that $G/K \simeq L$ and $G/L \simeq K$.

In using the three isomorphism theorems, it is often necessary to know the actual correspondence given by the isomorphism and not just the fact that the groups are isomorphic. The next six exercises give you training for this.

*15.5 Let $\phi : \mathbf{Z}_{12} \to \mathbf{Z}_3$ be the homomorphism such that $1\phi = 2$.

a) Find the kernel K of ϕ.

b) List the cosets in $\mathbf{Z}_{12}/K$, showing the elements in each coset.

c) Give the correspondence between $\mathbf{Z}_{12}/K$ and $\mathbf{Z}_3$ given by the map ψ described in Theorem 15.1.

*15.6 Let $\phi : \mathbf{Z}_{18} \to \mathbf{Z}_{12}$ be the homomorphism where $1\phi = 10$.

a) Find the kernel K of ϕ.

b) List the cosets in $\mathbf{Z}_{18}/K$, showing the elements in each coset.

c) Find the group $\mathbf{Z}_{18}\phi$.

d) Give the correspondence between $\mathbf{Z}_{18}/K$ and $\mathbf{Z}_{18}\phi$ given by the map ψ described in Theorem 15.1.

*15.7 In the group $\mathbf{Z}_{24}$, let $H = \langle 4 \rangle$ and $N = \langle 6 \rangle$.

a) List the elements in HN (which we might write $H + N$ for these additive groups) and in $H \cap N$.

b) List the cosets in HN/N, showing the elements in each coset.

c) List the cosets in $H/(H \cap N)$, showing the elements in each coset.

d) Give the correspondence between HN/N and $H/(H \cap N)$ described in the proof of Theorem 15.2.

*15.8 Repeat Exercise 15.7 for the group $\mathbf{Z}_{36}$ with $H = \langle 6 \rangle$ and $N = \langle 9 \rangle$.

*15.9 In the group $G = \mathbf{Z}_{24}$, let $H = \langle 4 \rangle$ and $K = \langle 8 \rangle$.

a) List the cosets in G/H, showing the elements in each coset.

b) List the cosets in G/K, showing the elements in each coset.

c) List the cosets in H/K, showing the elements in each coset.

d) List the cosets in $(G/K)/(H/K)$, showing the elements in each coset.

e) Give the correspondence between G/H and $(G/K)/(H/K)$ described in the proof of Theorem 15.3.

*15.10 Repeat Exercise 15.9 for the group $G = \mathbf{Z}_{36}$ with $H = \langle 9 \rangle$ and $K = \langle 18 \rangle$.

*15.11 Let G be $\mathbf{Z}_{36}$. Refer to the proof of Theorem 15.4. Let the subnormal series (15.1) be

$$\{0\} < \langle 12 \rangle < \langle 3 \rangle < \mathbf{Z}_{36}$$

and let the subnormal series (15.2) be

$$\{0\} < \langle 18 \rangle < \mathbf{Z}_{36}.$$

Find the chains (15.3) and (15.4) and exhibit the isomorphic factor groups as described in the proof. Write the chains (15.3) and (15.4) in the rectangular array shown in the text.

*15.12 Repeat Exercise 15.11 for the group $\mathbf{Z}_{24}$ with the subnormal series (15.1)

$$\{0\} < \langle 12 \rangle < \langle 4 \rangle < \mathbf{Z}_{24}$$

and (15.2)

$$\{0\} < \langle 6 \rangle < \langle 3 \rangle < \mathbf{Z}_{24}.$$

*15.13 Show that a subgroup K of a solvable group G is solvable. [Hint: Let $H_0 = \{e\} < H_1 < \cdots < H_n = G$ be a composition series for G. Show that the distinct groups among $K \cap H_i$ for $i = 0, \ldots, n$ form a composition series for K. Observe that

$$(K \cap H_i)/(K \cap H_{i-1}) \simeq [H_{i-1}(K \cap H_i)]/[H_{i-1}],$$

by Theorem 15.2, with $H = K \cap H_i$ and $N = H_{i-1}$, and that $H_{i-1}(K \cap H_i) \le H_i$.]

*15.14 Let $H_0 = \{e\} < H_1 < \cdots < H_n = G$ be a composition series for a group G. Let N be a normal subgroup of G, and suppose that N is a simple group. Show that the distinct groups among $H_0, H_i N$ for $i = 0, \ldots, n$ also form a composition series for G.

[*Hint:* H_iN is a group by Lemma 15.1. Show that $H_{i-1}N$ is normal in H_iN. By Theorem 15.2,

$$(H_iN)/(H_{i-1}N) \simeq H_i/[H_i \cap (H_{i-1}N)],$$

and the latter group is isomorphic to

$$[H_i/H_{i-1}]/[(H_i \cap (H_{i-1}N))/H_{i-1}],$$

by Theorem 15.3. But H_i/H_{i-1} is simple.]

*15.15 Let G be a group, and let $H_0 = \{e\} < H_1 < \cdots < H_n = G$ be a composition series for G. Let N be a normal subgroup of G, and let $\gamma: G \to G/N$ be the canonical map. Show that the distinct groups among $H_i\gamma$ for $i = 0, \ldots, n$ form a composition series for G/N. [*Hint:* Observe that the map

$$\psi: H_iN \to (H_i\gamma)/(H_{i-1}\gamma)$$

defined by

$$(h_in)\psi = ((h_in)\gamma)(H_{i-1}\gamma)$$

is a homomorphism with kernel $H_{i-1}N$. By Theorem 15.1,

$$(H_i\gamma)/(H_{i-1}\gamma) \simeq (H_iN)/(H_{i-1}N).$$

Proceed via Theorem 15.2, as shown in the hint for Exercise 15.14.]

*15.16 Prove that a homomorphic image of a solvable group is solvable. [*Hint:* Apply Exercise 15.15 to get a composition series for the homomorphic image. The hints of Exercises 15.14 and 15.15 then show how the factor groups of this composition series in the image look.]

*sixteen

Group
Action
on
a Set

*16.1 THE NOTION OF A
GROUP ACTION

You are now familiar with functions and Cartesian products, so we can take a more sophisticated view of a binary operation on a set S than we did in Chapter 1. A **binary operation on** S is a function mapping $S \times S$ into S. If we denote the function by $*$, then $(s_1, s_2)* = s_3$ is more conventionally expressed as $s_1 * s_2 = s_3$. The function $*$ gives us a rule for "multiplying" any element of S by an element of S to yield again an element of S.

More generally, for any sets A, B, and C, we can view a map $*: A \times B \to C$ as defining a "multiplication," where any element a of A times any element b of B has as value some element c of C. Of course, we write $a * b = c$, or simply $ab = c$. In this chapter, we will be concerned with the case where X is a set, G is a group, and we have a map $*: X \times G \to X$. We shall write $(x, g)*$ as $x * g$ or xg.

Definition Let X be a set and G a group. An **action of G on X** is a map $*: X \times G \to X$ such that

1. $xe = x$ for all $x \in X$,
2. $x(g_1 g_2) = (x g_1) g_2$ for all $x \in X$ and all $g_1, g_2 \in G$.

Under these conditions, X is a G-**set.**

Example 16.1 Let X be any set, and let S_X be the group of all permutations of X. Then X is an S_X-set, where for $x \in X$ and $\sigma \in S_X$, the action $x\sigma$ of σ on x is the effect of the permutation σ on x. Condition 2 is a consequence of the definition of permutation multiplication as function composition, and condition 1 is immediate from the definition of the identity permutation. In particular, $\{1, 2, 3, \ldots, n\}$ is an S_n-set.

Example 16.2 Every group G is itself a G-set, where the action on $g_1 \in G$ by $g_2 \in G$ is given by right multiplication. That is, $(g_1, g_2)* = g_1 g_2$. If H is a subgroup of G, we can also regard G as an H-set, where $(g, h)* = gh$. ■

Example 16.3 Let H be a subgroup of G. Then G is an H-set under conjugation, where $(g, h)* = h^{-1}gh$ for $g \in G$ and $h \in H$. Condition 1 is obvious, and for condition 2 note that $(g, h_1 h_2)* = (h_1 h_2)^{-1} g(h_1 h_2) = h_2^{-1}(h_1^{-1} gh_1)h_2 = ((g, h_1)*, h_2)*$. We always write this action of H on G by conjugation as $h^{-1}gh$. The abbreviation gh described before the definition would cause terrible confusion with the group operation of G. We will see in Chapter 18 that this action of H on G is very important in studying the structure of the group G.

Example 16.4 Let H be a subgroup of G, and let R_H be the set of all right cosets of H. Then R_H is a G-set, where the action of $g \in G$ on the right coset Hx is given by $(Hx)g = H(xg)$. A series of exercises shows that every G-set is isomorphic to one that may be formed using these right coset G-sets as building blocks. (See Exercises 16.12 through 16.15.)

Example 16.5 Let G be the group $D_4 = \{\rho_0, \rho_1, \rho_2, \rho_3, \mu_1, \mu_2, \delta_1, \delta_2\}$ of symmetries of the square, described in Example 4.2. In Fig. 16.1 we show the square with vertices 1, 2, 3, 4 as in Fig. 4.6. We also label the sides s_1, s_2, s_3, s_4, the diagonals d_1 and d_2, vertical and horizontal axes m_1 and m_2, the center point C, and midpoints P_i of the sides s_i. Recall that ρ_i corresponds to rotating the square counterclockwise through $\pi i/2$ radians, μ_i corresponds to flipping on the axis m_i, and δ_i to flipping on the diagonal d_i. We let

$$X = \{1, 2, 3, 4, s_1, s_2, s_3, s_4, m_1, m_2, d_1, d_2, C, P_1, P_2, P_3, P_4\}.$$

Then X can be regarded as a D_4-set in the obvious way. Table 16.1 describes completely the action of D_4 on X and is given to provide geometric illustrations of ideas to be introduced. Be sure you understand how this table is formed. ■

Table 16.1

	1	2	3	4	s_1	s_2	s_3	s_4	m_1	m_2	d_1	d_2	C	P_1	P_2	P_3	P_4
ρ_0	1	2	3	4	s_1	s_2	s_3	s_4	m_1	m_2	d_1	d_2	C	P_1	P_2	P_3	P_4
ρ_1	2	3	4	1	s_2	s_3	s_4	s_1	m_2	m_1	d_2	d_1	C	P_2	P_3	P_4	P_1
ρ_2	3	4	1	2	s_3	s_4	s_1	s_2	m_1	m_2	d_1	d_2	C	P_3	P_4	P_1	P_2
ρ_3	4	1	2	3	s_4	s_1	s_2	s_3	m_2	m_1	d_2	d_1	C	P_4	P_1	P_2	P_3
μ_1	2	1	4	3	s_1	s_4	s_3	s_2	m_1	m_2	d_2	d_1	C	P_1	P_4	P_3	P_2
μ_2	4	3	2	1	s_3	s_2	s_1	s_4	m_1	m_2	d_2	d_1	C	P_3	P_2	P_1	P_4
δ_1	3	2	1	4	s_2	s_1	s_4	s_3	m_2	m_1	d_1	d_2	C	P_2	P_1	P_4	P_3
δ_2	1	4	3	2	s_4	s_3	s_2	s_1	m_2	m_1	d_1	d_2	C	P_4	P_3	P_2	P_1

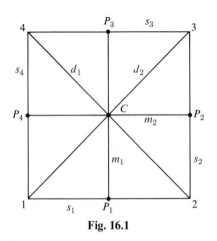

Fig. 16.1

*16.2 FIXED SETS AND ISOTROPY SUBGROUPS

Let X be a G-set. Let $x \in X$ and $g \in G$. It will be important to know when $xg = x$. We let

$$X_g = \{x \in X \mid xg = x\} \quad \text{and} \quad G_x = \{g \in G \mid xg = x\}.$$

Example 16.6 For the D_4-set X in Example 16.5, we have

$$X_{\rho_0} = X, \qquad X_{\rho_1} = \{C\}, \qquad X_{\mu_1} = \{s_1, s_3, m_1, m_2, C, P_1, P_3\}$$

Also, with $G = D_4$,

$$G_1 = \{\rho_0, \delta_2\}, \qquad G_{s_3} = \{\rho_0, \mu_1\}, \qquad G_{d_1} = \{\rho_0, \rho_2, \delta_1, \delta_2\}.$$

We leave the computation of the other X_σ and G_x to Exercises 16.1 and 16.2. ■

You probably noticed that the subsets G_x given in the preceding example were, in each case, subgroups of G. This is true in general.

Theorem 16.1   *Let X be a G-set. Then G_x is a subgroup of G for each $x \in X$.*

Proof.   Let $x \in X$ and let $g_1, g_2 \in G_x$. Then $xg_1 = x$ and $xg_2 = x$. Consequently, $x(g_1g_2) = (xg_1)g_2 = xg_2 = x$, so $g_1g_2 \in G_x$, and G_x is closed under the induced operation of G. Of course $xe = x$, so $e \in G_x$. If $g \in G_x$, then $xg = x$, so $x = xe = x(gg^{-1}) = (xg)g^{-1} = xg^{-1}$, and consequently $g^{-1} \in G_x$. Thus G_x is a subgroup of G. ∎

Definition   Let X be a G-set and let $x \in X$. The subgroup G_x is the *isotropy subgroup of* x.

*16.3 ORBITS

For the D_4-set X of Example 16.5 with action table in Table 16.1, the elements in the subset $\{1, 2, 3, 4\}$ are carried into elements of this same subset under action by D_4. Furthermore, each of the elements 1, 2, 3, and 4 is carried into all the other elements of the subset by the various elements of D_4. We proceed to show that every G-set X can be partitioned into subsets of this type.

Theorem 16.2   *Let X be a G-set. For $x_1, x_2 \in X$, let $x_1 \sim x_2$ if and only if there exists $g \in G$ such that $x_1g = x_2$. Then $\sim$ is an equivalence relation on X.*

Proof.   For each $x \in X$, we have $xe = x$, so $x \sim x$ and $\sim$ is reflexive.

Suppose $x_1 \sim x_2$, so $x_1g = x_2$ for some $g \in G$. Then $x_2g^{-1} = (x_1g)g^{-1} = x_1(gg^{-1}) = x_1e = x_1$, so $x_2 \sim x_1$, and $\sim$ is symmetric.

Finally, if $x_1 \sim x_2$ and $x_2 \sim x_3$, then $x_1g_1 = x_2$ and $x_2g_2 = x_3$ for some $g_1, g_2 \in G$. Then $x_1(g_1g_2) = (x_1g_1)g_2 = x_2g_2 = x_3$, so $x_1 \sim x_3$ and $\sim$ is transitive. ∎

Definition   Let X be a G-set. Each cell in the partition of the equivalence relation described in Theorem 16.2 is an *orbit in X under G.* If $x \in X$, the cell containing x is the *orbit of* x. We let this cell be xG.

The relationship between the orbits of X and the group structure of G lies at the heart of the applications that appear in the next two chapters. The following theorem gives this relationship. Recall that for a set X, we use $|X|$ for the number of elements in X, and $(G:H)$ is the index of a subgroup H in a group G.

Theorem 16.3   *Let X be a G-set and let $x \in X$. Then $|xG| = (G:G_x)$.*

Proof. We define a one-to-one map ϕ from xG onto the collection of right cosets of G_x in G. Let $x_1 \in xG$. Then there exists $g_1 \in G$ such that $xg_1 = x_1$. We define $x_1\phi$ to be the right coset G_xg_1 of G_x. We must show this map ϕ is well defined, independent of the choice of $g_1 \in G$. Suppose also $xg_1' = x_1$. Then $xg_1 = xg_1'$, so $(xg_1)g_1^{-1} = (xg_1')g_1^{-1}$, from which we deduce $x = x(g_1'g_1^{-1})$. Therefore $g_1'g_1^{-1} \in G_x$, so $g_1' \in G_xg_1$, and $G_xg_1 = G_xg_1'$. Thus the map ϕ is well defined.

To show the map ϕ is one to one, suppose $x_1, x_2 \in xG$, and $x_1\phi = x_2\phi$. Then there exist $g_1, g_2 \in G$ such that $x_1 = xg_1$, $x_2 = xg_2$, and $g_2 \in G_xg_1$. Then $g_2 = gg_1$ for some $g \in G_x$, so $x_2 = xg_2 = x(gg_1) = (xg)g_1 = xg_1 = x_1$. Thus ϕ is one to one.

Finally, we show each right coset of G_x in G is of the form $x_1\phi$ for some $x_1 \in xG$. Let G_xg_1 be a right coset. Then if $x_1 = xg_1$, we have $G_xg_1 = x_1\phi$. Thus ϕ maps xG onto the collection of right cosets. ∎

Example 16.7 Let X be the D_4-set in Example 16.5, with action table given by Table 16.1. With $G = D_4$, we have $1G = \{1, 2, 3, 4\}$ and $G_1 = \{\rho_0, \delta_2\}$. Since $|G| = 8$, we have $|1G| = (G : G_1) = 4$. ∎

You should remember not only the cardinality equation in Theorem 16.3 but also that the *elements of G carrying x into xg_1 are precisely the elements of the right coset G_xg_1.* This is clear from the proof.

Exercises

In Exercises 1 through 3, let

$$X = \{1, 2, 3, 4, s_1, s_2, s_3, s_4, m_1, m_2, d_1, d_2, C, P_1, P_2, P_3, P_4\}$$

be the D_4-set of Example 16.5 with action table in Table 16.1. Find the following, where $G = D_4$.

*16.1. The fixed sets X_σ for each $\sigma \in D_4$, that is, $X_{\rho_0}, X_{\rho_1}, \ldots, X_{\delta_2}$.

*16.2 The isotrophy subgroups G_x for each $x \in X$, that is, $G_1, G_2, \ldots, G_{P_3}, G_{P_4}$.

*16.3 The orbits in X under D_4.

*16.4 A nonempty G-set X is **transitive** if, for each $x_1, x_2 \in X$, there exists $g \in G$ such that $x_1g = x_2$. Characterize a transitive G-set in terms of its orbits.

*16.5 Let X be a G-set and let $S \subseteq X$. If $sG \subseteq S$ for all $s \in S$, then S is a **sub-G-set**. Characterize a sub-G-set of a G-set X in terms of orbits in $\bar{X}$ under G.

*16.6 Let X and Y be G-sets with the *same* group G. An **isomorphism** between G-sets X and Y is a map $\phi : X \to Y$ that is one to one, onto Y, and satisfies $(x\phi)g = (xg)\phi$ for all $x \in X$ and $g \in G$. Two G-sets are **isomorphic** if such an isomorphism between them exists. Let X be the D_4-set of Example 16.5.

a) Find two distinct orbits of X that are isomorphic sub-D_4-sets.

b) Show that the orbits $\{1, 2, 3, 4\}$ and $\{s_1, s_2, s_3, s_4\}$ are not isomorphic sub-D-sets. [*Hint:* Find an element of G that acts in an essentially different fashion on the two orbits.]

c) Are the orbits you gave for your answer to part (a) the only two different isomorphic sub-D_4-sets of X?

16.7 A group G acts **faithfully** on a G-set X if the identity is the only element of G that leaves every element of X fixed. Let X be the D_4-set in Example 16.5.

a) Does D_4 act faithfully on X?

b) Find all orbits in X on which D_4 acts faithfully as a sub-D_4-set.

*16.8 Let X be a G-set. Referring to Exercise 16.7, show that G acts faithfully on X if and only if no two distinct elements of G have the same action on each element of X.

†*16.9 Let X be a G-set and let $Y \subseteq X$. Let $G_Y = \{g \in G \mid yg = y \text{ for all } y \in Y\}$. Show G_Y is a subgroup of G, generalizing Theorem 16.1.

*16.10 Mark each of the following true or false.

___ a) Every G-set is also a group.

___ b) Each element of a G-set is left fixed by the identity of G.

___ c) If every element of a G-set is left fixed by the same element g of G, then g must be the identity e.

___ d) Let X be a G-set with $x_1, x_2 \in X$ and $g \in G$. If $x_1 g = x_2 g$, then $x_1 = x_2$.

___ e) Let X be a G-set with $x \in X$ and $g_1, g_2 \in G$. If $xg_1 = xg_2$, then $g_1 = g_2$.

___ f) Each orbit of a G-set X is a transitive sub-G-set.

___ g) Let X be a G-set and let $H \le G$. Then X can be regarded in a natural way as an H-set.

___ h) With reference to (g), the orbits in X under H are the same as the orbits in X under G.

___ i) If X is a G-set, then each element of G acts as a permutation of X.

___ j) Let X be a G-set and let $x \in X$. If G is finite, then $|G| = |xG| \cdot |G_x|$.

*16.11 Let G be the additive group of real numbers. Let the action of $\theta \in G$ on the real plane $\mathbf{R}^2$ be given by rotating the plane counterclockwise about the origin through θ radians. Let P be a point other than the origin in the plane.

a) Show $\mathbf{R}^2$ is a G-set.

b) Describe geometrically the orbit containing P.

c) Find the group G_P.

Exercises 12 through 15 show how all possible G-sets, up to isomorphism, can be formed from a group G.

*16.12 Let $\{X_i \mid i \in I\}$ be a disjoint collection of sets, so $X_i \cap X_j = \varnothing$ for $i \ne j$. Let each X_i be a G-set for the same group G.

a) Show that $\bigcup_{i \in I} X_i$ can be viewed in a natural way as a G-set, the **union** of the G-sets X_i.

b) Show that every G-set X is the union of its orbits.

*16.13 Let X be a transitive G-set, and let $x_0 \in X$. Show that X is isomorphic to the G-set R of all right cosets of G_{x_0}, described in Example 16.4. [*Hint:* For $x \in X$, suppose $x = x_0 g$, and define $\phi : X \to R$ by $x\phi = G_{x_0} g$. Be sure to show ϕ is well defined!]

*16.14 Let X_i for $i \in I$ be G-sets for the same group G, and suppose the sets X_i are not necessarily disjoint. Let $X'_i = \{(x, i) \mid x \in X_i\}$ for each $i \in I$. Then the sets X'_i are disjoint, and each can still be regarded as a G-set in an obvious way. (The elements of X_i have simply been tagged by i to distinguish them from the elements of X_j for $i \neq j$.) The G-set $\bigcup_{i \in I} X'_i$ is the **disjoint union** of the G-sets X_i. Using Exercises 16.12 and 16.13, show that every G-set is isomorphic to a disjoint union of right coset G-sets, as described in Example 16.4.

*16.15 The preceding exercises show that every G-set is isomorphic to a disjoint union of right coset G-sets. The question then arises whether right coset G-sets of distinct subgroups H and K of G can themselves be isomorphic. Note that the map defined in the hint of Exercise 16.13 depends on the choice of x_0 as "base point." If x_0 is replaced by $x_0 g_0$ and if $G_{x_0} \neq G_{x_0 g_0}$, then the collections R_H of right cosets of $H = G_{x_0}$ and R_K of right cosets of $K = G_{x_0 g_0}$ form distinct G-sets that must be isomorphic, since both R_H and R_K are isomorphic to X.

a) Let X be a transitive G-set and let $x_0 \in X$ and $g_0 \in G$. If $H = G_{x_0}$, describe $K = G_{x_0 g_0}$ in terms of H and g_0.

b) Based on (a), conjecture conditions on subgroups H and K of G such that the right coset G-sets of H and K are isomorphic.

c) Prove your conjecture in (b).

*16.16 Up to isomorphism, how many transitive Z_4 sets X are there? (Use the preceding exercises.) Give an example of each isomorphism type, listing an action table of each, as in Table 16.1. Take lowercase names a, b, c, and so on for the elements in the set X.

*16.17 Repeat Exercise 16.16 for the group Z_6.

*16.18 Repeat Exercise 16.16 for the group S_3. List the elements of S_3 in the order ι, (1 2 3), (1 3 2), (2 3), (1 3), (1 2).

Applications
of
G-Sets
to
Counting

This section presents an application of our work with G-sets to counting. Suppose, for example, you wish to count how many distinguishable ways the six faces of a cube can be marked with from one to six dots to form a die. The standard die is marked so that when placed on a table with the 1 on the bottom and the 2 toward the front, the 6 is on top, the 3 on the left, the 4 on the right, and the 5 on the back. Of course, other ways of marking the cube to give a distinguishably different die are possible.

Let us distinguish between the faces of the cube for the moment and call them the bottom, top, left, right, front, and back. Then the bottom can have any one of six marks from one dot to six dots, the top any one of the five remaining marks, and so on. There are $6! = 720$ ways the cube faces can be marked in all. Some markings yield the same die as others, in the sense that one marking can be carried into another by a rotation of the marked cube. For example, if the standard die described above is rotated $90°$ counterclockwise as you look down on it, then 3 will be on the front face rather than 2, but it is the same die.

There are twenty-four possible positions of a cube on a table, for any one of six faces can be placed down, and then any one of four to the front, giving $6 \cdot 4 = 24$ possible positions. Any position can be achieved from any other by a rotation of the die. These rotations form a group G, which is isomorphic to a subgroup of S_8 (see Exercise 4.10). We let X be the 720 possible ways of marking the cube and let G act on X by rotation of the cube. We consider two markings to give the same die if one can be carried into the other under action by an element of G, that is, by rotating the cube. In other words, we consider each *orbit* in X under G to correspond to a single die, and different orbits to give different dice. The determination of the number of distinguishable dice thus leads to the question of determining the number of orbits under G in a G-set X.

The following theorem gives a tool for determining the number of orbits in a G-set X under G. Recall that for each $g \in G$ we let X_g be the set of elements of X left fixed by g, so that $X_g = \{x \in X \mid xg = x\}$. Recall also that for each $x \in X$, we let $G_x = \{g \in G \mid xg = x\}$, and xG is the orbit of x under G.

Theorem 17.1 (Burnside)   *Let G be a finite group and X a finite G-set. If r is the number of orbits in X under G, then*

$$r \cdot |G| = \sum_{g \in G} |X_g|. \tag{17.1}$$

Proof.   We consider all pairs (x, g) where $xg = x$, and let N be the number of such pairs. For each $g \in G$ there are $|X_g|$ pairs having g as second member. Thus,

$$N = \sum_{g \in G} |X_g|. \tag{17.2}$$

On the other hand, for each $x \in X$ there are $|G_x|$ pairs having x as first member. Thus we also have

$$N = \sum_{x \in X} |G_x|.$$

By Theorem 16.3 we have $|xG| = (G : G_x)$. But we know that $(G : G_x) = |G|/|G_x|$, so we obtain $|G_x| = |G|/|xG|$. Then

$$N = \sum_{x \in X} \frac{|G|}{|xG|} = |G| \left(\sum_{x \in X} \frac{1}{|xG|} \right).$$

Now $1/|xG|$ has the same value for all x in the same orbit, and if we let $\mathcal{O}$ be any orbit, then

$$\sum_{x \in \mathcal{O}} \frac{1}{|xG|} = 1.$$

Thus we obtain

$$N = |G| \text{ (number of orbits in } X \text{ under } G) = |G| \cdot r. \qquad (17.3)$$

Comparison of Eq. (17.2) and Eq. (17.3) gives Eq. (17.1). ∎

Corollary   *If G is a finite group and X is a finite G-set, then*

$$\text{(number of orbits in } X \text{ under } G) = \frac{1}{|G|} \cdot \sum_{g \in G} |X_g|.$$

Proof.   The proof of this corollary follows immediately from the preceding theorem. ∎

Let us continue our computation of the number of distinguishable dice as our first example.

Example 17.1   We let X be the set of 720 different markings of faces of a cube using from one to six dots. Let G be the group of 24 rotations of the cube as discussed above. We saw that the number of distinguishable dice is the number of orbits in X under G. Now $|G| = 24$. For $g \in G$ where $g \neq e$, we have $|X_g| = 0$, because any rotation other than the identity changes any one of the 720 markings into a different one. However, $|X_e| = 720$ since the identity leaves all 720 markings fixed. Then by the corollary to Theorem 17.1,

$$\text{(number of orbits)} = \frac{1}{24} \cdot 720 = 30,$$

so there are thirty distinguishable dice. ∎

Of course the number of distinguishable dice could be counted without using the machinery of the preceding corollary, but by using elementary combinatorics as often taught in a freshman finite math course. In marking a cube to make a die, you can, by rotation if necessary, assume the face marked 1 is down. There are five choices for the top (opposite) face. By rotating the die as you look down on it, any one of the remaining four faces could be brought to the front position, so there are no different choices involved for the front face. But with respect to the number on the front face, there are $3 \cdot 2 \cdot 1$ possibilities for the remaining three side faces. Thus there are $5 \cdot 3 \cdot 2 \cdot 1 = 30$ possibilities in all.

The next two examples appear in some finite math texts and are easy to solve by elementary means. We use the corollary to Theorem 17.1 so that you have more practice thinking in terms of orbits.

Example 17.2   How many distinguishable ways can seven people be seated at a round table, where there is no distinguishable "head" to the table? Of course there are 7! ways to assign people to the different chairs. We take X

to be the 7! possible assignments. A rotation of people achieved by asking each person to move one place to the right results in the same arrangement. Such a rotation generates a cyclic group G of order 7, which we consider to act on X in the obvious way. Again, only the identity e leaves any arrangement fixed, and it leaves all 7! arrangements fixed. By the corollary to Theorem 17.1,

$$\text{(number of orbits)} = \frac{1}{7} \cdot 7! = 6! = 720. \ \blacksquare$$

Example 17.3   How many distinguishable necklaces (with no clasp) can be made using seven different-colored beads of the same size? Unlike the table in Example 17.2, the necklace can be turned over as well as rotated. Thus we consider the full dihedral group D_7 of order $2 \cdot 7 = 14$ as acting on the set X of 7! possibilities. Then the number of distinguishable necklaces is

$$\text{(number of orbits)} = \frac{1}{14} \cdot 7! = 360. \ \blacksquare$$

In using the corollary to Theorem 17.1, one has to compute $|G|$ and $|X_g|$ for each $g \in G$. In the examples and the exercises, $|G|$ will pose no real problem. Let us give an example where $|X_g|$ is not as trivial to compute as in the preceding examples. We will continue to assume you know very elementary combinatorics.

Example 17.4   Let us find the number of distinguishable ways the edges of an equilateral triangle can be painted if four different colors of paint are available, assuming only one color is used on a single edge, and the same color may be used on different edges.

Of course there are $4^3 = 64$ ways of painting the edges in all, since each of the three edges may be any one of four colors. We consider X to be the set of these 64 possible painted triangles. The group G acting on X is the group of symmetries of the triangle, which is isomorphic to S_3 and which we consider to be S_3. We use the notation for elements in S_3 given in Chapter 4. We need to compute $|X_g|$ for each of the six elements g in S_3.

$\|X_{\rho_0}\| = 64$	every painted triangle is left fixed by ρ_0.
$\|X_{\rho_1}\| = 4$	to be invariant under ρ_1, all edges must be the same color, and there are 4 possible colors.
$\|X_{\rho_2}\| = 4$	same reason as for ρ_1.
$\|X_{\mu_1}\| = 16$	the edges that are interchanged must be the same color (4 possibilities) and the other edge may also be any of the colors (times 4 possibilities).
$\|X_{\mu_2}\| = \|X_{\mu_3}\| = 16$	same reason as for μ_1.

Then

$$\sum_{g \in S_3} |X_g| = 64 + 4 + 4 + 16 + 16 + 16 = 120.$$

Thus

$$(\text{number of orbits}) = \frac{1}{6} \cdot 120 = 20,$$

and there are twenty distinguishable painted triangles. ■

Example 17.5 We repeat Example 17.4 with the assumption that a different color is used on each edge. The number of possible ways of painting the edges is then $4 \cdot 3 \cdot 2 = 24$, and we let X be the set of 24 possible painted triangles. Again, the group acting on X can be considered to be S_3. Since all edges are a different color, we see $|X_{\rho_0}| = 24$ while $|X_g| = 0$ for $g \neq \rho_0$. Thus

$$(\text{number of orbits}) = \frac{1}{6} \cdot 24 = 4,$$

so there are four distinguishable triangles. ■

Exercises

In each of the following exercises use the corollary to Theorem 17.1 to work the problem even though the answer may be obtained by more elementary methods.

*17.1 Find the number of orbits in $\{1, 2, 3, 4, 5, 6, 7, 8\}$ under the cyclic subgroup $\langle (1, 3, 5, 6) \rangle$ of S_8.

*17.2 Find the number of orbits in $\{1, 2, 3, 4, 5, 6, 7, 8\}$ under the subgroup of S_8 generated by $(1, 3)$ and $(2, 4, 7)$.

*17.3 Find the number of distinguishable tetrahedral dice that can be made using from one, two, three, and four dots on the faces of a regular tetrahedron, rather than a cube.

*17.4 Wooden cubes of the same size are to be painted a different color on each face to make children's blocks. How many distinguishable blocks can be made if eight colors of paint are available?

*17.5 Answer Exercise 17.4 if colors may be repeated on different faces at will. [*Hint:* The twenty-four rotations of a cube consist of the identity, nine that leave a pair of opposite faces invariant, eight that leave a pair of opposite vertices invariant, and six leaving a pair of opposite edges invariant.]

*17.6 Each of the eight corners of a cube is to be tipped with one of four colors, each of which may be used on from one to all eight corners. Find the number of distinguishable markings possible. (See the hint of Exercise 17.5.)

*17.7 Find the number of distinguishable ways the edges of a square of cardboard can be painted if six colors of paint are available and

a) no color is used more than once.

b) the same color can be used on any number of edges.

*17.8 Consider six straight wires of equal lengths with ends soldered together to form edges of a regular tetrahedron. Either a 50-ohm or 100-ohm resistor is to be inserted in the middle of each wire. Assume there are at least six of each type of resistor available. How many essentially different wirings are possible?

*17.9 A rectangular prism 2 ft. long with 1-ft.-square ends is to have each of its six faces painted with one of six possible colors. How many distinguishable painted prisms are possible if

a) no color is to be repeated on different faces,

b) each color may be used on any number of faces?

eighteen

Sylow Theorems

Theorem 9.3 gives us complete information about all finite abelian groups. The study of finite nonabelian groups is much more complicated. The Sylow theorems give us some important information about them.

We know the order of a subgroup of a finite group G must divide $|G|$. If G is abelian, then there exist subgroups of every order dividing $|G|$. We mentioned at the end of Chapter 11 that this is not true for nonabelian groups; A_4 can be shown to have no subgroup of order 6. The Sylow theorems assert that for a prime power dividing $|G|$, there is a subgroup of that prime-power order. They also give some information about the number of such subgroups.

Proofs of the Sylow theorems give us another application of action of a group on a set. This time, the set itself is formed from the group; in some instances the set is the group itself, sometimes it is a collection of cosets of a subgroup, and sometimes it is a collection of subgroups.

*18.1 *p*-Groups

The preceding chapter gave applications of an equation that counted the number of orbits in a finite G-set. Most of our results in this section flow from an equation that counts the number of elements in a finite G-set.

Let X be a finite G-set. Recall that for $x \in X$, the orbit of x in X under G is $xG = \{xg \mid g \in G\}$. Suppose there are r orbits in X under G, and let $\{x_1, x_2, \ldots, x_r\}$ contain one element from each orbit in X. Now every element of X is in precisely one orbit, so

$$|X| = \sum_{i=1}^{r} |x_i G|. \tag{18.1}$$

There may be some one-element orbits in X. Let $X_G = \{x \in X \mid xg = x$ for all $g \in G\}$. Thus X_G is precisely the union of the one-element orbits in X. Let us suppose there are s one-element orbits, where $0 \leq s \leq r$. Then $|X_G| = s$ and we may rewrite Eq. (18.1) as

$$|X| = |X_G| + \sum_{i=s+1}^{r} |x_i G|. \tag{18.2}$$

Most of the results of this chapter will flow from Eq. (18.2). We shall develop Sylow theory as in Hungerford [9], where credit is given to R. J. Nunke for the line of proof. The proof of Theorem 18.2 (Cauchy's theorem) is credited there to J. H. McKay.

Theorem 18.1, which follows, is not quite a counting theorem, but it does have a numerical conclusion. It counts modulo p. The theorem seems to be amazingly powerful. In the rest of the chapter, if we choose the correct set, the correct group action on it, and apply Theorem 18.1, what we want seems to fall right into our lap! Compared with older proofs, the arguments are extremely pretty and elegant.

Throughout this chapter, p will always be a prime integer.

Theorem 18.1 *Let G be a group of order p^n and let X be a finite G-set. Then $|X| \equiv |X_G| \pmod{p}$.*

Proof. In the notation of Eq. (18.2), we know that $|x_i G| = (G:G_{x_i})$ by Theorem 16.3. But $(G:G_{x_i})$ divides $|G|$, and consequently p divides $(G:G_{x_i})$ and thus divides $|x_i G|$ for $s + 1 \leq i \leq r$. Equation (18.2) then shows that $|X| - |X_G|$ is divisible by p, so $|X| \equiv |X_G| \pmod{p}$. ∎

Definition A group G is a *p-group* if every element in G has order a power of the prime p. A subgroup of a group G is a *p-subgroup of G* if the subgroup is itself a p-group.

Our goal in this chapter is to show that a finite group G has a subgroup of every prime-power order dividing $|G|$. As a first step, we prove Cauchy's theorem, which says that if p divides $|G|$, then G has a subgroup of order p.

Theorem 18.2 (Cauchy) *Let G be a finite group and let p divide $|G|$. Then G has an element of order p and, consequently, a subgroup of order p.*

Proof. We form the set X of all p-tuples $(g_1, g_2, \ldots, g_p)$ of elements of G having the property that the product of the coordinates in G is e. That is,

$$X = \{(g_1, g_2, \ldots, g_p) | g_i \in G \text{ and } g_1 g_2 \cdots g_p = e\}.$$

We claim p divides $|X|$. In forming a p-tuple in X, we may let $g_1, g_2, \ldots, g_{p-1}$ be any elements of G, and g_p is then uniquely determined as $(g_1 g_2 \cdots g_{p-1})^{-1}$. Thus $|X| = |G|^{p-1}$ and since p divides $|G|$, we see that p divides $|X|$.

Let σ be the cycle $(1, 2, 3, \ldots, p)$ in S_p. We let σ act on X by

$$(g_1, g_2, \ldots, g_p)\sigma = (g_{1\sigma}, g_{2\sigma}, \ldots, g_{p\sigma}) = (g_2, g_3, \ldots, g_p, g_1).$$

Note that $(g_2, g_3, \ldots, g_p, g_1) \in X$, for $g_1(g_2 g_3 \cdots g_p) = e$ implies that $g_1 = (g_2 g_3 \cdots g_p)^{-1}$, so $(g_2 g_3 \cdots g_p)g_1 = e$ also. Thus σ acts on X, and we consider the subgroup $\langle \sigma \rangle$ of S_p to act on X by iteration in the obvious way.

Now $|\langle \sigma \rangle| = p$, so we may apply Theorem 18.1, and we know that $|X| \equiv |X_{\langle \sigma \rangle}| \pmod{p}$. Since p divides $|X|$, it must be that p divides $|X_{\langle \sigma \rangle}|$ also. Let us examine $X_{\langle \sigma \rangle}$. Now $(g_1, g_2, \ldots, g_p)$ is left fixed by σ, and hence by $\langle \sigma \rangle$, if and only if $g_1 = g_2 = \cdots = g_p$. We know at least one element in $X_{\langle \sigma \rangle}$, namely $(e, e, \ldots, e)$. Since p divides $|X_{\langle \sigma \rangle}|$, there must be at least p elements in $X_{\langle \sigma \rangle}$. Hence there exists some element $a \in G$, $a \neq e$, such that $(a, a, \ldots, a) \in X_{\langle \sigma \rangle}$ and $a^p = e$, so a has order p. Of course $\langle a \rangle$ is a subgroup of G of order p. ∎

Corollary *Let G be a finite group. Then G is a p-group if and only if $|G|$ is a power of p.*

Proof. We leave the proof of this corollary to Exercise 18.5. ∎

*18.2 THE SYLOW THEOREMS

Let G be a group, and let $\mathscr{S}$ be the collection of all subgroups of G. We make $\mathscr{S}$ into a G-set by letting G act on $\mathscr{S}$ by conjugation. That is, if $H \in \mathscr{S}$ so $H \leq G$ and $g \in G$, then g acting on H yields the conjugate subgroup $g^{-1}Hg$. (To avoid confusion we will never write this action as Hg.) Now $H_G = \{g \in G | g^{-1}Hg = H\}$ is a subgroup of G by Theorem 16.1. Obviously H is a normal subgroup of H_G. Since H_G consists of *all* elements of G that leave H invariant under conjugation, H_G is the largest subgroup of G having H as a normal subgroup.

Definition   The subgroup H_G just discussed is the ***normalizer of H in G*** and will be denoted $N[H]$ from now on.

Lemma 18.1   *Let H be a p-subgroup of a finite group G. Then*

$$(N[H]:H) \equiv (G:H) \pmod{p}.$$

Proof.   Let $\mathscr{R}$ be the set of right cosets of H in G, and let H act on $\mathscr{R}$ by right translation, so that $(Hx)h = H(xh)$. Then $\mathscr{R}$ becomes an H-set. Note that $|\mathscr{R}| = (G:H)$.

Let us determine $\mathscr{R}_H$, that is, those right cosets left fixed by all elements of H. Now $Hx = (Hx)h$ if and only if $H = Hxhx^{-1}$, or if and only if $xhx^{-1} \in H$. Thus $Hx = (Hx)h$ for all $h \in H$ if and only if $xhx^{-1} = (x^{-1})^{-1}hx^{-1} \in H$ for all $h \in H$, or if and only if $x^{-1} \in N[H]$, or if and only if $x \in N[H]$. Thus the right cosets in $\mathscr{R}_H$ are those contained in $N[H]$. The number of such cosets $(N[H]:H)$, so $|\mathscr{R}_H| = (N[H]:H)$.

Since H is a p-group, it has order a power of p by the corollary to Theorem 18.2. Theorem 18.1 then tells us that $|\mathscr{R}| \equiv |\mathscr{R}_H| \pmod{p}$, that is, that $(G:H) \equiv (N[H]:H) \pmod{p}$.   ∎

Corollary   *Let H be a p-subgroup of a finite group G. If p divides $(G:H)$, then $N[H] \neq H$.*

Proof.   It follows from Lemma 18.1 that p divides $(N[H]:H)$, which must then be different from 1. Thus $H \neq N[H]$.   ∎

We are now ready for the first of the Sylow theorems, which asserts the existence of prime-power subgroups of G for any prime power dividing $|G|$.

Theorem 18.3 (*First Sylow Theorem*)   *Let G be a finite group and let $|G| = p^n m$ where $n \geq 1$ and where p does not divide m. Then*

1. *G contains a subgroup of order p^i for each i where $1 \leq i \leq n$,*
2. *every subgroup H of G of order p^i is a normal subgroup of a subgroup of order p^{i+1} for $1 \leq i < n$.*

Proof.   1. We know G contains a subgroup of order p by Cauchy's theorem (Theorem 18.2). We use an induction argument and show that the existence of a subgroup of order p^i for $i < n$ implies the existence of a subgroup of order p^{i+1}. Let H be a subgroup of order p^i. Since $i < n$, we see p divides $(G:H)$. By Lemma 18.1, we then know p divides $(N[H]:H)$. Since H is a normal subgroup of $N[H]$, we can form $N[H]/H$, and we see that p divides $|N[H]/H|$. By Cauchy's theorem, the factor group $N[H]/H$ has a subgroup K of order p. If $\gamma : N[H] \to N[H]/H$ is the canonical homomorphism, then

$K\gamma^{-1} = \{x \in N[H] \,|\, x\gamma \in K\}$ is a subgroup of $N[H]$, and hence of G. This subgroup contains H and is of order p^{i+1}.

2. We repeat the construction in part 1 and note that $H < K\gamma^{-1} \le N[H]$ where $|K\gamma^{-1}| = p^{i+1}$. Since H is normal in $N[H]$, it is of course normal in the possibly smaller group $K\gamma^{-1}$. ∎

Definition A **Sylow p-subgroup** P of a group G is a maximal p-subgroup of G, that is, a p-subgroup contained in no larger p-subgroup.

Let G be a finite group, where $|G| = p^n m$ as in Theorem 18.3. The theorem shows that the Sylow p-subgroups of G are precisely those subgroups of order p^n. If P is a Sylow p-subgroup, every conjugate $g^{-1}Pg$ of P is also a Sylow p-subgroup. The second Sylow theorem states that every Sylow p-subgroup can be obtained from P in this fashion, that is, any two Sylow p-subgroups are conjugate.

Theorem 18.4 (Second Sylow Theorem) *Let P_1 and P_2 be Sylow p-subgroups of a finite group G. Then P_1 and P_2 are conjugate subgroups of G.*

Proof. The trick here is to let one of the subgroups act on right cosets of the other, and use Theorem 18.1. Let $\mathcal{R}$ be the collection of right cosets of P_1, and let P_2 act on $\mathcal{R}$ by $(P_1 x)y = P_1(xy)$ for $y \in P_2$. Then $\mathcal{R}$ is a P_2-set. By Theorem 18.1, $|\mathcal{R}_{P_2}| \equiv |\mathcal{R}| \pmod{p}$, and $|\mathcal{R}| = (G : P_1)$ is not divisible by p, so $|\mathcal{R}_{P_2}| \ne 0$. Let $P_1 x \in \mathcal{R}_{P_2}$. Then $P_1 xy = P_1 x$ for all $y \in P_2$, so $P_1 xyx^{-1} = P_1$ for all $y \in P_2$. Thus $xyx^{-1} \in P_1$ for all $y \in P_2$, so $xP_2 x^{-1} \le P_1$. Since $|P_1| = |P_2|$ we must have $P_1 = x^{-1}P_2 x$, so P_1 and P_2 are indeed conjugate subgroups. ∎

The final Sylow theorem gives information on the number of Sylow p-subgroups. A few illustrations are given after the theorem, and many more are given in the next chapter.

Theorem 18.5 (Third Sylow Theorem) *If G is a finite group and p divides $|G|$, then the number of Sylow p-subgroups is congruent to 1 modulo p and divides $|G|$.*

Proof. Let P be one Sylow p-subgroup of G. Let $\mathcal{S}$ be the set of all Sylow p-subgroups and let P act on $\mathcal{S}$ by conjugation, so that $x \in P$ carries $T \in \mathcal{S}$ into $x^{-1}Tx$. By Theorem 18.1, $|\mathcal{S}| \equiv |\mathcal{S}_P| \pmod{p}$. Let us find $\mathcal{S}_P$. If $T \in \mathcal{S}_P$, then $x^{-1}Tx = T$ for all $x \in P$. Thus $P \le N[T]$. Of course $T \le N[T]$ also. Since P and T are both Sylow p-subgroups of G, they are also Sylow p-subgroups of $N[T]$. But then they are conjugate in $N[T]$ by Theorem 18.4. Since T is a normal subgroup of $N[T]$, it is its only conjugate in $N[T]$. Thus $T = P$. Then $\mathcal{S}_P = \{P\}$. Since $|\mathcal{S}| \equiv |\mathcal{S}_P| = 1 \pmod{p}$, we see the number of Sylow p-subgroups is congruent to 1 modulo p.

Now let G act on $\mathscr{S}$ by conjugation. Since all Sylow p-subgroups are conjugate, there is only one orbit in $\mathscr{S}$ under G. If $P \in \mathscr{S}$, then $G_P = N[P]$. Then $|\mathscr{S}| = |\text{orbit of } P| = (G : G_P)$ by Theorem 16.3. But $(G : G_P)$ is a divisor of $|G|$, so the number of Sylow p-subgroups divides $|G|$. ■

Example 18.1 The Sylow 2-subgroups of S_3 have order 2. The subgroups of order 2 of S_3 in Example 4.1 are

$$\{\rho_0 \quad \mu_1\}, \qquad \{\rho_0 \quad \mu_2\}, \qquad \{\rho_0 \quad \mu_3\}.$$

Note that there are three subgroups and that $3 \equiv 1 \pmod 2$. Also, 3 divides 6, the order of S_3. You can easily see that

$$\{\rho_0, \mu_1\} i_{\rho_1} = \{\rho_0, \mu_2\} \qquad \text{and} \qquad \{\rho_0, \mu_1\} i_{\rho_2} = \{\rho_0, \mu_3\}$$

where $xi_{\rho_j} = (\rho_j)^{-1} x \rho_j$, illustrating that they are all conjugate. ■

Example 18.2 Let us use the Sylow theorems to show that no group of order 15 is simple. Let G have order 15. We claim that G has a normal subgroup of order 5. By Theorem 18.3 G has at least one subgroup of order 5, and by Theorem 18.5 the number of such subgroups is congruent to 1 modulo 5 and divides 15. Since 1, 6, and 11 are the only positive numbers less than 15 that are congruent to 1 modulo 5, and since among these only the number 1 divides 15, we see that G has exactly one subgroup P of order 5. But for each $g \in G$, the inner automorphism i_g of G with $xi_g = g^{-1}xg$ maps P onto a subgroup $g^{-1}Pg$, again of order 5. Hence we must have $g^{-1}Pg = P$ for all $g \in G$, so P is a normal subgroup of G. Therefore, G is not simple. ■

We trust that Example 18.2 gives you some inkling of the power of Theorem 18.5. *Never underestimate a theorem that counts something, even modulo p.*

Exercises

*18.1 Fill in the blanks.

a) A Sylow 3-subgroup of a group of order 12 has order ___3___.

b) A Sylow 3-subgroup of a group of order 54 has order ___27___.

c) A group of order 24 must have either ___1___ or ___3___ Sylow 2-subgroups. (Use only the information given in Theorem 18.5.)

d) A group of order $255 = (3)(5)(17)$ must have either ___ or ___ Sylow 3-subgroups and ___ or ___ Sylow 5-subgroups. (Use only the information given in Theorem 18.5.)

*18.2 Find all Sylow 3-subgroups of S_4 and demonstrate that they are all conjugate.

*18.3 Let G be a finite group and let p divide $|G|$. Prove that if G has only one Sylow p-subgroup, it is a normal subgroup, so G is not simple.

*18.4 Show that every group of order 45 has a normal subgroup of order 9.

*18.5 Prove the corollary of Theorem 18.2.

†*18.6 Let G be a finite group and let p divide $|G|$. Let P be a Sylow p-subgroup of G. Show that $N[N[P]] = N[P]$. [*Hint:* Argue that P is the only Sylow p-subgroup of $N[N[P]]$, and use Theorem 18.4.]

*18.7 Mark each of the following true or false.

— a) Any two Sylow p-subgroups of a finite group are conjugate.

— b) Theorem 18.5 shows that a group of order 15 has only one Sylow 5-subgroup.

— c) Every Sylow p-subgroup of a finite group has order a power of p.

— d) Every p-subgroup of every finite group is a Sylow p-subgroup.

— e) Every finite abelian group has exactly one Sylow p-subgroup for each prime p dividing the order of G.

— f) The normalizer in G of a subgroup H of G is always a normal subgroup of G.

— g) If H is a subgroup of G, then H is always a normal subgroup of $N[H]$.

— h) A Sylow p-subgroup of a finite group G is normal in G if and only if it is the only Sylow p-subgroup of G.

— i) If G is an abelian group and H is a subgroup of G, then $N[H] = H$.

— j) A group of prime-power order p^n has no Sylow p-subgroup.

*18.8 Let G be a finite group and let p divide $|G|$. Let P be a Sylow p-subgroup of G and let H be any p-subgroup of G. Show there exists $g \in G$ such that $g^{-1}Hg \le P$.

*18.9 Find two Sylow 2-subgroups of S_4 and show that they are conjugate.

*18.10 Show that every group of order $(35)^3$ has a normal subgroup of order 125.

*18.11 Show that there are no simple groups of order $255 = (3)(5)(17)$.

*18.12 Show that there are no simple groups of order $p^r m$, where p is a prime and $m < p$.

*18.13 Let G be a finite group. Regard G as a G-set where G acts on itself by conjugation.

a) Show that G_G is the center $Z(G)$ of G. (See Section 14.3.)

b) Use Theorem 18.1 to show that the center of a finite nontrivial p-group is nontrivial.

*18.14 Show that a finite group of order p^n contains *normal* subgroups H_i for $0 \le i \le n$ such that $|H_i| = p^i$ and $H_i < H_{i+1}$ for $0 \le i < n$. [*Hint:* See Exercise 18.13, and get an idea from Section 14.3.]

*18.15 Show that a normal p-subgroup of a finite group is contained in every Sylow p-subgroup.

nineteen

Applications of the Sylow Theory

In this chapter we give several applications of the Sylow theorems. You should find it intriguing to see how easily certain facts about groups of particular orders can be deduced. However, you should realize that we are working only with groups of finite order and really making only a small dent in the general problem of determining the structure of all finite groups. If the order of a group has only a few factors, then the techniques illustrated in this chapter may be of some use in determining the structure of the group. This will be demonstrated further in Chapter 22, where we shall show how it is sometimes possible to describe all groups (up to isomorphism) of certain orders, even when some of the groups are not abelian. However, if the order of a finite group is highly composite, that is, has a large number of factors, the problem is in general much harder.

*19.1 APPLICATIONS TO p-GROUPS AND THE CLASS EQUATION

Theorem 19.1 *Every group of prime power order (that is, every finite p-group) is solvable.*

172

Proof. If G has order p^r, it is immediate from Theorem 18.3 that G has subgroups H_i of order p^i, each normal in a subgroup of order p^{i+1} for $1 \le i \le r$. Then

$$\{e\} = H_0 < H_1 < H_2 < \cdots < H_r = G$$

is a composition series, where the factor groups are of order p, and hence abelian and actually cyclic. Thus, G is solvable. ∎

The older proofs of the Sylow theorems used the *class equation*. The line of proof in Chapter 18 avoided explicit mention of the class equation, although Eq. (18.2) is a general form of it. We now develop the classic class equation so you will be familiar with it.

Let X be a finite G-set where G is a finite group. Then Eq. (18.2) tells us that

$$|X| = |X_G| + \sum_{i=s+1}^{r} |x_i G| \tag{19.1}$$

where x_i is an element in the ith orbit in X. Consider now the special case of Eq. (19.1) where $X = G$ and the action of G on G is by conjugation, so $g \in G$ carries $x \in X = G$ into $g^{-1}xg$. Then

$$X_G = \{x \in G \mid g^{-1}xg = x \text{ for all } g \in G\}$$
$$= \{x \in G \mid xg = gx \text{ for all } g \in G\} = Z(G),$$

the center of G. If we let $c = |Z(G)|$ and $n_i = |x_i G|$ in Eq. (19.1), then we obtain

$$|G| = c + n_{c+1} + \cdots + n_r \tag{19.2}$$

where n_i is the number of elements in the ith orbit of G under conjugation by itself. Note that n_i divides $|G|$ for $c + 1 \le i \le r$ since in Eq. (19.1) we know $|x_i G| = (G : G_{x_i})$, which is a divisor of $|G|$.

Definition Equation (19.2) is the *class equation of G*. Each orbit in G under conjugation by G is a *conjugate class in G*.

Example 19.1 It is easy to check that for S_3 of Example 4.1, the conjugate classes are

$$\{\rho_0\}, \qquad \{\rho_1, \rho_2\}, \qquad \{\mu_1, \mu_2, \mu_3\}.$$

The class equation of S_3 is

$$6 = 1 + 2 + 3. \quad ∎$$

For illustration of the use of the class equation, we prove a theorem that we asked you to prove in Exercise 18.13(b).

Theorem 19.2 *The center of a nontrivial p-group G is nontrivial.*

Proof. In Eq. (19.2) for G, each n_i divides $|G|$ for $c + 1 \leq i \leq r$, so p divides each n_i, and of course p divides $|G|$. Therefore p divides c. Now $e \in Z(G)$, so $c \geq 1$. Therefore $c \geq p$, and there exists some $a \in Z(G)$ where $a \neq e$. ∎

We turn now to a lemma on direct products that will be used in some of the following theorems.

Lemma 19.1 *Let G be a group containing normal subgroups H and K such that $H \cap K = \{e\}$ and $H \vee K = G$. Then G is isomorphic to $H \times K$.*

Proof. We show that the three conditions of Theorem 8.6 are satisfied. Since $H \cap K = \{e\}$ and $H \vee K = G$, we need show only that $hk = kh$ for $h \in H$ and $k \in K$. Consider the commutator $hkh^{-1}k^{-1}$. The grouping $(hkh^{-1})k^{-1}$ shows that the commutator is in K, since K is normal and $hkh^{-1} \in K$. Similarly, the grouping $h(kh^{-1}k^{-1})$ shows that the commutator is in H. Hence $hkh^{-1}k^{-1} \in (H \cap K)$, so $hkh^{-1}k^{-1} = e$, or $hk = kh$. Our lemma is proved. ∎

Theorem 19.3 *For a prime number p, every group G of order p^2 is abelian.*

Proof. If G is not cyclic, then every element except e must be of order p. Let a be such an element. Then the cyclic subgroup $\langle a \rangle$ of order p does not exhaust G. Also let $b \in G$ with $b \notin \langle a \rangle$. Then $\langle a \rangle \cap \langle b \rangle = \{e\}$, since an element c in $\langle a \rangle \cap \langle b \rangle$ with $c \neq e$ would generate both $\langle a \rangle$ and $\langle b \rangle$ giving $\langle a \rangle = \langle b \rangle$, contrary to construction. From Theorem 18.3, $\langle a \rangle$ is normal in some subgroup of order p^2 of G, that is, normal in all of G. Likewise $\langle b \rangle$ is normal in G. Now $\langle a \rangle \vee \langle b \rangle$ is a subgroup of G properly containing $\langle a \rangle$ and of order dividing p^2. Hence $\langle a \rangle \vee \langle b \rangle$ must be all of G. Thus the hypotheses of Lemma 19.1 are satisfied, and G is isomorphic to $\langle a \rangle \times \langle b \rangle$ and therefore abelian. ∎

*19.2 FURTHER APPLICATIONS

We turn now to a discussion of whether there exist simple groups of certain orders. We have seen that every group of prime order is simple. We also asserted that A_n is simple for $n \geq 5$ and that A_5 is the smallest simple group that is not of prime order. There was a famous conjecture of Burnside that every finite simple group of nonprime order must be of even order. It was a triumph when this was recently proved by Thompson and Feit [21].

Theorem 19.4 *If p and q are distinct primes with $p < q$, then every group G of order pq has a single subgroup of order q and this subgroup is normal in G. Hence G is not simple. If q is not congruent to 1 modulo p, then G is abelian and cyclic.*

Proof. Theorems 18.3 and 18.5 tell us that G has a Sylow q-subgroup and that the number of such subgroups is congruent to 1 modulo q and divides pq, and therefore must divide p. Since $p < q$, the only possibility is the number 1. Thus there is only one Sylow q-subgroup Q of G. This group Q must be normal in G, for under an inner automorphism it would be carried into a group of the same order, hence itself. Thus G is not simple.

Likewise, there is a Sylow p-subgroup P of G, and the number of these divides q and is congruent to 1 modulo p. This number must be either 1 or q. If q is not congruent to 1 modulo p, then the number must be 1 and P is normal in G. Let us assume that $q \not\equiv 1 \pmod{p}$. Since every element in Q other than e is of order q and every element in P other than e is of order p, we $Q \cap P = \{e\}$. Also $Q \vee P$ must be a subgroup of G properly containing Q and of order dividing pq. Hence $Q \vee P = G$ and by Lemma 19.1 G is isomorphic to $Q \times P$ or $\mathbf{Z}_q \times \mathbf{Z}_p$. Thus G is abelian and cyclic. ∎

We need another lemma for some of the counting arguments that follow.

Lemma 19.2 *If H and K are finite subgroups of a group G, then*

$$|HK| = \frac{(|H|)(|K|)}{|H \cap K|}.$$

Proof. Recall that $HK = \{hk | h \in H, \ k \in K\}$. Let $|H| = r$, $|K| = s$, and $|H \cap K| = t$. Clearly, HK has at most rs elements. However, it is possible for $h_1 k_1$ to equal $h_2 k_2$, for $h_1, h_2 \in H$ and $k_1, k_2 \in K$, that is, there may be some collapsing. If $h_1 k_1 = h_2 k_2$, then let

$$x = (h_2)^{-1} h_1 = k_2 (k_1)^{-1}.$$

Now $x = (h_2)^{-1} h_1$ shows that $x \in H$, and $x = k_2(k_1)^{-1}$ shows that $x \in K$. Hence $x \in (H \cap K)$, and

$$h_2 = h_1 x^{-1} \quad \text{and} \quad k_2 = x k_1.$$

On the other hand, if for $y \in (H \cap K)$ we let $h_3 = h_1 y^{-1}$ and $k_3 = y k_1$, then clearly $h_3 k_3 = h_1 k_1$, with $h_3 \in H$ and $k_3 \in K$. Thus each element $hk \in HK$ can be represented in the form $h_i k_i$, for $h_i \in H$ and $k_i \in K$, as many times as there are elements of $H \cap K$, that is, t times. Therefore, the number of elements in HK is rs/t. ∎

The preceding lemma is another result that counts something, so don't underestimate it. The lemma will be used in the following way: A finite group can't have subgroups H and K that are too large with intersections that are too small, or the order of HK would have to exceed the order of G, which is impossible. For example, a group of order 24 can't have two subgroups of orders 12 and 8 with an intersection of order 2.

The remainder of this chapter consists of several examples illustrating techniques of proving that all groups of certain orders are abelian or that they have nontrivial proper normal subgroups, that is, that they are not simple. We will use one fact we mentioned before only in the exercises. *A subgroup H of index 2 in a finite group G is always normal,* for by counting, we see that there are only the left cosets H itself and the coset consisting of all elements in G not in H. The right cosets are the same. Thus every right coset is a left coset, and by Theorem 12.1 and the following discussion, H is normal in G.

Example 19.2   No group of order p^r for $r > 1$ is simple, where p is a prime. For by Theorem 18.3 such a group G contains a subgroup of order p^{r-1} normal in a subgroup of order p^r, which must be all of G. Thus a group of order 16 is not simple; it has a normal subgroup of order 8. ∎

Example 19.3   Every group of order 15 is cyclic (hence abelian and not simple, since 15 is not a prime). This is because $15 = (5)(3)$, and 5 is not congruent to 1 modulo 3. By Theorem 19.4 we are done. ∎

Example 19.4   No group of order 20 is simple, for such a group G contains Sylow 5-subgroups in number congruent to 1 modulo 5 and a divisor of 20, hence only 1. This Sylow 5-subgroup is then normal, since all conjugates of it must be itself. ∎

Example 19.5   No group of order 30 is simple. We have seen that if there is only one Sylow p-subgroup for some prime p dividing 30, we are done. By Theorem 18.5 the possibilities for the number of Sylow 5-subgroups are 1 or 6, and those for Sylow 3-subgroups are 1 or 10. But if G has six Sylow 5-subgroups, then the intersection of any two is a subgroup of each of order dividing 5, and hence just $\{e\}$. Thus each contains four elements of order 5 that are in none of the others. Hence G must contain twenty-four elements of order 5. Similarly, if G has ten Sylow 3-subgroups, it has at least twenty elements of order 3. The two types of Sylow subgroups together would require at least forty-four elements in G. Thus there is a normal subgroup either of order 5 or of order 3. ∎

Example 19.6   No group of order 48 is simple. Indeed, we shall show that a group G of order 48 has a normal subgroup of either order 16 or order 8. By Theorem 18.5 G has either one or three Sylow 2-subgroups of order 16. If there is only one subgroup of order 16, it is normal in G, by a now familiar argument.

Suppose that there are three subgroups of order 16, and let H and K be two of them. Then $H \cap K$ must be of order 8, for if $H \cap K$ were of order ≤ 4 then by Lemma 19.2 HK would have at least $(16)(16)/4 = 64$ elements contradicting the fact that G has only forty-eight elements. Therefore, $H \cap K$

is normal in both H and K (being of index 2, or by Theorem 18.3). Hence the normalizer of $H \cap K$ contains both H and K and must have order a multiple > 1 of 16 and a divisor of 48, therefore 48. Thus $H \cap K$ must be normal in G. ■

Example 19.7 No group of order 36 is simple. Such a group G has either one and four subgroups of order 9. If there is only one such subgroup, it is normal in G. If there are four such subgroups, let H and K be two of them. As in Example 19.6, $H \cap K$ must have at least three elements, or HK would have to have eighty-one elements, which is impossible. Thus the normalizer of $H \cap K$ has as order a multiple > 1 of 9 and a divisor of 36; hence the order must be either 18 or 36. If the order is 18, the normalizer is then of index 2, and therefore is normal in G. If the order is 36, then $H \cap K$ is normal in G. ■

Example 19.8 Every group of order $255 = (3)(5)(17)$ is abelian (hence cyclic by Theorem 9.3 and not simple, since 255 is not a prime). By Theorem 18.5 such a group G has only one subgroup H of order 17. Then G/H has order 15 and is abelian by Example 19.3. By Theorem 12.6, we see that the commutator subgroup G' of G is contained in H. Thus as a subgroup of H, G' has either order 1 or 17. Theorem 18.5 also shows that G has either 1 or 85 subgroups of order 3 and either 1 or 51 subgroups of order 5. However, 85 subgroups of order 3 would require 170 elements of order 3, and 51 subgroups of order 5 would require 204 elements of order 5 in G; both together would then require 375 elements in G, which is impossible. Hence there is a subgroup K having either order 3 or order 5 and normal in G. Then G/K has either order $(5)(17)$ or order $(3)(17)$, and in either case Theorem 19.4 shows that G/K is abelian. Thus $G' \leq K$ and has order either 3, 5, or 1. Since $G' \leq H$ showed that G' has order 17 or 1, we conclude that G' has order 1. Hence $G' = \{e\}$, and $G/G' \simeq G$ is abelian. Theorem 9.3 then shows that G is cyclic. ■

Exercises

*19.1 By arguments similar to those used in the examples of this section, convince yourself that every group of order not a prime and less than 60 contains a nontrivial proper normal subgroup and hence is not simple. You need not write out the details. (The hardest cases were discussed in the examples.)

*19.2 Prove that every group of order $(5)(7)(47)$ is abelian and cyclic.

*19.3 Mark each of the following true or false.

___ a) Every group of order 159 is cyclic.

___ b) Every group of order 102 has a nontrivial proper normal subgroup.

___ c) Every solvable group is of prime-power order.

—   d) Every group of prime-power order is solvable.

—   e) It would become quite tedious to show that no group of nonprime order between 60 and 168 is simple by the methods illustrated in the text.

—   f) No group of order 21 is simple.

—   g) Every group of 125 elements has at least 5 elements that commute with every element in the group.

—   h) Every group of order 42 has a normal subgroup of order 7.

—   i) Every group of order 42 has a normal subgroup of order 8.

—   j) The only simple groups are the groups Z_p and A_n, where p is a prime and $n \neq 4$.

---

*19.4 Prove that no group of order 96 is simple.

*19.5 Prove that no group of order 160 is simple.

*19.6 Let D_4 be the group of symmetries of the square in Example 4.2.

a) Find the decomposition of D_4 into conjugate classes.

b) Write the class equation for D_4.

*19.7 This exercise determines the conjugate classes of S_n for every integer $n \geq 1$.

a) Show that if $\sigma = (a_1, a_2, \ldots, a_m)$ is a cycle in S_n and τ is any element of S_n, then
$\tau^{-1}\sigma\tau = (a_1\tau, a_2\tau, \ldots, a_m\tau)$.

b) Argue from (a) that any two cycles in S_n of the same length are conjugate.

c) Argue from (a) and (b) that a product of s disjoint cycles in S_n of lengths r_i for $i = 1, 2, \ldots, s$ is conjugate to every other product of s disjoint cycles of lengths r_i in S_n.

d) Show that the number of conjugate classes in S_n is $p(n)$, where $p(n)$ is the number of ways, neglecting the order of the summands, that n can be expressed as a sum of positive integers. The number $p(n)$ is the **number of partitions of** n.

e) Compute $p(n)$ for $n = 1, 2, 3, 4, 5, 6, 7$.

*19.8 Find the conjugate classes and the class equation for S_4. [*Hint:* Use Exercise 19.7.]

*19.9 Find the class equations for S_5 and S_6 [*Hint:* Use Exercise 19.7.]

*19.10 Show that the number of conjugate classes in S_n is also the number of different abelian groups (up to isomorphism) of order p^n, where p is a prime number. [*Hint:* Use Exercise 19.7.]

*19.11 Show that if $n > 2$, the center of S_n is the subgroup consisting of the identity permutation only. [*Hint:* Use Exercise 19.7.]

twenty

Free
Abelian
Groups

In this chapter we introduce the concept of free abelian groups and prove some results concerning them. The chapter concludes with a demonstration of the Fundamental Theorem of Finitely Generated Abelian Groups (Theorem 9.3).

*20.1 FREE ABELIAN GROUPS

You should review the notions of a generating set for a group G and a finitely generated group, as given at the start of Chapter 9. In this chapter we shall deal exclusively with abelian groups and use additive notations as follows:

0 for the identity, $+$ for the operation,

$$na = \underbrace{a + a + \cdots + a}_{n \text{ summands}}$$

$$-na = \underbrace{(-a) + (-a) + \cdots + (-a)}_{n \text{ summands}}$$

$\left. \right\}$ for $n \in \mathbf{Z}^+$ and $a \in G$.

$0a = 0$ for the first 0 in $\mathbf{Z}$ and the second in G.

179

We shall continue to use the symbol $\times$ for direct product of groups rather than change to direct sum notation.

Of course $\{(1, 0), (0, 1)\}$ is a generating set for the group $\mathbf{Z} \times \mathbf{Z}$ since $(n, m) = n(1, 0) + m(0, 1)$ for any (n, m) in $\mathbf{Z} \times \mathbf{Z}$. This generating set has the property that each element of $\mathbf{Z} \times \mathbf{Z}$ can be *uniquely* expressed in the form $n(1, 0) + m(0, 1)$. That is, the coefficients n and m in $\mathbf{Z}$ are unique.

Theorem 20.1 *Let X be a subset of a nonzero abelian group G. The following conditions on X are equivalent.*

1. *Each nonzero element a in G can be **uniquely** expressed in the form $a = n_1 x_1 + n_2 x_2 + \cdots + n_r x_r$ for $n_i \neq 0$ in $\mathbf{Z}$ and distinct x_i in X.*
2. *X generates G, and $n_1 x_1 + n_2 x_2 + \cdots + n_r x_r = 0$ for $n_i \in \mathbf{Z}$ and distinct $x_i \in X$ if and only if $n_1 = n_2 = \cdots = n_r = 0$.*

Proof. Suppose condition 1 is true. Since $G \neq \{0\}$, we have $X \neq \{0\}$. It follows from 1 that $0 \notin X$, for if $x_i = 0$ and $x_j \neq 0$, then $x_j = x_i + x_j$ which would contradict the uniqueness of the expression for x_j. From 1 surely X generates G, and of course $n_1 x_1 + n_2 x_2 + \cdots + n_r x_r = 0$ if $n_1 = n_2 = \cdots = n_r = 0$. Suppose that $n_1 x_1 + n_2 x_2 + \cdots + n_r x_r = 0$ with some $n_i \neq 0$; by dropping terms with zero coefficients and renumbering, we can assume all $n_i \neq 0$. Then

$$x_1 = x_1 + (n_1 x_1 + n_2 x_2 + \cdots + n_r x_r)$$
$$= (n_1 + 1)x_1 + n_2 x_2 + \cdots + n_r x_r,$$

which gives two ways of writing $x_1 \neq 0$, contradicting the uniqueness assumption in condition 1. Thus condition 1 implies condition 2.

We now show that condition 2 implies condition 1. Let $a \in G$. Since X generates G, we see a can be written in the form $a = n_1 x_1 + n_2 x_2 + \cdots + n_r x_r$. Suppose a has another such expression in terms of elements of X. By using some zero coefficients in the two expressions, we can assume they involve the same elements in X and are of the form

$$a = n_1 x_1 + n_2 x_2 + \cdots + n_r x_r$$
$$a = m_1 x_1 + m_2 x_2 + \cdots + m_r x_r.$$

Subtracting, we obtain

$$0 = (n_1 - m_1)x_1 + (n_2 - m_2)x_2 + \cdots + (n_r - m_r)x_r,$$

so $n_i - m_i = 0$ by condition 2, and $n_i = m_i$ for $i = 1, 2, \ldots, r$. Thus the coefficients are unique. ∎

Definition An abelian group having a nonempty generating set X satisfying the conditions described in Theorem 20.1 in a **free abelian group,** and X is a **basis** for the group.

Example 20.1   The group $\mathbf{Z} \times \mathbf{Z}$ is free abelian and $\{(1, 0), (0, 1)\}$ is a basis. Of course a basis for the free abelian group $\mathbf{Z} \times \mathbf{Z} \times \mathbf{Z}$ is $\{(1, 0, 0), (0, 1, 0), (0, 0, 1)\}$, and so on. Thus finite direct products of the group $\mathbf{Z}$ with itself are free abelian groups. ∎

Example 20.2   The group $\mathbf{Z}_n$ is not free abelian, for $nx = 0$ for every $x \in \mathbf{Z}_n$, and $n \neq 0$, which would contradict condition 2. ∎

Suppose a free abelian group G has a finite basis $X = \{x_1, x_2, \ldots, x_r\}$. If $a \in G$ and $a \neq 0$, then a has a *unique* expression of the form

$$a = n_1 x_1 + n_2 x_2 + \cdots + n_r x_r \qquad \text{for} \quad n_i \in \mathbf{Z}.$$

We define

$$\phi : G \to \underbrace{\mathbf{Z} \times \mathbf{Z} \times \cdots \times \mathbf{Z}}_{r \text{ factors}}$$

by $a\phi = (n_1, n_2, \ldots, n_r)$ and $0\phi = (0, 0, \ldots, 0)$. It is easy to prove that ϕ is an isomorphism. We leave the details to the exercises (see Exercise 20.1) and state the result as a theorem.

Theorem 20.2   *If G is a nonzero free abelian group with a basis of r elements, then G is isomorphic to $\mathbf{Z} \times \mathbf{Z} \times \cdots \times \mathbf{Z}$ for r factors.*

It is a fact that any two bases of a free abelian group G contain the same number of elements. We shall prove this only if G has a finite basis, although it is also true if every basis of G is infinite. The proof is really lovely; it gives an easy characterization of the number of elements in a basis in terms of the size of a factor group.

Theorem 20.3   *Let $G \neq \{0\}$ be a free abelian group with a finite basis. Then every basis of G is finite, and all bases have the same number of elements.*

Proof.   Let G have a basis $\{x_1, x_2, \ldots, x_r\}$. Then G is isomorphic to $\mathbf{Z} \times \mathbf{Z} \times \cdots \times \mathbf{Z}$ for r factors. Let $2G = \{2g \mid g \in G\}$. It is easily checked that $2G$ is a subgroup of G. Since $G \simeq \mathbf{Z} \times \mathbf{Z} \times \cdots \times \mathbf{Z}$ for r factors, we have

$$G/2G \simeq (\mathbf{Z} \times \mathbf{Z} \times \cdots \times \mathbf{Z})/(2\mathbf{Z} \times 2\mathbf{Z} \times \cdots \times 2\mathbf{Z}) \simeq \mathbf{Z}_2 \times \mathbf{Z}_2 \times \cdots \times \mathbf{Z}_2$$

for r factors. Thus $|G/2G| = 2^r$, so the number of elements in any finite basis X is $\log_2 |G/2G|$. Thus any two finite bases have the same number of elements.

It remains to show that G cannot also have an infinite basis. Let Y be any basis for G, and let $\{y_1, y_2, \ldots, y_s\}$ be distinct elements in Y. Let H be the subgroup of G generated by $\{y_1, y_2, \ldots, y_s\}$, and let K be the subgroup of G generated by the remaining elements of Y. Then it is easy to see that $G \simeq H \times K$, so $G/2G \simeq (H \times K)/(2H \times 2K) \simeq (H/2H) \times (K/2K)$. Since

$|H/2H| = 2^s$, we see $|G/2G| \geq 2^s$. Since we have $|G/2G| = 2^r$, we see that $s \leq r$. Then Y cannot be an infinite set, for we could take $s > r$. ∎

Definition   If G is a free abelian group, the **rank of** G is the number of elements in a basis for G. (All bases have the same number of elements.)

*20.2 PROOF OF THE FUNDAMENTAL THEOREM

We shall prove the fundamental theorem (Theorem 9.3) by showing that any finitely generated abelian group is isomorphic to a factor group of the form

$$(\mathbf{Z} \times \mathbf{Z} \times \cdots \times \mathbf{Z})/(d_1\mathbf{Z} \times d_2\mathbf{Z} \times \cdots \times d_s\mathbf{Z} \times \{0\} \times \cdots \times \{0\}),$$

where both "numerator" and "denominator" have n factors, and d_1 divides d_2, which divides $d_3 \cdots$, which divides d_s. The decomposition 2 of Theorem 9.3 will then follow easily.

To show that G is isomorphic to such a factor group, we will show that there is a homomorphism of $\mathbf{Z} \times \mathbf{Z} \times \cdots \times \mathbf{Z}$ onto G with kernel of the form $d_1\mathbf{Z} \times d_2\mathbf{Z} \times \cdots \times d_s\mathbf{Z} \times \{0\} \times \cdots \times \{0\}$. The result will then follow by Theorem 13.3. The following theorems give the details of the argument. Our purpose in these introductory paragraphs is to let you see where you are going as you read what follows.

Theorem 20.4   *Let G be a finitely generated abelian group with generating set* $\{a_1, a_2, \ldots, a_n\}$. *Let*

$$\phi: \underbrace{\mathbf{Z} \times \mathbf{Z} \times \cdots \times \mathbf{Z}}_{n \text{ factors}} \to G$$

be defined by $(h_1, h_2, \ldots, h_n)\phi = h_1a_1 + h_2a_2 + \cdots + h_na_n$. *Then* ϕ *is a homomorphism onto G.*

Proof.   From the meaning of h_ia_i for $h_i \in \mathbf{Z}$ and $a_i \in G$, you see at once that $[(h_1, \ldots, h_n) + (k_1, \ldots, k_n)]\phi = (h_1 + k_1, \ldots, h_n + k_n)\phi = (h_1 + k_1)a_1 + \ldots + (h_n + k_n)a_n = (h_1a_1 + k_1a_1) + \cdots + (h_na_n + k_na_n) = (h_1a_1 + \cdots + h_na_n) + (k_1a_1 + \cdots + k_na_n) = (h_1, \ldots, h_n)\phi + (k_1, \ldots, k_n)\phi$. Since $\{a_1, \ldots, a_n\}$ generates G, clearly the homorphism ϕ is onto G. ∎

We now prove a "replacement property" that makes it possible for us to adjust a basis.

Theorem 20.5   *If* $X = \{x_1, \ldots, x_r\}$ *is a basis for a free abelian group G and* $t \in \mathbf{Z}$, *then for* $i \neq j$, *the set* $Y = \{x_1, \ldots, x_{j-1}, x_j + tx_i, x_{j+1}, \ldots, x_r\}$ *is also a basis for G.*

Proof. Since $x_j = (-t)x_i + (1)(x_j + tx_i)$, you see that x_j can be recovered from Y, which thus also generates G. Suppose

$$n_1 x_1 + \cdots + n_{j-1} x_{j-1} + n_j (x_j + tx_i) + n_{j+1} x_{j+1} + \cdots + n_r x_r = 0.$$

Then

$$n_1 x_1 + \cdots + (n_i + n_j t)x_i + \cdots + n_j x_j + \cdots + n_r x_r = 0,$$

and since X is a basis, $n_1 = \cdots = n_i + n_j t = \cdots = n_j = \cdots = n_r = 0$. From $n_j = 0$ and $n_i + n_j t = 0$, it follows that $n_i = 0$ also, so $n_1 = \cdots = n_i = \cdots = n_j = \cdots = n_r = 0$, and condition 2 of Theorem 20.1 is satisfied. Thus Y is a basis. ∎

Example 20.3 A basis for $\mathbf{Z} \times \mathbf{Z}$ is $\{(1, 0), (0, 1)\}$. Another basis is $\{(1, 0), (4, 1)\}$ for $(4, 1) = 4(1, 0) + (0, 1)$. However, $\{(3, 0), (0, 1)\}$ is not a basis. For example, you can't express $(2, 0)$ in the form $n_1(3, 0) + n_2(0, 1)$ for n_1, $n_2 \in \mathbf{Z}$. Here $(3, 0) = (1, 0) + 2(1, 0)$, and a multiple of a basis element was added to *itself*, rather than to a *different* basis element. ∎

A free abelian group G of finite rank may have many bases. We show that if $K \leq G$, then K is also free abelian with rank not exceeding that of G. Equally important, there exist bases of G and K nicely related to each other.

Theorem 20.6 *Let G be a nonzero free abelian group of finite rank n, and let K be a nonzero subgroup of G. Then K is free abelian of rank $s \leq n$. Furthermore, there exists a basis $\{x_1, x_2, \ldots, x_n\}$ for G and positive integers $d_1, d_2, \ldots, d_s$ where d_i divides d_{i+1} for $i = 1, \ldots, s-1$, such that $\{d_1 x_1, d_2 x_2, \ldots, d_s x_s\}$ is a basis for K.*

Proof. We show that K has a basis of the described form, which will of course show it is free abelian of rank at most n. Suppose $Y = \{y_1, \ldots, y_n\}$ is a basis for G. All nonzero elements in K can be expressed in the form

$$k_1 y_1 + \cdots + k_n y_n,$$

where some $|k_i|$ is nonzero. Among *all* bases Y for G, select one Y_1 that yields the minimal such nonzero value $|k_i|$ as all nonzero elements of K are written in terms of the basis elements in Y_1. By renumbering the elements of Y_1 if necessary, we can assume there is $w_1 \in K$ such that

$$w_1 = d_1 y_1 + k_2 y_2 + \cdots + k_n y_n$$

where $d_1 > 0$ and d_1 is the minimal attainable coefficient as just described. Using the division algorithm write $k_j = d_1 q_j + r_j$ where $0 \leq r_j < d_1$ for $j = 2, \ldots, n$. Then

$$w_1 = d_1(y_1 + q_2 y_2 + \cdots + q_n y_n) + r_2 y_2 + \cdots + r_n y_n. \tag{20.1}$$

Now let $x_1 = y_1 + q_2 y_2 + \cdots + q_n y_n$. By Theorem 20.5 $\{x_1, y_2 \ldots, y_n\}$ is also a basis for G. From Eq. (20.1) and our choice of Y_1 for minimal coefficient d_1, you see that $r_2 = \cdots = r_n = 0$. Thus $d_1 x_1 \in K$.

We now consider bases for G of the form $\{x_1, y_2, \ldots, y_n\}$. Each element of K can be expressed in the form

$$h_1 x_1 + k_2 y_2 + \cdots + k_n y_n.$$

Since $d_1 x_1 \in K$, we can subtract a suitable multiple of $d_1 x_1$ and then using the minimality of d_1 to see that h_1 is a multiple of d_1, we see we actually have $k_2 y_2 + \cdots + k_n y_n$ in K. Among all such bases $\{x_1, y_2, \ldots, y_n\}$, we choose one Y_2 that leads to some $k_i \neq 0$ of minimal magnitude. (It is possible all k_i are always zero. In this case, K is generated by $d_1 x_1$ and we are done.) By renumbering the elements of Y_2 we can assume that there is $w_2 \in K$ such that

$$w_2 = d_2 y_2 + \cdots + k_n y_n$$

where $d_2 > 0$ and d_2 is minimal as just described. Exactly as in the preceding paragraph, we can modify our basis from $Y_2 = \{x_1, y_2, \ldots, y_n\}$ to a basis $\{x_1, x_2, y_3, \ldots, y_n\}$ for G where $d_1 x_1 \in K$ and $d_2 x_2 \in K$. Writing $d_2 = d_1 q + r$ for $0 \leq r < d_1$, we see that $\{x_1 + q x_2, x_2, y_3, \ldots, y_n\}$ is a basis for G, and $d_1 x_1 + d_2 x_2 = d_1(x_1 + q x_2) + r x_2$ is in K. By our minimal choice of d_1, we see $r = 0$, so d_1 divides d_2.

We now consider all bases of the form $\{x_1, x_2, y_3, \ldots, y_n\}$ for G and examine elements of K of the form $k_3 y_3 + \cdots + k_n y_n$. The pattern is clear. The process continues until we obtain a basis $\{x_1, x_2, \ldots, x_s, y_{s+1}, \ldots, y_n\}$ where the only element of K of the form $k_{s+1} y_{s+1} + \cdots + k_n y_n$ is zero, that is, all k_i are zero. We then let $x_{s+1} = y_{s+1}, \ldots, x_n = y_n$ and obtain a basis for G of the form described in the statement of Theorem 20.6. ∎

Theorem 20.7 *Every finitely generated abelian group is isomorphic to a group of the form*

$$\mathbf{Z}_{m_1} \times \mathbf{Z}_{m_2} \times \cdots \times \mathbf{Z}_{m_r} \times \mathbf{Z} \times \mathbf{Z} \times \cdots \times \mathbf{Z}$$

where m_i divides m_{i+1} for $i = 1, \ldots, r - 1$.

Proof. For the purposes of this proof, it will be convenient to use as notations $\mathbf{Z}/1\mathbf{Z} = \mathbf{Z}/\mathbf{Z} \simeq \mathbf{Z}_1 = \{0\}$. Let G be finitely generated by n elements. Let $F = \mathbf{Z} \times \mathbf{Z} \times \cdots \times \mathbf{Z}$ for n factors. Consider the homomorphism $\phi : F \to G$ of Theorem 20.4, and let K be the kernel of this homomorphism. Then there is a basis for F of the form $\{x_1, \ldots, x_n\}$ where $\{d_1 x_1, \ldots, d_s x_s\}$ is a basis for K and d_i divides d_{i+1} for $i = 1, \ldots, s - 1$. By Theorem 13.3, G

is isomorphic to F/K. But

$$F/K \simeq (\mathbf{Z} \times \mathbf{Z} \times \cdots \times \mathbf{Z})/(d_1\mathbf{Z} \times d_2\mathbf{Z} \times \cdots \times d_s\mathbf{Z} \times \{0\} \times \cdots \times \{0\})$$
$$\simeq \mathbf{Z}_{d_1} \times \mathbf{Z}_{d_2} \times \cdots \times \mathbf{Z}_{d_s} \times \mathbf{Z} \times \cdots \times \mathbf{Z}.$$

It is possible that $d_1 = 1$, in which case $\mathbf{Z}_{d_1} = \{0\}$ and can be dropped (up to isomorphism) from this product. Similarly, d_2 may be 1, and so on. We let m_1 be the first $d_i > 1$, m_2 be the next d_i, and so on, and our theorem follows at once. ■

We have demonstrated the toughest part of Theorem 9.3. Of course a prime-power decomposition exists since you can break the groups $\mathbf{Z}_{m_i}$ into prime-power factors. The only remaining part of Theorem 9.3 concerns the uniqueness of the betti number, of the torsion coefficients, and of the prime powers. The betti number appears as the rank of the free abelian group G/T, where T is the torsion subgroup of G. This rank is invariant by Theorem 20.3, which shows the uniqueness of the betti number. The uniqueness of the torsion coefficients and of prime powers is a bit more difficult to show. We give some exercises that indicate their uniqueness (see Exercises 20.12 through 20.20).

Exercises

*20.1 Complete the proof of Theorem 20.2.

*20.2 Find a basis $\{(a_1, a_2, a_3), (b_1, b_2, b_3), (c_1, c_2, c_3)\}$ for $\mathbf{Z} \times \mathbf{Z} \times \mathbf{Z}$ with all $a_i \neq 0$, all $b_i \neq 0$, and all $c_i \neq 0$. (Many answers are possible.)

*20.3 Is $\{(2, 1), (3, 1)\}$ a basis for $\mathbf{Z} \times \mathbf{Z}$? Prove your assertion.

*20.4 Is $\{(2, 1), (4, 1)\}$ a basis for $\mathbf{Z} \times \mathbf{Z}$? Prove your assertion.

*20.5 Find conditions on $a, b, c, d \in \mathbf{Z}$ for $\{(a, b), (c, d)\}$ to be a basis for $\mathbf{Z} \times \mathbf{Z}$. [Hint: Solve $x(a, b) + y(c, d) = \left\{ \begin{smallmatrix} (1, & 0) \\ (0, & 1) \end{smallmatrix} \right\}$ in $\mathbf{R}$, and see when the values actually lie in $\mathbf{Z}$.]

†*20.6 Show that a free abelian group contains no nonzero elements of finite order.

*20.7 Mark each of the following true or false.

— a) Every free abelian group is torsion free.

— b) Every finitely generated torsion-free abelian group is a free abelian group.

— c) There exists a free abelian group of every positive integer rank.

— d) A finitely generated abelian group is free abelian if its betti number equals the number of elements in some generating set.

— e) If X generates a free abelian group G and $X \subseteq Y \subseteq G$, then Y generates G.

— f) If X is a basis for a free abelian group G and $X \subseteq Y \subseteq G$, then Y is a basis for G.

— g) Every nonzero free abelian group has an infinite number of bases.

— h) Every free abelian group of rank at least 2 has an infinite number of bases.
— i) If K is a nonzero subgroup of a finitely generated free abelian group, then K is free abelian.
— j) If K is a nonzero subgroup of a finitely generated free abelian group, then G/K is free abelian.

---

*20.8 Show by example that it is possible for a proper subgroup of a free abelian group of finite rank r also to have rank r.

*20.9 Show that if G and G' are free abelian groups, then $G \times G'$ is free abelian.

*20.10 Show that free abelian groups of finite rank are precisely the finitely generated abelian groups containing no nonzero elements of finite order.

*20.11 Show that **Q** under addition is not a free abelian group. [*Hint:* Show that no two distinct rational numbers n/m and r/s could be contained in a set satisfying condition 2 of Theorem 20.1.]

Exercises 12 through 17 deal with showing the uniqueness of the prime powers appearing in the prime-power decomposition of the torsion subgroup T of a finitely generated abelian group.

*20.12 Let p be a fixed prime. Show that the elements of T having as order some power of p, together with zero, form a subgroup T_p of T.

*20.13 Show that in any prime-power decomposition of T, the subgroup T_p in the preceding exercise is isomorphic to the direct product of those cyclic factors of order some power of the prime p. [This reduces our problem to showing that the group T_p cannot have essentially different decompositions into products of cyclic groups.]

*20.14 Let G be any abelian group and let n be any positive integer. Show that $G[n] = \{x \in G \mid nx = 0\}$ is a subgroup of G. (In multiplicative notation, $G[n] = \{x \in G \mid x^n = e\}$.)

*20.15 Referring to Exercise 20.14, show that $\mathbf{Z}_{p^r}[p] \simeq \mathbf{Z}_p$ for any $r \geq 1$ and any prime p.

*20.16 Using Exercise 20.15, show

$$(\mathbf{Z}_{p^{r_1}} \times \mathbf{Z}_{p^{r_2}} \times \cdots \times \mathbf{Z}_{p^{r_m}})[p] \simeq \underbrace{\mathbf{Z}_p \times \mathbf{Z}_p \times \cdots \times \mathbf{Z}_p}_{m \text{ factors}}$$

provided each $r_i \geq 1$.

*20.17 Let G be a finitely generated abelian group and T_p the subgroup defined in Exercise 20.12. Suppose $T_p \simeq \mathbf{Z}_{p^{r_1}} \times \mathbf{Z}_{p^{r_2}} \times \cdots \times \mathbf{Z}_{p^{r_m}} \simeq \mathbf{Z}_{p^{s_1}} \mathbf{Z}_{p^{s_2}} \times \cdots \times \mathbf{Z}_{p^{s_n}}$ where $1 \leq r_1 \leq r_2 \leq \cdots \leq r_m$ and $1 \leq s_1 \leq s_2 \leq \cdots \leq s_n$. We need to show $m = n$ and $r_i = s_i$ for $i = 1, \ldots, n$ to complete the demonstration of uniqueness of the prime-power decomposition.

a) Use Exercise 20.16 to show that $n = m$.

b) Suppose $r_i = s_i$ for all $i < j$. Show $r_j = s_j$, which will complete the proof. [*Hint:* Suppose $r_j < s_j$. Consider the subgroup $p^{r_j}T_p = \{p^{r_j}x \mid x \in T_p\}$, and show that this subgroup would then have two prime-power decompositions involving different numbers of nonzero factors. Then argue that this is impossible by part (a) of this exercise.]

　　Let T be the torsion subgroup of a finitely generated abelian group. Suppose $T \simeq \mathbf{Z}_{m_1} \times \mathbf{Z}_{m_2} \times \cdots \times \mathbf{Z}_{m_r} \simeq \mathbf{Z}_{n_1} \times \mathbf{Z}_{n_2} \times \cdots \times \mathbf{Z}_{n_s}$ where m_i divides m_{i+1} for $i = 1, \ldots,$

$r - 1$ and n_j divides n_{j+1} for $n = 1, \ldots, s - 1$, and $m_1 > 1$ and $n_1 > 1$. We wish to show $r = s$ and $m_k = n_k$ for $k = 1, \ldots, r$, demonstrating the uniqueness of the torsion coefficients. This is done in Exercises 18 through 20.

*20.18 Indicate how a prime-power decomposition can be obtained from a torsion-coefficient decomposition. (Observe that the preceding exercises show the prime powers obtained are unique.)

*20.19 Argue from Exercise 20.18 that m_r and n_s can both be characterized as follows. Let $p_1, \ldots, p_t$ be the distinct primes dividing $|T|$, and let $p_1^{h_1}, \ldots, p_t^{h_t}$ be the highest powers of these primes appearing in the (unique) prime-power decomposition. Then $m_r = n_s = p_1^{h_1} p_2^{h_2} \cdots p_t^{h_t}$.

*20.20 Characterize m_{r-1} and n_{s-1}, showing that they are equal, and continue to show $m_{r-i} = n_{s-i}$ for $i = 1, \ldots, r - 1$, and that $r = s$.

*twenty-one

Free
Groups

In this chapter and the next we discuss a portion of group theory that is of great interest not only in algebra but in topology as well. In fact, an excellent and readable discussion of free groups and presentations of groups is found in Crowell and Fox [45, Chapters 3 and 4].

*21.1 WORDS AND REDUCED WORDS

Let A be any (not necessarily finite) set of elements a_i for $i \in I$. We think of A as an **alphabet** and of the a_i as **letters** in the alphabet. Any symbol of the form a_i^n with $n \in \mathbf{Z}$ is a **syllable** and a finite string w of syllables written in juxtaposition is a **word** We also introduce the **empty word** 1, which has no syllables.

Example 21.1 Let $A = \{a_1, a_2, a_3\}$. Then

$$a_1 a_3^{-4} a_2^{\,2} a_3, \qquad a_2^{\,3} a_2^{-1} a_3 a_1^{\,2} a_1^{-7}, \qquad \text{and} \qquad a_3^{\,2}$$

are all words, if we follow the convention of understanding that a_i^1 is the same as a_i. ∎

There are two natural types of modifications of certain words, the **elementary contractions.** The first type consists of replacing an occurrence of $a_i^m a_i^n$ in a word by a_i^{m+n}. The second type consists of replacing an occurrence of a_i^0 in a word by 1, that is, dropping it out of the word. By means of a finite number of elementary contractions, every word can be changed to a **reduced word,** one for which no more elementary contractions are possible. Note that these elementary contractions formally amount to the usual manipulations of integral exponents.

Example 21.2   The reduced form of the word $a_2^3 a_2^{-1} a_3 a_1^2 a_1^{-7}$ of Example 21.1 is $a_2^2 a_3 a_1^{-5}$. ∎

It should be said here once and for all that we are going to gloss over several points that some books spend pages proving, usually by quite complicated induction arguments broken down into many cases. For example, suppose you are given a word and wish to find its reduced form. There may be a variety of elementary contractions that could be performed first. How do you know that the reduced word you end up with is the same no matter in which order you perform the elementary contractions? The student will probably say this is obvious. Some authors spend considerable effort proving this. The author tends to agree here with the student. Proofs of this sort he regards as tedious, and they have never made him more comfortable about the situation. However, the author is the first to acknowledge that he is not a great mathematician. In deference to the fact that many mathematicians feel that these things do need considerable discussion, we shall mark an occasion when we just state such facts by the phrase, "It would seem obvious that," keeping the quotation marks.

*21.2 FREE GROUPS

Let the set of all reduced words formed from our alphabet A be $F[A]$. We now make $F[A]$ into a group in a natural way. For w_1 and w_2 in $F[A]$, define $w_1 \cdot w_2$ to be the reduced form of the word obtained by the juxtaposition $w_1 w_2$ of the two words.

Example 21.3   If
$$w_1 = a_2^3 a_1^{-5} a_3^2$$
and
$$w_2 = a_3^{-2} a_1^2 a_3 a_2^{-2},$$
then $w_1 \cdot w_2 = a_2^3 a_1^{-3} a_3 a_2^{-2}$. ∎

"It would seem obvious that" this operation of multiplication on $F[A]$ is well defined and associative. It is obvious that the empty word 1 acts as an identity element. "It would seem obvious that" given a reduced word $w \in F[A]$, if you form the word obtained by first writing the syllables of w in the opposite order and secondly replacing each a_i^n by a_i^{-n}, then the resulting word w^{-1} is a reduced word also, and

$$w \cdot w^{-1} = w^{-1} \cdot w = 1.$$

Definition The group $F[A]$ just described is the *free group generated by A*.

Look back at Theorem 9.1 and the definition preceding it to see that the present use of the term *generated* is consistent with the earlier use.

Starting with a group G and a generating set $\{a_i \mid i \in I\}$, one might ask if G is *free* on $\{a_i\}$, that is, if G is essentially the free group generated by $\{a_i\}$. We define precisely what this is to mean.

Definition If G is a group with a set $A = \{a_i\}$ of generators, and if G is isomorphic to $F[A]$ under a map $\phi : G \to F[A]$ such that $a_i \phi = a_i$, then G is *free on* $\{a_i\}$ and the a_i are *free generators of G*. A group is *free* if it is free on some nonempty set $\{a_i\}$.

Example 21.4 The only example of a free group that has occurred before is Z, which is free on one generator. Clearly every free group is infinite. ∎

You are referred to the literature for proofs of the next three theorems. We will not be using these results. They are stated simply to inform you of these interesting facts.

Theorem 21.1 *If a group G is free on $\{a_i\}$ and also on $\{b_j\}$, then the sets $\{a_i\}$ and $\{b_j\}$ have the same number of elements, that is, any two sets of free generators of a free group have the same cardinality.*

Definition If G is free on $\{a_i\}$, the number of elements in $\{a_i\}$ is the *rank of the free group G*.

Actually, the next theorem is pretty evident from Theorem 21.1.

Theorem 21.2 *Two free groups are isomorphic if and only if they have the same rank.*

Theorem 21.3 *A nontrivial proper subgroup of a free group is free.*

Example 21.5 Let $F[\{x, y\}]$ be the free group on $\{x, y\}$. Let

$$y_k = x^k y x^{-k}$$

for $k \geq 0$. You will have no difficulty convincing yourself that the y_k for $k \geq 0$ are free generators for the subgroup of $F[\{x, y\}]$ that they generate. This illustrates that although a subgroup of a free group is free, the rank of the subgroup may be much greater than the rank of the whole group! ∎

*21.3 HOMOMORPHISMS OF FREE GROUPS

Our work in this section will be concerned primarily with homomorphisms defined on a free group. The results here are simple and elegant.

Theorem 21.4   *Let G be generated by $\{a_i \mid i \in I\}$ and let G' be any group. If a'_i for $i \in I$ are any elements in G', not necessarily distinct, then there is at most one homomorphism $\phi : G \to G'$ such that $a_i\phi = a'_i$. If G is free on $\{a_i\}$, then there is exactly one such homomorphism.*

Proof.   Let ϕ be a homomorphism from G into G' such that $a_i\phi = a'_i$. Now by Theorem 9.1, for any $x \in G$ we have

$$x = \prod_j a_{i_j}{}^{n_j}$$

for some finite product of the generators a_i, where the a_{i_j} appearing in the product need not be distinct. Then since ϕ is a homomorphism, we must have

$$x\phi = \prod_j (a_{i_j}{}^{n_j})\phi = \prod_j (a'_{i_j})^{n_j}.$$

Thus a homomorphism is completely determined by its values on elements of a generating set. This shows that there is at most one homomorphism such that $a_i\phi = a'_i$.

Now suppose G is free on $\{a_i\}$, that is, $G = F[\{a_i\}]$. For

$$x = \prod_j a_{i_j}{}^{n_j}$$

in G, define $\psi : G \to G'$ by

$$x\psi = \prod_j (a'_{i_j})^{n_j}.$$

This map is well defined, since $F[\{a_i\}]$ consists precisely of reduced words; no two different formal products in $F[\{a_i\}]$ are equal. Since the rules for computation involving exponents in G' are formally the same as those involving exponents in G, it is clear that $(xy)\psi = (x\psi)(y\psi)$ for any elements x and y in G, so ψ is indeed a homomorphism. ∎

Perhaps we should have proved the first part of this theorem earlier, rather than having relegated it to the exercises. Note that the theorem states that *a homomorphism of a group is completely determined if you know its value on each element of a generating set.* This is really very obvious and could have been mentioned immediately after the definition of a homomorphism. In particular, a homomorphism of a cyclic group is completely determined by its value on any single generator of the group.

Theorem 21.5   *Every group G' is a homomorphic image of a free group G.*

Proof.   Let $G' = \{a_i'\}$, and let $\{a_i\}$ be a set with the same number of elements as G'. Let $G = F[\{a_i\}]$. Then by Theorem 21.4 there exists a homomorphism ψ mapping G into G' such that $a_i\psi = a_i'$. Clearly, the image of G under ψ is all of G'. ∎

*21.4 ANOTHER LOOK AT FREE ABELIAN GROUPS

It is important that you do not confuse the notion of a free group with the notion of a free abelian group. A free group on more than one generator is not abelian. In the preceding chapter, we defined a free abelian group as an abelian group that has a basis, that is, a generating set satisfying properties described in Theorem 20.1. There is another approach, via free groups, to free abelian groups. We now describe this approach.

Let $F[A]$ be the free group on the generating set A. We shall write F in place of $F[A]$ for the moment. Note that F is not abelian if A contains more than one element. Let F' be the commutator subgroup of F. Then F/F' is an abelian group, and it is pretty clear that F/F' is free abelian with basis $\{a + F' \mid a \in A\}$. If $a + F$ is renamed a, we can view F/F' as a free abelian group with basis A. This indicates how a free abelian group having a given set as basis can be constructed. Every free abelian group can be constructed in this fashion, up to isomorphism. That is, if G is free abelian with basis X, form the free group $F[X]$, form the factor group of $F[X]$ modulo its commutator subgroup, and you have a group isomorphic to G.

Theorems 21.1, 21.2, and 21.3 hold for free abelian groups as well as for free groups. In fact, the abelian version of Theorem 21.3 was proved for the finite rank case in Theorem 20.6. In contrast to Example 21.5 for free groups, it is true that for a free abelian group the rank of a subgroup is at most the rank of the entire group. Theorem 20.6 also showed this for the finite rank case.

Exercises

*21.1 Find the reduced form and the inverse of the reduced form of each of the following words.

a) $a^2b^{-1}b^3a^3c^{-1}c^4b^{-2}$

b) $a^2a^{-3}b^3a^4c^4c^2a^{-1}$

*21.2 Compute the products given in parts (a) and (b) of Exercise 21.1 in the case that $\{a, b, c\}$ is a set of generators forming a basis for a free abelian group. Find the inverses of these products.

*21.3 How many different homomorphisms are there of a free group of rank 2 into

a) $\mathbf{Z}_4$?   b) $\mathbf{Z}_6$?   c) S_3?

*21.4 How many different homomorphisms are there of a free group of rank 2 onto

a) $\mathbf{Z}_4$?   b) $\mathbf{Z}_6$?   c) S_3?

*21.5 How many different homomorphisms are there of a free abelian group of rank 2 into

a) $\mathbf{Z}_4$?   b) $\mathbf{Z}_6$?   c) S_3?

*21.6 How many different homomorphisms are there of a free abelian group of rank 2 onto

a) $\mathbf{Z}_4$?   b) $\mathbf{Z}_6$?   c) S_3?

†*21.7 Take one of the instances in this section in which the phrase "It would seem obvious that" was used and discuss your reaction in that instance.

*21.8 Mark each of the following true or false.

___ a) Every proper subgroup of a free group is a free group.

___ b) Every proper subgroup of every free abelian group is a free group.

___ c) A homomorphic image of a free group is a free group.

___ d) Every free abelian group has a basis.

___ e) The free abelian groups of finite rank are precisely the finitely generated abelian groups.

___ f) No free group is abelian.

___ g) No free abelian group is free.

___ h) No free abelian group of rank > 1 is free.

___ i) Any two free groups are isomorphic.

___ j) Any two free abelian groups of the same rank are isomorphic.

*21.9 Let G be a finitely generated abelian group with identity 0. A finite set $\{b_1, \ldots, b_n\}$, where $b_i \in G$, is a **basis for** G if $\{b_1, \ldots, b_n\}$ generates G and $\sum_{i=1}^{n} m_i b_i = 0$ if and only if $m_i b_i = 0$, where $m_i \in \mathbf{Z}$.

a) Show that $\{2, 3\}$ is not a basis for $\mathbf{Z}_4$. Find a basis for $\mathbf{Z}_4$.

b) Show that both $\{1\}$ and $\{2, 3\}$ are bases for $\mathbf{Z}_6$. (This shows that for a finitely generated abelian group G with torsion, the number of elements in a basis may vary, that is, it need not be an *invariant* of the group G.)

c) Is a basis for a free abelian group as we defined it in Chapter 20 a basis in the sense in which it is used in this exercise?

d) Show that every finite abelian group has a basis $\{b_1, \ldots, b_n\}$, where the order of b_i divides the order of b_{i+1}. You may use any theorems in the text, even if they were not proved.

In present-day expositions of algebra, a frequently used technique (particularly by the disciples of N. Bourbaki) for introducing a new algebraic entity is the following:

1. *Describe algebraic properties that this algebraic entity is to possess.*

2. *Prove that any two algebraic entities with these properties are isomorphic, that is, that these properties characterize this entity.*

3. *Show that at least one such entity exists.*

The next three exercises illustrate this technique for three algebraic entities, each of which the student has met before. So that we don't give away their identities, we use fictitious names for them in the first two exercises. The last part of these first two exercises asks you to give the usual name for the entity.

*21.10 Let G be any group. An abelian group G^* is a **blip group of** G if there exists a fixed homomorphism ϕ of G onto G^* such that each homomorphism ψ of G into an abelian group G' can be factored as $\psi = \phi\theta$, where θ is a homomorphism of G^* into G' (see Fig. 21.1).

a) Show that any two blip groups of G are isomorphic. [*Hint:* Let G_1^* and G_2^* be two blip groups of G. Then each of the fixed homomorphisms $\phi_1 : G \to G_1^*$ and $\phi_2 : G \to G_2^*$ can be factored via the other blip group according to the definition of a blip group, that is, $\phi_1 = \phi_2\theta_1$ and $\phi_2 = \phi_1\theta_2$. Show that θ_1 is an isomorphism of G_2^* onto G_1^*. Apply Exercise 13.16.]

b) Show for every group G that a blip group G^* of G exists.

c) What concept that we have introduced before corresponds to this idea of a blip group of G?

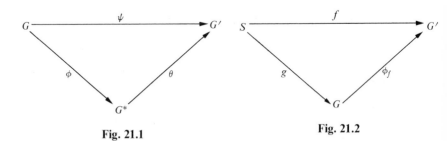

Fig. 21.1   Fig. 21.2

*21.11 Let S be any set. A group G together with a fixed function $g : S \to G$ constitutes a **blop group on** S if for each group G' and map $f : S \to G'$ there exists a *unique* homomorphism ϕ_f of G into G' such that $f = g\phi_f$ (see Fig. 21.2).

a) Let S be a fixed set. Show that if both G_1, together with $g_1 : S \to G_1$, and G_2, together with $g_2 : S \to G_2$, are blop groups on S, then G_1 and G_2 are isomorphic. [*Hint:* Show that g_1 and g_2 are one-to-one maps and that Sg_1 and Sg_2 generate G_1 and G_2 respectively. Then proceed in a way analogous to that given by the hint of Exercise 21.10.]

b) Let S be a set. Show that a blop group on S exists. You may use any theorems of the text.

c) What concept that we have introduced before corresponds to this idea of a blop group on S?

*21.12 Characterize a free abelian group by properties in a fashion similar to that used in Exercise 21.11.

*twenty-two

Group Presentations

*22.1 DEFINITION

Following most of the literature on group presentations, in this chapter we let 1 be the identity of a group. The idea of a *group presentation* is to form a group by giving a set of generators for the group and certain equations or relations that you want the generators to satisfy. You want the group to be as free as it possibly can be on the generators subject to these relations.

Example 22.1 Suppose G has generators x and y and is *free except for the relation* $xy = yx$, which we may express as $xyx^{-1}y^{-1} = 1$. Clearly, the condition $xy = yx$ is exactly what is needed to make G commutative, even though $xyx^{-1}y^{-1}$ is just one of the many possible commutators of $F[\{x, y\}]$. Thus G is free abelian on two generators and is isomorphic to $F[\{x, y\}]$ modulo its commutator subgroup. This commutator subgroup of $F[\{x, y\}]$ is the smallest normal subgroup containing $xyx^{-1}y^{-1}$, since any normal subgroup containing $xyx^{-1}y^{-1}$ gives rise to a factor group that is abelian, and thus contains the commutator subgroup by Theorem 12.6. ∎

The preceding example illustrates the general situation. Let $F[A]$ be a free group and suppose that you want to form a new group as much like $F[A]$ as it can be, subject to certain equations that you want satisfied. Any equation can be written in a form in which the right-hand side is 1. Thus we can consider the equations to be $r_i = 1$, where $r_i \in F[A]$. Clearly, if you require that $r_i = 1$, then you will have to have

$$x^{-1}(r_i{}^n)x = 1$$

for any $x \in F[A]$ and $n \in \mathbf{Z}$. Also any product of elements equal to 1 will again have to equal 1. Thus any finite product of the form

$$\prod_j x_j^{-1}(r_{i_j}{}^{n_j})x_j,$$

where the r_{i_j} need not be distinct, will have to equal 1 in the new group. It is very easy to check that the set of all these finite products is a normal subgroup R of $F[A]$. Thus any group looking as much as possible like $F[A]$, subject to the requirements $r_i = 1$, also has $r = 1$ for every $r \in R$. But $F[A]/R$ looks like $F[A]$ (remember that we multiply cosets by choosing representatives) except that R has been collapsed to form the identity 1. Thus the group we are after is (at least isomorphic to) $F[A]/R$. We can view this group as described by the generating set A and the set $\{r_i\}$.

Definition   Let A be a set and let $\{r_i\} \subseteq F[A]$. Let R be the least normal subgroup of $F[A]$ containing the r_i. An isomorphism ϕ of $F[A]/R$ onto a group G is a ***presentation of*** G. The sets A and $\{r_i\}$ give a ***group presentation***. The set A is the set of ***generators for the presentation*** and each r_i is a ***relator***. Each $r \in R$ is a ***consequence*** of $\{r_i\}$. An equation $r_i = 1$ is a ***relation***. A ***finite presentation*** is one in which both A and $\{r_i\}$ are finite sets.

This definition may seem complicated, but it really isn't. In Example 22.1 $\{x, y\}$ is our set of generators and $xyx^{-1}y^{-1}$ is the only relator. The equation $xyx^{-1}y^{-1} = 1$, or $xy = yx$, is a relation. This was an example of a finite presentation.

If a group presentation has generators x_j and relators r_i, we shall use the notations

$$(x_j : r_i) \qquad \text{or} \qquad (x_j : r_i = 1)$$

to denote the group presentation. We may refer to $F[\{x_j\}]/R$ as *the group with presentation* $(x_j : r_i)$.

*22.2 ISOMORPHIC PRESENTATIONS

Example 22.2   Consider the group presentation with

$$A = \{a\} \qquad \text{and} \qquad \{r_i\} = \{a^6\},$$

that is, the presentation

$$(a : a^6 = 1).$$

This group defined by one generator a, with the relation $a^6 = 1$, is clearly isomorphic to $\mathbf{Z}_6$.

Now consider the group defined by two generators a and b, with $a^2 = 1$, $b^3 = 1$, and $ab = ba$, that is, the group with presentation

$$(a, b : a^2, b^3, aba^{-1}b^{-1}).$$

The condition $a^2 = 1$ gives $a^{-1} = a$. Also $b^3 = 1$ gives $b^{-1} = b^2$. Thus every element in this group can be written as a product of nonnegative powers of a and b. The relation $aba^{-1}b^{-1} = 1$, that is, $ab = ba$, allows us to write first all the factors involving a and then the factors involving b. Hence every element of the group is equal to some $a^m b^n$. But then $a^2 = 1$ and $b^3 = 1$ show that there are just six distinct elements,

$$1, b, b^2, a, ab, ab^2.$$

Therefore this presentation also gives a group of order 6 that is abelian, and by Theorem 9.3 it must again be cyclic and isomorphic to $\mathbf{Z}_6$. ∎

The preceding example illustrates that different presentations may give isomorphic groups. When this happens, we have **isomorphism presentation.** To determine whether two presentations are isomorphic may be very hard. It has recently been shown (see Rabin [22]) that a number of such problems connected with this theory are not generally solvable, that is, there is no *routine* and well-defined way of discovering a solution in all cases. These unsolvable problems include the problem of deciding whether two presentations are isomorphic, whether a group given by a presentation is finite, free, abelian, or trivial, and the famous *word problem* of determining whether a given word r is a consequence of a given set of words $\{r_i\}$.

The importance of this material is indicated by our Theorem 21.5, which guarantees that *every group has a presentation.*

Example 22.3   Let us show that

$$(x, y : y^2 x = y, yx^2 y = x)$$

is a presentation of the trivial group of one element. We need only show that x and y are consequences of the relators $y^2 x y^{-1}$ and $yx^2 y x^{-1}$, or that $x = 1$ and $y = 1$ can be deduced from $y^2 x = y$ and $yx^2 y = x$. We illustrate both techniques.

As a consequence of y^2xy^{-1}, we get yx upon conjugation by y. From yx we deduce $x^{-1}y^{-1}$, and then $(x^{-1}y^{-1})(yx^2yx^{-1})$ gives xyx^{-1}. Conjugating xyx^{-1} by x, we get y. From y we get y^{-1}, and $y^{-1}(yx)$ is x.

Working with relations instead of relators, from $y^2x = y$ we deduce $yx = 1$ upon multiplication by y^{-1} on the left. Then substituting $yx = 1$ into $yx^2y = x$, that is, $(yx)(xy) = x$, we get $xy = x$. Then multiplying by x^{-1} on the left, we have $y = 1$. Substituting this in $yx = 1$, we get $x = 1$.

Both techniques amount to the same work, but it somehow seems more natural to most of us to work with relations. ∎

*22.3 APPLICATIONS

We conclude this chapter with two applications.

Example 22.4   Let us determine all groups of order 10 up to isomorphism. We know from Theorem 9.3 that every abelian group of order 10 is isomorphic to $\mathbf{Z}_{10}$. Suppose that G is nonabelian of order 10. By Sylow theory, G contains a normal subgroup H of order 5, and H must be cyclic. Let a be a generator of H. Then G/H is of order 2 and thus isomorphic to $\mathbf{Z}_2$. If $b \in G$ and $b \notin H$, we must then have $b^2 \in H$. Since every element of H except 1 has order 5, if b^2 were not equal to 1, then b^2 would have order 5, so b would have order 10. This would mean that G would be cyclic, contradicting our assumption that G is not abelian. Thus $b^2 = 1$. Finally, since H is a normal subgroup of G, $bHb^{-1} = H$, so in particular, $bab^{-1} \in H$. Since conjugation by b^{-1} is an automorphism of H, bab^{-1} must be another element of H of order 5, hence bab^{-1} equals a, a^2, a^3, or a^4. But $bab^{-1} = a$ would give $ba = ab$, and then clearly G would be abelian, since a and b generate G. Thus the possibilities for presentations of G are:

1. $(a, b : a^5 = 1, b^2 = 1, ba = a^2b)$,
2. $(a, b : a^5 = 1, b^2 = 1, ba = a^3b)$,
3. $(a, b : a^5 = 1, b^2 = 1, ba = a^4b)$.

Note that all three of these presentations can give groups of order at most 10, since the last relation $ba = a^ib$ enables us to express every product of a's and b's in G in the form a^sb^t. Then $a^5 = 1$ and $b^2 = 1$ show that the set

$$S = \{a^0b^0, a^1b^0, a^2b^0, a^3b^0, a^4b^0, a^0b^1, a^1b^1, a^2b^1, a^3b^1, a^4b^1\}$$

includes all elements of G.

It is not yet clear that all these elements in S are distinct, so that we have in all three cases a group of order 10. For example, the group presentation

$$(a, b : a^5 = 1, b^2 = 1, ba = a^2b)$$

gives a group in which, using the associative law, we have

$$a = b^2 a = (bb)a = b(ba) = b(a^2 b) = (ba)(ab)$$
$$= (a^2 b)(ab) = a^2(ba)b = a^2(a^2 b)b = a^4 b^2 = a^4.$$

Thus in this group, $a = a^4$, so $a^3 = 1$, which, together with $a^5 = 1$, yields $a^2 = 1$. But $a^2 = 1$, together with $a^3 = 1$, means that $a = 1$. Hence every element in the group with presentation

$$(a, b : a^5 = 1, b^2 = 1, ba = a^2 b)$$

is equal to either 1 or b, that is, this group is isomorphic to $\mathbf{Z}_2$. A similar study of

$$(bb)a = b(ba)$$

for

$$(a, b : a^5 = 1, b^2 = 1, ba = a^3 b)$$

shows that $a = a^4$ again, so this also yields a group isomorphic to $\mathbf{Z}_2$.

This leaves just

$$(a, b : a^5 = 1, b^2 = 1, ba = a^4 b)$$

as a candidate for a nonabelian group of order 10. In this case, it can be shown that all elements of S are distinct, so this presentation does give a nonabelian group G of order 10. How can we show that all elements in S represent distinct elements of G? The easy way is to observe that we know that there is at least one nonabelian group of order 10, the dihedral group D_5. Since G is the only remaining candidate, we must have $G \simeq D_5$. Another attack is as follows. Let us try to make S into a group by *defining* $(a^s b^t)(a^u b^v)$ to be $a^x b^y$, where x is the remainder of $s + u(4^t)$ when divided by 5, and y is the remainder of $t + v$ when divided by 2, in the sense of Lemma 6.1. In other words, we use the relation $ba = a^4 b$ as a guide in *defining* the product $(a^s b^t)(a^u b^v)$ of two elements of S. It is easy to see that $a^0 b^0$ acts as identity, and that given $a^u b^v$, we can determine t and s successively by letting

$$t \equiv -v \pmod 2$$

and then

$$s \equiv -u(4^t) \pmod 5,$$

giving $a^s b^t$, which is a left inverse for $a^u b^v$. We will then have a group structure on S if and only if the associative law holds. In Exercise 22.7 we ask you to carry out the straightforward computation for the associative law and to discover a condition for S to be a group under such a definition of multiplication. The criterion of the exercise in this case amounts to the valid congruence

$$4^2 \equiv 1 \pmod 5.$$

Thus we do get a group of order 10. Note that

$$2^2 \not\equiv 1 \pmod 5$$

and

$$3^2 \not\equiv 1 \pmod 5,$$

so Exercise 22.7 also shows that

$$(a, b : a^5 = 1, b^2 = 1, ba = a^2 b)$$

and

$$(a, b : a^5 = 1, b^2 = 1, ba = a^3 b)$$

do not give groups of order 10. ∎

Example 22.5 Let us determine all groups of order 8 up to isomorphism. We know the three abelian ones:

$$\mathbf{Z}_8, \qquad \mathbf{Z}_2 \times \mathbf{Z}_4, \qquad \mathbf{Z}_2 \times \mathbf{Z}_2 \times \mathbf{Z}_2.$$

Using generators and relations, we shall give presentations of the nonabelian groups.

Let G be nonabelian of order 8. Since G is nonabelian, it has no elements of order 8, so each element but the identity is of order either 2 or 4. If every element were of order 2, then for $a, b \in G$, we would have $(ab)^2 = 1$, that is, $abab = 1$. Then since $a^2 = 1$ and $b^2 = 1$ also, we would have

$$ba = a^2 bab^2 = a(ab)^2 b = ab,$$

contrary to our assumption that G is not abelian. Thus G has an element of order 4.

Let $\langle a \rangle$ be a subgroup of G of order 4. If $b \notin \langle a \rangle$, the cosets $\langle a \rangle$ and $b \langle a \rangle$ exhaust all of G. Hence a and b are generators for G and $a^4 = 1$. Since $\langle a \rangle$ is normal in G (by Sylow theory, or because it is of index 2), $G/\langle a \rangle$ is isomorphic to $\mathbf{Z}_2$ and we have $b^2 \in \langle a \rangle$. If $b^2 = a$ or $b^2 = a^3$, then b would be of order 8. Hence $b^2 = 1$ or $b^2 = a^2$. Finally, since $\langle a \rangle$ is normal, we have $bab^{-1} \in \langle a \rangle$, and since $b \langle a \rangle b^{-1}$ is a subgroup conjugate to $\langle a \rangle$ and hence isomorphic to $\langle a \rangle$, we see that bab^{-1} must be an element of order 4. Thus $bab^{-1} = a$ or $bab^{-1} = a^3$. If bab^{-1} were equal to a, then ba would equal ab, which would make G abelian. Hence $bab^{-1} = a^3$ so $ba = a^3 b$. Thus, we have two possibilities for G, namely

$$G_1 : (a, b : a^4 = 1, b^2 = 1, ba = a^3 b)$$

and

$$G_2 : (a, b : a^4 = 1, b^2 = a^2, ba = a^3 b).$$

Note that $a^{-1} = a^3$, and that b^{-1} is b in G_1 and b^3 in G_2. These facts, with the relation $ba = a^3 b$, enable us to express every element in G_i in the form $a^m b^n$, as in Examples 22.2 and 22.4. Since $a^4 = 1$ and either $b^2 = 1$ or

$b^2 = a^2$, the possible elements in each group are

$$1, a, a^2, a^3, b, ab, a^2b, a^3b.$$

Thus G_1 and G_2 each have order at most 8. That G_1 is a group of order 8 can be seen from Exercise 22.7. An argument similar to that used in Exercise 22.7 shows that G_2 has order 8 also.

Since $ba = a^3b \neq ab$, we see that both G_1 and G_2 are nonabelian. That the two groups are not isomorphic follows from the fact that a computation shows that G_1 has only two elements of order 4, namely a and a^3. On the other hand, in G_2 all elements but 1 and a^2 are of order 4. We leave the computations of the tables for these groups for you to perform in Exercise 22.3. To illustrate, suppose you wish to compute $(a^2b)(a^3b)$. Using $ba = a^3b$ repeatedly, we get

$$(a^2b)(a^3b) = a^2(ba)a^2b = a^5(ba)ab = a^8(ba)b = a^{11}b^2.$$

Then for G_1, we have

$$a^{11}b^2 = a^{11} = a^3,$$

but if we are in G_2, we get

$$a^{11}b^2 = a^{13} = a.$$

The group G_1 is the **octic group** and is nothing more than our old friend the group D_4 of symmetries of the square. The group G_2 is the **quaternion group**; the reason for the name will be explained in Section 25.4. ∎

Exercises

*22.1 Give a presentation of Z_4 involving one generator; involving two generators; involving three generators.

*22.2 Give a presentation of S_3 involving three generators.

*22.3 Give the tables for both the octic group

$$(a, b : a^4 = 1, b^2 = 1, ba = a^3b)$$

and the quaternion group

$$(a, b : a^4 = 1, b^2 = a^2, ba = a^3b).$$

In both cases, write the elements in the order $1, a, a^2, a^3, b, ab, a^2b, a^3b$. (Note that you don't have to compute *every* product. You know that these presentations give groups of order 8, and once you have computed enough products the rest are forced so that each row and each column of the table has each element exactly once.)

*22.4 Mark each of the following true or false.

— a) Every group has a presentation.

— b) Every group has many different presentations.

— c) Every group has two presentations that are not isomorphic.

— d) Every group has a finite presentation.

— e) Every group with a finite presentation is of finite order.

— f) Every cyclic group has a presentation with just one generator.

— g) Every conjugate of a relator is a consequence of the relator.

— h) Two presentations with the same number of generators are always isomorphic.

— i) In a presentation of an abelian group, the set of consequences of the relators contains the commutator subgroup of the free group on the generators.

— j) Every presentation of a free group has 1 as the only relator.

*22.5 Show that

$$(a, b : a^3 = 1, b^2 = 1, ba = a^2b)$$

gives a group of order 6. Show that it is nonabelian.

*22.6 Show that the presentation

$$(a, b : a^3 = 1, b^2 = 1, ba = a^2b)$$

of Exercise 22.5 gives (up to isomorphism) the only nonabelian group of order 6, and hence gives a group isomorphic to S_3.

*22.7 Let

$$S = \{a^i b^j \,|\, 0 \le i < m, 0 \le j < n\},$$

that is, S consists of all formal products $a^i b^j$ starting with $a^0 b^0$ and ending with $a^{m-1}b^{n-1}$. Let r be a positive integer, and define multiplication on S by

$$(a^s b^t)(a^u b^v) = a^x b^y,$$

where x is the remainder of $s + u(r^t)$ when divided by m, and y is the remainder of $t + v$ when divided by n, in the sense of Lemma 6.1.

a) Show that a necessary and sufficient condition for the associative law to hold and for S to be a group under this multiplication is that $r^n \equiv 1 \pmod{m}$.

b) Deduce from part (a) that the group presentation

$$(a, b : a^m = 1, b^n = 1, ba = a^r b)$$

gives a group of order mn if and only if $r^n \equiv 1 \pmod{m}$.

*22.8 Determine all groups of order 14 up to isomorphism. [*Hint:* Follow the outline of Example 22.4 and use Exercise 22.7, part (b).]

*22.9 Determine all groups of order 21 up to isomorphism. [*Hint:* Follow the outline of Example 22.4 and use Exercise 22.7, part (b). It may seem that there are two presentations giving nonabelian groups. Show that they are isomorphic.]

*22.10 Show that if $n = pq$, with p and q primes and $q > p$ and $q \equiv 1 \pmod p$, then there is exactly one nonabelian group (up to isomorphism) of order n. Assume (as will be proved later) that the $q-1$ nonzero elements of $\mathbf{Z}_q$ form a cyclic group $\mathbf{Z}_q^*$ under multiplication modulo q. [*Hint:* The solutions of $x^p \equiv 1 \pmod{q}$ form a cyclic subgroup of $\mathbf{Z}_q^*$ with elements $1, r, r^2, \ldots, r^{p-1}$. In the group with presentation $(a, b : a^q = 1, b^p = 1, ba = a^r b)$, we have $bab^{-1} = a^r$, so $b^j ab^{-j} = a^{(r^j)}$. Thus, since b^j generates $\langle b \rangle$ for $j = 1, \ldots, p-1$, this presentation is isomorphic to

$$(a, b^j : a^q = 1, (b^j)^p = 1, (b^j)a = a^{(r^j)}(b^j)),$$

so all the presentations $(a, b : a^q = 1, b^p = 1, ba = a^{(r^j)}b)$ are isomorphic.]

*22.11 Determine all groups (up to isomorphism) of order 12.

*22.12 Determine all groups (up to isomorphism) of order 30.

Rings
and
Fields

twenty-three

Rings

23.1 DEFINITION AND BASIC PROPERTIES

All our work thus far has been concerned with sets on which a single binary operation has been defined. Our familiar examples of sets of numbers show that a study of sets on which two binary operations have been defined should be of great importance. The most general system of this nature that we shall study is a ring.

Definition A *ring* $\langle R, +, \cdot \rangle$ is a set R together with two binary operations $+$ and $\cdot$, which we call addition and multiplication, defined on R such that the following axioms are satisfied:

$\mathscr{R}_1$. $\langle R, + \rangle$ is an abelian group.

$\mathscr{R}_2$. Multiplication is associative.

$\mathscr{R}_3$. For all $a, b, c \in R$, the **left distributive law,** $a(b + c) = (ab) + (ac)$, and the **right distributive law,** $(a + b)c = (ac) + (bc)$, hold.

206

Example 23.1 You are well aware that axioms $\mathcal{R}_1$, $\mathcal{R}_2$, and $\mathcal{R}_3$ for a ring hold in any subset of the complex numbers that is a group under addition and that is closed under multiplication. For example, $\langle \mathbf{Z}, +, \cdot \rangle$, $\langle \mathbf{Q}, +, \cdot \rangle$, $\langle \mathbf{R}, +, \cdot \rangle$, and $\langle \mathbf{C}, +, \cdot \rangle$ are all rings. ∎

We shall observe the usual convention that multiplication is performed before addition, so the left distributive law, for example, becomes

$$a(b + c) = ab + ac,$$

without the parentheses on the right side of the equation. Also, as a convenience analogous to our notation in group theory, we shall somewhat incorrectly refer to *a ring R* in place of *a ring $\langle R, +, \cdot \rangle$*, provided that no confusion will result. In particular, from now on $\mathbf{Z}$ will always be $\langle \mathbf{Z}, +, \cdot \rangle$, and $\mathbf{Q}$, $\mathbf{R}$, and $\mathbf{C}$ will also be the obvious rings. We may on occasion refer to $\langle R, + \rangle$ as *the additive group of the ring R.*

Example 23.2 Consider the cyclic group $\langle \mathbf{Z}_n, + \rangle$. If one defines for $a, b \in \mathbf{Z}_n$ the product ab to be the remainder of the usual product of integers when divided by n, it can be shown that $\langle \mathbf{Z}_n, +, \cdot \rangle$ is a ring. We shall feel free to use this fact. For example, in $\mathbf{Z}_{10}$ we have $(3)(7) = 1$. This operation on $\mathbf{Z}_n$ is **multiplication modulo** n. We do not check the ring axioms here, for they will later follow directly from some of the theory we have to develop anyway. ∎

From now on, $\mathbf{Z}_n$ will always be $\langle \mathbf{Z}_n, +, \cdot \rangle$. Continuing matters of notation, we shall always let 0 be the additive identity of a ring. The additive inverse of an element a of a ring is $-a$. We shall frequently have occasion to refer to a sum

$$a + a + \cdots + a$$

having n summands. We shall let this sum be $n \cdot a$. However, $n \cdot a$ is not to be construed as a multiplication of n and a in the ring, for the integer n may not be in the ring at all. If $n < 0$, we let

$$n \cdot a = (-a) + (-a) + \cdots + (-a)$$

for $|n|$ summands. Finally, we define

$$0 \cdot a = 0$$

for $0 \in \mathbf{Z}$ on the left side of the equation and $0 \in R$ on the right side. Actually the equation $0a = 0$ holds also for $0 \in R$ on both sides. The following theorem proves this and various other easy but important facts. Note the strong use of the distributive laws in the proof of this theorem. *These distributive laws are the only means we have of relating additive concepts to multiplicative concepts in a ring.*

Theorem 23.1 *If R is a ring with additive identity 0, then for any $a, b \in R$ we have*

1. $0a = a0 = 0$,
2. $a(-b) = (-a)b = -(ab)$,
3. $(-a)(-b) = ab$.

Proof. For condition 1, note that

$$a0 = a(0 + 0) = a0 + a0.$$

Then by the cancellation law for the additive group $\langle R, + \rangle$, we have $0 = a0$. Likewise,

$$0a = (0 + 0)a = 0a + 0a$$

implies that $0a = 0$. This proves condition 1.

In order to understand the proof of condition 2, you must remember that, by *definition*, $-(ab)$ is the element that when added to ab gives 0. Thus to show that $a(-b) = -(ab)$, you must show precisely that $a(-b) + ab = 0$. By the left distributive law,

$$a(-b) + ab = a(-b + b) = a0 = 0,$$

since $a0 = 0$ by condition 1. Likewise,

$$(-a)b + ab = (-a + a)b = 0b = 0.$$

For condition 3, note that

$$(-a)(-b) = -(a(-b))$$

by condition 2. Again by 2,

$$-(a(-b)) = -(-(ab)),$$

and $-(-(ab))$ is the element that when added to $-(ab)$ gives 0. This is ab by definition of $-(ab)$ and by the uniqueness of an inverse in a group. Thus, $(-a)(-b) = ab$. ∎

It is important that you *understand* the preceding proof. If you can't follow the logic and use of definitions, you will come to grief later. (Actually you may have already come to grief.) The theorem allows us to use our usual rules for signs.

We hope that you are beginning to realize that in the study of any sort of mathematical structure, an idea of basic importance is the concept of two systems being structurally identical, that is, one being just like the other except for names. In algebra this concept is always called *isomorphism*. The concept of two rings being just alike except for names of elements leads us, just as it did for groups, to the following definition.

Definition An *isomorphism ϕ of a ring R with a ring R'* is a one-to-one function mapping R onto R' such that for all $a, b \in R$,

1. $(a + b)\phi = a\phi + b\phi$,
2. $(ab)\phi = (a\phi)(b\phi)$.

The rings R and R' are then *isomorphic.*

Example 23.3 As abelian groups, $\langle \mathbf{Z}, + \rangle$ and $\langle 2\mathbf{Z}, + \rangle$ are isomorphic under the map $\phi : \mathbf{Z} \to 2\mathbf{Z}$, with $x\phi = 2x$ for $x \in \mathbf{Z}$. Note that $2\mathbf{Z}$ is closed under the usual multiplication and that $\langle 2\mathbf{Z}, +, \cdot \rangle$ is a ring. Here ϕ is *not* a ring isomorphism, for $(xy)\phi = 2xy$, while $(x\phi)(y\phi) = 2x2y = 4xy$. ∎

From now on, $n\mathbf{Z}$ will always be the ring $\langle n\mathbf{Z}, +, \cdot \rangle$.

23.2 MULTIPLICATIVE QUESTIONS; FIELDS

All the rings we have seen so far have a multiplication that is commutative. Many of them, such as $\mathbf{Z}$, $\mathbf{Q}$, and $\mathbf{R}$, also have a multiplicative identity 1. However, $2\mathbf{Z}$ does not have an identity element for multiplication. There are many rings in which multiplication is not commutative. The student familiar with a little matrix theory will see that the $n \times n$ matrices with entries in $\mathbf{Z}$ (or $\mathbf{Q}$, $\mathbf{R}$, or $\mathbf{C}$) form a ring under matrix addition and multiplication in which multiplication is not commutative if $n \geq 2$. These rings of matrices do have an identity element for multiplication. We treat them in more detail in Chapter 25.

It is evident that $\{0\}$, with $0 + 0 = 0$ and $(0)(0) = 0$, gives a ring. Here 0 acts as multiplicative as well as additive identity. By Theorem 23.1, this is the only case in which 0 could act as a multiplicative identity, for if $0a = a$, we can deduce that $a = 0$. We shall exclude this trivial case when speaking of a multiplicative identity in a ring, that is, whenever we speak of a multiplicative identity, we will assume that it is nonzero.

Definition A ring in which the multiplication is commutative is a *commutative ring.* A ring R with a multiplicative identity 1 such that $1x = x1 = x$ for all $x \in R$ is a *ring with unity.* A multiplicative identity in a ring is *unity.*

Theorem 23.2 *If R is a ring with unity, then this unity 1 is the only multiplicative identity.*

Proof. We proceed exactly as we did for groups. Let 1 and 1′ both be multiplicative identities in a ring R, and set up a competition. Regarding 1 as identity, we have

$$(1)(1') = 1'.$$

Regarding $1'$ as identity, we have

$$(1)(1') = 1.$$

Thus $1 = 1'$. ∎

If $R_1, R_2, \ldots, R_n$ are rings, we can form the set $R_1 \times R_2 \times \cdots \times R_n$ of all ordered n-tuples $(r_1, r_2, \ldots, r_n)$, where $r_i \in R_i$. Defining addition and multiplication of n-tuples by components (just as for groups), we see at once from the ring axioms in each component that the set of all these n-tuples forms a ring under addition and multiplication by components. The ring $R_1 \times R_2 \times \cdots \times R_n$ is the **direct product** of the rings R_i. Clearly, such a direct product is commutative or has unity if and only if each R_i is commutative or has unity, respectively.

In a ring R with unity, the set R^* of nonzero elements, if closed under the ring multiplication, will be a multiplicative group if inverses exist. A **multiplicative inverse** of an element a in a ring R with unity 1 is an element $a^{-1} \in R$ such that $aa^{-1} = a^{-1}a = 1$. Precisely as for groups, a multiplicative inverse for an element a in R is unique, if it exists at all (see Exercise 23.12). Theorem 23.1 shows that it would be hopeless to have a multiplicative inverse for 0 unless one wishes to regard the set $\{0\}$, where $0 + 0 = 0$ and $(0)(0) = 0$, as a ring with 0 as both additive and multiplicative identity. We have agreed to exclude this trivial case when speaking of a ring with unity. We are thus led to discuss the existence of multiplicative inverses for nonzero elements in a ring with unity. You are probably getting tired of all these definitions, but we have no choice.

Definition   Let R be a ring with unity. An element u in R is a ***unit of R*** if it has a multiplicative inverse in R. If every nonzero element of R is a unit, then R is a ***skew field*** or ***division ring***. A ***field*** is a commutative division ring.

Example 23.4 $\mathbf{Z}$ is not a field, since 2, for example, has no multiplicative inverse, so 2 is not a unit in $\mathbf{Z}$. The only units in $\mathbf{Z}$ are 1 and -1. Clearly, $\mathbf{Q}$ and $\mathbf{R}$ are fields. ∎

One has the natural concepts of a subring of a ring and a subfield of a field. A **subring of a ring** is a subset of the ring that is a ring under induced operations from the whole ring; a **subfield** is defined similarly for a subset of a field. In fact, let us say here once and for all that if we have a set, together with a certain specified type of algebraic structure on the set, the resulting conglomeration being a **glob** (group, ring, field, integral domain, vector space, and so on), then any subset of this set, together with a natural induced algebraic structure *that yields an algebraic structure of the same type*, is a **subglob.** If K and L are globs, we shall let $K \leq L$ denote that K is a subglob of L, and $K < L$ denote that $K \leq L$ but $K \neq L$.

Finally, we wish to caution you not to confuse our use of the words *unit* and *unity*. Unity is the multiplicative identity, while a unit is any element having a multiplicative inverse. Thus the multiplicative identity or unity is a unit, but not every unit is unity. For example, -1 is a unit in $\mathbf{Z}$, but -1 is not unity, that is, $-1 \neq 1$.

Exercises

23.1 State for which of the following sets the indicated operations of addition and multiplication are defined (the set is closed), and give a ring structure. If a ring is not formed, tell why this is the case.

a) $n\mathbf{Z}$ with the usual addition and multiplication

b) $\mathbf{Z}^+$ with the usual addition and multiplication

c) $\mathbf{Z} \times \mathbf{Z}$ with addition and multiplication by components

d) $2\mathbf{Z} \times \mathbf{Z}$ with addition and multiplication by components

e) $\{a + b\sqrt{2} \mid a, b \in \mathbf{Z}\}$ with the usual addition and multiplication

f) $\{a + b\sqrt{2} \mid a, b \in \mathbf{Q}\}$ with the usual addition and multiplication

g) The set of all pure imaginary complex numbers ri for $r \in \mathbf{R}$ with the usual addition and multiplication

23.2 For each part of Exercise 23.1 that gives a ring, state whether the ring is commutative, has unity, and is a field.

23.3 Describe all units in each of the following rings.

a) $\mathbf{Z}$ b) $\mathbf{Z} \times \mathbf{Z}$

c) $\mathbf{Z}_5$ d) $\mathbf{Q}$

e) $\mathbf{Z} \times \mathbf{Q} \times \mathbf{Z}$ f) $\mathbf{Z}_4$

†**23.4** Show that if U is the collection of all units in a ring $\langle R, +, \cdot \rangle$ with unity, then $\langle U, \cdot \rangle$ is a group. [*Warning:* Be sure to show that U is closed under multiplication.]

23.5 Show that $a^2 - b^2 = (a + b)(a - b)$ for all a and b in a ring R if and only if R is commutative.

23.6 Mark each of the following true or false.

___ a) Every field is also a ring.

___ b) Every ring has a multiplicative identity.

___ c) Every ring with unity has at least two units.

___ d) Every ring with unity has at most two units.

___ e) It is possible for a subset of some field to be a ring but not a subfield, under the induced operations.

___ f) The distributive laws for a ring are not very important.

___ g) Multiplication in a field is commutative.

___ h) The nonzero elements of a field form a group under the multiplication in the field.

— i) Addition in every ring is commutative.

— j) Every element in a ring has an additive inverse.

23.7 Let $(R, +)$ be an abelian group. Show that $(R, +, \cdot)$ is a ring if we define $ab = 0$ for all $a, b \in R$.

23.8 Show that the rings $2\mathbf{Z}$ and $3\mathbf{Z}$ are not isomorphic. Show that the fields $\mathbf{R}$ and $\mathbf{C}$ are not isomorphic.

23.9 (Freshman exponentiation) Let p be a prime. Show that in the ring Z_p you have $(a + b)^p = a^p + b^p$ for all $a, b \in Z_p$. [*Hint:* Observe that the usual binomial expansion for $(a + b)^n$ is valid in a *commutative* ring.]

23.10 Give an example of a ring with unity 1 that has a subring with unity $1' \neq 1$.

23.11 Show that the unity element in a subfield of a field must be the unity of the whole field, in contrast to Exercise 23.10 for rings.

23.12 Show that the multiplicative inverse of a unit in a ring with unity is unique.

23.13 An element a of a ring R is **nilpotent** if $a^n = 0$ for some $n \in Z^+$. Show that if a and b are nilpotent elements of a *commutative* ring, then $a + b$ is also nilpotent.

23.14 Show that a ring R has no nonzero nilpotent element if and only if 0 is the only solution of $x^2 = 0$ in R.

23.15 Show that a subset S of a ring R gives a subring of R if and only if the following hold:

$0 \in S$;

$(a - b) \in S$ for all $a, b \in S$;

$ab \in S$ for all $a, b \in S$.

23.16 a) Show that an intersection of subrings of a ring R is again a subring of R.
b) Show that an intersection of subfields of a field F is again a subfield of F.

23.17 Let R be a ring, and let a be a fixed element of R. Let $I_a = \{x \in R \mid ax = 0\}$. Show that I_a is a subring of R.

23.18 Let R be a ring, and let a be a fixed element of R. Let R_a be the subring of R that is the intersection of all subrings of R containing a (see Exercise 23.16). The ring R_a is the **subring of R generated by** a. Show that the abelian group $\langle R_a, + \rangle$ is generated (in the sense of Chapter 9) by $\{a^n \mid n \in Z^+\}$.

23.19 Consider $\langle S, +, \cdot \rangle$, where S is a set and $+$ and $\cdot$ are binary operations on S such that

$\langle S, + \rangle$ is a group,

$\langle S^*, \cdot \rangle$ is a group, where S^* consists of all elements of S except the additive identity,

$a(b + c) = (ab) + (ac)$ and $(a + b)c = (ac) + (bc)$ for all $a, b, c \in S$.

Show that $\langle S, +, \cdot \rangle$ is a division ring. [*Hint:* Apply the distributive laws to $(1 + 1)(a + b)$ to prove the commutativity of addition.]

23.20 A ring R is a **Boolean ring** if $a^2 = a$ for all $a \in R$. Show that every Boolean ring is commutative.

23.21 (For students having some knowledge of the laws of set theory.) For a set S let $\mathscr{P}(S)$ be the collection of all subsets of S. Define binary operations $+$ and $\cdot$ on $\mathscr{P}(S)$ by

$$A + B = (A \cup B) - (A \cap B) = \{x \mid x \in A \text{ or } x \in B \text{ but } x \notin (A \cap B)\}$$

and

$$A \cdot B = A \cap B$$

for $A, B \in \mathscr{P}(S)$.

a) Give the tables for $+$ and $\cdot$ for $\mathscr{P}(S)$, where $S = \{a, b\}$. [*Hint*: $\mathscr{P}(S)$ has four elements.]

b) Show that for *any* set S, $\langle \mathscr{P}(S), +, \cdot \rangle$ is a Boolean ring (see Exercise 23.20).

twenty-four

Integral Domains

While a careful treatment of polynomials is not given until Chapter 30, for purposes of motivation we shall make naive use of them in this section.

24.1 DIVISORS OF 0
AND CANCELLATION

One of the most important algebraic properties of our usual number system is that a product of two numbers can only be 0 if at least one of the factors is 0. The student uses this fact constantly, perhaps without realizing it. Suppose, for example, you are asked to solve the equation

$$x^2 - 5x + 6 = 0.$$

The first thing you do is to factor the left side:

$$x^2 - 5x + 6 = (x - 2)(x - 3).$$

Then you conclude that the only possible values for x are 2 and 3. Why? It is because if x is replaced by any number a, the product $(a - 2)(a - 3)$ of the resulting numbers is 0 if and only if either $a - 2 = 0$ or $a - 3 = 0$.

Example 24.1   Let us solve the equation $x^2 - 5x + 6 = 0$ in $\mathbf{Z}_{12}$. Now $x^2 - 5x + 6 = (x - 2)(x - 3)$ is still valid if we think of x as standing for any number in $\mathbf{Z}_{12}$. But in $\mathbf{Z}_{12}$, not only is $0a = a0 = 0$ for all $a \in \mathbf{Z}_{12}$, but also

$$(2)(6) = (6)(2) = (3)(4) = (4)(3) = (3)(8) = (8)(3)$$
$$= (4)(6) = (6)(4) = (4)(9) = (9)(4) = (6)(6) = (6)(8)$$
$$= (8)(6) = (6)(10) = (10)(6) = (8)(9) = (9)(8) = 0.$$

Thus our equation has not only 2 and 3 as solutions, but also 6 and 11, for $(6 - 2)(6 - 3) = (4)(3) = 0$ and $(11 - 2)(11 - 3) = (9)(8) = 0$ in $\mathbf{Z}_{12}$. ∎

These ideas are of such importance that we formalize them in a definition.

Definition   If a and b are two nonzero elements of a ring R such that $ab = 0$, then a and b are *divisors of* **0** (or **0** *divisors*). In particular, a is a *left divisor of* **0** and b is a *right divisor of* **0**.

In a commutative ring, every left divisor of 0 is also a right divisor of 0 and conversely. Thus there is no distinction between left and right divisors of 0 in a commutative ring.

Example 24.1 shows that in $\mathbf{Z}_{12}$ the elements 2, 3, 4, 6, 8, 9, and 10 are all divisors of 0. Note that these are exactly the numbers in $\mathbf{Z}_{12}$ that are not relatively prime to 12, that is, whose gcd with 12 is not 1. Our next theorem shows that this is an example of a general situation.

Theorem 24.1   *In the ring* $\mathbf{Z}_n$, *the divisors of* 0 *are precisely those elements that are not relatively prime to n.*

Proof.   Let $m \in \mathbf{Z}_n$, where $m \neq 0$, and let the gcd of m and n be $d \neq 1$. Then

$$m\left(\frac{n}{d}\right) = \left(\frac{m}{d}\right)n,$$

and $(m/d)n$ gives 0 as a multiple of n. Thus, $m(n/d) = 0$ in $\mathbf{Z}_n$, while neither m nor n/d is 0, so m is a divisor of 0.

On the other hand, suppose $m \in \mathbf{Z}_n$ is relatively prime to n. If for $s \in \mathbf{Z}_n$ we have $ms = 0$, then n divides the product ms of m and s as elements in the ring $\mathbf{Z}$. Since n has no factors > 1 in common with m, it must be that n divides s, so $s = 0$ in $\mathbf{Z}_n$. ∎

Corollary   *If p is a prime, then* $\mathbf{Z}_p$ *has no divisors of* 0.

Proof.   Proof of this corollary is immediate from Theorem 24.1. ∎

Another indication of the importance of the concept of 0 divisors is shown in the following theorem. Let R be a ring, and let a, b, $c \in R$. The **cancellation laws** hold in R if $ab = ac$, with $a \neq 0$, implies $b = c$, and $ba = ca$, with $a \neq 0$, implies $b = c$. These are multiplicative cancellation laws. Of course, the additive cancellation laws hold in R, since $\langle R, + \rangle$ is a group.

Theorem 24.2 *The cancellation laws hold in a ring R if and only if R has no left or right divisors of 0.*

Proof. Let R be a ring in which the cancellation laws hold, and suppose $ab = 0$ for some a, $b \in R$. We must show that either a or b is 0. If $a \neq 0$, then $ab = a0$ implies that $b = 0$ by the cancellation laws. Similarly, $b \neq 0$ implies that $a = 0$, so there can be no left or right divisors of 0 if the cancellation laws hold.

Conversely, suppose that R has no left or right divisors of 0, and suppose that $ab = ac$ with $a \neq 0$. Then

$$ab - ac = a(b - c) = 0.$$

Since $a \neq 0$, and since R has no left divisors of 0, we must have $b - c = 0$, so $b = c$. A similar argument shows that $ba = ca$, with $a \neq 0$, implies $b = c$. ∎

Suppose that R is a ring with no divisors of 0. Then an equation $ax = b$, with $a \neq 0$, in R can have at most one solution x in R, for if $ax_1 = b$ and $ax_2 = b$, then $ax_1 = ax_2$, and by Theorem 24.2 $x_1 = x_2$, since R has no divisors of 0. If R has unity 1 and a is a unit in R with multiplicative inverse a^{-1}, then clearly the solution x of $ax = b$ is $a^{-1}b$. In the case that R is commutative, in particular if R is a field, it is customary to denote $a^{-1}b$ and ba^{-1} (they are equal by commutativity) by the formal quotient b/a. This quotient notation must not be used in the event that R is not commutative, for then one does not know whether b/a denotes $a^{-1}b$ or ba^{-1}. In a field F it is usual to *define* a **quotient** b/a, where $a \neq 0$, as the solution x in F of the equation $ax = b$. This definition is consistent with our preceding remarks, and we shall be using this quotient notation when we work in a field. In particular, the multiplicative inverse of a nonzero element a of a field is $1/a$.

24.2 INTEGRAL DOMAINS

Definition An *integral domain* D is a commutative ring with unity containing no divisors of 0.

Thus, if the coefficients of a polynomial are from an integral domain, one can solve a polynomial equation in which the polynomial can be factored into linear factors in the usual fashion by setting each factor equal to 0.

In our hierarchy of algebraic structures, an integral domain belongs between a commutative ring with unity and a field, as we shall now show. Theorem 24.2 shows that the cancellation laws for multiplication hold in an intergral domain. We have seen that $\mathbf{Z}$ and $\mathbf{Z}_p$ for any prime p are integral domains, but $\mathbf{Z}_n$ is not an integral domain if n is not prime.

Theorem 24.3   *Every field F is an integral domain.*

Proof.   Let $a, b \in F$, and suppose that $a \neq 0$. Then if $ab = 0$, we have

$$\left(\frac{1}{a}\right)(ab) = \left(\frac{1}{a}\right)0 = 0.$$

But then

$$0 = \left(\frac{1}{a}\right)(ab) = \left[\left(\frac{1}{a}\right)a\right]b = 1b = b.$$

We have shown that $ab = 0$ with $a \neq 0$ implies that $b = 0$ in F, so there are no divisors of 0 in F. Of course, F is a commutative ring with unity, so our theorem is proved. ∎

Thus far the only fields we have seen are $\mathbf{Q}$, $\mathbf{R}$, and $\mathbf{C}$. The corollary of the next theorem will give us some fields of finite order!

Theorem 24.4   *Every finite integral domain is a field.*

Proof.   Let

$$0, 1, a_1, \ldots, a_n$$

be all the elements of a finite integral domain D. We need to show that for $a \in D$, where $a \neq 0$, there exists $b \in D$ such that $ab = 1$. Now consider

$$a1, aa_1, \ldots, aa_n.$$

We claim that all these elements of D are distinct, for $aa_i = aa_j$ implies that $a_i = a_j$, by the cancellation laws that hold in an integral domain. Also, since D has no 0 divisors, none of these elements is 0. Hence by counting, we find that $a1, aa_1, \ldots, aa_n$ are the elements $1, a_1, \ldots, a_n$ in some order, so that either $a1 = 1$, that is, $a = 1$, or $aa_i = 1$ for some i. Thus a has a multiplicative inverse. ∎

Corollary   *If p is a prime, then $\mathbf{Z}_p$ is a field.*

Proof.   Proof of this corollary follows immediately from the fact that $\mathbf{Z}_p$ is an integral domain and from Theorem 24.4. ∎

24.3 THE CHARACTERISTIC OF A RING

Let R be any ring. We might ask whether there is a positive integer n such that $n \cdot a = 0$ for all $a \in R$, where $n \cdot a$ means $a + a + \cdots + a$ for n summands, as explained in Section 23.1. For example, the integer m has this property for the ring $\mathbf{Z}_m$.

Definition   If for a ring R a positive integer n exists such that $n \cdot a = 0$ for all $a \in R$, then the least such positive integer is the *characteristic of the ring* R. If no such positive integer exists, then R is of *characteristic* 0.

We shall be using the concept of a characteristic chiefly for fields.

Example 24.2   The ring $\mathbf{Z}_n$ is of characteristic n, while $\mathbf{Z}$, $\mathbf{Q}$, $\mathbf{R}$, and $\mathbf{C}$ all have characteristic 0. ∎

Theorem 24.5   *If R is a ring with unity 1, then R has characteristic $n > 0$ if and only if n is the smallest positive integer such that $n \cdot 1 = 0$.*

Proof.   By definition, if R has characteristic $n > 0$, then $n \cdot a = 0$ for all $a \in R$, so in particular $n \cdot 1 = 0$.

Conversely, suppose that n is a positive integer such that $n \cdot 1 = 0$. Then for any $a \in R$, we have

$$n \cdot a = a + a + \cdots + a = a(1 + 1 + \cdots + 1) = a(n \cdot 1) = a0 = 0.$$

Our theorem follows directly. ∎

24.4 FERMAT'S THEOREM

We conclude this section with some elegant applications to number theory. It is easy to see that *for any field, the nonzero elements form a group under the field multiplication*. In particular, for $\mathbf{Z}_p$, the elements

$$1, 2, 3, \ldots, p - 1$$

form a group of order $p - 1$ under multiplication modulo p. Since the order of any element in a group divides the order of the group, we see for $a \neq 0$ and $a \in \mathbf{Z}_p$ that $a^{p-1} = 1$ in $\mathbf{Z}_p$. We shall see in detail later, for multiplication as well as addition, that $a \in \mathbf{Z}_p$ can be viewed as representing the coset $a + p\mathbf{Z}$ and that the product of cosets can be computed by multiplication modulo p of representatives in a fashion analogous to the way we compute sums. The collection $\mathbf{Z}/p\mathbf{Z}$ of these cosets becomes a ring isomorphic to $\mathbf{Z}_p$. Let us assume this now. This gives us immediately the so-called Little Theorem of Fermat.

Theorem 24.6 (Fermat) *If $a \in \mathbf{Z}$ and p is a prime not dividing a, then p divides $a^{p-1} - 1$, that is, $a^{p-1} \equiv 1 \pmod{p}$ for $a \not\equiv 0 \pmod{p}$.*

Corollary *If $a \in \mathbf{Z}$, then $a^p \equiv a \pmod{p}$ for any prime p.*

Proof. Proof of the corollary follows from Theorem 24.6, if $a \not\equiv 0 \pmod{p}$. If $a \equiv 0 \pmod{p}$, then both sides reduce to 0 modulo p. ∎

This corollary will be of great importance to us later in our work with finite fields.

Example 24.3 Let us compute the remainder of 8^{103} when divided by 13. Using Fermat's theorem, we have

$$8^{103} \equiv (8^{12})^8(8^7) \equiv (1^8)(8^7) \equiv 8^7 \equiv (-5)^7$$
$$\equiv (25)^3(-5) \equiv (-1)^3(-5) \equiv 5 \pmod{13}. \blacksquare$$

*24.5 EULER'S GENERALIZATION

Euler gave a generalization of Fermat's theorem. His generalization will follow at once from our next theorem.

Theorem 24.7 *The set G_n of nonzero elements of $\mathbf{Z}_n$ that are not 0 divisors forms a group under multiplication modulo n.*

Proof. First we must show that G_n is closed under multiplication modulo n. Let a, $b \in G_n$. If $ab \notin G_n$, then there would exist $c \neq 0$ in $\mathbf{Z}_n$ such that $(ab)c = 0$. Now $(ab)c = 0$ implies that $a(bc) = 0$. Since $b \in G_n$ and $c \neq 0$, we have $bc \neq 0$ by definition of G_n. But then $a(bc) = 0$ would imply that $a \notin G_n$ contrary to assumption. *Note that we have shown that for any ring the set of elements that are not divisors of 0 is closed under multiplication.* No structure of $\mathbf{Z}_n$ other than ring structure has been involved so far.

We now show that G_n is a group. Of course, multiplication modulo n is associative, and $1 \in G_n$. It remains to show that for $a \in G_n$, there is $b \in G_n$ such that $ab = 1$. Let

$$1, a_1, \ldots, a_r$$

be the elements of G_n. The elements

$$a1, aa_1, \ldots, aa_r$$

are all different, for if $aa_i = aa_j$, then $a(a_i - a_j) = 0$, and since $a \in G_n$ and thus is not a divisor of 0, we must have $a_i - a_j = 0$ or $a_i = a_j$. Therefore by counting, we find that either $a1 = 1$, or some aa_i must be 1, so a has a multiplicative inverse. ∎

Note that the only property of $\mathbf{Z}_n$ used in this last theorem, other than the fact that it was a ring with unity, was that it was finite. In both Theorem 24.4 and Theorem 24.7 we have (in essentially the same construction) employed a counting argument. *Counting arguments are often simple, but they are among the most powerful tools of all mathematics.*

Let $\varphi(n)$ be defined as the number of positive integers less than or equal to n and relatively prime to n. For example, if $n = 12$, the positive integers less than or equal to 12 and relatively prime to 12 are 1, 5, 7, and 11, so $\varphi(12) = 4$. By Theorem 24.1, $\varphi(n)$ is the number of elements of $\mathbf{Z}_n$ that are not divisors of 0. This function $\varphi: \mathbf{Z}^+ \to \mathbf{Z}^+$ is the **Euler phi-function.** We can now describe Euler's generalization of Fermat's theorem.

Theorem 24.8 (Euler) *If a is an integer relatively prime to n, then $a^{\varphi(n)} - 1$ is divisible by n, that is, $a^{\varphi(n)} \equiv 1 \pmod{n}$.*

Proof. If a is relatively prime to n, then the coset $a + n\mathbf{Z}$ of $n\mathbf{Z}$ containing a contains an integer $b < n$ and relatively prime to n. Using the fact (which will be proved later) that multiplication of these cosets by multiplication modulo n of representatives is well defined, we have

$$a^{\varphi(n)} \equiv b^{\varphi(n)} \pmod{n}.$$

But by Theorems 24.1 and 24.7, b can be viewed as an element of the multiplicative group G_n of order $\varphi(n)$ consisting of the $\varphi(n)$ elements of $\mathbf{Z}_n$ relatively prime to n. Thus

$$b^{\varphi(n)} \equiv 1 \pmod{n},$$

and our theorem follows. ∎

Exercises

24.1 Find all solutions of the equation $x^3 - 2x^2 - 3x = 0$ in $\mathbf{Z}_{12}$.

24.2 Solve the equation $3x = 2$ in the field $\mathbf{Z}_7$; in the field $\mathbf{Z}_{23}$.

24.3 Find the characteristic of each of the following rings.

a) $2\mathbf{Z}$ b) $\mathbf{Z} \times \mathbf{Z}$

c) $\mathbf{Z}_3 \times 3\mathbf{Z}$ d) $\mathbf{Z}_3 \times \mathbf{Z}_3$

e) $\mathbf{Z}_3 \times \mathbf{Z}_4$ f) $\mathbf{Z}_6 \times \mathbf{Z}_{15}$

24.4 Using Fermat's theorem, find the remainder of 3^{47} when divided by 23.

†**24.5** a) Show that 1 and $p - 1$ are the only elements of the field $\mathbf{Z}_p$ that are their own multiplicative inverse. [*Hint:* Consider the equation $x^2 - 1 = 0$.]

b) From part (a), deduce the half of *Wilson's theorem* that states that if p is a prime, then $(p - 1)! \equiv -1 \pmod{p}$. (The other half states that if $(n - 1)! \equiv -1 \pmod{n}$, then n is a prime.)

24.6 Mark each of the following true or false.

___ a) $n\mathbf{Z}$ has zero divisors if n is not prime.

___ b) Every field is an integral domain.

___ c) The characteristic of $n\mathbf{Z}$ is n.

___ d) As a ring, $\mathbf{Z}$ is isomorphic to $n\mathbf{Z}$ for all $n \geq 1$.

___ e) The cancellation law holds in any ring that is isomorphic to an integral domain.

___ f) Every integral domain of characteristic 0 is infinite.

___ g) The direct product of two integral domains is again an integral domain.

___ h) A divisor of zero in a commutative ring with unity can have no multiplicative inverse.

___ i) $n\mathbf{Z}$ is a subdomain of $\mathbf{Z}$.

___ j) $\mathbf{Z}$ is a subfield of $\mathbf{Q}$.

24.7 Find all solutions of the equation $x^2 + 2x + 2 = 0$ in $\mathbf{Z}_6$; of the equation $x^2 + 2x + 4 = 0$ in $\mathbf{Z}_6$.

24.8 An element a of a ring R is **idempotent** if $a^2 = a$. Show that a division ring contains exactly two idempotent elements.

24.9 Show that an intersection of subdomains of an integral domain D is again a subdomain of D.

24.10 Show that a finite ring R with unity and no divisors of 0 is a division ring. (It is actually a field, although commutativity is not easy to prove. See Theorem 25.5.) [*Note:* In your proof, to show that $a \neq 0$ is a unit, you must show that a "left multiplicative inverse" of $a \neq 0$ in R is also a "right multiplicative inverse."]

24.11 Let R be a ring that contains at least two elements. Suppose for each nonzero $a \in R$, there exists a unique $b \in R$ such that $aba = b$.

a) Show that R has no divisors of 0.

b) Show that $bab = b$.

c) Show that R has unity.

d) Show that R is a division ring.

24.12 Show that the characteristic of a subdomain of an integral domain D is equal to the characteristic of D.

24.13 Show that if D is an integral domain, then $\{n \cdot 1 \mid n \in \mathbf{Z}\}$ is a subdomain of D contained in every subdomain of D.

24.14 Show that the characteristic of an integral domain D must be either 0 or a prime p. [*Hint:* If the characteristic of D is mn, consider $(m \cdot 1)(n \cdot 1)$ in D.]

24.15 Use Fermat's theorem to show that for any positive integer n, $n^{37} - n$ is divisible by 383838. [*Hint:* $383838 = (37)(19)(13)(7)(3)(2)$.]

24.16 This exercise shows that every ring R can be enlarged (if necessary) to a ring S with unity, having the same characteristic as R. Let $S = R \times Z$ if R has characteristic 0, and $R \times Z_n$ if R has characteristic n. Let addition in S be the usual addition by components, and let multiplication be defined by

$$(r_1, n_1)(r_2, n_2) = (r_1 r_2 + n_1 \cdot r_2 + n_2 \cdot r_1, n_1 n_2)$$

where $n \cdot r$ has the meaning explained in Chapter 23.

a) Show that S is a ring.

b) Show that S has unity.

c) Show that S and R have the same characteristic.

d) Show that the map $\phi : R \to S$ given by $r\phi = (r, 0)$ for $r \in R$ is an isomorphism of R with a subring of S.

*24.17 Give the group multiplication table for the multiplicative group of those elements of Z_{12} relatively prime to 12. To which group of order 4 is it isomorphic?

*24.18 Make a table of the values of $\varphi(n)$ for $n \leq 30$.

*24.19 Use Euler's generalization of Fermat's theorem to find the remainder of 7^{1000} when divided by 24.

twenty-five

Some Noncommutative Examples

We shall be doing almost nothing with noncommutative rings and skew fields. So that you may realize that there are many important noncommutative rings occurring very naturally in algebra, in this chapter we give several examples of such rings.

*25.1 MATRICES OVER A FIELD

Let F be any field (say $\mathbf{Q}$, $\mathbf{R}$, or $\mathbf{C}$), and consider the set $M_2(F)$ of all 2×2 square arrays

$$(a_{ij}) = \begin{pmatrix} a_{11} & a_{12} \\ a_{21} & a_{22} \end{pmatrix},$$

where the a_{ij} are all in F. The first subscript i of a_{ij} indicates the *row* in which a_{ij} occurs in the square array, and the second subscript j indicates the *column*. Thus a_{12} is the element of F in the first row and second column of the square

array. Such a square array is a **2 × 2 matrix over** F. The set $M_n(F)$ of all
n × n matrices over F is similarly defined.

We define matrix addition on $M_2(F)$ by

$$\begin{pmatrix} a_{11} & a_{12} \\ a_{21} & a_{22} \end{pmatrix} + \begin{pmatrix} b_{11} & b_{12} \\ b_{21} & b_{22} \end{pmatrix} = \begin{pmatrix} a_{11} + b_{11} & a_{12} + b_{12} \\ a_{21} + b_{21} & a_{22} + b_{22} \end{pmatrix},$$

that is, by adding corresponding entries. After a few moments of thought,
you will see from the field axioms for F that $\langle M_2(F), + \rangle$ is an abelian group,
with additive identity

$$\begin{pmatrix} 0 & 0 \\ 0 & 0 \end{pmatrix},$$

and with

$$-\begin{pmatrix} a_{11} & a_{12} \\ a_{21} & a_{22} \end{pmatrix} = \begin{pmatrix} -a_{11} & -a_{12} \\ -a_{21} & -a_{22} \end{pmatrix}.$$

Matrix multiplication on $M_2(F)$ is defined by

$$\begin{pmatrix} a_{11} & a_{12} \\ a_{21} & a_{22} \end{pmatrix}\begin{pmatrix} b_{11} & b_{12} \\ b_{21} & b_{22} \end{pmatrix} = \begin{pmatrix} a_{11}b_{11} + a_{12}b_{21} & a_{11}b_{12} + a_{12}b_{22} \\ a_{21}b_{11} + a_{22}b_{21} & a_{21}b_{12} + a_{22}b_{22} \end{pmatrix}.$$

This multiplication looks difficult, and is best remembered by

$$(a_{ij})(b_{ij}) = (c_{ij}),$$

where

$$c_{rs} = \sum_{i=1}^{2} a_{ri}b_{is}.$$

With the analogous definition for a matrix multiplication in which the sum
goes from $i = 1$ to n and the obvious analogous definition of matrix addition,
everything we have done is valid for the set $M_n(F)$ of all $n \times n$ matrices over F.

Example 25.1   In $M_2(Q)$.

$$\begin{pmatrix} 2 & 1 \\ -3 & 4 \end{pmatrix} + \begin{pmatrix} -1 & 0 \\ 2 & -5 \end{pmatrix} = \begin{pmatrix} 1 & 1 \\ -1 & -1 \end{pmatrix}$$

and

$$\begin{pmatrix} 2 & 1 \\ -3 & 4 \end{pmatrix}\begin{pmatrix} -1 & 0 \\ 2 & -5 \end{pmatrix} = \begin{pmatrix} 0 & -5 \\ 11 & -20 \end{pmatrix}. \ \blacksquare$$

To show that $\langle M_n(F), +, \cdot \rangle$ is a ring, it remains to prove the associative
and distributive laws. We illustrate with the associative law for matrix
multiplication in $M_n(F)$. Using field properties of F and the definition of

matrix multiplication in $M_n(F)$, if d_{rs} is an entry in $(a_{ij})[(b_{ij})(c_{ij})]$, we have

$$d_{rs} = \sum_{k=1}^{n} a_{rk}\left(\sum_{j=1}^{n} b_{kj}c_{js}\right) = \sum_{j=1}^{n}\left(\sum_{k=1}^{n} a_{rk}b_{kj}\right)c_{js} = e_{rs},$$

where e_{rs} is the entry in the rth row and sth column of $[(a_{ij})(b_{ij})](c_{ij})$. The distributive laws are similarly proved. We consider the following theorem to have been proved.

Theorem 25.1   *If F is a field, then the set $M_n(F)$ of all $n \times n$ matrices with entries from F forms a ring under matrix addition and multiplication.*

These rings of matrices are considered in linear algebra. In this context, they can be viewed as corresponding to certain functions, and matrix multiplication, when viewed in this light, can be shown to be just function composition. Since function composition is always associative, this gives another, more elegant, demonstration of the associative law.

Note that $M_1(F)$ is isomorphic to F under the map $\phi : F \to M_1(F)$ given by $a\phi = (a)$ for $a \in F$. Remember that this section is supposed to be concerned with noncommutative rings. It is true that $M_n(F)$ is noncommutative if $n \geq 2$. Example 25.2 illustrates this for $M_2(F)$.

Example 25.2   Since every field F contains elements 0 and 1, $M_2(F)$ always has among its elements

$$\begin{pmatrix} 0 & 1 \\ 0 & 0 \end{pmatrix} \quad \text{and} \quad \begin{pmatrix} 0 & 0 \\ 0 & 1 \end{pmatrix}.$$

The definition of matrix multiplication shows that

$$\begin{pmatrix} 0 & 1 \\ 0 & 0 \end{pmatrix}\begin{pmatrix} 0 & 0 \\ 0 & 1 \end{pmatrix} = \begin{pmatrix} 0 & 1 \\ 0 & 0 \end{pmatrix},$$

while

$$\begin{pmatrix} 0 & 0 \\ 0 & 1 \end{pmatrix}\begin{pmatrix} 0 & 1 \\ 0 & 0 \end{pmatrix} = \begin{pmatrix} 0 & 0 \\ 0 & 0 \end{pmatrix}.$$

Thus $M_2(F)$ is noncommutative. Since

$$\begin{pmatrix} 0 & 0 \\ 0 & 0 \end{pmatrix}$$

is the additive identity, this example also shows that there exist divisors of 0 in $M_2(F)$. The same is true of $M_n(F)$ for $n \geq 2$. We leave as an exercise the demonstration that

$$\begin{pmatrix} 1 & 0 \\ 0 & 1 \end{pmatrix}$$

is unity in $M_2(F)$ (see Exercise 25.2). ■

*25.2 RINGS OF ENDOMORPHISMS

Let A be any abelian group. A homomorphism of A into itself is an **endomorphism of** A. Let the set of all endomorphisms of A be $Hom(A)$. Since the composition of two homomorphisms of A into itself is again such a homomorphism, we define multiplication on $Hom(A)$ by function composition, and thus multiplication is associative.

To define addition, for $\phi, \psi \in Hom(A)$, we have to describe the value of $(\phi + \psi)$ on each $a \in A$. Define

$$a(\phi + \psi) = (a\phi) + (a\psi).$$

Since

$$(a + b)(\phi + \psi) = (a + b)\phi + (a + b)\psi = (a\phi + b\phi) + (a\psi + b\psi)$$
$$= (a\phi + a\psi) + (b\phi + b\psi) = a(\phi + \psi) + b(\phi + \psi),$$

we see that $\phi + \psi$ is again in $Hom(A)$.

Since A is commutative, we have

$$a(\phi + \psi) = (a\phi) + (a\psi) = (a\psi) + (a\phi) = a(\psi + \phi)$$

for all $a \in A$, so $\phi + \psi = \psi + \phi$ and addition in $Hom(A)$ is commutative. The associativity of addition follows from

$$a[\phi + (\psi + \theta)] = a\phi + a(\psi + \theta) = a\phi + (a\psi + a\theta)$$
$$= (a\phi + a\psi) + a\theta = a(\phi + \psi) + a\theta = a[(\phi + \psi) + \theta].$$

If e is the additive identity of A, then the homomorphism 0 defined by

$$a0 = e$$

for $a \in A$ is clearly an additive identity in $Hom(A)$. Finally, for

$$\phi \in Hom(A),$$

$-\phi$ defined by

$$a(-\phi) = -(a\phi)$$

is in $Hom(A)$, since

$$(a + b)(-\phi) = -((a + b)\phi) = -(a\phi + b\phi)$$
$$= -(a\phi) + (-(b\phi)) = a(-\phi) + b(-\phi).$$

Clearly, $\phi + (-\phi) = 0$. Thus $\langle Hom(A), + \rangle$ is an abelian group.

Note that we have not yet used the fact that our functions are *homomorphisms* except to show that $\phi + \psi$ and $-\phi$ are again *homomorphisms*. Thus the set A^A of *all functions* from A into A is an abelian group under exactly the same definition of addition, and, of course, function composition again gives a nice associative multiplication in A^A. However, we do need the fact that these functions in $Hom(A)$ are homomorphisms now to prove the right distributive law in $Hom(A)$. Except for this right distributive law, $\langle A^A, +, \cdot \rangle$ satisfies all the axioms for a ring. Let ϕ, ψ, and θ be in $Hom(A)$, and let $a \in A$. Then

$$a[(\phi + \psi)\theta] = [a(\phi + \psi)]\theta = (a\phi + a\psi)\theta.$$

Since θ is a *homomorphism*,

$$(a\phi + a\psi)\theta = (a\phi)\theta + (a\psi)\theta = a(\phi\theta) + a(\psi\theta) = a(\phi\theta + \psi\theta).$$

Thus $(\phi + \psi)\theta = \phi\theta + \psi\theta$. The left distributive law causes no trouble, even in A^A, and follows from

$$a[\phi(\psi + \theta)] = a\phi(\psi + \theta) = (a\phi)\psi + (a\phi)\theta = a(\phi\psi) + a(\phi\theta) = a(\phi\psi + \phi\theta).$$

Thus we have proved the following theorem.

> **Theorem 25.2** The set $Hom(A)$ of all endomorphisms of an abelian group A forms a ring under homomorphism addition and homomorphism multiplication (function composition).

Again, to show applicability to this chapter, we should give an example showing that $Hom(A)$ need not be commutative. Since function composition is in general not commutative, this seems reasonable to expect. However, $Hom(A)$ may be commutative in some cases. In fact, $Hom(\langle \mathbf{Z}, + \rangle)$ is commutative. We ask the student to show this in Exercise 25.10.

Example 25.3 Consider the free abelian group $\langle \mathbf{Z} \times \mathbf{Z}, + \rangle$ discussed in Part I. We can specify an endomorphism of this free abelian group by giving its values on the generators $(1, 0)$ and $(0, 1)$ of the group. Define

$$\phi \in Hom(\langle \mathbf{Z} \times \mathbf{Z}, + \rangle)$$

by

$$(1, 0)\phi = (1, 0) \quad \text{and} \quad (0, 1)\phi = (1, 0).$$

Define ψ by

$$(1, 0)\psi = (0, 0) \quad \text{and} \quad (0, 1)\psi = (0, 1).$$

Naively, ϕ maps everything onto the first factor of $\mathbf{Z} \times \mathbf{Z}$, and ψ collapses the first factor. Thus

$$(n, m)(\phi\psi) = (n + m, 0)\psi = (0, 0),$$

while

$$(n, m)(\psi\phi) = (0, m)\phi = (m, 0).$$

Hence $\phi\psi \neq \psi\phi$. ∎

Example 25.4 Let F be a field and let $F[x]$ be the set of all formal polynomial expressions with coefficients in F. (We will discuss such polynomials in Chapter 30. Here we build on your intuition.) A typical element of $F[x]$

can be written in the form

$$a_0 + a_1 x + a_2 x^2 + \cdots + a_n x^n$$

where $a_0, a_1, a_2, \ldots, a_n \in F$. With the usual polynomial addition, $F[x]$ becomes an abelian group, so we can consider $Hom(F[x])$. One element of $Hom(F[x])$ acts on each polynomial in $F[x]$ by multiplying it by x. Let this endomorphism be X, so

$$(a_0 + a_1 x + a_2 x^2 + \cdots + a_n x^n)X = a_0 x + a_1 x^2 + a_2 x^3 + \cdots + a_n x^{n+1}.$$

Another element of $Hom(F[x])$ is formal differentiation with respect to x. (The familiar formula "the derivative of a sum is the sum of the derivatives" guarantees that differentiation is an endomorphism of $F[x]$.) Let Y be this endomorphism, so

$$(a_0 + a_1 x + a_2 x^2 + \cdots + a_n x^n)Y = a_1 + 2a_2 x + \cdots + na_n x^{n-1}.$$

In Exercise 25.12, we ask you to show that $XY - YX = 1$, where 1 is unity (the identity map) in $Hom(F[x])$. Thus $XY \neq YX$. Multiplication of polynomials in $F[x]$ by any element of F also gives an element of $Hom(F[x])$. The subring of $Hom(F[x])$ generated by X and Y and multiplications by elements of F is the Weyl Algebra and is important in quantum mechanics. ∎

*25.3 GROUP RINGS AND GROUP ALGEBRAS

Let $G = \{g_i \mid i \in I\}$ be any multiplicative group, and let R be any commutative ring with unity. Let $R(G)$ be the set of all *formal sums*

$$\sum_{i \in I} a_i g_i$$

for $a_i \in R$ and $g_i \in G$, *where all but a finite number of the a_i are* 0. Define the sum of two elements of $R(G)$ by

$$\left(\sum_{i \in I} a_i g_i \right) + \left(\sum_{i \in I} b_i g_i \right) = \sum_{i \in I} (a_i + b_i) g_i.$$

It is clear that $(a_i + b_i) = 0$ except for a finite number of indices i, so $\sum_{i \in I} (a_i + b_i) g_i$ is again in $R(G)$. It is immediate that $\langle R(G), + \rangle$ is an abelian group with additive identity $\sum_{i \in I} 0 g_i$.

Multiplication of two elements of $R(G)$ is defined by the use of the multiplications in G and R as follows:

$$\left(\sum_{i \in I} a_i g_i\right)\left(\sum_{i \in I} b_i g_i\right) = \sum_{i \in I}\left(\sum_{g_j g_k = g_i} a_j b_k\right) g_i.$$

Naively, we formally distribute the sum $\sum_{i \in I} a_i g_i$ over the sum $\sum_{i \in I} b_i g_i$ and rename a term $a_j g_j b_k g_k$ by $a_j b_k g_i$, where $g_j g_k = g_i$ in G. Since a_i and b_i are 0 for all but a finite number of i, the sum $\sum_{g_j g_k = g_i} a_j b_k$ contains only a finite number of nonzero summands $a_j b_k \in R$ and may thus be viewed as an element of R. Clearly, again at most a finite number of such sums $\sum_{g_j g_k = g_i} a_j b_k$ are nonzero. Thus multiplication is closed on $R(G)$.

The distributive laws follow at once from the definition of addition and the formal way we used distributivity to define multiplication. For the associativity of multiplication

$$\left(\sum_{i \in I} a_i g_i\right)\left[\left(\sum_{i \in I} b_i g_i\right)\left(\sum_{i \in I} c_i g_i\right)\right] = \left(\sum_{i \in I} a_i g_i\right)\left[\sum_{i \in I}\left(\sum_{g_j g_k = g_i} b_j c_k\right) g_i\right]$$

$$= \sum_{i \in I}\left(\sum_{g_h g_j g_k = g_i} a_h b_j c_k\right) g_i$$

$$= \left[\sum_{i \in I}\left(\sum_{g_h g_j = g_i} a_h b_j\right) g_i\right]\left(\sum_{i \in I} c_i g_i\right)$$

$$= \left[\left(\sum_{i \in I} a_i g_i\right)\left(\sum_{i \in I} b_i g_i\right)\right]\left(\sum_{i \in I} c_i g_i\right).$$

Thus we have proved the following theorem.

Theorem 25.3 *If G is any multiplicative group, then $\langle R(G), +, \cdot \rangle$ is a ring.*

If we rename the element $\sum_{i \in I} a_i g_i$ of $R(G)$, where $a_i = 0$ for $i \neq j$ and $a_j = 1$, by g_j, we see that $\langle R(G), \cdot \rangle$ can be considered to contain G naturally as a multiplicative subsystem. Thus, if G is not abelian, $R(G)$ will not be a commutative ring.

Definition The ring $R(G)$ defined above is the *group ring of G over R*. If F is a field, then $F(G)$ is the *group algebra of G over F*.

Example 25.5 Let us give the addition and multiplication tables for the group algebra $\mathbf{Z}_2(G)$, where $G = \{e, a\}$ is cyclic of order 2. The elements of $\mathbf{Z}_2(G)$ are

$$0e + 0a, \qquad 0e + 1a, \qquad 1e + 0a, \qquad \text{and} \qquad 1e + 1a.$$

Table 25.1

+	0	a	e	e + a
0	0	a	e	e + a
a	a	0	e + a	e
e	e	e + a	0	a
e + a	e + a	e	a	0

Table 25.2

	0	a	e	e + a
0	0	0	0	0
a	0	e	a	e + a
e	0	a	e	e + a
e + a	0	e + a	e + a	0

If we denote these elements in the obvious, natural way by

$$0, \quad a, \quad e, \quad \text{and} \quad e + a,$$

respectively, we get Tables 25.1 and 25.2.
For example, to see that $(e + a)(e + a) = 0$, we have

$$(1e + 1a)(1e + 1a) = (1 + 1)e + (1 + 1)a = 0e + 0a.$$

This example shows that a group algebra may have 0 divisors. Indeed, this is usually the case. ∎

*25.4 THE QUATERNIONS

You have not yet seen an example of a skew field. The *quaternions* of Hamilton are the standard example of a skew field; let us describe them.

Let the set $\mathcal{2}$ be $\mathbf{R} \times \mathbf{R} \times \mathbf{R} \times \mathbf{R}$. Now $\langle \mathbf{R} \times \mathbf{R} \times \mathbf{R} \times \mathbf{R}, + \rangle$ is a group under addition by components, the direct product of $\mathbf{R}$ under addition with itself four times. This gives the operation of addition on $\mathcal{2}$. Let us rename certain elements of $\mathcal{2}$. We shall let

$$1 = (1, 0, 0, 0), \quad i = (0, 1, 0, 0),$$
$$j = (0, 0, 1, 0), \quad \text{and} \quad k = (0, 0, 0, 1).$$

We furthermore agree to let

$$a_1 = (a_1, 0, 0, 0), \quad a_2 i = (0, a_2, 0, 0),$$
$$a_3 j = (0, 0, a_3, 0), \quad \text{and} \quad a_4 k = (0, 0, 0, a_4).$$

In view of our definition of addition, we then have

$$(a_1, a_2, a_3, a_4) = a_1 + a_2 i + a_3 j + a_4 k.$$

Thus

$$(a_1 + a_2 i + a_3 j + a_4 k) + (b_1 + b_2 i + b_3 j + b_4 k)$$
$$= (a_1 + b_1) + (a_2 + b_2)i + (a_3 + b_3)j + (a_4 + b_4)k.$$

To define multiplication on $\mathcal{Q}$, we start by defining

$$1a = a1 = a \qquad \text{for} \quad a \in \mathcal{Q},$$
$$i^2 = j^2 = k^2 = -1,$$

and

$$ij = k, \qquad jk = i, \qquad ki = j, \qquad ji = -k, \qquad kj = -i, \qquad \text{and} \qquad ik = -j.$$

You should note the similarity with the so-called cross product of vectors. These formulas are easy to remember if you think of the sequence

$$i, j, k, i, j, k.$$

The product from left to right of two adjacent elements is the next one to the right. The product from right to left of two adjacent elements is the negative of the next one to the left. One then defines a product to be what it must be to make the distributive laws hold, namely

$$(a_1 + a_2 i + a_3 j + a_4 k)(b_1 + b_2 i + b_3 j + b_4 k)$$
$$= (a_1 b_1 - a_2 b_2 - a_3 b_3 - a_4 b_4) + (a_1 b_2 + a_2 b_1 + a_3 b_4 - a_4 b_3)i$$
$$+ (a_1 b_3 - a_2 b_4 + a_3 b_1 + a_4 b_2)j$$
$$+ (a_1 b_4 + a_2 b_3 - a_3 b_2 + a_4 b_1)k.$$

Verification that $\mathcal{Q}$ is a skew field is now a tedious chore, some of which is assigned in an exercise. Since $ij = k$ and $ji = -k$, we see that multiplication is not commutative, so $\mathcal{Q}$ is definitely not a field. The only axiom that cannot be verified mechanically is the existence of a multiplicative inverse for $a = a_1 + a_2 i + a_3 j + a_4 k$, with not all $a_i = 0$. The student can check that

$$(a_1 + a_2 i + a_3 j + a_4 k)(a_1 - a_2 i - a_3 j - a_4 k) = a_1^2 + a_2^2 + a_3^2 + a_4^2.$$

If we let

$$|a|^2 = a_1^2 + a_2^2 + a_3^2 + a_4^2 \qquad \text{and} \qquad \bar{a} = a_1 - a_2 i - a_3 j - a_4 k,$$

we see that

$$\frac{\bar{a}}{|a|^2} = \frac{a_1}{|a|^2} - \left(\frac{a_2}{|a|^2}\right)i - \left(\frac{a_3}{|a|^2}\right)j - \left(\frac{a_4}{|a|^2}\right)k$$

is a multiplicative inverse for a. We consider that we have demonstrated the following theorem.

Theorem 25.4   *The quaternions $\mathcal{Q}$ form a skew field under addition and multiplication.*

Note that $G = \{\pm 1, \pm i, \pm j, \pm k\}$ is a group of order 8 under quaternion multiplication. In terms of generators and relations, this group is generated by i and j, where

$$i^4 = 1, \qquad j^2 = i^2 \qquad \text{and} \qquad ji = i^3 j.$$

Since we saw in Example 22.5 that G_2, with presentation

$$(a, b : a^4 = 1, b^2 = a^2, ba = a^3b),$$

is a group of order 8, we must have $G_2 \simeq G$. This explains why the group G_2 of Example 22.5 was called the *quaternion group*.

Algebra is not as rich in (strictly) skew fields as it is in fields. For example, there are no finite skew fields (that are not fields). This is the content of a famous theorem of Wedderburn, which we state without proof.

Theorem 25.5 (Wedderburn)   *A finite division ring is a field.*

Proof.   See the literature for proof of Wedderburn's theorem.

Exercises

*25.1 Compute

$$\begin{pmatrix} 3 & -4 \\ 1 & 5 \end{pmatrix} + \begin{pmatrix} 4 & 17 \\ 5 & -3 \end{pmatrix}$$

and

$$\begin{pmatrix} 3 & -4 \\ 1 & 5 \end{pmatrix}\begin{pmatrix} 4 & 17 \\ 5 & -3 \end{pmatrix}$$

in $M_2(Q)$.

*25.2 Show that

$$\begin{pmatrix} 1 & 0 \\ 0 & 1 \end{pmatrix}$$

is unity in $M_2(F)$. Describe the unity element in $M_n(F)$.

*25.3 Let ϕ be the element of $Hom(\langle \mathbf{Z} \times \mathbf{Z}, + \rangle)$ given in Example 25.3. This example showed that ϕ is a left divisor of 0. Show that ϕ is also a right divisor of 0.

*25.4 Let $G = \{e, a, b\}$ be a cyclic group of order 3 with identity element e. Write each of the following elements in the group algebra $\mathbf{Z}_5(G)$ in the form

$$re + sa + tb \qquad \text{for} \quad r, s, t \in \mathbf{Z}_5.$$

a) $(2e + 3a + 0b) + (4e + 2a + 3b)$

b) $(2e + 3a + 0b)(4e + 2a + 3b)$

c) $(3e + 3a + 3b)^4$

*25.5 Write the following elements of $\mathcal{Q}$ in the form $a_1 + a_2i + a_3j + a_4k$ for $a_i \in \mathbf{R}$.

√ a) $(i + 3j)(4 + 2j - k)$

b) $i^2j^3kji^5$

√ c) $(i + j)^{-1}$

d) $[(1 + 3i)(4j + 3k)]^{-1}$

*25.6 Mark each of the following true or false.
— a) $M_n(F)$ has no divisors of 0 for any n.
— b) Every nonzero element of $M_2(\mathbf{Z}_2)$ is a unit.
— c) $Hom(A)$ is always a ring with unity $\neq 0$ for every abelian group A.
— d) $Hom(A)$ is never a ring with unity $\neq 0$ for any abelian group A.
— e) The subset $Iso(A)$ of $Hom(A)$, consisting of the isomorphisms of A onto A, forms a subring of $Hom(A)$ for every abelian group A.
— f) $R(\langle \mathbf{Z}, + \rangle)$ is isomorphic to $\langle \mathbf{Z}, +, \cdot \rangle$ for every commutative ring R with unity.
— g) The group ring $R(G)$ of an abelian group G is a commutative ring for any commutative ring R with unity.
— h) The quaternions are a field.
— i) $\langle \mathscr{2}^*, \cdot \rangle$ is a group where $\mathscr{2}^*$ is the set of nonzero quaternions.
— j) No subring of $\mathscr{2}$ is a field.

*25.7 Show that the matrix

$$\begin{pmatrix} 0 & 0 \\ 0 & 1 \end{pmatrix}$$

in $M_2(F)$ is not only a left divisor of 0 as was shown in Example 25.2, but is also a right divisor of 0.

*25.8 Prove the left distributive law in $M_2(F)$.

*25.9 Show that $M_2(F)$ has at least six units for every field F. Exhibit these units. [*Hint:* F has at least two elements, 0 and 1.]

*25.10 Show that $Hom(\langle \mathbf{Z}, + \rangle)$ is naturally isomorphic to $\langle \mathbf{Z}, +, \cdot \rangle$ and that $Hom(\langle \mathbf{Z}_n, + \rangle)$ is naturally isomorphic to $\langle \mathbf{Z}_n, +, \cdot \rangle$.

*25.11 Show that $Hom(\langle \mathbf{Z}_2 \times \mathbf{Z}_2, + \rangle)$ is not isomorphic to $\langle \mathbf{Z}_2 \times \mathbf{Z}_2, +, \cdot \rangle$.

*25.12 Referring to Example 25.4, show that $XY - YX = 1$.

*25.13 Referring to the group S_3 given in Example 4.1, compute the product

$$(0\rho_0 + 1\rho_1 + 0\rho_2 + 0\mu_1 + 1\mu_2 + 1\mu_3)(1\rho_0 + 1\rho_1 + 0\rho_2 + 1\mu_1 + 0\mu_2 + 1\mu_3)$$

in the group algebra $\mathbf{Z}_2(S_3)$.

*25.14 If $G = \{e\}$, the group of one element, show that $R(G)$ is isomorphic to R for for any ring R.

*25.15 Find two subsets of $\mathscr{2}$ different from $\mathbf{C}$ and from each other, each of which is a field isomorphic to $\mathbf{C}$ under the induced addition and multiplication from $\mathscr{2}$.

*25.16 Show by an example that a polynomial equation of degree n may have more than n solutions in a skew field. [*Hint:* Consider $n = 2$ and the skew field $\mathscr{2}$.]

*25.17 Prove the associative law for multiplication in $\mathscr{2}$. (This should cure you of wanting to verify any other skew field axioms for $\mathscr{2}$.)

*25.18 Find the center of the group $\langle \mathscr{2}^*, \cdot \rangle$, where $\mathscr{2}^*$ is the set of nonzero quaternions.

twenty-six

The Field
of Quotients
# of an
Integral
Domain

If an integral domain is such that every nonzero element has a multiplicative inverse, then it is a field. However, many integral domains, such as the integers **Z**, do not form a field. This dilemma is not too serious. It is the purpose of this chapter to show that every integral domain can be regarded as being contained in a certain field, *a field of quotients of the integral domain.* This field will be a minimal field containing the integral domain in a sense that we shall describe. For example, the integers are contained in the field **Q**, whose elements can all be expressed as quotients of integers. Our construction of a field of quotients of an integral domain is exactly the same as the construction of the rational numbers from the integers, which the student may have seen in a course in foundations or advanced calculus. To follow this construction through is such a good exercise in the use of definitions and the concept of isomorphism that we discuss it in some detail, although to write out, or to read, every last detail would be tedious. We can be motivated at every step by the way **Q** can be formed from **Z**. Recall that the different representations of a rational number as a quotient of integers constituted our motivation for the discussion of equivalence relations in Section 0.3.

234

26.1 THE CONSTRUCTION

Let D be an integral domain that we desire to enlarge to a field of quotients F. A coarse outline of the steps we take is as follows:

1. Define what the elements of F are to be.
2. Define the binary operations of addition and multiplication on F.
3. Check all the field axioms to show that F is a field under these operations.
4. Show that F can be viewed as containing D as an integral subdomain.

Steps 1, 2, and 4 are very interesting, and step 3 is largely a boring chore. We proceed with the construction.

STEP 1.   Let D be a given integral domain, and form the Cartesian product

$$D \times D = \{(a, b) \mid a, b \in D\}.$$

We are going to think of an ordered pair (a, b) as representing a *formal quotient* a/b, that is, if $D = \mathbf{Z}$, the pair $(2, 3)$ will eventually represent the number $\frac{2}{3}$ for us. The pair $(2, 0)$ represents no element of $\mathbf{Q}$, and also in the general case we cut the set $D \times D$ down a bit. Let S be the subset of $D \times D$ given by

$$S = \{(a, b) \mid a, b \in D, b \neq 0\}.$$

Now S is still not going to be our field as is indicated by the fact that, with $D = \mathbf{Z}$, *different* pairs of integers such as $(2, 3)$ and $(4, 6)$ can represent the *same* rational number. We next define when two elements of S represent the same element of F or, as we shall say, when two elements of S are *equivalent*.

> **Definition**   Two elements (a, b) and (c, d) in S are **equivalent**, denoted by $(a, b) \sim (c, d)$, if and only if $ad = bc$.

Observe that this definition is reasonable, since the criterion for $(a, b) \sim (c, d)$ is an equation $ad = bc$ involving elements in D and concerning the known multiplication in D. Note also that for $D = \mathbf{Z}$, the criterion gives us our usual definition of *equality*, for example, $\frac{2}{3} = \frac{4}{6}$, since $(2)(6) = (3)(4)$. The rational number that we usually denote by $\frac{2}{3}$ can be thought of as the collection of *all* quotients of integers that reduce to, or are equivalent to, $\frac{2}{3}$.

> **Lemma 26.1**   *The relation $\sim$ between elements of the set S as just described is an equivalence relation.*

Proof.   We must check the three properties of an equivalence relation.

> *Reflexive.* $(a, b) \sim (a, b)$ since $ab = ba$, for multiplication in D is commutative.

Symmetric. If $(a, b) \sim (c, d)$, then $ad = bc$. Since multiplication in D is commutative, we deduce that $cb = da$, and consequently $(c, d) \sim (a, b)$.

Transitive. If $(a, b) \sim (c, d)$ and $(c, d) \sim (r, s)$, then $ad = bc$ and $cs = dr$. Using these relations and the fact that multiplication in D is commutative, we have

$$asd = sad = sbc = bcs = bdr = brd.$$

Now $d \neq 0$, and D is an integral domain, so cancellation is valid; this is a crucial step in the argument. Hence from $asd = brd$ we obtain $as = br$, so that $(a, b) \sim (r, s)$. ∎

It is worth comparing the preceding proof with the demonstration in Example 0.1. The steps are identical.

We now know, in view of Theorem 0.1, that $\sim$ gives a partition of S into equivalence classes. To avoid long bars over extended expressions, we shall let $[(a, b)]$, rather than $\overline{(a, b)}$, be the equivalence class of (a, b) in S under the relation $\sim$. We now finish step 1 by defining F to be the set of all equivalence classes $[(a, b)]$ for $(a, b) \in S$.

STEP 2.   The next lemma serves to define addition and multiplication in F. You should check to see that if $D = \mathbf{Z}$ and $[(a, b)]$ is viewed as $(a/b) \in \mathbf{Q}$ these definitions applied to $\mathbf{Q}$ give the usual operations.

Lemma 26.2   *For* $[(a, b)]$ *and* $[(c, d)]$ *in F, the equations*

$$[(a, b)] + [(c, d)] = [(ad + bc), bd]$$

and

$$[(a, b)][(c, d)] = [(ac, bd)]$$

give well-defined operations of addition and multiplication on F.

Proof.   Note first that if $[(a, b)]$ and $[(c, d)]$ are in F, then (a, b) and (c, d) are in S, so $b \neq 0$ and $d \neq 0$. Since D is an integral domain, $bd \neq 0$, so both $(ad + bc, bd)$ and (ac, bd) are in S. (Note the crucial use here of the fact that D has no divisors of 0.) This shows that the right-hand sides of the defining equations are at least in F.

It remains for us to show that these operations of addition and multiplication are well defined. That is, they were defined by means of representatives in S of elements of F. We must show that if different representatives in S are chosen, the same element of F will result. To this end, suppose that $(a_1, b_1) \in [(a, b)]$ and $(c_1, d_1) \in [(c, d)]$. We must show that

$$(a_1 d_1 + b_1 c_1, b_1 d_1) \in [(ad + bc, bd)]$$

and

$$(a_1 c_1, b_1 d_1) \in [(ac, bd)].$$

Now $(a_1, b_1) \in [(a, b)]$ means that $(a_1, b_1) \sim (a, b)$, that is,

$$a_1 b = b_1 a.$$

Similarly, $(c_1, d_1) \in [(c, d)]$ implies that

$$c_1 d = d_1 c.$$

Multiplying the first equation by $d_1 d$, the second by $b_1 b$, and adding the resulting equations, we obtain the following equation in D:

$$a_1 b d_1 d + c_1 d b_1 b = b_1 a d_1 d + d_1 c b_1 b.$$

Using various axioms for an integral domain, we see that

$$(a_1 d_1 + b_1 c_1) b d = b_1 d_1 (ad + bc),$$

so

$$(a_1 d_1 + b_1 c_1, b_1 d_1) \sim (ad + bc, bd),$$

giving $(a_1 d_1 + b_1 c_1, b_1 d_1) \in [(ad + bc, bd)]$. This takes care of addition in F. For multiplication in F, on multiplying the equations $a_1 b = b_1 a$ and $c_1 d = d_1 c$, we obtain

$$a_1 b c_1 d = b_1 a d_1 c,$$

so, using axioms of D, we get

$$a_1 c_1 b d = b_1 d_1 ac,$$

which implies that

$$(a_1 c_1, b_1 d_1) \sim (ac, bd).$$

Thus $(a_1 c_1, b_1 d_1) \in [(ac, bd)]$, which completes the proof. ∎

You should be sure you *understand* the meaning of the last lemma and the necessity for proving it. This completes our step 2.

STEP 3. Step 3 is pretty boring, but it is good for you to work through a few of these details. The reason for this is that you cannot work through them unless you *understand* what we have done. Thus working through them will contribute to your understanding of this construction. We outline the things that must be proved and prove a couple of them. The rest are left to the exercises.

1. Addition in F is commutative.

Proof. Now $[(a, b)] + [(c, d)]$ is by definition $[(ad + bc, bd)]$. Also $[(c, d)] + [(a, b)]$ is by definition $[(cb + da, db)]$. We need to show that $(ad + bc, bd) \sim (cb + da, db)$. This is clear, since $ad + bc = cb + da$ and $bd = db$, by the axioms of D. ∎

2. Addition is associative.

3. $[(0, 1)]$ is an identity for addition in F.

4. $[(-a, b)]$ is an additive inverse for $[(a, b)]$ in F.
5. Multiplication in F is associative.
6. Multiplication in F is commutative.
7. The distributive laws hold in F.
8. $[(1, 1)]$ is a multiplicative identity in F.
9. If $[(a, b)] \in F$ is not the additive identity, then $a \neq 0$ in D and $[(b, a)]$ is a multiplicative inverse for $[(a, b)]$.

Proof.   Let $[(a, b)] \in F$. If $a = 0$, then

$$a1 = b0 = 0,$$

so

$$(a, b) \sim (0, 1),$$

that is, $[(a, b)] = [(0, 1)]$. But $[(0, 1)]$ is the additive identity by part 3. Thus if $[(a, b)]$ is not the additive identity in F, we have $a \neq 0$, so it makes sense to talk about $[(b, a)]$ in F. Now $[(a, b)][(b, a)] = [(ab, ba)]$. But in D we have $ab = ba$, or $(ab)1 = (ba)1$, so

$$(ab, ba) \sim (1, 1).$$

Thus

$$[(a, b)][(b, a)] = [(1, 1)],$$

and $[(1, 1)]$ is the multiplicative identity by part 8. ∎

This completes step 3.

STEP 4.   It remains for us to show that F can be regarded as containing D. To do this, we show that there is an isomorphism i of D with a subdomain of F. Then if we rename the image of D under i using the names of the elements of D, we will be done. The next lemma gives us this isomorphism.

Lemma 26.3   *The map $i : D \to F$ given by $ai = [(a, 1)]$ is an isomorphism of D with a subdomain of F.*

Proof.   For a and b in D, we have

$$(a + b)i = [(a + b, 1)].$$

Also,

$$(ai) + (bi) = [(a, 1)] + [(b, 1)] = [(a1 + 1b, 1)] = [(a + b, 1)],$$

so $(a + b)i = (ai) + (bi)$. Furthermore,

$$(ab)i = [(ab, 1)],$$

while

$$(ai)(bi) = [(a, 1)][(b, 1)] = [(ab, 1)],$$

so $(ab)i = (ai)(bi)$.

It remains for us to show only that i is one to one. If $ai = bi$, then

$$[(a, 1)] = [(b, 1)],$$

so $(a, 1) \sim (b, 1)$ giving $a1 = 1b$, that is,

$$a = b.$$

Thus i is an isomorphism of D with Di, and, of course, Di is then a subdomain of F. ∎

Since $[(a, b)] = [(a, 1)][(1, b)] = [(a, 1)]/[(b, 1)] = ai/bi$ clearly holds in F, we have now proved the following theorem.

Theorem 26.1 *Any integral domain D can be enlarged to (or embedded in) a field F such that every element of F can be expressed as a quotient of two elements of D. (Such a field F is a **field of quotients** of D.)*

26.2 UNIQUENESS

We said in the beginning that F could be regarded in some sense as a minimal field containing D. This is pretty obvious, since every field containing D must contain all elements a/b for every $a, b \in D$ with $b \neq 0$. The next theorem will show that every field containing D contains a subfield of quotients of D, and that any two fields of quotients of D are isomorphic.

Theorem 26.2 *Let F be a field of quotients of D and let L be any field containing D. Then there exists a map $\psi : F \to L$ that gives an isomorphism of F with a subfield of L such that $a\psi = a$ for $a \in D$.*

Proof. The lattice and mapping diagram in Fig. 26.1 may help you to visualize the situation for this theorem.

An element of F is of the form $a/_F b$ where $/_F$ denotes the quotient of $a \in D$ by $b \in D$ regarded as elements of F. We of course want to map $a/_F b$

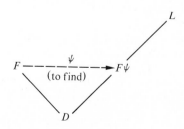

Fig. 26.1

onto $a/_L b$ where $/_L$ denotes the quotient of elements in L. This is so trivial that it seems as though you must be cheating somewhere when you define the mapping ψ, but let's do it anyway.

We must define $\psi : F \to L$, and we start by defining

$$a\psi = a \qquad \text{for} \quad a \in D.$$

Every $x \in F$ is a quotient $a/_F b$ of some two elements a and b, $b \neq 0$, of D. Let us attempt to define ψ by

$$(a/_F b)\psi = (a\psi)/_L(b\psi).$$

We must first show that this map ψ is sensible and well defined. Since ψ is the identity on D, for $b \neq 0$ we have $b\psi \neq 0$, so our definition of $(a/_F b)\psi$ as $(a\psi)/_L(b\psi)$ makes sense. If $a/_F b = c/_F d$ in F, then $ad = bc$ in D, so $(ad)\psi = (bc)\psi$. But since ψ is the identity on D,

$$(ad)\psi = (a\psi)(d\psi) \qquad \text{and} \qquad (bc)\psi = (b\psi)(c\psi).$$

Thus

$$(a\psi)/_L(b\psi) = (c\psi)/_L(d\psi)$$

in L, so ψ is well defined.

The equations

$$(xy)\psi = (x\psi)(y\psi)$$

and

$$(x + y)\psi = x\psi + y\psi$$

follow easily from the definition of ψ on F and from the fact that ψ is the identity on D.

If $(a/_F b)\psi = (c/_F d)\psi$, we have

$$(a\psi)/_L(b\psi) = (c\psi)/_L(d\psi)$$

so

$$(a\psi)(d\psi) = (b\psi)(c\psi).$$

Since ψ is the identity on D, we then deduce that $ad = bc$, so $a/_F b = c/_F d$. Thus ψ is one to one.

By definition, $a\psi = a$ for $a \in D$. ∎

Corollary   *Every field L containing an integral domain D contains a field of quotients of D.*

Proof. In the proof of Theorem 26.2, every element of the subfield $F\psi$ of L is a quotient in L of elements of D. ∎

Corollary   *Any two fields of quotients of an integral domain D are isomorphic.*

Proof. Suppose in Theorem 26.2 that L is a field of quotients of D, so that every element x of L can be expressed in the form $a/_L b$ for a, $b \in D$. Then L is the field $F\psi$ of the proof of Theorem 26.2 and is thus isomorphic to F. ∎

Exercises

26.1 Describe the field F of quotients of the integral subdomain

$$D = \{n + mi \,|\, n, m \in \mathbf{Z}\}$$

of $\mathbf{C}$. "Describe" means give the elements of $\mathbf{C}$ that make up the field of quotients of D in $\mathbf{C}$.

26.2 Describe (in the sense of Exercise 26.1) the field F of quotients of the integral subdomain $D = \{n + m\sqrt{2} \,|\, n, m \in \mathbf{Z}\}$ of $\mathbf{R}$.

26.3 Show by an example that a field F' of quotients of a proper subdomain D' of an integral domain D may also be a field of quotients for D.

†**26.4** Prove part 7 of step 3. You may assume any preceding part of step 3.

26.5 Mark each of the following true or false.

— a) $\mathbf{Q}$ is a field of quotients of $\mathbf{Z}$.

— b) $\mathbf{R}$ is a field of quotients of $\mathbf{Z}$.

— c) $\mathbf{R}$ is a field of quotients of $\mathbf{R}$.

— d) $\mathbf{C}$ is a field of quotients of $\mathbf{R}$.

— e) If D is a field, then any field of quotients of D is isomorphic to D.

— f) The fact that D has no divisors of 0 was used strongly several times in the construction of a field F of quotients of the integral domain D.

— g) Every element of an integral domain D is a unit in a field F of quotients of D.

— h) Every nonzero element of an integral domain D is a unit in a field F of quotients of D.

— i) A field of quotients F' of a subdomain D' of an integral domain D can be regarded as a subfield of some field of quotients of D.

— j) Every field of quotients of $\mathbf{Z}$ is isomorphic to $\mathbf{Q}$.

26.6 Prove part 2 of step 3. You may assume any preceding part of step 3.

26.7 Prove part 3 of step 3. You may assume any preceding part of step 3.

26.8 Prove part 4 of step 3. You may assume any preceding part of step 3.

26.9 Prove part 5 of step 3. You may assume any preceding part of step 3.

26.10 Prove part 6 of step 3. You may assume any preceding part of step 3.

26.11 Let R be a commutative ring, and let $T \neq \{0\}$ be a nonempty subset of R closed under multiplication and containing no divisors of 0. Starting with $R \times T$ and otherwise exactly following the construction in this chapter, one can show that the ring R can be enlarged to a *partial ring of quotients* $Q(R, T)$. Think about this for fifteen minutes or so; look back over the construction and see why things still work. In particular, show the following:

a) $Q(R \ T)$ has unity even if R does not.

b) In $Q(R, T)$, every nonzero element of T is a unit.

26.12 Prove from Exercise 26.11 that every commutative ring containing an element a that is not a divisor of 0 can be enlarged to a commutative ring with unity. Compare with Exercise 24.16.

26.13 With reference to Exercise 26.11, how many elements are there in the ring $Q(Z_4, \{1, 3\})$?

26.14 With reference to Exercise 26.11, describe the ring $Q(Z, \{2^n | n \in Z^+\})$ by describing a subring of **R** to which it is isomorphic.

26.15 With reference to Exercise 26.11, describe the ring $Q(3Z, \{6^n | n \in Z^+\})$ by describing a subring of **R** to which it is isomorphic.

26.16 With reference to Exercise 26.11, suppose we drop the condition that T have no divisors of zero and just require that nonempty $T \neq \{0\}$ be closed under multiplication. The attempt to enlarge R to a commutative ring with unity in which every nonzero element of T is a unit must fail if T contains an element a that is a divisor of 0 for a divisor of 0 cannot also be a unit. Try to discover where a construction parallel to that in the text but starting with $R \times T$ first runs into trouble. In particular, for $R = Z_6$ and $T = \{1, 2, 4\}$, illustrate the first difficulty encountered. [*Hint:* It is in step 1.]

twenty-seven

Our Basic Goal

This chapter is expository and is designed to give you the proper perspective on the unstarred portion of the rest of this text. There are no exercises.

In the following two chapters we shall be concerned with topics in ring theory that are analogous to the material on factor groups and homomorphisms for group theory. However, our aim in developing these analogous concepts is quite different from our aims in group theory. In group theory we used (partly in starred sections) the concepts of factor groups and homomorphisms to study the structure of a given group and to determine the types of group structures of certain orders that could exist. With apologies to the professional mathematician, we explain now, in terms with which you are familiar and that you think you understand, our purpose in developing this analogous machinery for rings.

> *Every single bit of the unstarred material in the rest of this text is aimed at finding and studying solutions of polynomial equations such as*

$$x^2 + x - 6 = 0.$$

Let us take a moment to talk about this aim in the light of mathematical history.

We start way back with the Pythagorean school of mathematics of about 525 B.C. The Pythagoreans asserted with an almost fanatical fervor that all distances are **commensurable,** that is, that given distances a and b there should exist a unit of distance u and integers n and m such that $a = (n)(u)$ and $b = (m)(u)$. In terms of numbers, then, thinking of u as being one unit of distance, they maintained that all numbers are integers. This idea of commensurability can be rephrased according to our ideas as an assertion that all numbers are rational, for if a and b are rational numbers then each is an integral multiple of the reciprocal of the least common multiple of their denominators. For example, if $a = \frac{7}{12}$ and $b = \frac{19}{15}$, then $a = (35)(\frac{1}{60})$ and $b = (76)(\frac{1}{60})$.

The Pythagoreans knew, of course, what is now called the *Pythagorean theorem*, that is, that for a right triangle with legs of lengths a and b and a hypotenuse of length c, one has

$$a^2 + b^2 = c^2.$$

They also had to grant the existence of a hypotenuse of a right triangle having two legs of equal length, say one unit each. The hypotenuse of such a right triangle would, as we know, have to have a length of $\sqrt{2}$. Imagine then their consternation, dismay, and even fury when one of their society— according to some stories it was Pythagoras himself—came up with the embarrassing fact that is stated in our terminology in the following theorem.

> **Theorem 27.1**   *The equation $x^2 = 2$ has no solutions in rational numbers. Thus $\sqrt{2}$ is not a rational number.*

Proof.   Suppose that m/n for $m, n \in \mathbf{Z}$ is a rational number such that $(m/n)^2 = 2$. Then

$$m^2 = 2n^2,$$

where both m^2 and $2n^2$ are integers. Since m^2 and $2n^2$ are the same integer and since 2 is a factor of $2n^2$, we see that 2 must be one of the factors of m^2. But as a square, m^2 has as factors the factors of m repeated twice. Thus m must have two factors 2. Then $2n^2$ must have two factors 2, so n^2 must have 2 as a factor. But then as a square, n^2 must have two factors 2. Therefore $2n^2$ has three factors 2. But then m^2 must have three factors 2 and hence four factors 2 as a square. But then $2n^2$ must have at least four factors 2 and hence .... By bouncing back and forth from m^2 to $2n^2$ in this fashion you get yourself into an impossible situation. We have shown by contradiction that $2 \neq (m/n)^2$ for $m, n \in \mathbf{Z}$. This argument can be made much more concise and elegant, but we are in a reckless mood. ∎

Thus the Pythagoreans ran right into the question of a solution of a polynomial equation, $x^2 - 2 = 0$. We refer the student to Shanks [35, Chapter 3], for a lively and totally delightful account of this Pythagorean dilemma and its significance in mathematics.

In our motivation of the definition of a group, we commented on the necessity of having negative numbers, so that equations such as $x + 2 = 0$ might have solutions. The introduction of negative numbers caused a certain amount of consternation in some philosophical circles. One can visualize 1 apple, 2 apples, and even $\frac{13}{11}$ apples, but how can you point to anything and say that it is -17 apples? Finally, consideration of the equation $x^2 + 1 = 0$ led to the introduction of the number *i*. The very name of an "imaginary number" given to *i* shows how this number was regarded. Even today many students are led by this name to regard *i* with some degree of suspicion. The negative numbers were introduced to you at such an early stage in your mathematical development that you accept them without question.

So much for history. We reiterate:

The unstarred portion of the rest of this text is devoted to finding and studying solutions of polynomial equations.

You first met polynomials in high school freshman algebra. The first problem there was to learn how to add, multiply, and factor polynomials. Then, in both freshman algebra and in the second course in algebra in high school, considerable emphasis was placed on solving polynomial equations. These topics are exactly those with which we shall be concerned. The difference is that while in high school only polynomials with real number coefficients were considered, *we shall be doing our work for polynomials with coefficients from any arbitrary given field.*

As we said above, the next two chapters deal with factor rings and ring homomorphisms, material formally analogous to our work on factor groups and group homomorphisms. Polynomials won't even be mentioned. Then we shall introduce polynomials and demonstrate how the idea of solving a polynomial equation can be phrased in terms of the language of homomorphisms. Don't be scared by the terminology at that point. Remember always:

We shall just be doing high school algebra in a more general context.

And then, with amazing ease, elegance, beauty, and sophistication, as a bright star suddenly appears in your possibly drab mathematical life, we shall achieve our

Basic goal: *To show that given any polynomial equation of degree* ≥ 1, *where the coefficients of the polynomial may be from any field, there exists a solution of the equation.*

If you feel that all this fuss is ridiculous, just think back in history. *This is the culmination of more than 2000 years of mathematical endeavor in working with polynomial equations.* After achieving our *basic goal*, we shall spend the rest of our time studying the nature of these solutions of polynomial equations. We emphasize again that you need have no fear in approaching this material. *We shall be dealing with familiar topics of high school algebra. This work should seem much more natural to you than group theory.*

In conclusion, we remark that the machinery of factor rings and ring homomorphisms is not really necessary in order for us to achieve our *basic goal*. For a direct demonstration, see Artin [26, p. 29]. However, factor rings and ring homomorphisms are fundamental ideas that you should grasp, and our *basic goal* will follow very easily once we have mastered them. We will further use these concepts effectively in the study of properties of solutions of polynomial equations.

twenty-eight

Quotient Rings and Ideals

28.1 INTRODUCTION

This chapter begins a study of rings analogous to the material on groups contained in Chapters 11, 12, and 13, concerning factor groups and homomorphisms. Since $\langle R, + \rangle$ is a group for every ring R, indeed $\langle R, + \rangle$ is an abelian group, the additive part of this theory is already done. We shall have to concern ourselves only with its multiplicative aspects. To make the analogy with the situation for groups as clear as possible, we shall develop this theory according to the same plan that we used for groups. This will give you another chance to master it.

Let R be a ring. We are concerned with studying a partition of R into disjoint subsets or cells such that these cells themselves can be viewed as elements in a ring where *both* addition and multiplication of the cells are the induced operations from R. That is, we wish to define *both* operations by choosing representatives from the cells, adding or multiplying these representatives in R, and *defining* the sum or product of the cells to be the cell in which the sum or product of the representatives is found. *Both* these operations must be *well defined*, that is, independent of the choices of representatives from the cells.

> **Theorem 28.1 (Analog of Theorem 11.1)**   *If a ring R can be partitioned into cells with both the induced operations described above well defined, and if the cells form a ring under these induced operations, then the cell containing the additive identity 0 of R must be an additive subgroup N of the additive group $\langle R, + \rangle$. Furthermore, N must have the additional property that for all $r \in R$ and $n \in N$, both rn $\in$ N and nr $\in$ N. We express this last condition as $rN \subseteq N$ and $Nr \subseteq N$.*

Proof.   Viewing $\langle R, + \rangle$ as a group under addition, we know from Theorem 11.1 that N must be an additive subgroup of $\langle R, + \rangle$. Indeed, we know from work following Theorem 11.1 that N must be a normal additive subgroup of $\langle R, + \rangle$, but since addition in R is commutative, every additive subgroup of $\langle R, + \rangle$ is normal in $\langle R, + \rangle$.

We need only show that for $r \in R$ we have $rN \subseteq N$ and $Nr \subseteq N$. Let $r \in R$. Now r is in some cell $A \subseteq R$, and by our hypothesis that the induced cell multiplication is well defined we can compute the products AN and NA by choosing any representatives. Let us choose $r \in A$ and $0 \in N$. Then AN and NA are the cells containing $r0 = 0r = 0$, that is, $AN = NA = N$. Therefore, for all possible representatives $n \in N$, we have $rn \in N$ and $nr \in N$. ∎

28.2 CRITERIA FOR THE EXISTENCE OF A COSET RING

> **Theorem 28.2 (Analog of Theorem 11.2)**   *If a ring R can be partitioned into cells with both induced operations well defined and giving a ring, then the cells must be precisely the left (and also the right) cosets with respect to addition of the additive subgroup $\langle N, + \rangle$ of $\langle R, + \rangle$, where N is the cell containing 0.*

Proof.   Proof of Theorem 28.2 is immediate from Theorem 11.2 if we consider just $\langle R, + \rangle$, an additive group, and forget about multiplication. ∎

Of course, Theorem 11.3 holds here also, that is, these cosets are disjoint. Since addition is commutative, the left coset $r + N$ is the same as the right coset $N + r$.

> **Lemma 28.1 (Analog of Lemma 12.1)**   *If $\langle N, + \rangle$ is an additive subgroup of $\langle R, + \rangle$ for a ring R, and if the induced operations of addition and multiplication on cosets $r + N$ for $r \in R$ are well defined, that is, inde-*

pendent of the choice of representatives, then the collection of these cosets r + N is a ring under these induced coset operations.

Proof. Lemma 12.1 shows us that the cosets form a group under the induced addition. The ring axioms involving multiplication follow at once since we also compute products by choosing representatives, and since the ring axioms hold in R. For example, the left distributive law

$$(r_1 + N)[(r_2 + N) + (r_3 + N)]$$
$$= [(r_1 + N)(r_2 + N)] + [(r_1 + N)(r_3 + N)]$$

follows when we choose $r_i \in (r_i + N)$ as representatives and observe that the required

$$r_1(r_2 + r_3) \in [(r_1 r_2 + r_1 r_3) + N]$$

is true, since $r_1(r_2 + r_3) = (r_1 r_2 + r_1 r_3)$ by the left distributive law in R. The other distributive law and the associative law for multiplication follow similarly. ∎

Theorem 28.3 (Analog of Theorem 12.1) *If $\langle N, + \rangle$ is an additive subgroup of the additive group $\langle R, + \rangle$ of a ring R, then the operations of induced addition and multiplication are both well defined on the cosets r + N for r ∈ R if and only if rN ⊆ N and Nr ⊆ N for all r ∈ R.*

Proof. The fact that if the operations are well defined, then $rN \subseteq N$ and $Nr \subseteq N$ follows immediately from Lemma 28.1 and Theorem 28.1.

Suppose that $\langle N, + \rangle$ is an additive subgroup of $\langle R, + \rangle$ such that for all $r \in R$ we have $rN \subseteq N$ and $Nr \subseteq N$. Coset addition is well defined, by Theorem 12.1 applied to the subgroup $\langle N, + \rangle$ of $\langle R, + \rangle$. For multiplication, we must show that a product $(r_1 + N)(r_2 + N)$, computed by choosing representatives, is well defined. It is no loss of generality to take as two representatives of $r_1 + N$ the elements r_1 itself and $r_1 + n_1$ for $n_1 \in N$. Similarly, let r_2 and $r_2 + n_2$ be two elements of $r_2 + N$. We must show that $(r_1 + n_1)(r_2 + n_2)$ is in the same coset $r_1 r_2 + N$ as $r_1 r_2$. Now

$$(r_1 + n_1)(r_2 + n_2) = r_1 r_2 + n_1 n_2 + n_1 r_2 + r_1 n_2,$$

by the distributive laws in R. The assumption that $rN \subseteq N$ and $Nr \subseteq N$ implies that $r_1 n_2 \in N$ and $n_1 r_2 \in N$. Viewing n_1 as an element of R, we have $n_1 n_2 \in n_1 N$ and $n_1 N \subseteq N$, so $n_1 n_2 \in N$ also. Then since $\langle N, + \rangle$ is an additive subgroup of $\langle R, + \rangle$, we see that $(n_1 r_2 + r_1 n_2 + n_1 n_2) \in N$, so $(r_1 + n_1)(r_2 + n_2) \in (r_1 r_2 + N)$. ∎

It is now clear that these special additive subgroups $\langle N, + \rangle$ of a ring R having the property that $rN \subseteq N$ and $Nr \subseteq N$ for all $r \in R$ are going to play a role of basic importance in ring theory, analogous to the role of a normal subgroup in a group. Note that the conditions $rN \subseteq N$ and $Nr \subseteq N$ imply in particular, for $r \in N$, that N is closed under the multiplication of R. Thus, we can consider N with the induced addition and multiplication from R to be a subring of R. Not every subring N of every ring R satisfies the conditions $rN \subseteq N$ and $Nr \subseteq N$ however. For example, $\mathbf{Q} \leq \mathbf{R}$, but $\pi\mathbf{Q} \nsubseteq \mathbf{Q}$.

28.3 IDEALS AND QUOTIENT RINGS

Definition (Analog of the definition of a normal subgroup)   An additive subgroup $\langle N, + \rangle$ of a ring R satisfying $rN \subseteq N$ and $Nr \subseteq N$ for all $r \in R$ is an *ideal* (or *two-sided ideal*) *of* R. A subgroup $\langle N, + \rangle$ of R satisfying $rN \subseteq N$ for all $r \in R$ is a *left ideal of* R, and one satisfying $Nr \subseteq N$ for all $r \in R$ is a *right ideal of* R.

When we refer to an *ideal*, it will always be assumed to be a two-sided ideal. Left or right ideals that are not two-sided will not concern us. Let us reiterate for emphasis:

An ideal is to a ring as a normal subgroup is to a group.

Definition (Analog of the definition of a factor group)   If N is an ideal in a ring R, then the ring of cosets $r + N$ under the induced operations is the *quotient ring,* or *factor ring,* or *residue class ring of R modulo N,* and is denoted by R/N. The cosets are *residue classes modulo N.*

Example 28.1   Consider the ring $\mathbf{Z}$. The only additive subgroups of $\langle \mathbf{Z}, + \rangle$ are the subgroups $n\mathbf{Z}$, as we have seen. Clearly, if r is any integer and $m \in n\mathbf{Z}$, then $rm = mr$ is again a multiple of n, that is, if $m = ns$, then $rm = mr = n(sr)$, and $n(sr) \in n\mathbf{Z}$. Thus, $n\mathbf{Z}$ is an ideal, and the cosets $a + n\mathbf{Z}$ of $n\mathbf{Z}$ form a ring $\mathbf{Z}/n\mathbf{Z}$ under the induced operations of addition and multiplication. ∎

In Example 23.2, we defined for a and b in $\mathbf{Z}_n$ the product ab modulo n to be the remainder when divided by n of the usual product of a and b in $\mathbf{Z}$. The map $\phi : \mathbf{Z}_n \to \mathbf{Z}/n\mathbf{Z}$ given by

$$a\phi = a + n\mathbf{Z}$$

is clearly a one-to-one onto map such that $(a + b)\phi = a\phi + b\phi$ and $(ab)\phi = (a\phi)(b\phi)$. Then $\mathbf{Z}_n$ under addition and multiplication modulo n may be viewed as $\mathbf{Z}/n\mathbf{Z}$ renamed and consequently is a ring under this addition and multiplication. We used this fact in the preceding sections with the comment that it would be justified in a general situation later. *This is the justification.* We shall often identify $\mathbf{Z}/n\mathbf{Z}$ with $\mathbf{Z}_n$ by means of this isomorphism ϕ, which is a natural or canonical isomorphism.

Example 28.2   As was shown in the corollary of Theorem 24.4, $\mathbf{Z}_p$, which is isomorphic to $\mathbf{Z}/p\mathbf{Z}$, is a field for p a prime. *Thus a quotient ring of an integral domain may be a field.* We shall say more about this in the next chapter. ∎

Example 28.3   The subset $N = \{0, 3\}$ of $\mathbf{Z}_6$ is easily seen to be an ideal of $\mathbf{Z}_6$, and $\mathbf{Z}_6/N$ has three elements, $0 + N, 1 + N$, and $2 + N$. These obviously add and multiply in such a fashion as to show that $\mathbf{Z}_6/N \simeq \mathbf{Z}_3$ under the correspondence

$$(0 + N) \leftrightarrow 0, \qquad (1 + N) \leftrightarrow 1, \qquad (2 + N) \leftrightarrow 2. \blacksquare$$

Example 28.4   The ring $\mathbf{Z} \times \mathbf{Z}$ is not an integral domain, for

$$(0, 1)(1, 0) = (0, 0),$$

showing that $(0, 1)$ and $(1, 0)$ are 0 divisors. If $N = \{(0, n) \mid n \in \mathbf{Z}\}$, it is clear that N is an ideal of $\mathbf{Z} \times \mathbf{Z}$, and that $(\mathbf{Z} \times \mathbf{Z})/N$ is isomorphic to $\mathbf{Z}$ under the correspondence $[(m, 0) + N] \leftrightarrow m$, the residue classes being of the form $(m, 0) + N$, where $m \in \mathbf{Z}$. *Thus a quotient ring of a ring may be an integral domain, even though the original ring is not.* We shall say more about this also in the next section. ∎

The preceding examples may indicate to you the tremendous importance of the concepts of ideals and quotient rings. Every ring R has two ideals, the **improper ideal** R and the **trivial ideal** $\{0\}$. For these ideals, the factor rings are R/R, which has only one element, and $R/\{0\}$, which is clearly isomorphic to R. These are uninteresting cases. Just as for a subgroup of a group, a **proper nontrivial ideal** of a ring R is an ideal N of R such that $N \neq R$ and $N \neq \{0\}$.

While quotient rings of rings and integral domains may be of great interest, as the above examples indicate, the corollary of the next and last theorem of this section shows that a quotient ring of a field is really not of much additional use.

Theorem 28.4   *If R is a ring with unity, and N is an ideal of R containing a unit, then $N = R$.*

Proof.   Let N be an ideal of R, and suppose that $u \in N$ for some unit u in R. Then the condition $rN \subseteq N$ for all $r \in R$ implies, if we take $r = u^{-1}$ and $u \in N$, that $1 = u^{-1}u$ is in N. But then $rN \subseteq N$ for all $r \in R$ implies that $r1 = r$ is in N for all $r \in R$, so $N = R$. ∎

Corollary   *A field contains no proper nontrivial ideals.*

Proof.   Since every nonzero element of a field is a unit, it follows at once from Theorem 28.4 that an ideal of a field F is either $\{0\}$ or all of F. ∎

Exercises

28.1 Find all ideals N of $\mathbf{Z}_{12}$. In each case compute $\mathbf{Z}_{12}/N$, that is, find a known ring to which the quotient ring is isomorphic.

28.2 Give addition and multiplication tables for $2\mathbf{Z}/8\mathbf{Z}$. Are $2\mathbf{Z}/8\mathbf{Z}$ and $\mathbf{Z}_4$ isomorphic rings?

28.3 Find a subring of the ring $\mathbf{Z} \times \mathbf{Z}$ that is not an ideal of $\mathbf{Z} \times \mathbf{Z}$.

†28.4 A student is asked to prove that a quotient ring of a ring R modulo an ideal N is commutative if and only if $(rs - sr) \in N$ for all $r, s \in R$. The student starts out:

Assume R/N is commutative. Then $rs = sr$ for all $r, s \in R/N$.

a) Why does the instructor reading this expect nonsense from there on?

b) What should the student have written?

c) Prove the assertion. (Note the "if and only if.")

28.5 Mark each of the following true or false.

— a) $\mathbf{Q}$ is an ideal in $\mathbf{R}$.

— b) Every ideal in a ring is a subring of the ring.

— c) Every subring of every ring is an ideal of the ring.

— d) Every quotient ring of every commutative ring is again a commutative ring.

— e) The rings $\mathbf{Z}/4\mathbf{Z}$ and $\mathbf{Z}_4$ are isomorphic.

— f) An ideal N in a ring with unity R is all of R if and only if $1 \in N$.

— g) The concept of an ideal is to the concept of a ring as the concept of a normal subgroup is to the concept of a group.

— h) $\mathbf{Z}_4$ is an ideal of $4\mathbf{Z}$.

— i) If a ring R has 0 divisors, then every quotient ring of R has 0 divisors.

— j) $\mathbf{Z}$ is an ideal in $\mathbf{Q}$.

28.6 Show that a factor ring of a field is either the trivial ring of one element or is isomorphic to the field.

28.7 Show that if R is a ring with unity and N is an ideal of R such that $N \neq R$, then R/N is a ring with unity $\neq 0$.

28.8 Let R be a commutative ring and let $a \in R$. Show that $I_a = \{x \in R \mid ax = 0\}$ is an ideal of R.

28.9 Show that an intersection of ideals of a ring R is again an ideal of R.

28.10 Determine all ideals of $\mathbf{Z} \times \mathbf{Z}$.

28.11 An element a of a ring R is **nilpotent** if $a^n = 0$ for some $n \in \mathbf{Z}^+$. Show that the collection of all nilpotent elements in a commutative ring R is an ideal, the **radical of** R.

28.12 Referring to the definition given in Exercise 28.11, find the radical of the ring $\mathbf{Z}_{12}$ and observe that it is one of the ideals of $\mathbf{Z}_{12}$ found in Exercise 28.1. What is the radical of $\mathbf{Z}$? of $\mathbf{Z}_{32}$?

28.13 Referring to Exercise 28.11, show that if N is the radical of a commutative ring R, then R/N has as radical the trivial ideal $\{0 + N\}$.

28.14 Let R be a commutative ring and N an ideal of R. Referring to Exercise 28.11, show that if every element of N is nilpotent and the radical of R/N is R/N, then the radical of R is R.

28.15 Let R be a commutative ring and N an ideal of R. Show that the set $\sqrt{N}$ of all $a \in R$, such that $a^n \in N$ for some $n \in \mathbf{Z}^+$, is an ideal of R, the **radical of** N. Is this terminology consistent with that in Exercise 28.11?

28.16 Referring to Exercise 28.15, show by examples that for proper ideals N of a commutative ring R,

a) $\sqrt{N}$ need not equal N,   b) $\sqrt{N}$ may equal N.

28.17 What is the relation of the ideal $\sqrt{N}$ of Exercise 28.15 to the radical of R/N (see Exercise 28.11)? Word your answer carefully.

There is a sort of arithmetic of ideals in a commutative ring. The next three exercises define sum, product, and quotient of ideals.

28.18 If A and B are ideals of a ring R, the **sum** $A + B$ **of** A **and** B is defined by

$$A + B = \{a + b \mid a \in A, b \in B\}.$$

a) Show that $A + B$ is an ideal.   b) Show that $A \subseteq A + B$ and $B \subseteq A + B$.

28.19 Let A and B be ideals of a ring R. The **product** AB **of** A **and** B is defined by

$$AB = \left\{ \sum_{i=1}^{n} a_i b_i \,\middle|\, a_i \in A, b_i \in B, n \in \mathbf{Z}^+ \right\}.$$

a) Show that AB is an ideal in R.   b) Show that $AB \subseteq (A \cap B)$.

28.20 Let A and B be ideals of a commutative ring R. The **quotient** $A:B$ **of** A **by** B is defined by

$$A : B = \{r \in R \mid rb \in A \text{ for all } b \in B\}.$$

Show that $A : B$ is an ideal of R.

***28.21** Show that for a field F, the set of all matrices of the form

$$\begin{pmatrix} a & b \\ 0 & 0 \end{pmatrix}$$

for $a, b \in F$ is a right ideal but not a left ideal of $M_2(F)$.

***28.22** Show that the matrix ring $M_2(\mathbf{Z}_2)$ is a **simple ring,** that is, $M_2(\mathbf{Z}_2)$ has no proper nontrivial ideals.

twenty-nine

Homomorphisms of Rings

29.1 DEFINITION AND ELEMENTARY PROPERTIES

We keep our discussion of homomorphisms of rings parallel to that in Chapter 13 for homomorphisms of groups.

Definition (Analog of the definition of a group homomorphism) A map ϕ of a ring R into a ring R' is a **homomorphism** if

$$(a + b)\phi = a\phi + b\phi$$

and

$$(ab)\phi = (a\phi)(b\phi)$$

for all elements a and b in R.

Theorem 29.1 (Analog of Theorem 13.1) *If N is an ideal of a ring R, then the canonical map $\gamma: R \to R/N$ given by $a\gamma = a + N$ for $a \in R$ is a homomorphism.*

254

Proof. Theorem 13.1 applied to $\langle R, + \rangle$ as an additive group with $\langle N, + \rangle$ a normal subgroup shows that

$$(a + b)\gamma = a\gamma + b\gamma.$$

Also,

$$(ab)\gamma = ab + N = (a + N)(b + N) = (a\gamma)(b\gamma). \quad \blacksquare$$

Definition (Analog of the definition of the kernel of a group homomorphism) The *kernel of a homomorphism* ϕ of a ring R into a ring R' is the set of all elements of R mapped onto the additive identity $0'$ of R' by ϕ.

Theorem 29.2 (*Analog of Theorem 13.2*) *Let ϕ be a homomorphism of a ring R into a ring R'. If 0 is the additive identity in R, then $0\phi = 0'$ is the additive identity in R', and if $a \in R$, then $(-a)\phi = -(a\phi)$. If S is a subring of R, then $S\phi$ is a subring of R', and S an ideal of R implies that $S\phi$ is an ideal of $R\phi$. Going the other way, if S' is a subring of R', then $S'\phi^{-1}$ is a subring of R, and S' an ideal of $R\phi$ implies that $S'\phi^{-1}$ is an ideal of R. Finally, if R has unity 1 and $1\phi \neq 0'$, then $1\phi = 1'$ is unity for $R\phi$. Loosely, subrings correspond to subrings, ideals to ideals, and rings with unity to rings with unity under a ring homomorphism.*

Proof. Let ϕ be a homomorphism of a ring R into a ring R'. Since, in particular, ϕ can be viewed as a group homomorphism of $\langle R, + \rangle$ into $\langle R', +' \rangle$, Theorem 13.2 tells us that $0\phi = 0'$ is the additive identity of R' and that $(-a)\phi = -(a\phi)$.

Theorem 13.2 also tells us that if S is a subring of R, then, considering the additive group $\langle S, + \rangle$, we find that $\langle S\phi, +' \rangle$ is a subgroup of $\langle R', +' \rangle$. If $s_1\phi$ and $s_2\phi$ are two elements of $S\phi$, then

$$(s_1\phi)(s_2\phi) = (s_1 s_2)\phi$$

and $(s_1 s_2)\phi \in S\phi$, so $S\phi$ is closed under multiplication, and is then a subring of R'. If S is an ideal of R, then for $s \in S$ and $r \in R$ we have

$$(r\phi)(s\phi) = (rs)\phi$$

and $(rs)\phi \in S\phi$. Also,

$$(s\phi)(r\phi) = (sr)\phi$$

and $(sr)\phi \in S\phi$. Thus $S\phi$ is an ideal in $R\phi$.

Going the other way, Theorem 13.2 again shows that if S' is a subring of R', then $\langle S'\phi^{-1}, + \rangle$ is a subgroup of $\langle R, + \rangle$. If $a\phi \in S'$ and $b\phi \in S'$, then

$$(ab)\phi = (a\phi)(b\phi)$$

and $[(a\phi)(b\phi)] \in S'$, so $S'\phi^{-1}$ is closed under multiplication, and is then a subring of R. If S' is an ideal of $R\phi$, then for $a \in S'\phi^{-1}$ and any $r \in R$ we have

$$(ra)\phi = (r\phi)(a\phi)$$

and $[(r\phi)(a\phi)] \in S'$, so $ra \in S'\phi^{-1}$. Similarly, $ar \in S'\phi^{-1}$. Thus $S'\phi^{-1}$ is an ideal of R.

Finally, if R has unity 1, then for all $r \in R$,

$$r\phi = (1r)\phi = (r1)\phi = (1\phi)(r\phi) = (r\phi)(1\phi),$$

so $1' = 1\phi$ is a multiplicative identity for $R\phi$. If $1' \neq 0'$, then $1'$ is unity for $R\phi$. ∎

In Theorem 29.2, it is important to note that if $\phi: R \rightarrow R'$ is a ring homomorphism and A is an ideal of R, then $A\phi$ is not necessarily an ideal of R', but only an ideal of $R\phi$. For example, let $\phi: \mathbf{Z} \rightarrow \mathbf{Z} \times \mathbf{Z}$ be defined by $n\phi = (n, n)$. Now $2\mathbf{Z}$ is an ideal of $\mathbf{Z}$. However, $(2\mathbf{Z})\phi = \{(2n, 2n) | n \in \mathbf{Z}\}$ is not an ideal of $\mathbf{Z} \times \mathbf{Z}$, since $(1, 2)(4, 4) = (4, 8) \notin (2\mathbf{Z})\phi$.

Theorem 29.2 shows in particular that for a homomorphism $\phi: R \rightarrow R'$, the kernel $K = \{0'\}\phi^{-1}$ is an ideal of R.

Theorem 29.3 (Fundamental Homomorphism Theorem; Analog of Theorem 13.3) *Let ϕ be a homomorphism of a ring R into a ring R' with kernel K. Then $R\phi$ is a ring and there is a canonical isomorphism of $R\phi$ with R/K.*

Proof. Theorem 29.2 shows that $R\phi$ is a ring. Let $(a + K) \in R/K$, and define the map $\psi: R/K \rightarrow R\phi$ by

$$(a + K)\psi = a\phi.$$

Theorem 13.3 shows that ψ is well defined, one to one and onto with

$$[(a + K) + (b + K)]\psi = (a + K)\psi + (b + K)\psi.$$

Now

$$[(a + K)(b + K)]\psi = (ab + K)\psi = (ab)\phi = (a\phi)(b\phi)$$
$$= [(a + K)\psi][(b + K)\psi].$$

Thus ψ is a ring isomorphism.

Again ψ is canonical in the sense that if $\gamma: R \rightarrow R/K$ is the canonical map, then $\phi = \gamma\psi$. ∎

29.2 MAXIMAL AND PRIME IDEALS

We now take up the question of when a quotient ring of a ring is a field and when it is an integral domain. The analogy with groups in Chapter 13 can be stretched a bit further to cover the case in which the quotient ring is a field.

Definition (Analog of the definition of a maximal normal subgroup)   A *maximal ideal of a ring* R is an ideal M different from R such that there is no proper ideal N of R properly containing M.

Theorem 29.4 (Analog of Theorem 13.4)   *Let R be a commutative ring with unity. Then M is a maximal ideal of R if and only if R/M is a field.*

Proof.   Suppose M is a maximal ideal in R. It is easy to see that if R is a commutative ring with unity, then R/M is also a commutative ring with unity if $M \neq R$, which is the case if M is maximal. Let $(a + M) \in R/M$, with $a \notin M$, so that $a + M$ is not the additive identity of R/M. We must show that $a + M$ has a multiplicative inverse in R/M. Let

$$N = \{ra + m \mid r \in R, m \in M\}.$$

Then $\langle N, + \rangle$ is a group, for

$$(r_1 a + m_1) + (r_2 a + m_2) = (r_1 + r_2)a + (m_1 + m_2),$$

and the latter is clearly in N, while also,

$$0 = 0a + 0 \qquad \text{and} \qquad -(ra + m) = (-r)a + (-m).$$

Now

$$r_1(ra + m) = (r_1 r)a + r_1 m$$

shows that $r_1(ra + m) \in N$ for $r_1 \in R$, and since R is a commutative ring, $(ra + m)r_1 \in N$ also. Thus N is an ideal. But

$$a = 1a + 0.$$

shows that $a \in N$, and for $m \in M$,

$$m = 0a + m$$

shows that $M \subseteq N$. Hence N is an ideal of R properly containing M, since $a \in N$ and $a \notin M$. Since M is maximal, we must have $N = R$. In particular, $1 \in N$. Then by definition of N, there is $b \in R$ and $m \in M$ such that $1 = ba + m$. Therefore,

$$1 + M = ba + M = (b + M)(a + M),$$

so $b + M$ is a multiplicative inverse of $a + M$.

Conversely, suppose that R/M is a field. By Theorem 29.2, if N is any ideal of R such that $M \subset N \subset R$ and γ is the canonical homomorphism of R onto R/M, then $N\gamma$ is an ideal of R/M with $\{(0 + M)\} \subset N\gamma \subset R/M$, contrary to the corollary of Theorem 28.4, which states that the field R/M contains no proper nontrivial ideals. Hence if R/M is a field, M is maximal. ∎

Corollary   *A commutative ring with unity is a field if and only if it has no proper nontrivial ideals.*

Proof. The corollary of Theorem 28.4 shows that a field has no proper nontrivial ideals.

Conversely, if a commutative ring R with unity has no proper nontrivial ideals, then $\{0\}$ is a maximal ideal and $R/\{0\}$, which is isomorphic to R, is a field by Theorem 29.4. ■

We now turn to the question of the characterization of the ideals $N \neq R$ for a commutative ring R with unity such that R/N is an integral domain. The answer here is rather obvious. The factor ring R/N will be an integral domain if and only if $(a + N)(b + N) = N$ implies that either

$$a + N = N \qquad \text{or} \qquad b + N = N.$$

This is exactly the statement that R/N has no divisors of 0, since the coset N plays the role of 0 in R/N. Looking at representatives, we see that this condition amounts to saying that $ab \in N$ implies that either $a \in N$ or $b \in N$.

Example 29.1   The ideals of $\mathbf{Z}$ are of the form $n\mathbf{Z}$. We have seen that $\mathbf{Z}/n\mathbf{Z} \simeq \mathbf{Z}_n$ and that $\mathbf{Z}_n$ is an integral domain if and only if n is a prime. Thus the ideals $n\mathbf{Z}$ such that $\mathbf{Z}/n\mathbf{Z}$ is an integral domain are of the form $p\mathbf{Z}$, where p is a prime. Of course, $\mathbf{Z}/p\mathbf{Z}$ is actually a field, so that $p\mathbf{Z}$ is a maximal ideal of $\mathbf{Z}$. Note that for a product rs of integers to be in $p\mathbf{Z}$, the prime p must divide either r or s. The role of prime integers in this example makes the use of the word *prime* in the next definition more reasonable. ■

Definition   An ideal $N \neq R$ in a commutative ring R is a ***prime ideal*** if $ab \in N$ implies that either $a \in N$ or $b \in N$ for all $a, b \in R$.

Our remarks preceding Example 29.1 constitute a proof of the following theorem.

Theorem 29.5   *Let R be a commutative ring with unity, and let $N \neq R$ be an ideal in R. Then R/N is an integral domain if and only if N is a prime ideal in R.*

Corollary   *Every maximal ideal in a commutative ring R with unity is a prime ideal.*

Proof.   If M is maximal in R, then R/M is a field, hence an integral domain, and therefore M is a prime ideal by Theorem 29.5. ■

The material that has just been presented regarding maximal and prime ideals is very important and we shall be using it quite a lot. You should keep the main ideas well in mind, even though you may have become lost in the fascinating proof of Theorem 29.4. That is, you must know and understand the definitions of maximal and prime ideals and must remember the following facts that we have demonstrated.

For a commutative ring R with unity:

1. *An ideal M of R is maximal if and only if R/M is a field.*
2. *An ideal N of R is prime if and only if R/N is an integral domain.*
3. *Every maximal ideal of R is a prime ideal.*

29.3 PRIME FIELDS

Let R be any ring with unity 1. Recall that by $n \cdot 1$ we mean $1 + 1 + \cdots + 1$ for n summands for $n > 0$, and $(-1) + (-1) + \cdots + (-1)$ for $|n|$ summands for $n < 0$, while $n \cdot 1 = 0$ for $n = 0$.

Theorem 29.6 *If R is a ring with unity 1, then the map $\phi : \mathbf{Z} \to R$ given by*

$$n\phi = n \cdot 1$$

for $n \in \mathbf{Z}$ is a homomorphism of $\mathbf{Z}$ into R.

Proof. It is obvious that

$$(n + m)\phi = (n + m) \cdot 1 = (n \cdot 1) + (m \cdot 1) = n\phi + m\phi.$$

The distributive laws in R show that

$$\underbrace{(1 + 1 + \cdots + 1)}_{n \text{ summands}} \underbrace{(1 + 1 + \cdots + 1)}_{m \text{ summands}} = \underbrace{(1 + 1 + \cdots + 1)}_{nm \text{ summands}}.$$

Thus $(n \cdot 1)(m \cdot 1) = (nm) \cdot 1$ for $n, m > 0$. Similar arguments with the distributive laws shows that for all $n, m \in \mathbf{Z}$, we have

$$(n \cdot 1)(m \cdot 1) = (nm) \cdot 1.$$

Thus

$$(nm)\phi = (nm) \cdot 1 = (n \cdot 1)(m \cdot 1) = (n\phi)(m\phi). \ \blacksquare$$

Corollary *If R is a ring with unity and characteristic $n > 1$, then R contains a subring isomorphic to $\mathbf{Z}_n$. If R has characteristic 0, then R contains a subring isomorphic to $\mathbf{Z}$.*

Proof. The map $\phi : \mathbf{Z} \to R$ given by $m\phi = m \cdot 1$ for $m \in \mathbf{Z}$ is a homomorphism by Theorem 29.6. The kernel must be an ideal in $\mathbf{Z}$. All ideals in $\mathbf{Z}$ are of the form $s\mathbf{Z}$ for some $s \in \mathbf{Z}$. By Theorem 24.5, it is clear that if R has characteristic $n > 0$, then the kernel of ϕ is $n\mathbf{Z}$. Then the image $\mathbf{Z}\phi \le R$ is isomorphic to $\mathbf{Z}/n\mathbf{Z} \simeq \mathbf{Z}_n$. If the characteristic of R is 0, then $m \cdot 1 \ne 0$ for all $m \ne 0$, so the kernel of ϕ is $\{0\}$. Thus, the image $\mathbf{Z}\phi \le R$ is isomorphic to $\mathbf{Z}$. $\blacksquare$

Theorem 29.7 *A field F is either of prime characteristic p and contains a subfield isomorphic to $\mathbf{Z}_p$ or of characteristic 0 and contains a subfield isomorphic to $\mathbf{Q}$.*

Proof.   If the characteristic of F is not 0, the above corollary shows that F contains a subring isomorphic to $\mathbf{Z}_n$. Then n must be a prime p, or F would have 0 divisors. If F is of characteristic 0, then F must contain a subring isomorphic to $\mathbf{Z}$. In this case the corollaries of Theorem 26.2 show that F must contain a field of quotients of this subring, and that this field of quotients must be isomorphic to $\mathbf{Q}$. ∎

Thus every field contains either a subfield isomorphic to $\mathbf{Z}_p$ for some prime p or a subfield isomorphic to $\mathbf{Q}$. These fields $\mathbf{Z}_p$ and $\mathbf{Q}$ are the fundamental building blocks on which all fields rest.

Definition   The fields $\mathbf{Z}_p$ and $\mathbf{Q}$ are *prime fields*.

Exercises

29.1 Describe all ring homomorphisms of $\mathbf{Z}$ into $\mathbf{Z}$. [*Hint:* By group theory, a homomorphism of a ring R is determined by its values on an additive generating set of the group $\langle R, + \rangle$. Describe the homomorphisms by giving their values on the generator 1 of $\langle \mathbf{Z}, + \rangle$.]

29.2 Describe all ring homomorphisms of $\mathbf{Z} \times \mathbf{Z}$ into $\mathbf{Z}$. [See the hint of Exercise 29.1.]

29.3 Give an example of a ring homomorphism $\phi : R \to S$ where R has unity 1_R and $1_R \phi \neq 0$, but $1_R \phi$ is not unity for S.

29.4 Find all prime ideals and all maximal ideals of $\mathbf{Z}_{12}$.

29.5 Find a maximal ideal of $\mathbf{Z} \times \mathbf{Z}$. Find a prime ideal of $\mathbf{Z} \times \mathbf{Z}$ that is not maximal. Find a proper ideal of $\mathbf{Z} \times \mathbf{Z}$ that is not prime.

†**29.6** Prove directly from the definitions of maximal and prime ideals that every maximal ideal of a commutative ring R with unity is a prime ideal. [*Hint:* Suppose M is maximal in R, $ab \in M$, and $a \notin M$. Consider the ideal of R generated by a and M.]

29.7 Mark each of the following true or false.

___ a) The concept of a ring homomorphism is closely connected with the idea of a factor ring.

___ b) A homomorphism is to a ring as an isomorphism is to a group.

___ c) A ring homomorphism is a one-to-one map if and only if the kernel is 0.

___ d) In a sense, a field is to the theory of commutative rings with unity as a simple group is to the theory of groups.

___ e) The kernel of a ring homomorphism is an ideal of the whole ring.

___ f) Every prime ideal of every commutative ring with unity is a maximal ideal.

___ g) Every maximal ideal of every commutative ring with unity is a prime ideal.

— h) **Q** is its own prime subfield.

— i) The prime subfield of **C** is **R**.

— j) Every field contains a prime field as a subfield.

29.8 Describe all ring homomorphisms of $\mathbf{Z} \times \mathbf{Z}$ into $\mathbf{Z} \times \mathbf{Z}$. [See the hint of Exercise 29.1.]

29.9 Show that each homomorphism of a field is either one to one or maps everything onto 0.

29.10 Show that if R, R', and R'' are rings, and if $\phi:R \to R'$ and $\psi:R' \to R''$ are homomorphisms, then the composite function $\phi\psi:R \to R''$ is a homomorphism. (Use Exercise 13.14.)

29.11 Let R be a commutative ring with unity of prime characteristic p. Show that the map $\phi_p:R \to R$ given by $a\phi = a^p$ is a homomorphism (the **Frobenius homomorphism**).

29.12 Let R and S be rings and let $\phi:R \to S$ be a ring homomorphism such that $R\phi \neq \{0\}$. Show that if R has unity 1_R and S has no 0 divisors, then $1_R\phi$ is unity for S.

29.13 Show that N is a maximal ideal in a ring R if and only if R/N is a simple ring, that is, it has no proper nontrivial ideals. (Compare with Theorem 13.4.)

29.14 The corollary of Theorem 29.6 tells us that every ring with unity contains a subring isomorphic to either $\mathbf{Z}$ or some $\mathbf{Z}_n$. Is it possible that a ring with unity may simultaneously contain two subrings isomorphic to $\mathbf{Z}_n$ and $\mathbf{Z}_m$ for $n \neq m$? Is it possible that a ring with unity may simultaneously contain two subrings isomorphic to the fields $\mathbf{Z}_p$ and $\mathbf{Z}_q$ for two different primes p and q? If it is impossible, prove it. If it is possible, give illustrations. (This is related to Exercise 23.10.)

29.15 Following the idea of Exercise 29.14, is it possible for an integral domain to contain two subrings isomorphic to $\mathbf{Z}_p$ and $\mathbf{Z}_q$ for $p \neq q$ and p and q both prime? Give reasons or an illustration. (This is related to Exercise 23.11.)

29.16 Let R and R' be rings and let N and N' be ideals of R and R', respectively. Let ϕ be a homomorphism of R into R'. Show that ϕ induces a natural homomorphism $\phi_*:R/N \to R'/N'$ if $N\phi \subseteq N'$. (Use Exercise 13.17.)

29.17 Let ϕ be a homomorphism of a ring R with unity onto a ring R'. Let u be a unit in R. Show that $u\phi$ is a unit in R' if and only if u is not in the kernel of ϕ.

*29.18 (Second Isomorphism Theorem for Rings) Let M and N be ideals of a ring R and let

$$M + N = \{m + n \,|\, m \in M, n \in N\}.$$

Show that $M + N$ is an ideal of R, and that $(M + N)/N$ is naturally isomorphic to $M/(M \cap N)$. (Use Theorem 15.2.)

*29.19 (Third Isomorphism Theorem for Rings) Let M and N be ideals of a ring R such that $M \leq N$. Show that there is a natural isomorphism of R/N with $(R/M)/(N/M)$. (Use Theorem 15.3.)

*29.20 Show that $\phi:\mathbf{C} \to M_2(\mathbf{R})$ given by

$$(a + bi)\phi = \begin{pmatrix} a & b \\ -b & a \end{pmatrix}$$

for $a, b \in \mathbf{R}$ is an isomorphism of **C** into $M_2(\mathbf{R})$.

*29.21 Let R be a ring with unity and let $Hom(\langle R, + \rangle)$ be the ring of endomorphisms of $\langle R, + \rangle$ as described in Section 25.2. Let $a \in R$, and let $\rho_a : R \to R$ be given by

$$x\rho_a = xa$$

for $x \in R$.

a) Show that ρ_a is an endomorphism of $\langle R, + \rangle$.

b) Show that $R' = \{\rho_a | a \in R\}$ is a subring of $Hom(\langle R, + \rangle)$.

c) Prove the analog of Cayley's Theorem for R by showing that R' of (b) is isomorphic to R.

thirty

Rings
of
Polynomials

30.1 POLYNOMIALS IN AN INDETERMINATE

You probably have a pretty workable idea of what constitutes a *polynomial in x with coefficients in a ring R*. You know how to add and multiply such objects, you have done this for years, and you know what is meant by *degree of a polynomial*. If you are indeed this fortunate, we suggest that you proceed at once to the statement of Theorem 30.1, skip the proof as trivial, and start reading after the proof. Oh yes, we should say that we are considering x an *indeterminate* rather than a variable.

Our problem is twofold: to explain what a polynomial is, and to explain what x is. If we just define a polynomial with coefficients in a ring R to be a *finite formal sum*

$$\sum_{i=0}^{n} a_i x^i = a_0 + a_1 x + \cdots + a_n x^n,$$

where $a_i \in R$, we get ourselves into a bit of trouble. For surely $0 + a_1 x$ and $0 + a_1 x + 0x^2$ are different as formal sums, but we want to regard them as the same polynomial. Perhaps the easy way out is to define a polynomial as an *infinite formal sum*

$$\sum_{i=0}^{\infty} a_i x^i = a_0 + a_1 x + \cdots + a_n x^n + \cdots,$$

where $a_i = 0$ for all but a finite number of values of i. Now there is no problem of having more than one formal sum represent what we wish to consider a single polynomial.

This brings us to the question of what x is. The *elegant* thing to do is to throw x away altogether. A polynomial $a_0 + a_1 x + \cdots + a_n x^n + \cdots$ is completely determined by its sequence of coefficients

$$a_0, a_1, \ldots, a_n, \ldots,$$

which doesn't involve the problem child x. A polynomial could be *elegantly defined* as such a sequence. Why carry x around when you don't even need it? Mathematicians have simply become used to x, that is why. It has seniority, tenure, or whatever it is that things have in the WFMN (World Federation of Mathematical Notations). If your ring has unity 1 and you must have x, the *elegant* thing to do is to define x to be the sequence

$$0, 1, 0, 0, 0, \ldots.$$

But then x would probably lodge a protest with the Grievance Committee at such treatment. Since its role is that of excess baggage anyway, let's not fuss too much about what it really is. Let's agree just to call it an **indeterminate.** You have to admit, that is a pretty good term for something you find hard to analyze. Perhaps *indeterminable* would be better. One thing is sure, it isn't 0 or 2 or any other number. *Thus from now on, we shall never be writing expressions such as $x = 0$ or $x = 2$.*

We are now about ready to define a polynomial with coefficients in a ring R as an infinite formal sum

$$\sum_{i=0}^{\infty} a_i x^i = a_0 + a_1 x + \cdots + a_n x^n + \cdots,$$

where $a_i \in R$ and $a_i = 0$ for all but a finite number of values of i. But with $R = \mathbf{Z}$, $2 + x^2$ would not be a polynomial! We would always have to write

$$2 + 0x + 1x^2 + 0x^3 + 0x^4 + \cdots$$

Thus we shall change the notation a bit after the definition.

Definition   Let R be a ring. A **polynomial $f(x)$ with coefficients** in R is an infinite formal sum

$$\sum_{i=0}^{\infty} a_i x^i = a_0 + a_1 x + \cdots + a_n x^n + \cdots,$$

where $a_i \in R$ and $a_i = 0$ for all but a finite number of values of i. The a_i are **coefficients of $f(x)$**. If for some $i > 0$ it is true that $a_i \neq 0$, the largest such value of i is the **degree of $f(x)$**. If no such $i > 0$ exists, then $f(x)$ is of **degree zero**.

Let us agree that if $f(x) = a_0 + a_1 x + \cdots + a_n x^n + \cdots$ has $a_i = 0$ for $i > n$, then we may denote $f(x)$ by $a_0 + a_1 x + \cdots + a_n x^n$. Also, if some $a_i = 1$, we may drop this from the formal sum, so that we shall consider, for example, $2 + x$ to be the polynomial $2 + 1x$ with coefficients in $\mathbf{Z}$. Finally, we shall agree that we may omit altogether from the formal sum any term $0x^i$, or a_0, if $a_0 = 0$ but not all $a_i = 0$. Thus 0, 2, x, and $2 + x^2$ are all polynomials with coefficients in $\mathbf{Z}$. An element of R is a **constant polynomial**.

Addition and multiplication of polynomials with coefficients in a ring R are defined in a way formally familiar to you. If

$$f(x) = a_0 + a_1 x + \cdots + a_n x^n + \cdots$$

and

$$g(x) = b_0 + b_1 x + \cdots + b_n x^n + \cdots,$$

then for polynomial addition, we have

$$f(x) + g(x) = c_0 + c_1 x + \cdots + c_n x^n + \cdots,$$

where $c_n = a_n + b_n$, and for polynomial multiplication, we have

$$f(x)g(x) = d_0 + d_1 x + \cdots + d_n x^n + \cdots,$$

where $d_n = \sum_{i=0}^{n} a_i b_{n-i}$. It is clear that again both c_i and d_i are 0 for all but a finite number of values of i, so these definitions make sense. Note that $\sum_{i=0}^{n} a_i b_{n-i}$ need not equal $\sum_{i=0}^{n} b_i a_{n-i}$ if R is not commutative. With these definitions of addition and multiplication, we have the following theorem.

Theorem 30.1   *The set $R[x]$ of all polynomials in an indeterminate x with coefficients in a ring R is a ring under polynomial addition and multiplication. If R is commutative, then so is $R[x]$, and if R has unity 1, then 1 is also unity for $R[x]$.*

Proof.   That $\langle R[x], + \rangle$ is an abelian group is obvious. The associative law for multiplication and the distributive laws are straightforward, but slightly cumbersome, computations. We illustrate by proving the associative law.

Applying ring axioms to $a_i, b_j, c_k \in R$, we obtain

$$\left[\left(\sum_{i=0}^{\infty} a_i x^i\right)\left(\sum_{j=0}^{\infty} b_j x^j\right)\right]\left(\sum_{k=0}^{\infty} c_k x^k\right) = \left[\sum_{n=0}^{\infty}\left(\sum_{i=0}^{n} a_i b_{n-i}\right) x^n\right]\left(\sum_{k=0}^{\infty} c_k x^k\right)$$

$$= \sum_{s=0}^{\infty}\left[\sum_{n=0}^{s}\left(\sum_{i=0}^{n} a_i b_{n-i}\right) c_{s-n}\right] x^s$$

$$= \sum_{s=0}^{\infty}\left(\sum_{i+j+k=s} a_i b_j c_k\right) x^s$$

$$= \sum_{s=0}^{\infty}\left[\sum_{m=0}^{s} a_{s-m}\left(\sum_{j=0}^{m} b_j c_{m-j}\right)\right] x^s$$

$$= \left(\sum_{i=0}^{\infty} a_i x^i\right)\left[\sum_{m=0}^{\infty}\left(\sum_{j=0}^{m} b_j c_{m-j}\right) x^m\right]$$

$$= \left(\sum_{i=0}^{\infty} a_i x^i\right)\left[\left(\sum_{j=0}^{\infty} b_j x^j\right)\left(\sum_{k=0}^{\infty} c_k x^k\right)\right].$$

Whew!! The distributive laws are similarly proved.

The comments prior to the statement of the theorem show that $R[x]$ is a commutative ring if R is commutative, and a unity 1 in R is obviously also unity for $R[x]$, in view of the definition of multiplication in $R[x]$. ∎

Thus $\mathbf{Z}[x]$ is the ring of polynomials in the indeterminate x with integral coefficients, $\mathbf{Q}[x]$ the ring of polynomials in x with rational coefficients, and so on.

Example 30.1 In $\mathbf{Z}_2[x]$, we have

$$(x + 1)^2 = (x + 1)(x + 1) = x^2 + (1 + 1)x + 1 = x^2 + 1.$$

Still working in $\mathbf{Z}_2[x]$, we obtain

$$(x + 1) + (x + 1) = (1 + 1)x + (1 + 1) = 0x + 0 = 0. \quad ∎$$

If R is a ring and x and y are two indeterminates, then we can form the ring $(R[x])[y]$, that is, the ring of polynomials in y with coefficients that are polynomials in x. It is pretty obvious, but somewhat tedious to prove carefully, that $(R[x])[y]$ is naturally isomorphic to $(R[y])[x]$. Every polynomial in y with coefficients that are polynomials in x can be naturally rewritten as a polynomial in x with coefficients that are polynomials in y. We shall identify these rings by means of this natural isomorphism and shall consider this ring $R[x, y]$ the **ring of polynomials in two indeterminates x and y with** coefficients in R. The **ring** $R[x_1, \ldots, x_n]$**of polynomials in the** n **indeterminates** x_i **with coefficients in** R is similarly defined.

We leave as an exercise the easy proof that if D is an integral domain then so is $D[x]$. In particular, if F is a field, then $F[x]$ is an integral domain. Note that $F[x]$ is not a field, for x is not a unit in $F[x]$. That is, there is no polynomial $f(x) \in F[x]$ such that $xf(x) = 1$. By Theorem 26.1, one can construct the field of quotients $F(x)$ of $F[x]$. Any element in $F(x)$ can be represented as a quotient $f(x)/g(x)$ of two polynomials in $F[x]$ with $g(x) \neq 0$. We similarly define $F(x_1, \ldots, x_n)$ to be the field of quotients of $F[x_1, \ldots, x_n]$. This field $F(x_1, \ldots, x_n)$ is the **field of rational functions in** n **indeterminates over** F. These fields play a very important role in algebraic geometry.

30.2 THE EVALUATION HOMOMORPHISMS

We are now ready to proceed to show, as we promised in Chapter 27, how the machinery of homomorphisms and factor rings can be used to study what the student has always referred to as "solving a polynomial equation." Let E and F be fields, with F a subfield of E, that is, $F \leq E$. The next theorem asserts the existence of very important homomorphisms of $F[x]$ into E. *These homomorphisms will be the fundamental tools for all the rest of our work.* If you really comprehend them and the homomorphism theory in the preceding section, you are in pretty good shape for the rest of the course.

Theorem 30.2 (The Evaluation Homomorphisms for Field Theory)   *Let F be a subfield of a field E, let α be any element of E, and let x be an indeterminate. The map $\phi_\alpha : F[x] \to E$ defined by*

$$(a_0 + a_1 x + \cdots + a_n x^n)\phi_\alpha = a_0 + a_1\alpha + \cdots + a_n\alpha^n$$

for $(a_0 + a_1 x + \cdots + a_n x^n) \in F[x]$ is a homomorphism of $F[x]$ into E. Also, $x\phi_\alpha = \alpha$, and ϕ_α maps F isomorphically by the identity map, that is, $a\phi_\alpha = a$ for $a \in F$. The homomorphism ϕ_α is **evaluation at** α.

Proof.   The lattice and mapping diagram in Fig. 30.1 may help you to visualize this situation. The dashed lines indicate an element of the set. The theorem is really an immediate consequence of our definitions of addition and multiplication in $F[x]$. It is clear that the map ϕ_α is well defined, that is, independent of our representation of $f(x) \in F[x]$ as a finite sum

$$a_0 + a_1 x + \cdots + a_n x^n.$$

Such a finite sum representing $f(x)$ can be changed only by insertion or deletion of terms $0x^i$, which clearly does not affect the value of $(f(x))\phi_\alpha$.

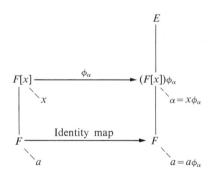

Fig. 30.1

If $f(x) = a_0 + a_1 x + \cdots + a_n x^n$, $g(x) = b_0 + b_1 x + \cdots + b_m x^m$, and $h(x) = f(x) + g(x) = c_0 + c_1 x + \cdots + c_r x^r$, then

$$(f(x) + g(x))\phi_\alpha = (h(x))\phi_\alpha = c_0 + c_1 \alpha + \cdots + c_r \alpha^r,$$

while

$$(f(x))\phi_\alpha + (g(x))\phi_\alpha = (a_0 + a_1 \alpha + \cdots + a_n \alpha^n) + (b_0 + b_1 \alpha + \cdots + b_m \alpha^m).$$

Since by definition of polynomial addition we have $c_i = a_i + b_i$, we see that

$$(f(x) + g(x))\phi_\alpha = (f(x))\phi_\alpha + (g(x))\phi_\alpha.$$

Turning to multiplication, we see that if

$$f(x)g(x) = d_0 + d_1 x + \cdots + d_s x^s,$$

then

$$(f(x)g(x))\phi_\alpha = d_0 + d_1 \alpha + \cdots + d_s \alpha^s,$$

while

$$[(f(x))\phi_\alpha][(g(x))\phi_\alpha] = (a_0 + a_1 \alpha + \cdots + a_n \alpha^n)(b_0 + b_1 \alpha + \cdots + b_m \alpha^m).$$

Since by definition of polynomial multiplication, $d_j = \sum_{i=0}^{j} a_i b_{j-i}$, we see that

$$(f(x)g(x))\phi_\alpha = [(f(x))\phi_\alpha][(g(x))\phi_\alpha].$$

Thus ϕ_α is a homomorphism.

The very definition of ϕ_α applied to a constant polynomial $a \in F[x]$ where $a \in F$, gives $a\phi_\alpha = a$, so ϕ_α maps F isomorphically by the identity map. Again by definition of ϕ_α, we have $x\phi_\alpha = (1x)\phi_\alpha = 1\alpha = \alpha$. ∎

We point out that this theorem is valid with the identical proof if F and E are merely commutative rings with unity rather than fields. However we shall be interested only in the case in which they are fields.

It is hard to overemphasize the importance of this simple theorem for us. It is the very foundation for all of our further work in field theory. It is so simple that it could justifiably be called an *observation* rather than a theorem. Perhaps we shouldn't have tried to write out the proof. The polynomial notation makes it look so complicated that you may be fooled into thinking it is a difficult theorem.

Example 30.2 Let F be $\mathbf{Q}$ and E be $\mathbf{R}$ in Theorem 30.2, and consider the evaluation homomorphism $\phi_0:\mathbf{Q}[x] \to \mathbf{R}$. Here

$$(a_0 + a_1x + \cdots + a_nx^n)\phi_0 = a_0 + a_10 + \cdots + a_n0^n = a_0.$$

Thus every polynomial is mapped onto its constant term. Note that the kernel of ϕ_0 is the ideal N of all polynomials with constant term 0. By Theorem 29.3, the image $(\mathbf{Q}[x])\phi_0 = \mathbf{Q}$ is naturally isomorphic to $\mathbf{Q}[x]/N$. A coset element of $\mathbf{Q}[x]/N$ consists precisely of all polynomials having a given fixed constant term. ∎

Example 30.3 Let F be $\mathbf{Q}$ and E be $\mathbf{R}$ in Theorem 30.2, and consider the evaluation homomorphism $\phi_2:\mathbf{Q}[x] \to \mathbf{R}$. Here

$$(a_0 + a_1x + \cdots + a_nx^n)\phi_2 = a_0 + a_12 + \cdots + a_n2^n.$$

Note that

$$(x^2 + x - 6)\phi_2 = 2^2 + 2 - 6 = 0.$$

Thus $x^2 + x - 6$ is in the kernel N of ϕ_2. Of course,

$$x^2 + x - 6 = (x - 2)(x + 3),$$

and, if you like, the reason that $(x^2 + x - 6)\phi_2 = 0$ is that $(x - 2)\phi_2 = 2 - 2 = 0$. We shall see later that N is precisely the ideal of all polynomials of the form $(x - 2)f(x)$ for $f(x) \in \mathbf{Q}[x]$. Here the image of $\mathbf{Q}[x]$ under ϕ_2 is again just $\mathbf{Q}$, and again by Theorem 29.3, $\mathbf{Q}[x]/N$ is naturally isomorphic to $\mathbf{Q}$. ∎

Example 30.4 Let F be $\mathbf{Q}$ and E be $\mathbf{C}$ in Theorem 30.2, and consider the evaluation homomorphism $\phi_i:\mathbf{Q}[x] \to \mathbf{C}$. Here

$$(a_0 + a_1x + \cdots + a_nx^n)\phi_i = a_0 + a_1i + \cdots + a_ni^n$$

and $x\phi_i = i$. Note that

$$(x^2 + 1)\phi_i = i^2 + 1 = 0,$$

so $x^2 + 1$ is in the kernel N of ϕ_i. We shall see later that N is precisely the ideal of all polynomials of the form $(x^2 + 1)f(x)$ for $f(x) \in \mathbf{Q}[x]$. By Theorem 29.3, the image $(\mathbf{Q}[x])\phi_i$ is naturally isomorphic to $\mathbf{Q}[x]/N$. We shall see later that this ring $(\mathbf{Q}[x])\phi_i$ consists of all complex numbers of the form $q_1 + q_2i$ for $q_1, q_2 \in \mathbf{Q}$, and is a subfield of $\mathbf{C}$. ∎

Example 30.5 Let F be $\mathbf{Q}$ and let E be $\mathbf{R}$ in Theorem 30.2, and consider the evaluation homomorphism $\phi_\pi : \mathbf{Q}[x] \to \mathbf{R}$. Here

$$(a_0 + a_1 x + \cdots + a_n x^n)\phi_\pi = a_0 + a_1\pi + \cdots + a_n\pi^n.$$

It can be proved that $a_0 + a_1\pi + \cdots + a_n\pi^n = 0$ if and only if $a_i = 0$ for $i = 0, 1, \ldots, n$. Thus, the kernel of ϕ_π is $\{0\}$, and ϕ_π is an isomorphic mapping. This shows that all *formal polynomials in π with rational coefficients* form a ring isomorphic to $\mathbf{Q}[x]$ in a natural way with $x\phi_\pi = \pi$. ∎

30.3 THE NEW APPROACH

We now complete the connection between our new ideas and the classical concept of solving a polynomial equation. Rather than speak of *solving a polynomial equation*, we shall refer to *finding a zero of a polynomial*.

Definition Let F be a subfield of a field E, and let α be an element of E. Let $f(x) = a_0 + a_1 x + \cdots + a_n x^n$ be in $F[x]$, and let $\phi_\alpha : F[x] \to E$ be the evaluation homomorphism of Theorem 30.2. Let $f(\alpha)$ denote

$$(f(x))\phi_\alpha = a_0 + a_1\alpha + \cdots + a_n\alpha^n.$$

If $f(\alpha) = 0$, then α is a *zero of* $f(x)$.

In terms of this definition, we can rephrase the classical problem of finding all real solutions of the polynomial equation $r^2 + r - 6 = 0$ by letting $F = \mathbf{Q}$ and $E = \mathbf{R}$ and *finding all $\alpha \in \mathbf{R}$ such that*

$$(x^2 + x - 6)\phi_\alpha = 0,$$

that is, finding all zeros of $x^2 + x - 6$ in $\mathbf{R}$. Both problems have the same answer, since

$$\{\alpha \in \mathbf{R} \,|\, (x^2 + x - 6)\phi_\alpha = 0\} = \{r \in \mathbf{R} \,|\, r^2 + r - 6 = 0\} = \{2, -3\}.$$

It probably seems to you that we have merely succeeded in making a simple problem seem quite complicated. *What we have done is to phrase the problem in the language of mappings, and we can now use all the mapping machinery that we have developed for its solution.* Remember our *basic goal*, which can now be rephrased as follows.

Basic goal: To show that for a field F, every nonconstant polynomial $f(x) \in F[x]$ has a zero.

Let us cheat and take a little peek ahead and see how our *basic goal* might be accomplished. If $f(x)$ has no zero in F, we have to *manufacture* in some fashion a field E containing F such that there is an α in E with $f(\alpha) = (f(x))\phi_\alpha = 0$. How can we build E? Now E is to contain the image $(F[x])\phi_\alpha$ of $F[x]$ under our evaluation homomorphism ϕ_α. Remember that by Theorem 29.3 $(F[x])\phi_\alpha$ is isomorphic to $F[x]/(\text{kernel of } \phi_\alpha)$. This suggests that we try to form E by forming a factor ring $F[x]/N$ for some suitable ideal N in $F[x]$. Of course to make $F[x]/N$ a field, N must be a *maximal ideal* of $F[x]$, by Theorem 29.4. Thus our task in the next section, as our final step toward the achievement of our *basic goal*, will be to examine the nature of ideals in $F[x]$. This is how all the machinery we have developed concerning factor rings and homomorphisms comes into play. There is a saying: "Don't bring a cannon on stage unless you intend to fire it." We brought the cannon onto the stage in Chapters 28 and 29, and we are now priming it for firing.

Exercises

30.1 Find the sum and product of $f(x) = 2x^3 + 4x^2 + 3x + 2$ and $g(x) = 3x^4 + 2x + 4$, given that $f(x), g(x) \in \mathbf{Z}_5[x]$.

30.2 Consider the element

$$f(x, y) = (3x^3 + 2x)y^3 + (x^2 - 6x + 1)y^2 + (x^4 - 2x)y + (x^4 - 3x^2 + 2)$$

of $(\mathbf{Q}[x])[y]$. Write $f(x, y)$ as it would appear if viewed as an element of $(\mathbf{Q}[y])[x]$.

30.3 Let $F = E = \mathbf{Z}_7$ in Theorem 30.2. Evaluate each of the following for the indicated evaluation homomorphism $\phi_a : \mathbf{Z}_7[x] \to \mathbf{Z}_7$.

a) $(x^2 + 3)\phi_2$

b) $(2x^3 - x^2 + 3x + 2)\phi_0$

c) $[(x^4 + 2x)(x^3 - 3x^2 + 3)]\phi_3$

d) $[(x^3 + 2)(4x^2 + 3)(x^7 + 3x^2 + 1)]\phi_5$

30.4 Consider the evaluation homomorphism $\phi_5 : \mathbf{Q}[x] \to \mathbf{R}$. Find six elements in the kernel of ϕ_5.

30.5 Find all zeros in $\mathbf{Z}_5$ of $(x^5 + 3x^3 + x^2 + 2x) \in \mathbf{Z}_5[x]$. [*Hint:* There are only five candidates. Try them.]

†**30.6** Prove that if D is an integral domain, then $D[x]$ is an integral domain.

30.7 Mark each of the following true or false.

— a) The polynomial $(a_nx^n + \cdots + a_1x + a_0) \in R[x]$ is zero if and only if $a_i = 0$, for $i = 0, 1, \ldots, n$.

— b) If R is a commutative ring, then $R[x]$ is commutative.

— c) If D is an integral domain, then $D[x]$ is an integral domain.

— d) If R is a ring containing divisors of zero, then $R[x]$ has divisors of zero.

— e) If R is a ring and $f(x)$ and $g(x)$ in $R[x]$ are of degrees 3 and 4, respectively, then $f(x)g(x)$ may be of degree 8 in $R[x]$.

— f) If R is any ring and $f(x)$ and $g(x)$ in $R[x]$ are of degrees 3 and 4, respectively, then $f(x)g(x)$ is always of degree 7.

— g) If F is a subfield of a field E and $a \in E$ is a zero of $f(x) \in F[x]$, then a is a zero of $h(x) = f(x)g(x)$ for all $g(x) \in F[x]$.

— h) The evaluation homomorphism ϕ_α of Theorem 30.2 is an extension of the injection map $i : F \to E$ to $F[x]$, where $(a)i = a$ for $a \in F$.

— i) If F is a subfield of a field E and $f(x) \in F[x]$, then the set of all zeros of $f(x)$ in E is an ideal of E.

— j) If F is a subfield of a field E and $\alpha \in E$, then the set of all $f(x) \in F[x]$ such that $f(\alpha) = 0$ is an ideal of $F[x]$.

30.8 Prove the left distributive law for $R[x]$, where R is a ring and x is an indeterminate.

30.9 Let F be a field and let D be the formal polynomial differentiation map, so that

$$D(a_0 + a_1 x + a_2 x^2 + \cdots + a_n x^n) = a_1 + 2a_2 x + \cdots + n a_n x^{n-1}.$$

(We write the map D on the left to conform with the usual convention in analysis.)

a) Show that $D : F[x] \to F[x]$ is a group automorphism of $\langle F[x], + \rangle$. Is D a ring automorphism?

b) Find the kernel of D.

c) Find the image of $F[x]$ under D.

30.10 Let $\phi_a : \mathbf{Z}_5[x] \to \mathbf{Z}_5$ be an evaluation homomorphism as in Theorem 30.2. Use Fermat's theorem to evaluate $(x^{231} + 3x^{117} - 2x^{53} + 1)\phi_3$.

30.11 Use Fermat's theorem to find all zeros in $\mathbf{Z}_5$ of

$$2x^{219} + 3x^{74} + 2x^{57} + 3x^{44}.$$

30.12 Let D be an integral domain and x an indeterminate.

a) Describe the units in $D[x]$.

b) Find the units in $\mathbf{Z}[x]$.

c) Find the units in $\mathbf{Z}_7[x]$.

30.13 Let F be a subfield of a field E.

a) Define an *evaluation homomorphism*

$$\phi_{\alpha_1, \ldots, \alpha_n} : F[x_1, \ldots, x_n] \to E \qquad \text{for} \quad \alpha_i \in E,$$

stating the analog of Theorem 30.2.

b) With $E = F = \mathbf{Q}$, compute $(x_1{}^2 x_2{}^3 + 3x_1{}^4 x_2)\phi_{-3,2}$.

c) Define the concept of a *zero of a polynomial* $f(x_1, \ldots, x_n) \in F[x_1, \ldots, x_n]$ in a way analogous to the definition in the text of a zero of $f(x)$.

30.14 Let F be a field. Show that all polynomials with constant term $a_0 = 0$ form an ideal $\langle x \rangle$ of $F[x]$.

30.15 Let F be a field, and let $\langle x \rangle$ be the ideal in $F[x]$ defined in Exercise 30.14. Show that $F[x]/\langle x \rangle$ is a field isomorphic to F by

a) showing directly that each residue class in $F[x]/\langle x \rangle$ contains exactly one element of F that may be chosen as a representative for computing in $F[x]/\langle x \rangle$,

b) considering the evaluation homomorphism $\phi_0 : F[x] \to F$ as defined in Theorem 30.2, showing that $\langle x \rangle$ is the kernel of ϕ_0 and applying Theorem 29.3.

30.16 Let R be a ring, and let R^R be the set of all functions mapping R into R. For $\phi, \psi \in R^R$, define the sum $\phi + \psi$ by

$$r(\phi + \psi) = (r\phi) + (r\psi)$$

and the product $\phi \cdot \psi$ by

$$r(\phi \cdot \psi) = (r\phi)(r\psi)$$

for $r \in R$. Note that $\cdot$ is *not* function composition. Show that $\langle R^R, +, \cdot \rangle$ is a ring.

30.17 Referring to Exercise 30.16, let F be a field. An element ϕ of F^F is a **polynomial function on** F, if there exists $f(x) \in F[x]$ such that $a\phi = f(a)$ for all $a \in F$.

a) Show that the set P_F of all polynomial functions on F forms a subring of F^F.

b) Show that the ring P_F is not necessarily isomorphic to $F[x]$. [*Hint:* Show that if F is a finite field, P_F and $F[x]$ don't even have the same number of elements.]

30.18 Refer to Exercises 30.16 and 30.17 for the following questions.

a) How many elements are there in $\mathbf{Z}_2^{\mathbf{Z}_2}$? in $\mathbf{Z}_3^{\mathbf{Z}_3}$?

b) Classify $\langle \mathbf{Z}_2^{\mathbf{Z}_2}, + \rangle$ and $\langle \mathbf{Z}_3^{\mathbf{Z}_3}, + \rangle$ by Theorem 9.3, the Fundamental Theorem of Finitely Generated Abelian Groups.

c) Show that if F is a finite field, then $F^F = P_F$. [*Hint:* Of course, $P_F \subseteq F^F$. Let F have as elements $a_1, \ldots, a_n$. Note that if

$$f_i(x) = c(x - a_1) \cdots (x - a_{i-1})(x - a_{i+1}) \cdots (x - a_n),$$

then $f_i(a_j) = 0$ for $j \neq i$, and the value $f_i(a_i)$ can be controlled by the choice of $c \in F$. Use this to show that every function on F is a polynomial function.]

thirty-one

Factorization of Polynomials Over a Field

31.1 THE DIVISION ALGORITHM IN $F[x]$

We indicated at the end of the last section that the final step for the achievement of our *basic goal* would be the study of ideal structure in polynomial rings $F[x]$, where F is a field. Of particular interest is the characterization of maximal ideals of $F[x]$ to determine when $F[x]/N$ is a field. It may seem odd that we have to study factorization of polynomials to do this, but we can make this seem reasonable in two ways. First, our main purpose is to study zeros of polynomials, and suppose that $f(x) \in F[x]$ factors in $F[x]$, so that $f(x) = g(x)h(x)$ for $g(x), h(x) \in F[x]$. If $F \leq E$ and $\alpha \in E$, then $f(\alpha) = 0$ if and only if either $g(\alpha) = 0$ or $h(\alpha) = 0$. Thus the attempt to find a zero of $f(x)$ is reduced to the problem of finding a zero of a factor of $f(x)$. Second, we wish to find maximal ideals of $F[x]$. Now any maximal ideal M is also a prime ideal. Therefore, if $f(x) = g(x)h(x)$ and $f(x) \in M$ for a maximal ideal M, then we must have either $g(x) \in M$ or $h(x) \in M$. These two considerations thus suggest the study of factorization of polynomials in $F[x]$.

The following theorem is the basic tool for our work in this section. The student should note the similarity with Lemma 6.1, the division algorithm for **Z**, the importance of which has been amply demonstrated. Division algorithms will be treated in a more general setting in starred Chapter 33.

Theorem 31.1 (Division Algorithm for $F[x]$).   *Let*

$$f(x) = a_n x^n + a_{n-1}x^{n-1} + \cdots + a_0$$

and

$$g(x) = b_m x^m + b_{m-1}x^{m-1} + \cdots + b_0$$

be two elements of $F[x]$, with a_n and b_m both nonzero elements of F and $m > 0$. Then there are unique polynomials $q(x)$ and $r(x)$ in $F[x]$ such that $f(x) = g(x)q(x) + r(x)$, with the degree of $r(x)$ less than $m = $ degree $g(x)$.

Proof.   Consider the set $S = \{ f(x) - g(x)s(x) \mid s(x) \in F[x]\}$. Let $r(x)$ be an element of minimal degree in S. Then

$$f(x) = g(x)q(x) + r(x)$$

for some $q(x) \in F[x]$. We must show that the degree of $r(x)$ is less than m. Suppose that

$$r(x) = c_t x^t + c_{t-1}x^{t-1} + \cdots + c_0,$$

with $c_j \in F$ and $c_t \neq 0$ if $t \neq 0$. If $t \geq m$, then

$$f(x) - q(x)g(x) - (c_t/b_m)x^{t-m}g(x) = r(x) - (c_t/b_m)x^{t-m}g(x), \qquad (31.1)$$

and the latter is of the form

$$r(x) - (c_t x^t + \text{terms of lower degree}),$$

which is a polynomial of degree lower than t, the degree of $r(x)$. However, the polynomial in Eq. (31.1) can be written in the form

$$f(x) - g(x)[q(x) + (c_t/b_m)x^{t-m}],$$

so it is in S, contradicting the fact that $r(x)$ was selected to have minimal degree in S. Thus the degree of $r(x)$ is less than $m = $ degree $g(x)$.

     For uniqueness, if

$$f(x) = g(x)q_1(x) + r_1(x)$$

and

$$f(x) = g(x)q_2(x) + r_2(x),$$

then subtracting we have

$$g(x)[q_1(x) - q_2(x)] = r_2(x) - r_1(x).$$

Since the degree of $r_2(x) - r_1(x)$ is less than the degree of $g(x)$, this can only hold if $q_1(x) - q_2(x) = 0$ or $q_1(x) = q_2(x)$. Then we must have $r_2(x) - r_1(x) = 0$ or $r_1(x) = r_2(x)$. ∎

You can compute the polynomials $q(x)$ and $r(x)$ of Theorem 31.1 by *long division* just as you divided polynomials in $\mathbf{R}[x]$ in high school. After the following corollaries, we give some computational examples.

You have no doubt seen in high school algebra the first corollary for the special case $\mathbf{R}[x]$. We phrase our proof in terms of the mapping (homomorphism) approach described in Chapter 30.

> **Corollary 1**   An element $a \in F$ is a zero of $f(x) \in F[x]$ if and only if $x - a$ is a factor of $f(x)$ in $F[x]$.

Proof.   Suppose that for $a \in F$ we have $f(a) = 0$. By Theorem 31.1, there exist $q(x), r(x) \in F[x]$ such that

$$f(x) = (x - a)q(x) + r(x),$$

where the degree of $r(x)$ is less than 1. Then we must have $r(x) = c$ for $c \in F$, so

$$f(x) = (x - a)q(x) + c.$$

Applying our evaluation homomorphism $\phi_a : F[x] \to F$ of Theorem 30.2, we find

$$0 = f(a) = 0q(a) + c,$$

so it must be that $c = 0$. Then $f(x) = (x - a)q(x)$, so $x - a$ is a factor of $f(x)$.

Conversely, if $x - a$ is a factor of $f(x)$ in $F[x]$, where $a \in F$, then applying our evaluation homomorphism ϕ_a to $f(x) = (x - a)q(x)$, we have $f(a) = 0q(a) = 0$. ∎

The next corollary should certainly also look familiar.

> **Corollary 2**   A nonzero polynomial $f(x) \in F[x]$ of degree n can have at most n zeros in a field F.

Proof.   The preceding corollary shows that if $a_1 \in F$ is a zero of $f(x)$, then

$$f(x) = (x - a_1)q_1(x),$$

where of course the degree of $q_1(x)$ is $n - 1$. A zero $a_2 \in F$ of $q_1(x)$ then results in a factorization

$$f(x) = (x - a_1)(x - a_2)q_2(x).$$

Continuing this process, we arrive at

$$f(x) = (x - a_1) \cdots (x - a_r)q_r(x),$$

where $q_r(x)$ has no further zeros in F. Clearly, $r \leq n$. Also, if $b \neq a_i$ for $i = 1, \ldots, r$ and $b \in F$, then

$$f(b) = (b - a_1) \cdots (b - a_r)q_r(b) \neq 0,$$

since F has no divisors of 0 and none of $b - a_i$ or $q_r(b)$ are 0 by construction. Hence the a_i for $i = 1, \ldots, r \leq n$ are all the zeros in F of $f(x)$. ■

Example 31.1 Let us work with polynomials in $\mathbf{Z}_5[x]$ and divide

$$f(x) = x^4 - 3x^3 + 2x^2 + 4x - 1$$

by $g(x) = x^2 - 2x + 3$ to find $q(x)$ and $r(x)$ of Theorem 31.1. The long division should be easy for you to follow, but remember that we are in $\mathbf{Z}_5[x]$, so, for example, $4x - (-3x) = 2x$.

$$
\begin{array}{r}
x^2 - x - 3 \\
x^2 - 2x + 3 \overline{\smash{\big)}\ x^4 - 3x^3 + 2x^2 + 4x - 1} \\
\underline{x^4 - 2x^3 + 3x^2} \\
-x^3 - x^2 + 4x \\
\underline{-x^3 + 2x^2 - 3x} \\
-3x^2 + 2x - 1 \\
\underline{-3x^2 + x - 4} \\
x + 3
\end{array}
$$

Thus

$$x^4 - 3x^3 + 2x^2 + 4x - 1 = (x^2 - 2x + 3)(x^2 - x - 3) + (x + 3),$$

so $q(x) = x^2 - x - 3$, and $r(x) = x + 3$. ■

Example 31.2 Working again in $\mathbf{Z}_5[x]$, note that 1 is a zero of

$$(x^4 + 3x^3 + 2x + 4) \in \mathbf{Z}_5[x].$$

Thus by Corollary 1 of Theorem 31.1, we should be able to factor $x^4 + 3x^3 + 2x + 4$ into $(x - 1)q(x)$ in $\mathbf{Z}_5[x]$. Let us try again by long division.

$$
\begin{array}{r}
x^3 + 4x^2 + 4x + 1 \\
x - 1 \overline{\smash{\big)}\ x^4 + 3x^3 + 2x + 4} \\
\underline{x^4 - x^3} \\
4x^3 \\
\underline{4x^3 - 4x^2} \\
4x^2 + 2x \\
\underline{4x^2 - 4x} \\
x + 4 \\
\underline{x - 1} \\
0
\end{array}
$$

Thus $x^4 + 3x^3 + 2x + 4 = (x - 1)(x^3 + 4x^2 + 4x + 1)$ in $\mathbf{Z}_5[x]$. Since 1 is seen to be a zero of $x^3 + 4x^2 + 4x + 1$ also, we can divide this polynomial by $x - 1$ and get

$$
\begin{array}{r}
x^2 + 4 \\
x - 1 \overline{\smash{\big)}\ x^3 + 4x^2 + 4x + 1} \\
\underline{x^3 - x^2} \\
0\ + 4x + 1 \\
\underline{4x - 4} \\
0
\end{array}
$$

Since $x^2 + 4$ still has 1 as a zero, we can divide again by $x - 1$ and get

$$
\begin{array}{r}
x + 1 \\
x - 1 \overline{\smash{\big)}\ x^2 + 4} \\
\underline{x^2 - x} \\
x + 4 \\
\underline{x - 1} \\
0
\end{array}
$$

Thus $x^4 + 3x^3 + 2x + 4 = (x - 1)^3(x + 1)$ in $\mathbf{Z}_5[x]$. ∎

The technique of *synthetic division* is undoubtedly valid in $F[x]$, if you can remember how it goes.

31.2 IRREDUCIBLE POLYNOMIALS

Our next definition singles out a type of polynomial in $F[x]$ that will be of utmost importance to us. Again, you are probably already familiar with the concept. You see, we really *are* doing high school algebra in a more general setting.

> **Definition**   A nonconstant polynomial $f(x) \in F[x]$ is **irreducible over F** or is an **irreducible polynomial in $F[x]$**, if $f(x)$ cannot be expressed as a product $g(x)h(x)$ of two polynomials $g(x)$ and $h(x)$ in $F[x]$ both of lower degree than the degree of $f(x)$.

Note that the preceding definition concerns the concept *irreducible over F* and not just the concept *irreducible*. A polynomial $f(x)$ may be irreducible over F, but may not be irreducible if viewed over a larger field E containing F. We illustrate this.

Example 31.3   Theorem 27.1 showed that $x^2 - 2$ viewed in $\mathbf{Q}[x]$ has no zeros in $\mathbf{Q}$. This shows that $x^2 - 2$ is irreducible over $\mathbf{Q}$, for a factorization $x^2 - 2 = (ax + b)(cx + d)$ for $a, b, c, d \in \mathbf{Q}$ would clearly give rise to zeros of $x^2 - 2$ in $\mathbf{Q}$. However, $x^2 - 2$ viewed in $\mathbf{R}[x]$ is not irreducible over $\mathbf{R}$, for $x^2 - 2$ factors in $\mathbf{R}[x]$ into $(x - \sqrt{2})(x + \sqrt{2})$. ∎

It is worthwhile for the student to remember that *the units in $F[x]$ are precisely the nonzero elements of F.* Thus we could have defined an irreducible polynomial $f(x)$ as a nonconstant polynomial such that in any factorization $f(x) = g(x)h(x)$ in $F[x]$, either $g(x)$ or $h(x)$ is a unit. This viewpoint will be elaborated in the next starred chapters, which deal with factorization in rings more general than $F[x]$.

Example 31.4   Let us show that $f(x) = x^3 + 3x + 2$ viewed in $\mathbf{Z}_5[x]$ is irreducible over $\mathbf{Z}_5$. If $x^3 + 3x + 2$ factored in $\mathbf{Z}_5[x]$ into polynomials of lower degree, then there would exist at least one linear factor of $f(x)$ of the form $x - a$ for some $a \in \mathbf{Z}_5$. But then $f(a)$ would be zero, by Corollary 1 of Theorem 31.1. However, $f(0) = 2$, $f(1) = 1$, $f(-1) = -2$, $f(2) = 1$, and $f(-2) = -2$, showing that $f(x)$ has no zeros in $\mathbf{Z}_5$. Thus $f(x)$ is irreducible over $\mathbf{Z}_5$. This test for irreducibility by finding zeros works nicely for quadratic and cubic polynomials over a finite field with a small number of elements. ∎

Irreducible polynomials will play a very important role in our work from now on. The problem of determining whether a given $f(x) \in F[x]$ is irreducible over F may be difficult. We now give some criteria for irreducibility that are useful in certain cases. One technique for determining irreducibility of quadratic and cubic polynomials was illustrated in Examples 31.3 and 31.4. We formalize it in a theorem.

Theorem 31.2   *Let $f(x) \in F[x]$, and let $f(x)$ be of degree 2 or 3. Then $f(x)$ is reducible over F if and only if it has a zero in F.*

Proof.   If $f(x)$ is reducible so that $f(x) = g(x)h(x)$, where the degree of $g(x)$ and the degree of $h(x)$ are both less than the degree of $f(x)$, then since $f(x)$ is either quadratic or cubic, either $g(x)$ or $h(x)$ is of degree 1. If, say, $g(x)$ is of degree 1, then except for a possible factor in F, $g(x)$ is of the form $x - a$. Then $g(a) = 0$, which implies that $f(a) = 0$, so $f(x)$ has a zero in F.

Conversely, Corollary 1 of Theorem 31.1 shows that if $f(a) = 0$ for $a \in F$, then $x - a$ is a factor of $f(x)$, so $f(x)$ is reducible. ∎

We turn to some conditions for irreducibility over $\mathbf{Q}$ of polynomials in $\mathbf{Q}[x]$. The most important condition that we shall give is contained in the next theorem. We shall not prove this theorem here; it is proved in a more general situation in starred Chapter 32 (see the corollary of Lemma 32.6).

Theorem 31.3   *If $f(x) \in \mathbf{Z}[x]$, then $f(x)$ factors into a product of two polynomials of lower degrees r and s in $\mathbf{Q}[x]$ if and only if it has such a factorization with polynomials of the same degrees r and s in $\mathbf{Z}[x]$.* ∎

Proof.   See the corollary of Lemma 32.6.

Corollary   *If $f(x) = x^n + a_{n-1}x^{n-1} + \cdots + a_0$ is in $\mathbf{Z}[x]$ with $a_0 \neq 0$, and if $f(x)$ has a zero in $\mathbf{Q}$, then it has a zero m in $\mathbf{Z}$, and m must divide a_0.*

Proof.   If $f(x)$ has a zero a in $\mathbf{Q}$, then $f(x)$ has a linear factor $x - a$ in $\mathbf{Q}[x]$ by Corollary 1 of Theorem 31.1. But then by Theorem 31.3, $f(x)$ has a factorization with a linear factor in $\mathbf{Z}[x]$, so for some $m \in \mathbf{Z}$ we must have

$$f(x) = (x - m)(x^{n-1} + \cdots + a_0/m).$$

Thus a_0/m is in $\mathbf{Z}$, so m divides a_0. ∎

Example 31.5   This corollary of Theorem 31.3 gives us another proof of the irreducibility of $x^2 - 2$ over $\mathbf{Q}$, for $x^2 - 2$ factors nontrivially in $\mathbf{Q}[x]$ if and only if it has a zero in $\mathbf{Q}$ by Theorem 31.2. By the corollary of Theorem 31.3, it has a zero in $\mathbf{Q}$ if and only if it has a zero in $\mathbf{Z}$, and moreover the only possibilities are the divisors ± 1 and ± 2 of 2. A check shows that none of these numbers is a zero of $x^2 - 2$. ∎

Example 31.6   Let us use Theorem 31.3 to show that

$$f(x) = x^4 - 2x^2 + 8x + 1$$

viewed in $\mathbf{Q}[x]$ is irreducible over $\mathbf{Q}$. If $f(x)$ has a linear factor in $\mathbf{Q}[x]$, then it has a zero in $\mathbf{Z}$, and by the corollary of Theorem 31.3, this zero would have to be a divisor in $\mathbf{Z}$ of 1, that is, either ± 1. But $f(1) = 8$, and $f(-1) = -8$, so such a factorization is impossible.

  If $f(x)$ factors into two quadratic factors in $\mathbf{Q}[x]$, then by Theorem 31.3 it has a factorization

$$(x^2 + ax + b)(x^2 + cx + d)$$

in $\mathbf{Z}[x]$. Equating coefficients of powers of x, we find that we must have

$$bd = 1, \qquad ad + bc = 8, \qquad ac + b + d = -2, \qquad \text{and} \qquad a + c = 0$$

for integers a, b, c, $d \in \mathbf{Z}$. From $bd = 1$, we see that either $b = d = 1$ or $b = d = -1$. If $b = d = 1$, then $ac + b + d = -2$ implies that $ac = -4$, which, with $a + c = 0$, gives $a = -c = \pm 2$. If $b = d = -1$, then $ac + b + d = -2$ implies that $ac = 0$, so $a = -c = 0$. No combination of the numbers arising from these possibilities can make $ad + bc$ as large as 8. Thus a factorization into two quadratic polynomials is also impossible, and $f(x)$ is irreducible over $\mathbf{Q}$. ∎

  We conclude our irreducibility criteria with the famous Eisenstein test for irreducibility. An additional and very useful criterion is given in Exercise 31.20.

Theorem 31.4 (Eisenstein) *Let* $p \in \mathbf{Z}$ *be a prime. Suppose that* $f(x) = a_n x^n + \cdots + a_0$ *is in* $\mathbf{Z}[x]$, *and* $a_n \not\equiv 0 \pmod{p}$, *but* $a_i \equiv 0 \pmod{p}$ *for* $i < n$, *with* $a_0 \not\equiv 0 \pmod{p^2}$. *Then* $f(x)$ *is irreducible over* $\mathbf{Q}$.

Proof. By Theorem 31.3 we need only show that $f(x)$ does not factor into polynomials of lower degree in $\mathbf{Z}[x]$. If

$$f(x) = (b_r x^r + \cdots + b_0)(c_s x^s + \cdots + c_0)$$

is a factorization in $\mathbf{Z}[x]$, with $b_r \neq 0, c_s \neq 0$ and $r, s < n$, then $a_0 \not\equiv 0 \pmod{p^2}$ implies that not both b_0 and c_0 are congruent to 0 modulo p. Suppose that $b_0 \not\equiv 0 \pmod{p}$ and $c_0 \equiv 0 \pmod{p}$. Now $a_n \not\equiv 0 \pmod{p}$ implies that $b_r, c_s \not\equiv 0 \pmod{p}$, since $a_n = b_r c_s$. Let m be the smallest value of k such that $c_k \not\equiv 0 \pmod{p}$. Then

$$a_m = b_0 c_m + b_1 c_{m-1} + \cdots + b_{m-i} c_i$$

for some i, $0 \leq i < m$. Now b_0 and c_m neither congruent to 0 modulo p and $c_{m-1}, \ldots, c_i$ all congruent to 0 modulo p implies that $a_m \not\equiv 0$ modulo p, so $m = n$. Consequently, $s = n$, contradicting our assumption that $s < n$, that is, that our factorization was nontrivial. ∎

Note that if we take $p = 2$, the Eisenstein criterion gives us still another proof of the irreducibility of $x^2 - 2$ over $\mathbf{Q}$.

Example 31.7 Taking $p = 3$, we see by Theorem 31.4 that

$$25x^5 - 9x^4 + 3x^2 - 12$$

is irreducible over $\mathbf{Q}$. ∎

Corollary *The cyclotomic polynomial*

$$\Phi_p(x) = \frac{x^p - 1}{x - 1} = x^{p-1} + x^{p-2} + \cdots + x + 1$$

is irreducible over $\mathbf{Q}$ *for any prime* p.

Proof. Again by Theorem 31.3, we need only consider factorizations in $\mathbf{Z}[x]$. Let

$$g(x) = \Phi_p(x + 1) = \frac{(x + 1)^p - 1}{(x + 1) - 1} = \frac{x^p + \binom{p}{1} x^{p-1} + \cdots + px}{x}.$$

Then

$$g(x) = x^{p-1} + \binom{p}{1} x^{p-2} + \cdots + p$$

satisfies the Eisenstein criterion for the prime p and is thus irreducible over $\mathbf{Q}$. But clearly if $\Phi_p(x) = h(x)r(x)$ were a nontrivial factorization of $\Phi_p(x)$

in $\mathbf{Z}[x]$, then

$$\Phi_p(x + 1) = g(x) = h(x + 1)r(x + 1)$$

would give a nontrivial factorization of $g(x)$ in $\mathbf{Z}[x]$. Thus $\Phi_p(x)$ must also be irreducible over $\mathbf{Q}$. ∎

31.3 IDEAL STRUCTURE IN $F[x]$

We give the next definition for a general commutative ring R with unity, although we are only interested in the case $R = F[x]$, where F is a field. It is clear that for a commutative ring R with unity and $a \in R$, the set $\{ra \mid r \in R\}$ is an ideal in R that contains the element a.

> **Definition** If R is a commutative ring with unity and $a \in R$, the ideal $\{ra \mid r \in R\}$ of all multiples of a is the ***principal ideal generated by*** a and is denoted by $\langle a \rangle$. An ideal N of R is a ***principal ideal*** if $N = \langle a \rangle$ for some $a \in R$.

Example 31.8 The ideal $\langle x \rangle$ in $F[x]$ clearly consists of all polynomials in $F[x]$ having zero constant term. ∎

The next theorem is another easy but very important application of Theorem 31.1, which is our basic tool for this whole section. You might be prepared to prove this theorem if your instructor ever asks for proofs. The proof of this theorem is to the division algorithm in $F[x]$ as the proof that a subgroup of a cyclic group is cyclic is to the division algorithm in $\mathbf{Z}$.

> **Theorem 31.5** *If F is a field, every ideal in $F[x]$ is principal.*

Proof. Let N be an ideal of $F[x]$. If $N = \{0\}$, then $N = \langle 0 \rangle$. Suppose that $N \neq \{0\}$, and let $g(x)$ be a nonzero element of N of minimal degree. If the degree of $g(x)$ is 0, then $g(x) \in F$ and is a unit, so $N = F[x] = \langle 1 \rangle$ by Theorem 28.4, so N is principal. If the degree of $g(x)$ is ≥ 1, let $f(x)$ be any element of N. Then by Theorem 31.1, $f(x) = g(x)q(x) + r(x)$, where (degree $r(x)$) < (degree $g(x)$). Now $f(x) \in N$ and $g(x) \in N$ implies that $f(x) - g(x)q(x) = r(x)$ is in N by definition of an ideal. Since $g(x)$ was a nonzero element of minimal degree in N, we must have $r(x) = 0$. Thus $f(x) = g(x)q(x)$ and $N = \langle g(x) \rangle$. ∎

We can now achieve our aim of characterizing the maximal ideals of $F[x]$.

> **Theorem 31.6** *An ideal $\langle p(x) \rangle \neq \{0\}$ of $F[x]$ is maximal if and only if $p(x)$ is irreducible over F.*

Proof. Suppose that $\langle p(x) \rangle \neq \{0\}$ is a maximal ideal of $F[x]$. Then $\langle p(x) \rangle \neq F[x]$, so $p(x) \notin F$. Let $p(x) = f(x)g(x)$ be a factorization of $p(x)$ in $F[x]$. Since $\langle p(x) \rangle$ is a maximal ideal and hence also a prime ideal, $(f(x)g(x)) \in \langle p(x) \rangle$ implies that $f(x) \in \langle p(x) \rangle$, or $g(x) \in \langle p(x) \rangle$, that is, either $f(x)$ or $g(x)$ has $p(x)$ as a factor. But then we can't have the degrees of both $f(x)$ and $g(x)$ less than the degree of $p(x)$. This shows that $p(x)$ is irreducible over F.

Conversely, if $p(x)$ is irreducible over F, suppose that N is an ideal such that $\langle p(x) \rangle \subseteq N \subseteq F[x]$. Now N is a principal ideal by Theorem 31.5, so $N = \langle g(x) \rangle$ for some $g(x) \in N$. Then $p(x) \in N$ implies that $p(x) = g(x)q(x)$ for some $q(x) \in F[x]$. But $p(x)$ is irreducible, which implies that either $g(x)$ or $q(x)$ is of degree 0. If $g(x)$ is of degree 0, that is, a nonzero constant in F, then $g(x)$ is a unit in $F[x]$, so $\langle g(x) \rangle = N = F[x]$. If $q(x)$ is of degree 0, then $q(x) = c$, where $c \in F$, and $g(x) = (1/c)p(x)$ is in $\langle p(x) \rangle$, so $N = \langle p(x) \rangle$. Thus $\langle p(x) \rangle \subset N \subset F[x]$ is impossible, so $\langle p(x) \rangle$ is maximal. ∎

Example 31.9   Example 31.4 shows that $x^3 + 3x + 2$ is irreducible in $\mathbf{Z}_5[x]$. Thus $\mathbf{Z}_5[x]/\langle x^3 + 3x + 2 \rangle$ is a field. Similarly, Theorem 27.1 shows that $x^2 - 2$ is irreducible in $\mathbf{Q}[x]$, so $\mathbf{Q}[x]/\langle x^2 - 2 \rangle$ is a field. We shall examine such fields in more detail later. ∎

31.4 UNIQUENESS OF FACTORIZATION IN $F[x]$

We shall demonstrate that polynomials in $F[x]$ can be factored into a product of irreducible polynomials in $F[x]$ in an essentially unique way, as you would guess from your high school work in $\mathbf{R}[x]$. We shall consider unique factorization in more general situations in the following starred chapter. Since the result for $F[x]$ can be obtained cheaply at this point, we decided to include this special case here. We shall be dealing exclusively with fields in the remaining unstarred portion of this text, so this special case is all that we shall need for our work.

For $f(x)$, $g(x) \in F[x]$ we say that $g(x)$ **divides** $f(x)$ **in** $F[x]$ if there exists $q(x) \in F[x]$ such that $f(x) = g(x)q(x)$.

Theorem 31.7   *Let $p(x)$ be an irreducible polynomial in $F[x]$. If $p(x)$ divides $r(x)s(x)$ for $r(x)$, $s(x) \in F[x]$, then either $p(x)$ divides $r(x)$ or $p(x)$ divides $s(x)$.*

Proof. Suppose that $p(x)$ divides $r(x)s(x)$. Then $r(x)s(x) \in \langle p(x) \rangle$, which is maximal by Theorem 31.6. Therefore, $\langle p(x) \rangle$ is a prime ideal by the corollary to Theorem 29.5. Hence $r(x)s(x) \in \langle p(x) \rangle$ implies that either $r(x) \in \langle p(x) \rangle$, giving $p(x)$ divides $r(x)$, or $s(x) \in \langle p(x) \rangle$, giving $p(x)$ divides $s(x)$. ∎

Corollary *If $p(x)$ is irreducible in $F[x]$ and $p(x)$ divides the product $r_1(x) \cdots r_n(x)$ for $r_i(x) \in F[x]$, then $p(x)$ divides $r_i(x)$ for at least one i.*

Proof. Using mathematical induction, we find that this is immediate from Theorem 31.7. ■

Theorem 31.8 *If F is a field, then every nonconstant polynomial $f(x) \in F[x]$ can be factored in $F[x]$ into a product of irreducible polynomials, the irreducible polynomials being unique except for order and for unit (that is, nonzero constant) factors in F.*

Proof. Let $f(x) \in F[x]$ be a nonconstant polynomial. If $f(x)$ is not irreducible, then $f(x) = g(x)h(x)$, with the degree of $g(x)$ and the degree of $h(x)$ both less than the degree of $f(x)$. If $g(x)$ and $h(x)$ are both irreducible, we stop here. If not, at least one of them factors into polynomials of lower degree. Continuing this process (an induction argument really), we arrive at a factorization

$$f(x) = p_1(x)p_2(x) \cdots p_r(x),$$

where $p_i(x)$ is irreducible.

It remains for us to show uniqueness. Suppose that

$$f(x) = p_1(x)p_2(x) \cdots p_r(x) = q_1(x)q_2(x) \cdots q_s(x)$$

are two factorizations of $f(x)$ into irreducible polynomials. Then by the corollary to Theorem 31.7, $p_1(x)$ divides some $q_j(x)$, let us assume $q_1(x)$. Since $q_1(x)$ is irreducible,

$$q_1(x) = u_1 p_1(x),$$

where $u_1 \neq 0$, but u_1 is in F and thus is a unit. Then substituting $u_1 p_1(x)$ for $q_1(x)$ and canceling, we get

$$p_2(x) \cdots p_r(x) = u_1 q_2(x) \cdots q_s(x).$$

By a similar argument, say $q_2(x) = u_2 p_2(x)$, so

$$p_3(x) \cdots p_r(x) = u_1 u_2 q_3(x) \cdots q_s(x).$$

Continuing in this manner, we eventually arrive at

$$1 = u_1 u_2 \cdots u_r q_{r+1}(x) \cdots q_s(x).$$

Clearly, this is only possible if $s = r$, so that this equation is actually $1 = u_1 u_2 \cdots u_r$. Thus the irreducible factors $p_i(x)$ and $q_j(x)$ were the same except possibly for order and unit factors. ■

Example 31.10 Example 31.2 shows that the factorization of $x^4 + 3x^3 + 2x + 4$ in $\mathbf{Z}_5[x]$ is $(x - 1)^3(x + 1)$. These irreducible factors in $\mathbf{Z}_5[x]$ are only defined up to units in $\mathbf{Z}_5[x]$, that is, nonzero constants in $\mathbf{Z}_5$. For example, $(x - 1)^3(x + 1) = (x - 1)^2(2x - 2)(3x + 3)$. ∎

Exercises

31.1 Let $f(x) = x^6 + 3x^5 + 4x^2 - 3x + 2$ and $g(x) = x^2 + 2x - 3$ be in $\mathbf{Z}_7[x]$. Find $q(x)$ and $r(x)$ in $\mathbf{Z}_7[x]$ such that $f(x) = g(x)q(x) + r(x)$, with (degree $r(x)$) < 2.

31.2 Let $f(x) = x^6 + 3x^5 + 4x^2 - 3x + 2$ and $g(x) = 3x^2 + 2x - 3$ be in $\mathbf{Z}_7[x]$. Find $q(x)$ and $r(x)$ in $\mathbf{Z}_7[x]$ such that $f(x) = g(x)q(x) + r(x)$, with (degree $r(x)$) < 2.

31.3 The polynomial $x^4 + 4$ can be factored into linear factors in $\mathbf{Z}_5[x]$. Find this factorization.

31.4 Is $x^3 + 2x + 3$ an irreducible polynomial of $\mathbf{Z}_5[x]$? Why? Express it as a product of irreducible polynomials of $\mathbf{Z}_5[x]$.

31.5 Show that $f(x) = x^2 + 8x - 2$ is irreducible over $\mathbf{Q}$. Is $f(x)$ irreducible over $\mathbf{R}$? over $\mathbf{C}$?

31.6 Repeat Exercise 31.5 with $g(x) = x^2 + 6x + 12$ in place of $f(x)$.

31.7 a) Show that $x^3 + 3x^2 - 8$ is irreducible over $\mathbf{Q}$.

b) Show that $x^4 - 22x^2 + 1$ is irreducible over $\mathbf{Q}$.

†**31.8** Prove that if F is a field, every proper prime ideal of $F[x]$ is maximal.

31.9 Mark each of the following true or false.

— a) $x - 2$ is irreducible over $\mathbf{Q}$.

— b) $3x - 6$ is irreducible over $\mathbf{Q}$.

— c) $x^2 - 3$ is irreducible over $\mathbf{Q}$.

— d) $x^2 + 3$ is irreducible over $\mathbf{Z}_7$.

— e) If F is a field, the units of $F[x]$ are precisely the nonzero elements of F.

— f) If F is a field, the units of $F(x)$ are precisely the nonzero elements of F.

— g) A polynomial $f(x)$ of degree n with coefficients in a field F can have at most n zeros in F.

— h) A polynomial $f(x)$ of degree n with coefficients in a field F can have at most n zeros in any given field E such that $F \leq E$.

— i) Every ideal of $F[x]$ is principal.

— j) Every principal ideal in $F[x]$ is a maximal ideal.

31.10 Determine which of the following polynomials in $\mathbf{Z}[x]$ satisfy an Eisenstein criterion for irreducibility over $\mathbf{Q}$.

a) $x^2 - 12$ b) $8x^3 + 6x^2 - 9x + 24$

c) $4x^{10} - 9x^3 + 24x - 18$ d) $2x^{10} - 25x^3 + 10x^2 - 30$

31.11 Show that the polynomial $x^p + a$ in $\mathbf{Z}_p[x]$ is not irreducible for any $a \in \mathbf{Z}_p$.

31.12 Find all odd prime numbers p such that $x + 2$ is a factor of $x^4 + x^3 + x^2 - x + 1$ in $\mathbf{Z}_p[x]$.

31.13 Is $2x^3 + x^2 + 2x + 2$ an irreducible polynomial in $Z_5[x]$? Why? Express it as a product of irreducible polynomials in $Z_5[x]$.

31.14 Find all zeros of $6x^4 + 17x^3 + 7x^2 + x - 10$ in Q. (This is a tedious high school algebra problem. *You* might use a bit of analytic geometry and calculus and make a graph, or use Newton's method to see which are the best candidates for zeros.)

31.15 If F is a field and $a \neq 0$ is a zero of $f(x) = a_0 + a_1 x + \cdots + a_n x^n$ in $F[x]$, show that $1/a$ is a zero of $a_n + a_{n-1}x + \cdots + a_0 x^n$.

31.16 Find all irreducible polynomials of degree 2 or 3 in $Z_2[x]$ and $Z_3[x]$.

31.17 Is $Q[x]/\langle x^2 - 5x + 6 \rangle$ a field? Why? What about $Q[x]/\langle x^2 - 6x + 6 \rangle$?

31.18 Let F be a field and $f(x), g(x) \in F[x]$. Show that $f(x)$ divides $g(x)$ if and only if $g(x) \in \langle f(x) \rangle$.

31.19 Let F be a field and let $f(x), g(x) \in F[x]$. Show that

$$N = \{r(x)f(x) + s(x)g(x) \mid r(x), s(x) \in F[x]\}$$

is an ideal of $F[x]$. Show that if $f(x)$ and $g(x)$ have different degrees and $N \neq F[x]$, then $f(x)$ and $g(x)$ cannot both be irreducible over F.

31.20 Let $\sigma_m : Z \to Z_m$ be the natural homomorphism given by $a\sigma_m =$ (the remainder of a when divided by m) for $a \in Z$.

a) Show that $\bar{\sigma}_m : Z[x] \to Z_m[x]$ given by

$$(a_0 + a_1 x + \cdots + a_n x^n)\bar{\sigma}_m = a_0\sigma_m + (a_1\sigma_m)x + \cdots + (a_n\sigma_m)x^n$$

is a homomorphism of $Z[x]$ onto $Z_m[x]$.

b) Show that if $f(x) \in Z[x]$ has degree n and $(f(x))\bar{\sigma}_m$ does not factor in $Z_m[x]$ into two polynomials of degree less than n, then $f(x)$ is irreducible in $Q[x]$.

c) Use part (b) to show that $x^3 + 17x + 36$ is irreducible in $Q[x]$. [*Hint:* Try a prime value of m that simplifies the coefficients.]

31.21 Let F be a field, and let S be any subset of $F \times F \times \cdots \times F$ for n factors. Show that the set of all $f(x_1, \ldots, x_n) \in F[x_1, \ldots, x_n]$ that are zero at every $(a_1, \ldots, a_n) \in S$ (see Exercise 30.13) forms an ideal in $F[x_1, \ldots, x_n]$. This is of importance in algebraic geometry.

thirty-two

Unique Factorization Domains

*32.1 INTRODUCTION

This chapter and the next two chapters are concerned with integral domains and the problem of factorization of elements in an integral domain. The integral domain **Z** is our standard example of an integral domain in which there is unique factorization into primes (irreducibles). Section 31.4 showed that for a field F, $F[x]$ is also such an integral domain with unique factorization. In order to discuss analogous ideas in an arbitrary integral domain, we shall give several definitions, some of which are repetitions of earlier ones. It is nice to have them all in one place for reference.

Definition   Let D be an integral domain and a, $b \in D$. If there exists $c \in D$ such that $b = ac$, then a **divides** b (or a **is a factor of** b), denoted by $a|b$.

Definition   An element u of an integral domain D is a **unit of** D if u divides 1, that is, if u has a multiplicative inverse in D. Two elements $a, b \in D$ are **associates in** D if $a = bu$, where u is a unit in D.

In Exercise 32.7, we ask you to show that this criterion for a and b to be associates is an equivalence relation on D.

Example 32.1   The only units in **Z** are 1 and -1. Thus the only associates of 26 in **Z** are 26 and -26.   ∎

It should be noted that while our condition $a = bu$ for elements $a, b \in D$ to be associates is not formally symmetric, if $a = bu$, then $b = au^{-1}$, where u^{-1} exists and is a unit of D, since u is a unit of D.

> **Definition**   A nonzero element p that is not a unit of an integral domain D is an *irreducible of D* if in any factorization $p = ab$ in D either a or b is a unit.

It is easily seen that an associate of an irreducible is again an irreducible.

> **Definition**   An integral domain D is a *unique factorization domain* (abbreviated UFD) if the following conditions are satisfied:
>
> 1. Every element of D that is neither 0 nor a unit can be factored into a product of a finite number of irreducibles.
> 2. If $p_1 \cdots p_r$ and $q_1 \cdots q_s$ are two factorizations of the same element of D into irreducibles, then $r = s$ and the q_j can be renumbered so that p_i and q_i are associates.

Example 32.2   Theorem 31.8 shows that for a field F, $F[x]$ is a UFD. Also you know that **Z** is a UFD; we have made frequent use of this fact, although we have never proved it. For example, in **Z** we have

$$24 = (2)(2)(3)(2) = (-2)(-3)(2)(2).$$

Here 2 and -2 are associates, as are 3 and -3. Thus except for order and associates, the irreducible factors in these two factorizations of 24 are the same.   ∎

After just one more definition we can describe what we wish to achieve in this section.

> **Definition**   An integral domain D is a *principal ideal domain* (abbreviated PID) if every ideal in D is a principal ideal.

Our purpose in this chapter is to prove two exceedingly important theorems:

1. Every PID is a UFD. (Theorem 32.2)
2. If D is a UFD, then $D[x]$ is a UFD. (Theorem 32.3)

Note that the fact that $F[x]$ is a UFD, where F is a field (by Theorem 31.8) illustrates both theorems. For by Theorem 31.5, $F[x]$ is a PID. Also, since F has no nonzero elements that are not units, F satisfies (in a vacuous sense that bothers some students) our definition for a UFD. Thus Theorem 32.3 would give another proof that $F[x]$ is a UFD, except for the fact that we shall actually use Theorem 31.8 quite often in proving Theorem 32.3. In the following chapter we shall study properties of a certain special class of UFDs, the *Euclidean domains*.

Let us proceed to prove the two theorems. The author has always disliked the proof of the second. Let's put it off as long as possible and prove Theorem 32.2 first.

*32.2 EVERY PID IS A UFD

The steps leading up to Theorem 31.8 and its proof indicate the way for our proof of Theorem 32.2. Much of the material will be repetitive. We inefficiently handled the special case of $F[x]$ separately in Theorem 31.8, since it was easy and was the only case we shall need for our field theory in general.

To prove that an integral domain D is a UFD, it is necessary to show that both conditions 1 and 2 of the definition of a UFD are satisfied. For our special case of $F[x]$ in Theorem 31.8, condition 1 was very easy and resulted from an argument that in a factorization of a polynomial of degree >0 into a product of two nonconstant polynomials, the degree of each factor was less than the degree of the original polynomial. Thus you couldn't keep on factoring indefinitely without running into unit factors, that is, polynomials of degree 0. For the general case of a PID, it is harder to show that this is so. We now turn to this problem. We shall need one more set-theoretic concept.

Definition   If $\{A_i | i \in I\}$ is a collection of sets, then the **union** $\bigcup_{i \in I} A_i$ *of the sets* A_i is the set of all x such that $x \in A_i$ for at least one $i \in I$.

Lemma 32.1 (Ascending Chain Condition for a PID)   *Let D be a PID. If $N_1 \subseteq N_2 \subseteq \cdots$ is a monotonic ascending chain of ideals N_i, then there exists a positive integer r such that $N_r = N_s$ for all $s \geq r$. Equivalently, every strictly ascending chain of ideals (all inclusions proper) in a PID is of finite length. That is, the **ascending chain condition** (ACC) holds for ideals in a PID.*

Proof.   Let $N_1 \subseteq N_2 \subseteq \cdots$ be a monotonic ascending chain of ideals N_i in D. Let $N = \bigcup_i N_i$. Clearly, $N \subseteq D$. We claim that N is an ideal in D. Let $a, b \in N$. Then there are ideals N_{i_1} and N_{i_2}, with $a \in N_{i_1}$ and $b \in N_{i_2}$. Now either $N_{i_1} \subseteq N_{i_2}$ or $N_{i_2} \subseteq N_{i_1}$; let us assume that $N_{i_1} \subseteq N_{i_2}$, so both a and b are in N_{i_2}. This implies that $a \pm b$ and ab are in N_{i_2}, so $a \pm b$ and ab are in N. Taking $a = 0$, we see that $b \in N$ implies $-b \in N$. Clearly, $0 \in N$. Thus N is a subring of D. For $a \in N$ and $d \in D$, we must have $a \in N_{i_1}$ for some N_{i_1}.

Then since N_{i_1} is an ideal, $da = ad$ is in N_{i_1}. Therefore, $da \in \bigcup_i N_i$, that is, $da \in N$. Hence N is an ideal.

Now as an ideal in D that is a PID, $N = \langle c \rangle$ for some $c \in D$. Since $N = \bigcup_i N_i$, we must have $c \in N_r$ for some $r \in \mathbf{Z}^+$. For $s \geq r$, we have

$$\langle c \rangle \subseteq N_r \subseteq N_s \subseteq N = \langle c \rangle.$$

Thus $N_r = N_s$ for $s \geq r$.

The equivalence with the ACC is obvious. ∎

In what follows, it will be useful to remember that for $a, b \in D$,

$\langle a \rangle \subseteq \langle b \rangle$ if and only if b divides a and

$\langle a \rangle = \langle b \rangle$ if and only if a and b are associates.

The first property is obvious. For the second property we conclude that $a = bd_1$ and $b = ad_2$ for some $d_1, d_2 \in D$. Then $a = ad_1d_2$. The case $a = 0$ is trivial, and for $a \neq 0$ we then have $1 = d_1d_2$. Thus d_1 and d_2 are units, so a and b are associates.

We can now prove condition 1 of the definition of a UFD for an integral domain that is a PID.

Theorem 32.1 Let D be a PID. *Every element that is neither 0 nor a unit in D is a product of irreducibles.*

Proof. Let $a \in D$, where a is neither 0 nor a unit. We first show that a has at least one irreducible factor. If a is an irreducible, we are done. If a is not an irreducible, then $a = a_1b_1$, where neither a_1 nor b_1 is a unit. Now

$$\langle a \rangle \subset \langle a_1 \rangle,$$

for $\langle a \rangle \subseteq \langle a_1 \rangle$ is obvious, and if $\langle a \rangle = \langle a_1 \rangle$, then a and a_1 would be associates and b_1 would be a unit, contrary to construction. Continuing this procedure then, starting now with a_1, we arrive at a strictly ascending chain of ideals

$$\langle a \rangle \subset \langle a_1 \rangle \subset \langle a_2 \rangle \subset \cdots.$$

By the ACC in Lemma 32.1, this chain terminates with some $\langle a_r \rangle$, and a_r must then be irreducible. Thus a has an irreducible factor a_r.

By what we have just proved, for an element a that is neither 0 nor a unit in D, either a is irreducible or $a = p_1c_1$ for p_1 an irreducible, and c_1 not a unit. By an argument similar to the one just made, in the latter case we can conclude that $\langle a \rangle \subset \langle c_1 \rangle$. If c_1 is not irreducible, then $c_1 = p_2c_2$ for an irreducible p_2 with c_2 not a unit. Continuing, we get a strictly ascending chain of ideals

$$\langle a \rangle \subset \langle c_1 \rangle \subset \langle c_2 \rangle \subset \cdots.$$

This chain must terminate, by the ACC in Lemma 32.1, with some $c_r = q_r$ that is an irreducible. Then $a = p_1p_2 \cdots p_rq_r$. ∎

This completes our demonstration of condition 1 of the definition of a UFD. Let us turn to condition 2. Our arguments here are parallel to those leading to Theorem 31.8. The results we encounter along the way are of some interest in themselves.

Lemma 32.2 (*Generalization of Theorem 31.6*) *An ideal $\langle p \rangle$ in a PID is maximal if and only if p is an irreducible.*

Proof. Let $\langle p \rangle$ be a maximal ideal of D, a PID. Suppose that $p = ab$ in D. Clearly, $\langle p \rangle \subseteq \langle a \rangle$. Suppose that $\langle a \rangle = \langle p \rangle$. Then a and p would be associates, so b must be a unit. If $\langle a \rangle \neq \langle p \rangle$, then we must have $\langle a \rangle = \langle 1 \rangle = D$, since $\langle p \rangle$ is maximal. But then a and 1 are associates, so a is a unit. Thus, if $p = ab$, either a or b must be a unit. Hence p is an irreducible of D.

Conversely, suppose that p is an irreducible in D. Then if $\langle p \rangle \subseteq \langle a \rangle$, we must have $p = ab$. Now if a is a unit, then $\langle a \rangle = \langle 1 \rangle = D$. If a is not a unit, then b must be a unit, so there exists $u \in D$ such that $bu = 1$. Then $pu = abu = a$, so $\langle a \rangle \subseteq \langle p \rangle$, and we have $\langle a \rangle = \langle p \rangle$. Thus $\langle p \rangle \subseteq \langle a \rangle$ implies that either $\langle a \rangle = D$ or $\langle a \rangle = \langle p \rangle$, and $\langle p \rangle \neq D$ or p would be a unit. Hence $\langle p \rangle$ is a maximal ideal. ∎

Lemma 32.3 (*Generalization of Theorem 31.7*) *In a PID, if an irreducible p divides ab, then either $p|a$ or $p|b$.*

Proof. Let D be a PID and suppose that for an irreducible p in D we have $p|ab$. Then $(ab) \in \langle p \rangle$. Since every maximal ideal in D is a prime ideal by the corollary to Theorem 29.5, $(ab) \in \langle p \rangle$ implies that either $a \in \langle p \rangle$ or $b \in \langle p \rangle$, giving either $p|a$ or $p|b$. ∎

Corollary *If p is an irreducible in a PID and p divides the product $a_1 a_2 \cdots a_n$ for $a_i \in D$, then $p|a_i$ for at least one i.*

Proof. Proof of this corollary is immediate from Lemma 32.3 if we use mathematical induction. ∎

Definition A nonzero nonunit element p of an integral domain D with the property that $p|ab$ implies either $p|a$ or $p|b$ is a **prime**.

Lemma 32.3 focused our attention on the defining property of a prime. In Exercise 32.5, we ask you to show that a prime in an integral domain is always an irreducible and that in a UFD an irreducible is also a prime. Thus the concepts of prime and irreducible coincide in a UFD. Example 32.3 will exhibit an integral domain containing some irreducibles that are not primes, so the concepts do not coincide in every domain.

Example 32.3 Let F be a field. Let D be the subdomain of $F[x, y]$ generated by F, x^3, xy, and y^3. Then x^3, xy, and y^3 are irreducibles in D, but

$$(x^3)(y^3) = (xy)(xy)(xy).$$

Since xy divides x^3y^3 but not x^3 or y^3, we see that xy is not a prime. Simila
arguments show that neither x^3 nor y^3 is a prime. ∎

The defining property of a prime is precisely what is needed to establis
the uniqueness property 2 in the definition of a UFD. We now complet
the proof of Theorem 32.2 by demonstrating the property 2 for a PID.

Theorem 32.2 (Generalization of Theorem 31.8)   *Every* PID *is a* UFD

Proof.   Theorem 32.1 shows that if D is a PID, then each $a \in D$, where a
neither 0 nor a unit, has a factorization

$$a = p_1 p_2 \cdots p_r$$

into irreducibles. It remains for us to show uniqueness. Let

$$a = q_1 q_2 \cdots q_s$$

be another such factorization into irreducibles. Then we have $p_1 | (q_1 q_2 \cdots q_s$
which implies that $p_1 | q_{j_1}$ for some j_1 by the corollary of Lemma 32.3. B
changing the order of the q_j if necessary, we can assume that $j_1 = 1$ or $p_1 | q$
Then $q_1 = p_1 u_1$, and since p_1 is an irreducible, u_1 is a unit, so p_1 and q_1 ar
associates. We have then

$$p_1 p_2 \cdots p_r = p_1 u_1 q_2 \cdots q_s,$$

so by the cancellation law in D,

$$p_2 \cdots p_r = u_1 q_2 \cdots q_s.$$

Continuing this process, starting with p_2 and so on, we finally arrive at

$$1 = u_1 u_2 \cdots u_r q_{r+1} \cdots q_s.$$

Since the q_j are irreducibles, we must have $r = s$. ∎

Many algebra texts start by proving the following corollary of Theorer
32.2. We have assumed that you were familiar with this corollary and use
it freely in our other work.

Corollary (Fundamental Theorem of Arithmetic)   *The integral doma*
Z *is a UFD.*

Proof. We have seen that all ideals in **Z** are of the form $n\mathbf{Z} = \langle n \rangle$ for $n \in \mathbf{Z}$
Thus **Z** is a PID, and Theorem 32.2 applies. ∎

It is worth noting that the proof that **Z** is a PID was really way back i
the corollary of Theorem 6.2. We proved Theorem 6.2 by using the divisio
algorithm for **Z** exactly as we proved, in Theorem 31.5, that $F[x]$ is a PI
by using the division algorithm for $F[x]$. In Chapter 33 we shall examin
this parallel more closely.

*32.3 IF D IS A UFD, THEN $D[x]$ IS A UFD

We proceed now with the proof of Theorem 32.3, our second main result for this chapter. The reason we don't like the proofs here is that they are all so naively obvious but so messy to write down. The idea of the argument is as follows. Let D be a UFD. We can form a field of quotients F of D. Then $F[x]$ is a UFD by Theorem 31.8, and we shall show that we can recover a factorization for $f(x) \in D[x]$ from its factorization in $F[x]$. It will be necessary to compare the irreducibles in $F[x]$ with those in $D[x]$, of course. This approach, which we prefer as more intuitive than some more efficient modern ones, is essentially due to Gauss.

Definition   Let D be a UFD. A nonconstant polynomial

$$f(x) = a_0 + a_1 x + \cdots + a_n x^n$$

in $D[x]$ is ***primitive*** if the only common divisors of all the a_i are units of D.

Example 32.4   In $\mathbf{Z}[x]$, $4x^2 + 3x + 2$ is primitive, but $4x^2 + 6x + 2$ is not, since 2, a nonunit in $\mathbf{Z}$, is a common divisor of 4, 6, and 2. ∎

Clearly, every nonconstant irreducible in $D[x]$ must be a primitive polynomial.

> **Lemma 32.4**   *If D is a UFD, then for every nonconstant $f(x) \in D[x]$ we have $f(x) = (c)g(x)$, where $c \in D$, $g(x) \in D[x]$, and $g(x)$ is primitive. The element c is unique up to a unit factor in D and is the **content of** $f(x)$. Also $g(x)$ is unique up to a unit factor in D.*

Proof.   Let $f(x) \in D[x]$ be given, where $f(x)$ is a nonconstant polynomial. Since D is a UFD, each coefficient of $f(x)$ can be factored into a finite product of irreducibles in D, uniquely up to order and associates. Imagine each coefficient of $f(x)$ to be so factored.

The number c to be constructed is really a greatest common divisor in D of the coefficients of $f(x)$. To form c, we proceed as follows. If p is a particular irreducible dividing every coefficient in $f(x)$, replace every occurrence of every associate of p in a factorization of a coefficient by pu for some unit u. Continuing this procedure for another irreducible q appearing in the factorization of some coefficient of $f(x)$, and so on, we arrive eventually at a factorization of the coefficients of $f(x)$ in which each irreducible p_i appearing in the factorization of one coefficient and dividing all coefficients actually appears in the factorization of all coefficients, but no other associate of p_i appears in the factorization of any coefficient. Let $c = \prod_i p_i^{v_i}$, where the product is taken over all irreducibles p_i appearing in the factorizations of all the coefficients in this adjusted factorization, and where v_i is the greatest integer such that $p_i^{v_i}$ divides all the coefficients. Clearly, we have $f(x) = (c)g(x)$, wnere $c \in D$, $g(x) \in D[x]$, and $g(x)$ is primitive by construction.

For uniqueness, if also $f(x) = (d)h(x)$ for $d \in D$, $h(x) \in D[x]$, and $h(x)$ primitive, then each irreducible factor of c must divide d and conversely. By setting $(c)g(x) = (d)h(x)$ and canceling irreducible factors of c into d, we arrive at $(u)g(x) = (v)h(x)$ for a unit $u \in D$. But then obviously we must also have v a unit of D or we would be able to cancel irreducible factors of v into u. Thus u and v are both units, so c is unique up to a unit factor. From $f(x) = (c)g(x)$, we see that the primitive polynomial $g(x)$ is also unique up to a unit factor. ∎

The preceding lemma illustrates what we meant when we said that the proofs here are obvious but messy to write down carefully. Surely this lemma falls into the "intuitively obvious" category.

Example 32.5   In $\mathbf{Z}[x]$,

$$4x^2 + 6x - 8 = (2)(2x^2 + 3x - 4),$$

where $2x^2 + 3x - 4$ is primitive. ∎

Lemma 32.5 (Gauss)   *If D is a* UFD, *then a product of two primitive polynomials in D[x] is again primitive.*

Proof.   Let

$$f(x) = a_0 + a_1 x + \cdots + a_n x^n$$

and

$$g(x) = b_0 + b_1 x + \cdots + b_m x^m$$

be primitive in $D[x]$, and let $h(x) = f(x)g(x)$. Let p be an irreducible in D. Then p does not divide all a_i and p does not divide all b_j, since $f(x)$ and $g(x)$ are primitive. Let a_r be the first coefficient of $f(x)$ not divisible by p, that is, $p \mid a_i$ for $i < r$, but $p \nmid a_r$ (that is, p does not divide a_r). Similarly, let $p \mid b_j$ for $j < s$, but $p \nmid b_s$. The coefficient of x^{r+s} in $h(x) = f(x)g(x)$ is

$$c_{r+s} = (a_0 b_{r+s} + \cdots + a_{r-1} b_{s+1}) + a_r b_s + (a_{r+1} b_{s-1} + \cdots + a_{r+s} b_0).$$

Now $p \mid a_i$ for $i < r$ implies that

$$p \mid (a_0 b_{r+s} + \cdots + a_{r-1} b_{s+1}),$$

and also $p \mid b_j$ for $j < s$ implies that

$$p \mid (a_{r+1} b_{s-1} + \cdots + a_{r+s} b_0).$$

But p does not divide a_r or b_s, so p does not divide $a_r b_s$, and consequently p does not divide c_{r+s}. This shows that given any irreducible $p \in D$, there is some coefficient of $f(x)g(x)$ not divisible by p. Thus $f(x)g(x)$ is primitive. ∎

Corollary *If D is a UFD, then a finite product of primitive polynomials in $D[x]$ is again primitive.*

Proof. Proof of this corollary follows from Lemma 32.5 by induction. ∎

Now let D be a UFD and let F be a field of quotients of D. By Theorem 31.8, $F[x]$ is a UFD. As we said earlier, we shall show that $D[x]$ is a UFD by carrying a factorization in $F[x]$ of $f(x) \in D[x]$ back into one in $D[x]$. The next lemma relates the nonconstant irreducibles of $D[x]$ to those of $F[x]$. This is the last important step.

Lemma 32.6 *Let D be a UFD and let F be a field of quotients of D. Let $f(x) \in D[x]$, where $(degree\ f(x)) > 0$. If $f(x)$ is an irreducible in $D[x]$, then $f(x)$ is also an irreducible in $F[x]$. Also, if $f(x)$ is primitive in $D[x]$ and irreducible in $F[x]$, then $f(x)$ is irreducible in $D[x]$.*

Proof. Suppose that a nonconstant $f(x) \in D[x]$ factors into polynomials of lower degree in $F[x]$, that is,

$$f(x) = r(x)s(x)$$

for $r(x), s(x) \in F[x]$. Then since F is a field of quotients of D, each coefficient in $r(x)$ and $s(x)$ is of the form a/b for some $a, b \in D$. By clearing denominators, we can get

$$(d)f(x) = r_1(x)s_1(x)$$

for $d \in D$, and $r_1(x)$, $s_1(x) \in D[x]$, where the degrees of $r_1(x)$ and $s_1(x)$ are the degrees of $r(x)$ and $s(x)$, respectively. By Lemma 32.4, $f(x) = (c)g(x)$, $r_1(x) = (c_1)r_2(x)$, and $s_1(x) = (c_2)s_2(x)$ for primitive polynomials $g(x)$, $r_2(x)$, and $s_2(x)$, and $c, c_1, c_2 \in D$. Then

$$(dc)g(x) = (c_1 c_2)r_2(x)s_2(x),$$

and by Lemma 32.5, $r_2(x)s_2(x)$ is primitive. By the uniqueness part of Lemma 32.4, $c_1 c_2 = dcu$ for some unit u in D. But then

$$(dc)g(x) = (dcu)r_2(x)s_2(x),$$

so

$$f(x) = (c)g(x) = (cu)r_2(x)s_2(x).$$

We have shown that if $f(x)$ factors nontrivially in $F[x]$, then $f(x)$ factors nontrivially into polynomials of the same degrees in $D[x]$. Thus, if $f(x) \in D[x]$ is irreducible in $D[x]$, it must be irreducible in $F[x]$.

It is obvious that a nonconstant $f(x) \in D[x]$ that is primitive in $D[x]$ and irreducible in $F[x]$ is also irreducible in $D[x]$, since $D[x] \subseteq F[x]$. ∎

Lemma 32.6 shows that *if D is a UFD the irreducibles in $D[x]$ are precisely the irreducibles in D, together with the nonconstant primitive polynomials that are irreducible in $F[x]$, where F is a field of quotients of $D[x]$.*

The preceding lemma is really very important in its own right. This is indicated by the following corollary, a special case of which was our Theorem 31.3. (We admit that it does not seem very sensible to call a special case of a corollary of a lemma a theorem. The label you assign to a result depends somewhat on the context in which it appears.)

Corollary *If D is a UFD and F is a field of quotients of D, then a nonconstant $f(x) \in D[x]$ factors into a product of two polynomials of lower degrees r and s in $F[x]$ if and only if it has a factorization into polynomials of the same degrees r and s in $D[x]$.*

Proof.   It was shown in the proof of Lemma 32.6 that if $f(x)$ factors into a product of two polynomials of lower degree in $F[x]$, then it has a factorization into polynomials of the same degrees in $D[x]$ (see the next to last sentence of the first paragraph of the proof)
The converse is obvious, since $D[x] \subseteq F[x]$. ∎

We are now prepared to prove our main theorem. We shall repeat the construction of the proof of Lemma 32.6 again; it is the core of this whole argument.

Theorem 32.3 *If D is a UFD, then $D[x]$ is a UFD.*

Proof.   Let $f(x) \in D[x]$, where $f(x)$ is neither 0 nor a unit. If $f(x)$ is of degree 0, we are done, since D is a UFD. Suppose that (degree $f(x)$) > 0, and let us view $f(x)$ as an element in $F[x]$, where F is a field of quotients of D. By Theorem 31.8, $f(x) = p_1(x) \cdots p_r(x)$ in $F[x]$, where $p_i(x)$ is irreducible in $F[x]$. Since F is a field of quotients of D, each coefficient in each $p_i(x)$ is of the form a/b for some $a, b \in D$. Clearing all denominators in the usual fashion, we arrive at

$$(d)f(x) = q_1(x) \cdots q_r(x)$$

for d, $q_i(x) \in D[x]$. Since each $p_i(x)$ was irreducible in $F[x]$, we see that $q_i(x)$, which is $p_i(x)$ multiplied by a *unit in F*, is also irreducible in $F[x]$. By Lemma 32.4, $f(x) = (c)g(x)$ and $q_i(x) = (c_i)q_i'(x)$ in $D[x]$ for $g(x)$ and $q_i'(x)$ primitive. Then

$$(dc)g(x) = (c_1 \cdots c_r)q_1'(x) \cdots q_r'(x),$$

where by Lemma 32.5, the product $q_1'(x) \cdots q_r'(x)$ is primitive. **By the** uniqueness part of Lemma 32.4, we see that

$$c_1 \cdots c_r = dcu$$

for some unit u in D. Then

$$(dc)g(x) = (dcu)q_1'(x) \cdots q_r'(x),$$

so

$$f(x) = (c)g(x) = (cu)q_1'(x) \cdots q_r'(x).$$

Now cu can be factored into irreducibles in D. Also $q_1'(x), \ldots, q_r'(x)$ are irreducible in $D[x]$, since they are primitive and irreducible in $F[x]$ by construction. Thus we have shown that we can factor $f(x)$ into a product of irreducibles in $D[x]$.

The uniqueness of a factorization of $f(x) \in D[x]$ is clear for $f(x) \in D$ that is neither 0 nor a unit. If (degree $f(x)$) > 0, we can view any factorization of $f(x)$ into irreducibles in $D[x]$ as a factorization in $F[x]$ into units (that is, the factors in D) are irreducible polynomials in $F[x]$ by Lemma 32.6. By Theorem 31.8, these polynomials are unique, except for possible constant factors in F. But as an irreducible in $D[x]$, each polynomial of degree > 0 appearing in the factorization of $f(x)$ in $D[x]$ is primitive. By the uniqueness part of Lemma 32.4, this shows that these polynomials are unique in $D[x]$ up to unit factors, that is, associates. The product of the irreducibles in D in the factorization of $f(x)$ is the content of $f(x)$, which is again unique up to a unit factor by Lemma 32.4. Thus all irreducibles in $D[x]$ appearing in the factorization are unique up to order and associates. ∎

Corollary *If F is a field and $x_1, \ldots, x_n$ are indeterminates, then $F[x_1, \ldots, x_n]$ is a UFD.*

Proof. By Theorem 38.1, $F[x_1]$ is a UFD. By Theorem 32.3, so is $(F[x_1])[x_2] = F[x_1, x_2]$. Continuing in this procedure, we see (by induction) that $F[x_1, \ldots, x_n]$ is a UFD. ∎

We have seen that a PID is a UFD. The corollary of Theorem 32.3 makes it easy for us to give an example that shows that *not every* UFD *is a* PID.

Example 32.6 Let F be a field and let x and y be indeterminates. Then $F[x, y]$ is a UFD by the corollary of Theorem 32.3. Consider the set N of all polynomials in x and y in $F[x, y]$ having constant term 0. Clearly, N is an ideal, but not a principal ideal. Thus $F[x, y]$ is not a PID. ∎

Another example of a UFD that is not a PID is $Z[x]$, as shown in Exercise 33.5.

Exercises

*32.1 State which of the following elements are irreducibles of the indicated integral domains.

a) 5 in Z

b) -17 in Z

c) 14 in Z

d) $2x - 3$ in $Z[x]$

e) $2x - 10$ in $Z[x]$

f) $2x - 3$ in $Q[x]$

g) $2x - 10$ in $Q[x]$

h) $2x - 10$ in $Z_{11}[x]$

*32.2 If possible, give four different associates of $2x - 7$ viewed as an element of $Z[x]$; of $Q[x]$; of $Z_{11}[x]$.

*32.3 Factor the polynomial $4x^2 - 4x + 8$ into a product of irreducibles viewing it as an element of the integral domain $\mathbf{Z}[x]$; of the integral domain $\mathbf{Q}[x]$; of the integral domain $\mathbf{Z}_{11}[x]$.

*32.4 Express each of the given polynomials as the product of its content with a primitive polynomial in the indicated UFD.

a) $18x^2 - 12x + 48$ in $\mathbf{Z}[x]$ b) $18x^2 - 12x + 48$ in $\mathbf{Q}[x]$

c) $2x^2 - 3x + 6$ in $\mathbf{Z}[x]$ d) $2x^2 - 3x + 6$ in $\mathbf{Z}_7[x]$

†*32.5 Prove the following.

a) If p is a prime in an integral domain D, then p is an irreducible.

b) If q is an irreducible in a UFD, then q is a prime.

*32.6 Mark each of the following true or false.

—   a) Every field is a UFD.

—   b) Every field is a PID.

—   c) Every PID is a UFD.

—   d) Every UFD is a PID.

—   e) $\mathbf{Z}[x]$ is a UFD.

—   f) Any two irreducibles in any UFD are associates.

—   g) If D is a PID, then $D[x]$ is a PID.

—   h) If D is a UFD, then $D[x]$ is a UFD.

—   i) In any UFD, if $p \mid a$ for an irreducible p, then p itself appears in every factorization of a.

—   j) A UFD has no divisors of 0.

---

*32.7 For an integral domain D, show that the relation $a \sim b$ if a is an associate of b (that is, if $a = bu$ for u a unit in D) is an equivalence relation on D.

*32.8 Let D be a UFD. Describe the irreducibles in $D[x]$ in terms of the irreducibles in D and the irreducibles in $F[x]$, where F is a field of quotients of D.

*32.9 Let D be an integral domain. Exercise 23.4 showed that $\langle U, \cdot \rangle$ is a group where U is the set of units of D. Show that the set $D^* - U$ of nonunits of D excluding 0 is closed under multiplication. Is this set a group under the multiplication of D?

*32.10 Lemma 32.6 states that if D is a UFD with a field of quotients F, then a nonconstant irreducible $f(x)$ of $D[x]$ is also an irreducible of $F[x]$. Show by an example that a $g(x) \in D[x]$ that is an irreducible of $F[x]$ need not be an irreducible of $D[x]$.

*32.11 All our work in this section was restricted to integral domains. Taking the same definitions in Section 32.1 but for a commutative ring with unity, consider factorizations into irreducibles in $\mathbf{Z} \times \mathbf{Z}$. What can happen? Consider in particular $(1, 0)$.

*32.12 Let D be a UFD. Show that a nonconstant divisor of a primitive polynomial in $D[x]$ is again a primitive polynomial.

*32.13 Show that in a PID, every ideal is contained in a maximal ideal. [Hint: Use Lemma 32.1.]

*32.14 Factor $x^3 - y^3$ into irreducibles in $\mathbf{Q}[x, y]$ and prove that each of the factors is irreducible.

There are several other concepts often considered that are similar in character to the ascending chain condition on ideals in a ring. The following three exercises concern some of these concepts.

*32.15 Let R be any ring. The **ascending chain condition** (ACC) **for ideals** holds in R if every strictly increasing sequence $N_1 \subset N_2 \subset N_3 \subset \cdots$ of ideals in R is of finite length. The **maximum condition** (MC) **for ideals** holds in R if every nonempty set S of ideals in R contains an ideal not properly contained in any other ideal of the set S. The **finite basis condition** (FBC) **for ideals** holds in R if for each ideal N in R, there is a finite set $B_N = \{b_1, \ldots, b_n\} \subseteq N$ such that N is the intersection of all ideals of R containing B_N. The set B_N is a **finite basis for** N.

Show that for every ring R, the conditions ACC, MC, and FBC are equivalent.

*32.16 Let R be any ring. The **descending chain condition** (DCC) **for ideals** holds in R if every strictly decreasing sequence $N_1 \supset N_2 \supset N_3 \supset \cdots$ of ideals in R is of finite length. The **minimum condition** (mC) **for ideals** holds in R if given any set S of ideals of R, there is an ideal of S that does not properly contain any other ideal in the set S.

Show that for every ring, the conditions DCC and mC are equivalent.

*32.17 Give an example of a ring in which ACC holds but DCC does not hold. (See Exercises 32.15 and 32.16.)

*thirty-three

Euclidean Domains

*33.1 INTRODUCTION AND DEFINITION

We have remarked several times on the importance of division algorithms. Our first contact with them was the *division algorithm for* **Z** (Lemma 6.1). This algorithm was immediately used to prove the important theorem that a subgroup of a cyclic group is cyclic, that is, has a single generator. The *division algorithm for* $F[x]$ appeared in Theorem 31.1 and was used in a completely analogous way to show that $F[x]$ is a PID, that is, that every ideal in $F[x]$ has a single generator. Now a modern technique of mathematics is to take some clearly related situations and to try to bring them under one roof by abstracting the important ideas common to them. It should be evident to the student how the following definition is an illustration of this technique. Let us see what we can develop by starting with the existence of a fairly general division algorithm in an integral domain.

Definition A *Euclidean valuation on an integral domain D* is a function v mapping the nonzero elements of D into the nonnegative integers such

300

that the following conditions are satisfied:

1. For all $a, b \in D$ with $b \neq 0$, there exist q and r in D such that $a = bq + r$, where either $r = 0$ or $v(r) < v(b)$.
2. For all $a, b \in D$, where neither a nor b is 0, $v(a) \leq v(ab)$.

An integral domain D is a **Euclidean domain** if there exists a Euclidean valuation on D.

The importance of condition 1 is clear from our discussion. The importance of condition 2 is that it will enable us to characterize the units of a Euclidean domain D.

Example 33.1 The integral domain $\mathbf{Z}$ is a Euclidean domain, for the valuation v defined by $v(n) = |n|$ for $n \neq 0$ in $\mathbf{Z}$ is a Euclidean valuation on $\mathbf{Z}$. Condition 1 holds by Lemma 6.1, and condition 2 is obvious. ∎

Example 33.2 If F is a field, then $F[x]$ is a Euclidean domain, for the valuation v defined by $v(f(x)) = (\text{degree } f(x))$ for $f(x) \in F[x]$ and $f(x) \neq 0$ is a Euclidean valuation. Condition 1 holds by Theorem 31.1, and condition 2 is obvious. ∎

Of course, we should give some examples of Euclidean domains other than these familiar ones that motivated the definition. We shall do this in the next chapter. In view of our opening remarks, you must surely be waiting for the following theorem.

Theorem 33.1 *Every Euclidean domain is a* PID.

Proof. Let D be a Euclidean domain with a Euclidean valuation v, and let N be an ideal in D. If $N = \{0\}$, then $N = \langle 0 \rangle$ and N is principal. Suppose that $N \neq \{0\}$. Then there exists $b \neq 0$ in N. Let us choose b such that $v(b)$ is minimal among all $v(n)$ for $n \in N$. We claim that $N = \langle b \rangle$. Let $a \in N$. Then by condition 1 for a Euclidean domain, there exist q and r in D such that

$$a = bq + r,$$

where either $r = 0$ or $v(r) < v(b)$. Now $r = a - bq$ and $a, b \in N$, so that $r \in N$, since N is an ideal. Thus $v(r) < v(b)$ is impossible by our choice of b. Hence $r = 0$, so $a = bq$. Since a was any element of N, we see that $N = \langle b \rangle$. ∎

Corollary *A Euclidean domain is a* UFD.

Proof. By Theorem 33.1, a Euclidean domain is a PID and by Theorem 32.2, a PID is a UFD. ∎

Finally, we should mention that while a Euclidean domain is a PID by Theorem 33.1, not every PID is a Euclidean domain. Examples of PIDs that are not Euclidean are not easily found, however.

*33.2 ARITHMETIC IN EUCLIDEAN DOMAINS

We shall now investigate some properties of Euclidean domains related to their multiplicative structure. It should be made clear that the arithmetic structure of a Euclidean domain is *intrinsic to the domain* and is not affected in any way by a Euclidean valuation v on the domain. A Euclidean valuation is merely a useful tool for possibly throwing some light on this arithmetic structure of the domain. The arithmetic structure of a domain D is completely determined by the set D and the two binary operations $+$ and $\cdot$ on D.

Let D be a Euclidean domain with a Euclidean valuation v. We can use property 2 of a Euclidean valuation to characterize the units of D.

Theorem 33.2   *For a Euclidean domain with a Euclidean valuation v, $v(1)$ is minimal among all $v(a)$ for nonzero $a \in D$, and $u \in D$ is a unit if and only if $v(u) = v(1)$.*

Proof.   Condition 2 for v tells us at once that for $a \neq 0$,

$$v(1) \leq v(1a) = v(a).$$

On the other hand, if u is a unit in D, then

$$v(u) \leq v(uu^{-1}) = v(1).$$

Thus

$$v(u) = v(1)$$

for a unit u in D.

Conversely, suppose that a nonzero $u \in D$ is such that $v(u) = v(1)$. Then by the division algorithm, there exist q and r in D such that

$$1 = uq + r,$$

where either $r = 0$ or $v(r) < v(u)$. But since $v(u) = v(1)$ is minimal over all $v(d)$ for nonzero $d \in D$, $v(r) < v(u)$ is impossible. Hence $r = 0$ and $1 = uq$, so u is a unit. ∎

Example 33.3   For **Z** with $v(n) = |n|$, the minimum of $v(n)$ for nonzero $n \in \mathbf{Z}$ is 1. Clearly, 1 and -1 are the only elements of **Z** with $v(n) = 1$. Of course, 1 and -1 are exactly the units of **Z**. ∎

Example 33.4   For $F[x]$ with $v(f(x)) = (\text{degree } f(x))$ for $f(x) \neq 0$, the minimum value of $v(f(x))$ for all nonzero $f(x) \in F[x]$ is 0. The nonzero polynomials of degree 0 are exactly the nonzero elements of F, and these are precisely the units of $F[x]$. ∎

You should realize that everything we prove here holds in *every* Euclidean domain, in particular in **Z** and $F[x]$. We are going to prove some nice, classical results about greatest common divisors in a Euclidean domain.

You surely have a naive idea of what a greatest common divisor (gcd) of two elements a and b in a UFD should be. You simply take a and b and factor them both, adjusting the factorizations by units, so that if any irreducible divides both a and b, it either appears in both factorizations, and no other of its associates appears, or it does not appear in either factorization, that is, one of its associates appears instead. We then obtain a gcd of a and b by multiplying together all irreducibles appearing in both factorizations, taking each irreducible to the highest power for which it divides both a and b. That is sloppily stated, but you probably get the idea. We have used the concept freely for $\mathbf{Z}$ in group theory already. Since an irreducible appearing in a factorization is only defined up to a unit factor, we see that a gcd must also be defined only up to a unit factor in a UFD. It is for this reason that we say "a" gcd, rather than "the" gcd, of a and b. It is customary to give the following more elegant definition of a gcd in a UFD. It should be obvious to you that for a UFD, this definition describes the same concept we have just discussed.

Definition   Let D be a UFD. An element $d \in D$ is a **greatest common divisor** (abbreviated gcd) **of elements a and b in D** if $d \mid a$, $d \mid b$, and also $c \mid d$ for all c dividing both a and b.

Example 33.5   In $\mathbf{Z}$, a gcd of 18 and 48 is 6. Another one is -6. In $\mathbf{Q}[x]$, a gcd of $x^2 - 2x + 1$ and $x^2 + x - 2$ is $x - 1$. Another one is $2(x - 1)$, since 2 is a unit in $\mathbf{Q}[x]$. Still another one is $(15/13)(x - 1)$. However, in $\mathbf{Z}[x]$, the only gcd's of $x^2 - 2x + 1$ and $x^2 + x - 2$ are $x - 1$ and $-(x - 1)$, for 1 and -1 are the only units in $\mathbf{Z}[x]$. ∎

As indicated after Example 33.4, one can show that any a and b in a UFD have a gcd by factoring a and b into irreducibles. (Actually, a gcd exists for any number of elements from a UFD. The proof of Lemma 32.4 shows in more detail how such a gcd can be constructed from factorizations into irreducibles.) Every PID is a UFD, so of course gcd's also exist in a PID. There is a nice proof for PIDs, not using factorization, that we would like to have you see.

Theorem 33.3   *If D is a PID and a and b are nonzero elements of D, then there exists a gcd of a and b. Furthermore, each gcd of a and b can be expressed in the form $\lambda a + \mu b$ for some $\lambda, \mu \in D$.*

Proof.   Consider the set

$$N = \{ra + sb \mid r, s \in D\}.$$

Since

$$(r_1 a + s_1 b) \pm (r_2 a + s_2 b) = (r_1 \pm r_2)a + (s_1 \pm s_2)b$$

and

$$t(ra + sb) = (tr)a + (ts)b$$

for $t \in D$, it is immediate that N is an ideal of D. Now $N = \langle d \rangle$ for some $d \in D$. Then $d \mid (ra + sb)$ for all r, $s \in D$, and taking first $s = 0$ with $r = 1$ and then $r = 0$ with $s = 1$, we see that $d \mid a$ and $d \mid b$. Also if $c \mid a$ and $c \mid b$, then $c \mid (ra + sb)$ for all $ra + sb$, that is, $c \mid n$ for all $n \in N$. Hence $c \mid d$. Thus d is a gcd of a and b.

For d as just constructed, $d \in N$ implies that there exist λ, $\mu \in D$ such that $d = \lambda a + \mu b$. But the definition of a gcd shows that if d_1 is also a gcd of a and b, then $d \mid d_1$ and $d_1 \mid d$. Thus

$$d_1 = vd = (v\lambda)a + (v\mu)b = \lambda_1 a + \mu_1 b. \quad \blacksquare$$

The preceding proof was extremely elegant, but not at all constructive. Of course we can find a gcd of a and b if we can factor them into irreducibles, but such factorizations can be very tough to do. However, if a UFD is actually Euclidean, and we know a Euclidean valuation, there is an easy constructive way to find gcd's, as the next theorem shows.

Theorem 33.4 (Euclidean Algorithm)   *Let D be a Euclidean domain with a Euclidean valuation v, and let a and b be nonzero elements of D. Let r_1 be as in condition 1 for a Euclidean valuation, that is,*

$$a = bq_1 + r_1,$$

where either $r_1 = 0$ or $v(r_1) < v(b)$. If $r_1 \neq 0$, let r_2 be such that

$$b = r_1 q_2 + r_2,$$

where either $r_2 = 0$ or $v(r_2) < v(r_1)$. In general, let r_{i+1} be such that

$$r_{i-1} = r_i q_{i+1} + r_{i+1},$$

where either $r_{i+1} = 0$ or $v(r_{i+1}) < v(r_i)$. Then the sequence $r_1, r_2, \ldots$ must terminate with some $r_s = 0$. If $r_1 = 0$, then b is a gcd of a and b. If $r_1 \neq 0$ and r_s is the first $r_i = 0$, then a gcd of a and b is r_{s-1}.

Proof.   Since $v(r_i) < v(r_{i-1})$ and $v(r_i)$ is a nonnegative integer, it is clear that after some finite number of steps we must arrive at some $r_s = 0$.

If $r_1 = 0$, then $a = bq_1$, and obviously b is a gcd of a and b. Suppose $r_1 \neq 0$. Then if $d \mid a$ and $d \mid b$, we have

$$d \mid (a - bq_1),$$

so $d \mid r_1$. However, if $d_1 \mid r_1$ and $d_1 \mid b$, then

$$d_1 \mid (bq_1 + r_1),$$

so $d_1 \mid a$. Thus the set of common divisors of a and b is the same set as the set of common divisors of b and r_1. By a similar argument, if $r_2 \neq 0$, the set of common divisors of b and r_1 is the same set as the set of common divisors of r_1 and r_2. Continuing this process, we see finally that the set of common divisors of a and b is the same set as the set of common divisors of r_{s-2} and r_{s-1}, where r_s is the first r_i equal to 0. Thus a gcd of r_{s-2} and r_{s-1} is also a gcd of a and b. But the equation

$$r_{s-2} = q_s r_{s-1} + r_s = q_s r_{s-1}$$

shows that a gcd of r_{s-2} and r_{s-1} is r_{s-1}. ∎

Example 33.6 Let us illustrate the Euclidean algorithm for the Euclidean valuation $||$ on $\mathbf{Z}$ by computing a gcd of 22,471 and 3,266. We just apply the division algorithm over and over again, and the last nonzero remainder is a gcd. We label the numbers obtained as in Theorem 33.4 to further illustrate the statement and proof of the theorem. The computations are easily checked.

$$a = 22{,}471$$
$$b = 3{,}266$$

$$
\begin{array}{ll}
22{,}471 = (3{,}266)6 + 2{,}875 & r_1 = 2{,}875 \\
3{,}266 = (2{,}875)1 + 391 & r_2 = 391 \\
2{,}875 = (391)7 + 138 & r_3 = 138 \\
391 = (138)2 + 115 & r_4 = 115 \\
138 = (115)1 + 23 & r_5 = 23 \\
115 = (23)5 + 0 & r_6 = 0
\end{array}
$$

Thus $r_5 = 23$ is a gcd of 22,471 and 3,266. We found a gcd without factoring! This is important, for sometimes it is very difficult to find a factorization of an integer into primes. ∎

Example 33.7 Note that the division algorithm 1 of the definition of a Euclidean valuation says nothing about r being "positive." In computing a gcd in $\mathbf{Z}$ by the Euclidean algorithm for $||$, as in Example 33.6, it is surely to our interest to make $|r_i|$ as small as possible in each division. Thus, repeating Example 33.6, it would be more efficient to write

$$a = 22{,}471$$
$$b = 3{,}266$$

$$
\begin{array}{ll}
22{,}471 = (3{,}266)7 - 391 & r_1 = -391 \\
3{,}266 = (391)8 + 138 & r_2 = 138 \\
391 = (138)3 - 23 & r_3 = -23 \\
138 = (23)6 + 0 & r_4 = 0
\end{array}
$$

The fact that we can change the sign of r_i from negative to positive when we wish is due to the fact that the divisors of r_i and $-r_i$ are the same. ∎

Exercises

*33.1 State which of the given functions v are Euclidean valuations for the given integral domains.

a) The function v for $\mathbf{Z}$ given by $v(n) = n^2$ for nonzero $n \in \mathbf{Z}$

b) The function v for $\mathbf{Z}[x]$ given by $v(f(x)) = $ (degree of $f(x)$) for nonzero $f(x) \in \mathbf{Z}[x]$

c) The function v for $\mathbf{Z}[x]$ given by $v(f(x)) = $ (the absolute value of the coefficient of the highest degree nonzero term of $f(x)$) for nonzero $f(x) \in \mathbf{Z}[x]$

d) The function v for $\mathbf{Q}$ given by $v(a) = a^2$ for nonzero $a \in \mathbf{Q}$

e) The function v for $\mathbf{Q}$ given by $v(a) = 50$ for nonzero $a \in \mathbf{Q}$

*33.2 Find a gcd of 49,349 and 15,555 in $\mathbf{Z}$.

*33.3 Find a gcd of

$$x^{10} - 3x^9 + 3x^8 - 11x^7 + 11x^6 - 11x^5 + 19x^4 - 13x^3 + 8x^2 - 9x + 3$$

and

$$x^6 - 3x^5 + 3x^4 - 9x^3 + 5x^2 - 5x + 2$$

in $\mathbf{Q}[x]$.

*33.4 By referring to Example 33.7 of the text, actually express the gcd 23 in the form $\lambda(22,471) + \mu(3,266)$ for λ, $\mu \in \mathbf{Z}$. [*Hint:* From the next to the last line of the computation in Example 33.7, $23 = (138)3 - 391$. From the line before that, $138 = 3,266 - (391)8$, so substituting, you get $23 = [3,266 - (391)8]3 - 391$, and so on. That is, work your way back up to actually find values for λ and μ.]

*33.5 Let us consider $\mathbf{Z}[x]$.

a) Is $\mathbf{Z}[x]$ a UFD? Why?

b) Show that $\{a + xf(x) | a \in 2\mathbf{Z}, f(x) \in \mathbf{Z}[x]\}$ is an ideal in $\mathbf{Z}[x]$.

c) Is $\mathbf{Z}[x]$ a PID? (Consider part (b).)

d) Is $\mathbf{Z}[x]$ a Euclidean domain? Why?

†*33.6 Let D be a Euclidean domain and let v be a Euclidean valuation on D. Show that if a and b are associates in D, then $v(a) = v(b)$.

*33.7 Mark each of the following true or false.

___ a) Every Euclidean domain is a PID.

___ b) Every PID is a Euclidean domain.

___ c) Every Euclidean domain is a UFD.

___ d) Every UFD is a Euclidean domain.

___ e) A gcd of 2 and 3 in $\mathbf{Q}$ is $\frac{1}{2}$.

___ f) The Euclidean algorithm gives a constructive method for finding a gcd of two integers.

___ g) If v is a Euclidean valuation on a Euclidean domain D, then $v(1) \leq v(a)$ for all nonzero $a \in D$.

___ h) If v is a Euclidean valuation on a Euclidean domain D, then $v(1) < v(a)$ for all nonzero $a \in D$, $a \neq 1$.

— i) If v is a Euclidean valuation on a Euclidean domain D, then $v(1) < v(a)$ for all nonzero nonunits $a \in D$.

— j) For any field F, $F[x]$ is a Euclidean domain.

---

*33.8 Does the choice of a particular Euclidean valuation v on a Euclidean domain D influence the arithmetic structure of D in any way? Explain.

*33.9 Following the idea of Exercise 33.4 and referring to Exercise 33.2, express the positive gcd of 49,349 and 15,555 in $\mathbf{Z}$ in the form $\lambda(49{,}349) + \mu(15{,}555)$ for λ, $\mu \in \mathbf{Z}$.

*33.10 Let D be a Euclidean domain and let v be a Euclidean valuation on D. Show that for nonzero $a, b \in D$, one has $v(a) < v(ab)$ if and only if b is not a unit of D. [*Hint:* Argue from Exercise 33.6 that $v(a) < v(ab)$ implies that b is not a unit of D. Using the Euclidean algorithm, show that $v(a) = v(ab)$ implies $\langle a \rangle = \langle ab \rangle$. Conclude that if b is not a unit, then $v(a) < v(ab)$.]

*33.11 Prove or disprove the following statement: If v is a Euclidean valuation on a Euclidean domain D, then $\{a \in D \mid v(a) > v(1)\}$ is an ideal of D.

*33.12 Show that every field is a Euclidean domain.

*33.13 Let v be a Euclidean valuation on a Euclidean domain D.

a) Show that if $s \in \mathbf{Z}$ such that $s + v(1) > 0$, then $\eta : D^* \to \mathbf{Z}$ defined by $\eta(a) = v(a) + s$ for nonzero $a \in D$ is a Euclidean valuation on D. As usual, D^* is the set of nonzero elements of D.

b) Show that for $r \in \mathbf{Z}^+$, $\lambda : D^* \to \mathbf{Z}$ given by $\lambda(a) = r(v(a))$ for nonzero $a \in D$ is a Euclidean valuation on D.

c) Show that there exists a Euclidean valuation μ on D such that $\mu(1) = 1$ and $\mu(a) > 100$ for all nonzero nonunits $a \in D$.

*33.14 Let D be a UFD. An element c in D is a **least common multiple** (abbreviated lcm) **of two elements** a **and** b **in** D if $a \mid c, b \mid c$, and if c divides every element of D that is divisible by both a and b. Show that every two nonzero elements a and b of a Euclidean domain D have an lcm in D. [*Hint:* Show that all common multiples, in the obvious sense, of both a and b form an ideal of D.]

*33.15 Use the last statement in Theorem 33.3 to show that two nonzero elements $r, s \in \mathbf{Z}$ generate the group $\langle \mathbf{Z}, + \rangle$ if and only if r and s, viewed as integers in the domain $\mathbf{Z}$, are **relatively prime,** that is, have a gcd of 1.

*33.16 Using the last statement in Theorem 33.3, show that for nonzero a, b, $n \in \mathbf{Z}$, the congruence $ax \equiv b \pmod{n}$ has a solution in $\mathbf{Z}$ if a and n are relatively prime.

*33.17 Generalize Exercise 33.16 by showing that for nonzero $a, b, n \in \mathbf{Z}$, the congruence $ax \equiv b \pmod{n}$ has a solution in $\mathbf{Z}$ if and only if the positive gcd of a and n in $\mathbf{Z}$ divides b. Interpret this result in the ring $\mathbf{Z}_n$.

*33.18 Following the idea of Exercises 33.4 and 33.17, outline a constructive method for finding a solution in $\mathbf{Z}$ of the congruence $ax \equiv b \pmod{n}$ for nonzero a, b, $n \in \mathbf{Z}$, the congruence does have a solution. Use this method to find a solution of the congruence $22x \equiv 18 \pmod{42}$.

*thirty-four

Gaussian Integers and Norms

*34.1 GAUSSIAN INTEGERS

We should give an example of a Euclidean domain different from $\mathbf{Z}$ and $F[x]$.

Definition   A **Gaussian integer** is a complex number $a + bi$, where $a, b \in \mathbf{Z}$. For a Gaussian integer $\alpha = a + bi$, the **norm** $N(\alpha)$ *of* α is $a^2 + b^2$.

We shall let $\mathbf{Z}[i]$ be the set of all Gaussian integers. The following lemma gives some basic properties of the norm function N on $\mathbf{Z}[i]$ and leads to a demonstration that the function v defined by $v(\alpha) = N(\alpha)$ for nonzero $\alpha \in \mathbf{Z}[i]$ is a Euclidean valuation on $\mathbf{Z}[i]$. Note that the Gaussian integers include all the **rational integers,** that is, all the elements of $\mathbf{Z}$.

Lemma 34.1   *In* $\mathbf{Z}[i]$, *the following properties of the norm function N hold for all* $\alpha, \beta \in \mathbf{Z}[i]$:

1. $N(\alpha) \geq 0$.
2. $N(\alpha) = 0$ *if and only if* $\alpha = 0$.
3. $N(\alpha\beta) = N(\alpha)N(\beta)$.

308

Proof. If we let $\alpha = a_1 + a_2i$ and $\beta = b_1 + b_2i$, these results are all either obvious or straightforward computations. We leave the proof of these properties as an exercise (see Exercise 34.8). ∎

Lemma 34.2 $\mathbf{Z}[i]$ *is an integral domain.*

Proof. It is obvious that $\mathbf{Z}[i]$ is a commutative ring with unity. We show that there are no divisors of 0. Let $\alpha, \beta \in \mathbf{Z}[i]$. Using Lemma 34.1, if $\alpha\beta = 0$, then

$$N(\alpha)N(\beta) = N(\alpha\beta) = N(0) = 0.$$

Thus $\alpha\beta = 0$ implies that $N(\alpha) = 0$ or $N(\beta) = 0$. By Lemma 34.1 again, this implies that either $\alpha = 0$ or $\beta = 0$. Thus $\mathbf{Z}[i]$ has no divisors of 0, so $\mathbf{Z}[i]$ is an integral domain. ∎

Of course, since $\mathbf{Z}[i]$ is a subring of $\mathbf{C}$, where $\mathbf{C}$ is the field of complex numbers, it is really obvious that $\mathbf{Z}[i]$ has no 0 divisors. We gave the argument of Lemma 34.2 to illustrate the use of the multiplicative property 3 of the norm function N and to avoid going outside of $\mathbf{Z}[i]$ in our argument.

Theorem 34.1 *The function v given by $v(\alpha) = N(\alpha)$ for nonzero $\alpha \in \mathbf{Z}[i]$ is a Euclidean valuation on $\mathbf{Z}[i]$. Thus $\mathbf{Z}[i]$ is a Euclidean domain.*

Proof. Note that for $\beta = b_1 + b_2i \neq 0$, $N(b_1 + b_2i) = b_1{}^2 + b_2{}^2$, so $N(\beta) \geq 1$. Then for all $\alpha, \beta \neq 0$ in $\mathbf{Z}[i]$, $N(\alpha) \leq N(\alpha)N(\beta) = N(\alpha\beta)$. This proves condition 2 for a Euclidean valuation.

It remains to prove the division algorithm, condition 1, for N. Let $\alpha, \beta \in \mathbf{Z}[i]$, with $\alpha = a_1 + a_2i$ and $\beta = b_1 + b_2i$, where $\beta \neq 0$. We must find σ and ρ in $\mathbf{Z}[i]$ such that $\alpha = \beta\sigma + \rho$, where either $\rho = 0$ or $N(\rho) < N(\beta) = b_1{}^2 + b_2{}^2$. Let us put $\sigma = q_1 + q_2i$, where q_1 and q_2 are *rational integers* in $\mathbf{Z}$ to be determined. Then ρ will have to have the form

$$\rho = (a_1 + a_2i) - (b_1 + b_2i)(q_1 + q_2i)$$
$$= (a_1 - b_1q_1 + b_2q_2) + (a_2 - b_1q_2 - b_2q_1)i.$$

We have to try to find *rational integers* q_1 and q_2 such that

$$N(\rho) = (a_1 - b_1q_1 + b_2q_2)^2 + (a_2 - b_1q_2 - b_2q_1)^2 < b_1{}^2 + b_2{}^2$$

that is, such that

$$\frac{(a_1 - b_1q_1 + b_2q_2)^2}{b_1{}^2 + b_2{}^2} + \frac{(a_2 - b_1q_2 - b_2q_1)^2}{b_1{}^2 + b_2{}^2} < 1.$$

Now you will recall that

$$\frac{(a_1 - b_1q_1 + b_2q_2)^2}{b_1{}^2 + b_2{}^2}$$

is exactly the square of the distance d in the Euclidean plane from a point (q_1, q_2) to the line l with equation $a_1 - b_1 X + b_2 Y = 0$. Similarly,

$$\frac{(a_2 - b_1 q_2 - b_2 q_1)^2}{b_1{}^2 + b_2{}^2}$$

is the square of the distance d' from (q_1, q_2) to the line l' with equation $a_2 - b_2 X - b_1 Y = 0$. Note that l is perpendicular to l'. Let P be the point of intersection of these two lines as shown in Fig. 34.1. From this figure, we see that $d^2 + (d')^2$ is the square of the distance from (q_1, q_2) to P. Thus we must show that there is a point (q_1, q_2) *with integral coordinates* and with a distance from P whose square is less than 1. Since P is contained within or on the boundary of some square of unit side such that both coordinates of each vertex are integers, it is clear that if (q_1, q_2) is chosen to be the point with integer coordinates *closest* to P, its distance from P can be at most half the length of a diagonal of the square, that is, at most $\sqrt{2}/2$ (see Fig. 34.2). Thus the square of this distance from P is at most $\frac{1}{2}$, which is less than 1. ∎

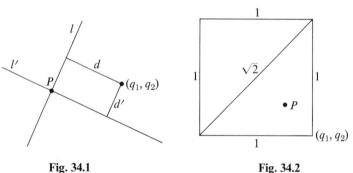

Fig. 34.1   Fig. 34.2

We could have proved the division algorithm for the function N purely algebraically; the algebraic proof is easier, shorter to describe, and more useful in applying the Euclidean algorithm for N (see Exercises 34.4 and 34.12). But we confess a certain fondness for our geometric argument. Besides, it is nice to have variety in the proofs in a text. We leave an algebraic proof to the exercises (see Exercise 34.11).

Example 34.1 We can now apply all our results of Chapter 33 to $\mathbf{Z}[i]$. In particular, since $N(1) = 1$, the units of $\mathbf{Z}[i]$ are exactly the $\alpha = a_1 + a_2 i$ with $N(\alpha) = a_1{}^2 + a_2{}^2 = 1$. From the fact that a_1 and a_2 are integers, it follows that the only possibilities are $a_1 = \pm 1$ with $a_2 = 0$, or $a_1 = 0$ with $a_2 = \pm 1$. Thus the units of $\mathbf{Z}[i]$ are ± 1 and $\pm i$. One can also use the Euclidean algorithm to compute a gcd of two nonzero elements. We leave such computations to the exercises. Finally, note that while 5 is an irreducible in $\mathbf{Z}$, 5 is no longer an irreducible in $\mathbf{Z}[i]$, for $5 = (1 + 2i)(1 - 2i)$, and neither $1 + 2i$ nor $1 - 2i$ is a unit. ∎

*34.2 MULTIPLICATIVE NORMS

Let us point out again that for an integral domain D, *the arithmetic concepts of irreducibles and units are intrinsic to the integral domain itself* and are not affected in any way by a valuation or norm that may be defined on the domain. However, as the preceding chapter and our work thus far in this chapter show, a suitably defined valuation or norm may be of help in determining the arithmetic structure of D. This is strikingly illustrated in *algebraic number theory*, where for a domain of *algebraic integers* one considers many different valuations of the domain, each doing its part in helping to determine the arithmetic structure of the domain. In a domain of algebraic integers, one has essentially one valuation for each irreducible (up to associates), and each such valuation gives information concerning the behavior in the integral domain of the irreducible to which it corresponds. This is an example of the importance of studying properties of elements in an algebraic structure by means of mappings associated with them. We shall be doing this for zeros of polynomials in the following sections.

Let us study integral domains that have a multiplicative norm satisfying the properties of N on $\mathbf{Z}[i]$ given in Lemma 34.1.

Definition   Let D be an integral domain. A ***multiplicative norm*** N on D is a function mapping D into the integers $\mathbf{Z}$ such that the following conditions are satisfied:
1. $N(\alpha) \geq 0$ for all $\alpha \in D$.
2. $N(\alpha) = 0$ if and only if $\alpha = 0$.
3. $N(\alpha\beta) = N(\alpha)N(\beta)$ for all $\alpha, \beta \in D$.

Theorem 34.2   *If D is an integral domain with a multiplicative norm N, then $N(1) = 1$ and $N(u) = 1$ for every unit u in D. If, furthermore, every α such that $N(\alpha) = 1$ is a unit in D, then an element π in D, with $N(\pi) = p$ for a prime $p \in \mathbf{Z}$, is an irreducible of D.*

Proof.   Let D be an integral domain with a multiplicative norm N. Then

$$N(1) = N((1)(1)) = N(1)N(1)$$

shows that $N(1) = 1$. Also, if u is a unit in D, then

$$1 = N(1) = N(uu^{-1}) = N(u)N(u^{-1}).$$

Since $N(u)$ is a nonnegative integer, this implies that $N(u) = 1$.

Now suppose that the units of D are *exactly* the elements of norm 1. Let $\pi \in D$ be such that $N(\pi) = p$, where p is a prime in $\mathbf{Z}$. Then if $\pi = \alpha\beta$, we have

$$p = N(\pi) = N(\alpha)N(\beta),$$

so either $N(\alpha) = 1$ or $N(\beta) = 1$. By assumption, this means that either α or β is a unit of D. Thus π is an irreducible of D. ∎

Example 34.2   On $\mathbf{Z}[i]$, the function N defined by $N(a + bi) = a^2 + b^2$ gives a multiplicative norm in the sense of our definition. We saw that the function v given by $v(\alpha) = N(\alpha)$ for nonzero $\alpha \in \mathbf{Z}[i]$ is a Euclidean valuation on $\mathbf{Z}[i]$, so the units are precisely the elements α of $\mathbf{Z}[i]$ with $N(\alpha) = N(1) = 1$. Thus the second part of Theorem 34.2 applies in $\mathbf{Z}[i]$. We saw in Example 34.1 that 5 is not an irreducible in $\mathbf{Z}[i]$, for $5 = (1 + 2i)(1 - 2i)$. Since $N(1 + 2i) = N(1 - 2i) = 1^2 + 2^2 = 5$ and 5 is a prime in $\mathbf{Z}$, we see from Theorem 34.2 that $1 + 2i$ and $1 - 2i$ are both irreducibles in $\mathbf{Z}[i]$. ∎

As an application of multiplicative norms, we shall now give another example of an integral domain that is *not* a UFD. We saw one example in Example 32.3. The following is the standard illustration.

Example 34.3   Let $\mathbf{Z}[\sqrt{-5}] = \{a + ib\sqrt{5} \,|\, a, b \in \mathbf{Z}\}$. As a subset of the complex numbers closed under addition, subtraction, and multiplication, and containing 0 and 1, $\mathbf{Z}[\sqrt{-5}]$ is an integral domain. Define N on $\mathbf{Z}[\sqrt{-5}]$ by

$$N(a + b\sqrt{-5}) = a^2 + 5b^2.$$

(Here $\sqrt{-5} = i\sqrt{5}$.) Clearly, $N(\alpha) \geq 0$ and $N(\alpha) = 0$ if and only if $\alpha = a + b\sqrt{-5} = 0$. That $N(\alpha\beta) = N(\alpha)N(\beta)$ is a straightforward computation that we leave to the exercises (see Exercise 34.9). Let us find all candidates for units in $\mathbf{Z}[\sqrt{-5}]$ by finding all elements α in $\mathbf{Z}[\sqrt{-5}]$ with $N(\alpha) = 1$. If $\alpha = a + b\sqrt{-5}$, and $N(\alpha) = 1$, we must have $a^2 + 5b^2 = 1$ for *integers* a and b. This is only possible if $b = 0$ and $a = \pm 1$. Hence ± 1 are the only candidates for units. Since ± 1 are units, they are then precisely the units in $\mathbf{Z}[\sqrt{-5}]$.

Now in $\mathbf{Z}[\sqrt{-5}]$, we have $21 = (3)(7)$ and also

$$21 = (1 + 2\sqrt{-5})(1 - 2\sqrt{-5}).$$

If we can show that 3, 7, $1 + 2\sqrt{-5}$, and $1 - 2\sqrt{-5}$ are all irreducibles in $\mathbf{Z}[\sqrt{-5}]$, we will then know that $\mathbf{Z}[\sqrt{-5}]$ cannot be a UFD, since neither 3 nor 7 is $\pm(1 + 2\sqrt{-5})$.

Suppose that $3 = \alpha\beta$. Then

$$9 = N(3) = N(\alpha)N(\beta)$$

shows that we must have $N(\alpha) = 1$, 3, or 9. If $N(\alpha) = 1$, then α is a unit. If $\alpha = a + b\sqrt{-5}$, then $N(\alpha) = a^2 + 5b^2$, and for no choice of integers a and b is $N(\alpha) = 3$..If $N(\alpha) = 9$, then $N(\beta) = 1$, so β is a unit. Thus from $3 = \alpha\beta$, we can conclude that either α or β is a unit. Therefore, 3 is an irreducible in $\mathbf{Z}[\sqrt{-5}]$. A similar argument shows that 7 is also an irreducible in $\mathbf{Z}[\sqrt{-5}]$.

If $1 + 2\sqrt{-5} = \gamma\delta$, we have

$$21 = N(1 + 2\sqrt{-5}) = N(\gamma)N(\delta),$$

so $N(\gamma) = 1, 3, 7,$ or 21. We have seen that there is no element of $\mathbf{Z}[\sqrt{-5}]$ of norm 3 or 7. Thus either $N(\gamma) = 1$, and γ is a unit, or $N(\gamma) = 21$, so $N(\delta) = 1$, and δ is a unit. Therefore, $1 + 2\sqrt{-5}$ is an irreducible in $\mathbf{Z}[\sqrt{-5}]$. A parallel argument shows that $1 - 2\sqrt{-5}$ is also an irreducible in $\mathbf{Z}[\sqrt{-5}]$.

In summary, we have shown that

$$\mathbf{Z}[\sqrt{-5}] = \{a + ib\sqrt{5} \,|\, a, b \in \mathbf{Z}\}$$

is an integral domain but not a UFD. In particular, there are two *different* factorizations

$$21 = 3 \cdot 7 = (1 + 2\sqrt{-5})(1 - 2\sqrt{-5})$$

of 21 into irreducibles. These irreducibles cannot be primes, for the property of a prime enables one to prove uniqueness of factorization (see the proof of Theorem 32.2). ∎

Exercises

*34.1 Factor each of the following Gaussian integers into a product of irreducibles in $\mathbf{Z}[i]$. [*Hint:* Since an irreducible factor of $\alpha \in \mathbf{Z}[i]$ must have norm > 1 and dividing $N(\alpha)$, there are only a finite number of Gaussian integers $a + bi$ to consider as possible irreducible factors of a given α. Divide α by each of them in $\mathbf{C}$, and see for which ones the quotient is again in $\mathbf{Z}[i]$.]

a) 5   b) 7   c) $4 + 3i$ d) $6 - 7i$

*34.2 Show that 6 does not factor uniquely (up to associates) into irreducibles in $\mathbf{Z}[\sqrt{-5}]$. Exhibit two different factorizations.

*34.3 Consider $\alpha = 7 + 2i$ and $\beta = 3 - 4i$ in $\mathbf{Z}[i]$. Find σ and ρ in $\mathbf{Z}[i]$ such that

$$\alpha = \beta\sigma + \rho \quad \text{with} \quad N(\rho) < N(\beta).$$

[*Hint:* Use the construction in the hint of Exercise 34.11.]

*34.4 Use a Euclidean algorithm in $\mathbf{Z}[i]$ to find a gcd of $8 + 6i$ and $5 - 15i$ in $\mathbf{Z}[i]$. [*Hint:* Use the construction in the hint of Exercise 34.11.]

†*34.5 Let D be an integral domain with a multiplicative norm N such that $N(\alpha) = 1$ for $\alpha \in D$ if and only if α is a unit of D. Let π be such that $N(\pi)$ is minimal among all $N(\beta) > 1$ for $\beta \in D$. Show that π is an irreducible of D.

*34.6 Mark each of the following true or false.

— a) $\mathbf{Z}[i]$ is a PID.

— b) $\mathbf{Z}[i]$ is a Euclidean domain.

— c) Every integer in $\mathbf{Z}$ is a Gaussian integer.

— d) Every complex number is a Gaussian integer.

— e) A Euclidean algorithm holds in $\mathbf{Z}[i]$.

— f) A multiplicative norm on an integral domain is sometimes an aid in finding irreducibles of the domain.

— g) If N is a multiplicative norm on an integral domain D, then $N(u) = 1$ for every unit u of D.

— h) If F is a field, then the function N defined by $N(f(x)) = (\text{degree of } f(x))$ is a multiplicative norm on $F[x]$.

— i) If F is a field, then the function defined by $N(f(x)) = 2^{(\text{degree of } f(x))}$ for $f(x) \neq 0$ and $N(0) = 0$ is a multiplicative norm on $F[x]$ according to our definition.

— j) $\mathbf{Z}[\sqrt{-5}]$ is an integral domain but not a UFD.

---

*34.7 Show that $1 + i$ is an irreducible of $\mathbf{Z}[i]$. [Hint: Apply Theorem 34.2.] (For a description of all Gaussian primes, see Pollard [34].)

*34.8 Prove Lemma 34.1.

*34.9 Prove that N of Example 34.3 is multiplicative, that is, that $N(\alpha\beta) = N(\alpha)N(\beta)$ for $\alpha, \beta \in \mathbf{Z}[\sqrt{-5}]$.

*34.10 Let D be an integral domain with a multiplicative norm N such that $N(\alpha) = 1$ for $\alpha \in D$ if and only if α is a unit of D. Show that every nonzero nonunit of D has a factorization into irreducibles in D.

*34.11 Prove algebraically that the division algorithm holds in $\mathbf{Z}[i]$ for v given by $v(\alpha) = N(\alpha)$ for nonzero $\alpha \in \mathbf{Z}[i]$. [Hint: For α and β in $\mathbf{Z}[i]$ with $\beta \neq 0$, $\alpha/\beta = r + si$ in $\mathbf{C}$ for $r, s \in \mathbf{Q}$. Let q_1 and q_2 be rational integers in $\mathbf{Z}$ as close as possible to the rational numbers r and s respectively. Show that for $\sigma = q_1 + q_2 i$ and $\rho = \alpha - \beta\sigma$, one has $N(\rho) < N(\beta)$, by showing that

$$N(\rho)/N(\beta) = |(\alpha/\beta) - \sigma|^2 < 1.$$

Here $| \ |$ is the usual absolute value for elements of $\mathbf{C}$.]

*34.12 Use a Euclidean algorithm in $\mathbf{Z}[i]$ to find a gcd of $16 + 7i$ and $10 - 5i$ in $\mathbf{Z}[i]$. [Hint: Use the construction in the hint of Exercise 34.11.]

*34.13 Let $\langle \alpha \rangle$ be a nonzero principal ideal in $\mathbf{Z}[i]$.

a) Show that $\mathbf{Z}[i]/\langle \alpha \rangle$ is a finite ring. [Hint: Use the division algorithm.]

b) Show that if π is an irreducible of $\mathbf{Z}[i]$, then $\mathbf{Z}[i]/\langle \pi \rangle$ is a field.

c) Referring to (b), find the order and characteristic of each of the following fields.

   i) $\mathbf{Z}[i]/\langle 3 \rangle$ ii) $\mathbf{Z}[i]/\langle 1 + i \rangle$ iii) $\mathbf{Z}[i]/\langle 1 + 2i \rangle$

*34.14 Let $n \in \mathbf{Z}^+$ be square free, that is, not divisible by the square of any prime integer. Let $\mathbf{Z}[\sqrt{-n}] = \{a + ib\sqrt{n} \mid a, b \in \mathbf{Z}\}$.

a) Show that the norm N, defined by $N(\alpha) = a^2 + nb^2$ for $\alpha = a + ib\sqrt{n}$, is a multiplicative norm on $\mathbf{Z}[\sqrt{-n}]$.

b) Show that $N(\alpha) = 1$ for $\alpha \in \mathbf{Z}[\sqrt{-n}]$ if and only if α is a unit of $\mathbf{Z}[\sqrt{-n}]$.

c) Show that every nonzero $\alpha \in \mathbf{Z}[\sqrt{-n}]$ that is not a unit has a factorization into irreducibles in $\mathbf{Z}[\sqrt{-n}]$. [*Hint:* Use (b).]

*34.15 Repeat Exercise 34.14 for $\mathbf{Z}[\sqrt{n}] = \{a + b\sqrt{n} \,|\, a, \ b \in \mathbf{Z}\}$, with N defined by $N(\alpha) = |a^2 - nb^2|$ for $\alpha = a + b\sqrt{n}$ in $\mathbf{Z}[\sqrt{n}]$.

*34.16 Show by a construction analogous to that given in the hint of Exercise 34.11 that the division algorithm holds in the integral domains $\mathbf{Z}[\sqrt{-2}]$, $\mathbf{Z}[\sqrt{2}]$, and $\mathbf{Z}[\sqrt{3}]$ for $\nu(\alpha) = N(\alpha)$ for nonzero α in one of these domains (see Exercises 34.14 and 34.15). (Thus these domains are Euclidean. See Hardy and Wright [28] for a discussion of which domains $\mathbf{Z}[\sqrt{n}]$ and $\mathbf{Z}[\sqrt{-n}]$ are Euclidean.)

thirty-five

Introduction to Extension Fields

35.1 OUR BASIC GOAL ACHIEVED

We are now in a position to achieve our *basic goal*, which, loosely stated, is to show that every nonconstant polynomial has a zero. This will be stated more precisely and proved in Theorem 35.1. We first introduce some new terminology for some old ideas.

Definition A field E is an *extension field of a field* F if $F \leq E$.

Thus **R** is an extension field of **Q**, and **C** is an extension field of both **R** and **Q**. As in the study of groups, it will often be convenient to use lattice diagrams to picture extension fields, the largest field being on top. We illustrate this in Fig. 35.1. A configuration where there is just one single column of fields, as at the left-hand side of Fig. 35.1, is often referred to, without any precise definition, as a *tower of fields.* We shall use this term freely.

Now for our *basic goal!* This great and important result follows easily and elegantly from the techniques we now have at our disposal. Find a nice

316

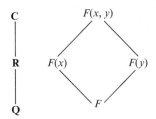

Fig. 35.1

quiet place where you can have a quarter hour to read it, digest it, and marvel to your heart's content without interruption. We write out the proof for you in our usual hideous detail. (A graduate text would give it about three lines at this point).

> **Theorem 35.1 (Kronecker) (Basic Goal)**   *Let F be a field and let f(x) be a nonconstant polynomial in F[x]. Then there exists an extension field E of F and an $\alpha \in E$ such that $f(\alpha) = 0$.*

Proof.   By Theorem 31.8, $f(x)$ has a factorization in $F[x]$ into polynomials that are irreducible over F. Let $p(x)$ be an irreducible polynomial in such a factorization. It is clearly sufficient to find an extension field E of F containing an element α such that $p(\alpha) = 0$.

By Theorem 31.6, $\langle p(x) \rangle$ is a maximal ideal in $F[x]$, so $F[x]/\langle p(x) \rangle$ is a field. We claim that F can be identified with a subfield of $F[x]/\langle p(x) \rangle$ in a natural way by use of the map $\psi : F \rightarrow F[x]/\langle p(x) \rangle$ given by

$$a\psi = a + \langle p(x) \rangle$$

for $a \in F$. This map is one to one, for if $a\psi = b\psi$, that is, if $a + \langle p(x) \rangle = b + \langle p(x) \rangle$ for some $a, b \in F$, then $(a - b) \in \langle p(x) \rangle$, so $a - b$ must be a multiple of the polynomial $p(x)$, which is of degree ≥ 1. Now $a, b \in F$ implies that $a - b$ is in F. Thus we must have $a - b = 0$, so $a = b$. We defined addition and multiplication in $F[x]/\langle p(x) \rangle$ by choosing any representatives, so we may choose $a \in (a + \langle p(x) \rangle)$. It is then clear that this map ψ is an isomorphism of F into $F[x]/\langle p(x) \rangle$. We identify F with $\{a + \langle p(x) \rangle \,|\, a \in F\}$ by means of this isomorphism. Thus we shall view $E = F[x]/\langle p(x) \rangle$ as an extension field of F. We have now manufactured our desired extension field E of F. It remains for us to show that E contains a zero of $p(x)$.

Let us set

$$\alpha = x + \langle p(x) \rangle,$$

so $\alpha \in E$. Consider the evaluation homomorphism $\phi_\alpha : F[x] \rightarrow E$, given by Theorem 30.2. If $p(x) = a_0 + a_1 x + \cdots + a_n x^n$, where $a_i \in F$, then we have

$$p(x)\phi_\alpha = a_0 + a_1(x + \langle p(x) \rangle) + \cdots + a_n(x + \langle p(x) \rangle)^n$$

in $E = F[x]/\langle p(x)\rangle$. *But we can compute in* $F[x]/\langle p(x)\rangle$ *by choosing representatives, and* x *is a representative of the coset* $\alpha = x + \langle p(x)\rangle$. Therefore,

$$p(\alpha) = (a_0 + a_1 x + \cdots + a_n x^n) + \langle p(x)\rangle$$
$$= p(x) + \langle p(x)\rangle = \langle p(x)\rangle = 0$$

in $F[x]/\langle p(x)\rangle$. We have found an element α in $E = F[x]/\langle p(x)\rangle$ such that $p(\alpha) = 0$, and therefore $f(\alpha) = 0$. ∎

We illustrate the construction involved in the proof of Theorem 35.1 by two examples.

Example 35.1   Let $F = \mathbf{R}$, and let $f(x) = x^2 + 1$, which is well known to have no zeros in $\mathbf{R}$ and thus is irreducible over $\mathbf{R}$ by Theorem 31.2. Then $\langle x^2 + 1\rangle$ is a maximal ideal in $\mathbf{R}[x]$, so $\mathbf{R}[x]/\langle x^2 + 1\rangle$ is a field. Identifying $r \in \mathbf{R}$ with $r + \langle x^2 + 1\rangle$ in $\mathbf{R}[x]/\langle x^2 + 1\rangle$, we can view $\mathbf{R}$ as a subfield of $E = \mathbf{R}[x]/\langle x^2 + 1\rangle$. Let

$$\alpha = x + \langle x^2 + 1\rangle.$$

Computing in $\mathbf{R}[x]/\langle x^2 + 1\rangle$, we find

$$\alpha^2 + 1 = (x + \langle x^2 + 1\rangle)^2 + (1 + \langle x^2 + 1\rangle)$$
$$= (x^2 + 1) + \langle x^2 + 1\rangle = 0.$$

Thus α is a zero of $x^2 + 1$. We shall identify $\mathbf{R}[x]/\langle x^2 + 1\rangle$ with $\mathbf{C}$ at the close of this chapter. ∎

Example 35.2   Let $F = \mathbf{Q}$, and let $f(x) = x^4 - 5x^2 + 6$. This time $f(x)$ factors in $\mathbf{Q}[x]$ into $(x^2 - 2)(x^2 - 3)$, both factors being irreducible over $\mathbf{Q}$, as we have seen. We can start with $x^2 - 2$ and construct an extension field E of $\mathbf{Q}$ containing α such that $\alpha^2 - 2 = 0$, or we can construct an extension field K of $\mathbf{Q}$ containing an element β such that $\beta^2 - 3 = 0$. The construction in either case is just as in Example 35.1. ∎

35.2 ALGEBRAIC AND TRANSCENDENTAL ELEMENTS

As we said before, the rest of the unstarred portion of this text is devoted to the study of zeros of polynomials. We commence this study by putting an element of an extension field E of a field F into one of two categories.

Definition   An element α of an extension field E of a field F is *algebraic over* F if $f(\alpha) = 0$ for some nonzero $f(x) \in F[x]$. If α is not algebraic over F, then α is *transcendental over* F.

Example 35.3   **C** is an extension field of **Q**. Since $\sqrt{2}$ is a zero of $x^2 - 2$, $\sqrt{2}$ is an algebraic element over **Q**. Also, i is an algebraic element over **Q**, being a zero of $x^2 + 1$. ∎

Example 35.4   It is well known (but not easy to prove) that the real numbers π and e are transcendental over **Q**. Here e is the base for the natural logarithms. ∎

Just as we do not speak simply of an *irreducible polynomial*, but rather of an *irreducible polynomial over* F, similarly we don't speak simply of an *algebraic element*, but rather of an *element algebraic over* F. The following illustration shows the reason for this.

Example 35.5   The real number π is transcendental over **Q**, as we stated in Example 35.4. However, π is algebraic over **R**, for it is a zero of $(x - \pi) \in$ **R**$[x]$. ∎

Example 35.6   It is easy to see that the real number $\sqrt{1 + \sqrt{3}}$ is algebraic over **Q**. For if $\alpha = \sqrt{1 + \sqrt{3}}$, then $\alpha^2 = 1 + \sqrt{3}$, so $\alpha^2 - 1 = \sqrt{3}$ and $(\alpha^2 - 1)^2 = 3$. Therefore $\alpha^4 - 2\alpha^2 - 2 = 0$, so α is a zero of $x^4 - 2x^2 - 2$, which is in **Q**$[x]$. ∎

To connect these ideas with those of number theory, we give the following definition.

Definition   An element of **C** that is algebraic over **Q** is an ***algebraic number***. A ***transcendental number*** is an element of **C** that is transcendental over **Q**.

There is an extensive and elegant theory of algebraic numbers.

The next theorem gives a useful characterization of algebraic and transcendental elements over F in an extension field E of F. It also illustrates the importance of our evaluation homomorphisms ϕ_α. *Note that once more we are phrasing our concepts in terms of mappings.*

Theorem 35.2   *Let E be an extension field of a field F and let $\alpha \in E$. Let $\phi_\alpha : F[x] \to E$ be the evaluation homomorphism of $F[x]$ into E such that $a\phi_\alpha = a$ for $a \in F$ and $x\phi_\alpha = \alpha$. Then α is transcendental over F if and only if ϕ_α is an isomorphism mapping $F[x]$ into E, that is, if and only if ϕ_α is a one-to-one map.*

Proof.   Now α is transcendental over F if and only if $f(\alpha) \neq 0$ for all nonconstant $f(x) \in F[x]$, which is true (by definition) if and only if $f(x)\phi_\alpha \neq 0$ for all nonconstant $f(x) \in F[x]$, which is true if and only if the kernel of ϕ_α is $\{0\}$, that is, if and only if ϕ_α is an isomorphism mapping $F[x]$ into E. ∎

35.3 THE IRREDUCIBLE POLYNOMIAL
FOR α OVER F

Consider the extension field **R** of **Q**. We know that $\sqrt{2}$ is algebraic over **Q**, being a zero of $x^2 - 2$. Of course, $\sqrt{2}$ is also a zero of $x^3 - 2x$ and of $x^4 - 3x^2 + 2 = (x^2 - 2)(x^2 - 1)$. All these other polynomials having $\sqrt{2}$ as a zero were multiples of $x^2 - 2$. The next theorem shows that this is an illustration of a general situation. This theorem plays a central role in our later work.

> **Theorem 35.3**   Let E be an extension field of F, and let $\alpha \in E$, where α is algebraic over F. Then there is an irreducible polynomial $p(x) \in F[x]$ such that $p(\alpha) = 0$. This irreducible polynomial $p(x)$ is uniquely determined up to a constant factor in F and is a polynomial of minimal degree ≥ 1 in $F[x]$ having α as a zero. If $f(\alpha) = 0$ for $f(x) \in F[x]$, with $f(x) \neq 0$, then $p(x)$ divides $f(x)$.

Proof.   Let ϕ_α be the evaluation homomorphism of $F[x]$ into E, given by Theorem 30.2. The kernel of ϕ_α is an ideal and by Theorem 31.5 it must be a principal ideal generated by some element $p(x) \in F[x]$. Clearly, $\langle p(x) \rangle$ consists precisely of those elements of $F[x]$ having α as a zero. Thus, if $f(\alpha) = 0$ for $f(x) \neq 0$, then $f(x) \in \langle p(x) \rangle$, so $p(x)$ divides $f(x)$. Clearly, $p(x)$ is a polynomial of minimal degree ≥ 1 having α as a zero, and any other such polynomial of the same degree as $p(x)$ must be of the form $(a)p(x)$ for some $a \in F$.

It only remains for us to show that $p(x)$ is irreducible. If $p(x) = r(x)s(x)$ were a factorization of $p(x)$ into polynomials of lower degree, then $p(\alpha) = 0$ would imply that $r(\alpha)s(\alpha) = 0$, so either $r(\alpha) = 0$ or $s(\alpha) = 0$, since E is a field. This would contradict the fact that $p(x)$ is of minimal degree ≥ 1 such that $p(\alpha) = 0$. Thus $p(x)$ is irreducible. ∎

By multiplying by a suitable constant in F, we can assume that the coefficient of the highest power of x appearing in $p(x)$ of Theorem 35.3 is 1. Such a polynomial having the coefficient of the highest power of x appearing equal to 1 is a **monic polynomial**.

> **Definition**   Let E be an extension field of a field F, and let $\alpha \in E$ be algebraic over F. The unique monic polynomial $p(x)$ of Theorem 35.3 is the *irreducible polynomial for α over F* and will be denoted by irr(α, F). The degree of irr(α, F) is the *degree of α over F*, denoted by deg(α, F).

Example 35.7   Clearly, irr($\sqrt{2}$, **Q**) $= x^2 - 2$. Referring to Example 35.6, we see that for $\alpha = \sqrt{1 + \sqrt{3}}$ in **R**, α is a zero of $x^4 - 2x^2 - 2$, which is in **Q**[x]. Since $x^4 - 2x^2 - 2$ is irreducible over **Q** (by Eisenstein with $p = 2$, or by application of the technique of Example 31.6), we see that

$$\text{irr}(\sqrt{1 + \sqrt{3}}, \mathbf{Q}) = x^4 - 2x^2 - 2.$$

Thus $\sqrt{1 + \sqrt{3}}$ is algebraic of degree 4 over $\mathbf{Q}$. ∎

Just as we must speak of an element α as *algebraic over F* rather than simply as *algebraic*, we must speak of the *degree of α over F* rather than the *degree of α*. To take a trivial illustration, $\sqrt{2} \in \mathbf{R}$ is algebraic of degree 2 over $\mathbf{Q}$ but algebraic of degree 1 over $\mathbf{R}$, for $\text{irr}(\sqrt{2}, \mathbf{R}) = x - \sqrt{2}$.

We hope that you are impressed with the beauty and elegance of this theory. You should appreciate that its easy development here is due to the machinery of homomorphisms and ideal theory that we now have at our disposal. Note especially our constant use of the evaluation homomorphisms ϕ_α.

35.4 SIMPLE EXTENSIONS

Let E be an extension field of a field F, and let $\alpha \in E$. Let ϕ_α be the evaluation homomorphism of $F[x]$ into E with $a\phi_\alpha = a$ for $a \in F$ and $x\phi_\alpha = \alpha$, as in Theorem 30.2. We consider two cases.

CASE I.   *Suppose α is algebraic over F.* Then as in Theorem 35.3, the kernel of ϕ_α is $\langle \text{irr}(\alpha, F) \rangle$ and by Theorem 31.6, $\langle \text{irr}(\alpha, F) \rangle$ is a maximal ideal of $F[x]$. Therefore, $F[x]/\langle \text{irr}(\alpha, F) \rangle$ is a field and is isomorphic to the image $(F[x])\phi_\alpha$ in E. This subfield $(F[x])\phi_\alpha$ of E is clearly the smallest subfield of E containing F and α. We shall denote this field by $F(\alpha)$.

CASE II.   *Suppose α is transcendental over F.* Then by Theorem 35.2, ϕ_α is an isomorphism mapping $F[x]$ into E. Thus in this case $(F[x])\phi_\alpha$ is *not* a field but an integral domain that we shall denote by $F[\alpha]$. By Corollary 1 of Theorem 26.2, E contains a field of quotients of $F[\alpha]$, which is clearly the smallest subfield of E containing F and α. As in Case I, we denote this field by $F(\alpha)$.

Example 35.8   Since π is transcendental over $\mathbf{Q}$, the field $\mathbf{Q}(\pi)$ is isomorphic to the field $\mathbf{Q}(x)$ of rational functions over $\mathbf{Q}$ in the indeterminate x. Thus from a structural viewpoint, an element that is transcendental over a field F behaves as though it were an indeterminate over F. ∎

Definition   An extension field E of a field F is a *simple extension of F* if $E = F(\alpha)$ for some $\alpha \in E$.

Many important results appear throughout this section. You see, we have now developed so much machinery that results are starting to pour out of our efficient plant at an alarming rate. The next theorem gives us insight into the nature of the field $F(\alpha)$ in the case where α is algebraic over F.

Theorem 35.4   *Let E be a simple extension $F(\alpha)$ of a field F, and let α be algebraic over F. Let the degree of* irr(α, F) *be $n \geq 1$. Then every element β of $E = F(\alpha)$ can be uniquely expressed in the form*

$$\beta = b_0 + b_1\alpha + \cdots + b_{n-1}\alpha^{n-1},$$

where the b_i are in F.

Proof.   For the usual evaluation homomorphism ϕ_α, every element of

$$F(\alpha) = (F[x])\phi_\alpha$$

is of the form $(f(x))\phi_\alpha = f(\alpha)$, a formal polynomial in α with coefficients in F. Let

$$\text{irr}(\alpha, F) = p(x) = x^n + a_{n-1}x^{n-1} + \cdots + a_0.$$

Then $p(\alpha) = 0$, so

$$\alpha^n = -a_{n-1}\alpha^{n-1} - \cdots - a_0.$$

This equation in $F(\alpha)$ can be used to express every monomial α^m for $m \geq n$ in terms of powers of α that are less than n. For example,

$$\alpha^{n+1} = \alpha\alpha^n = -a_{n-1}\alpha^n - a_{n-2}\alpha^{n-1} - \cdots - a_0\alpha$$
$$= -a_{n-1}(-a_{n-1}\alpha^{n-1} - \cdots - a_0) - a_{n-2}\alpha^{n-1} - \cdots - a_0\alpha.$$

Thus, if $\beta \in F(\alpha)$, β can be expressed in the required form

$$\beta = b_0 + b_1\alpha + \cdots + b_{n-1}\alpha^{n-1}.$$

For uniqueness, if

$$b_0 + b_1\alpha + \cdots + b_{n-1}\alpha^{n-1} = b'_0 + b'_1\alpha + \cdots + b'_{n-1}\alpha^{n-1}$$

for $b'_i \in F$, then

$$(b_0 - b'_0) + (b_1 - b'_1)x + \cdots + (b_{n-1} - b'_{n-1})x^{n-1} = g(x)$$

is in $F[x]$ and $g(\alpha) = 0$. Also, the degree of $g(x)$ is less than the degree of irr(α, F). Since irr(α, F) is a nonzero polynomial of minimal degree in $F[x]$ having α as a zero, we must have $g(x) = 0$. Therefore, $b_i - b'_i = 0$, so

$$b_i = b'_i,$$

and the uniqueness of the b_i is established. ∎

We give an impressive example illustrating Theorem 35.4.

Example 35.9   The polynomial $p(x) = x^2 + x + 1$ in $Z_2[x]$ is irreducible over Z_2 by Theorem 31.2, since neither element 0 nor element 1 of Z_2 is a

Table 35.1

+	0	1	α	$1 + \alpha$
0	0	1	α	$1 + \alpha$
1	1	0	$1 + \alpha$	α
α	α	$1 + \alpha$	0	1
$1 + \alpha$	$1 + \alpha$	α	1	0

Table 35.2

	0	1	α	$1 + \alpha$
0	0	0	0	0
1	0	1	α	$1 + \alpha$
α	0	α	$1 + \alpha$	1
$1 + \alpha$	0	$1 + \alpha$	1	α

zero of $p(x)$. By Theorem 35.1, we know that there is an extension field E of $\mathbf{Z}_2$ containing a zero α of $x^2 + x + 1$. By Theorem 35.4, $\mathbf{Z}_2(\alpha)$ has as elements $0 + 0\alpha$, $1 + 0\alpha$, $0 + 1\alpha$, and $1 + 1\alpha$, that is, 0, 1, α, and $1 + \alpha$. *This gives us a new finite field of four elements!* The addition and multiplication tables for this field are shown in Tables 35.1 and 35.2. For example, to compute $(1 + \alpha)(1 + \alpha)$ in $\mathbf{Z}_2(\alpha)$, one observes that since $p(\alpha) = \alpha^2 + \alpha + 1 = 0$, then,

$$\alpha^2 = -\alpha - 1 = \alpha + 1.$$

Therefore,

$$(1 + \alpha)(1 + \alpha) = 1 + \alpha + \alpha + \alpha^2 = 1 + \alpha^2 = 1 + \alpha + 1 = \alpha.$$

Finally, we can use Theorem 35.4 to fulfill our promise of Example 35.1 and show that $\mathbf{R}[x]/\langle x^2 + 1 \rangle$ is isomorphic to the field $\mathbf{C}$ of complex numbers. We saw in Example 35.1 that we can view $\mathbf{R}[x]/\langle x^2 + 1 \rangle$ as an extension field of $\mathbf{R}$. Let

$$\alpha = x + \langle x^2 + 1 \rangle.$$

Then $\mathbf{R}(\alpha) = \mathbf{R}[x]/\langle x^2 + 1 \rangle$ and consists of all elements of the form $a + b\alpha$ for $a, b \in \mathbf{R}$, by Theorem 35.4. But since $\alpha^2 + 1 = 0$, we see that α plays the role of $i \in \mathbf{C}$, and $a + b\alpha$ plays the role of $(a + bi) \in \mathbf{C}$. Thus $\mathbf{R}(\alpha) \simeq \mathbf{C}$. *This is the elegant algebraic way to construct* $\mathbf{C}$ *from* $\mathbf{R}$.

Exercises

35.1 For each of the given numbers $\alpha \in \mathbf{C}$, show that α is algebraic over $\mathbf{Q}$ by finding $f(x) \in \mathbf{Q}[x]$ such that $f(\alpha) = 0$.

a) $1 + \sqrt{2}$

b) $\sqrt{2} + \sqrt{3}$

c) $1 + i$

d) $\sqrt{1 + \sqrt[3]{2}}$

e) $\sqrt{\sqrt[3]{2} - i}$

35.2 For each of the given algebraic numbers $\alpha \in \mathbf{C}$, find irr$(\alpha, \mathbf{Q})$ and deg$(\alpha, \mathbf{Q})$. Be prepared to prove that your polynomials are irreducible over $\mathbf{Q}$ if challenged to do so.

a) $\sqrt{3 - \sqrt{6}}$

b) $\sqrt{(\frac{1}{3}) + \sqrt{7}}$

c) $\sqrt{2} + i$

35.3 Classify each of the given $\alpha \in \mathbf{C}$ as algebraic or transcendental over the given field F. If α is algebraic over F, find deg(α, F).

a) $\alpha = i, F = \mathbf{Q}$

b) $\alpha = 1 + i, F = \mathbf{R}$

c) $\alpha = \sqrt{\pi}, F = \mathbf{Q}$

d) $\alpha = \sqrt{\pi}, F = \mathbf{R}$

e) $\alpha = \sqrt{\pi}, F = \mathbf{Q}(\pi)$

f) $\alpha = \pi^2, F = \mathbf{Q}$

g) $\alpha = \pi^2, F = \mathbf{Q}(\pi)$

h) $\alpha = \pi^2, F = \mathbf{Q}(\pi^3)$

35.4 Refer to Example 35.9 of the text. The polynomial $x^2 + x + 1$ has a zero α in $\mathbf{Z}_2(\alpha)$ and thus must factor into a product of linear factors in $(\mathbf{Z}_2(\alpha))[x]$. Find this factorization. [*Hint:* Divide $x^2 + x + 1$ by $x - \alpha$ by long division, using the fact that $\alpha^2 = \alpha + 1$.]

35.5 Let E be an extension field of F, and let $\alpha, \beta \in E$. Suppose α is transcendental over F but algebraic over $F(\beta)$. Show that β is algebraic over $F(\alpha)$.

†**35.6** Let E be an extension field of a finite field F, where F has q elements. Let $\alpha \in E$ be algebraic over F of degree n. Prove that $F(\alpha)$ has q^n elements.

35.7 a) Show that the polynomial $x^2 + 1$ is irreducible in $\mathbf{Z}_3[x]$.

b) Let α be a zero of $x^2 + 1$ in an extension field of $\mathbf{Z}_3$. As in Example 35.9, give the multiplication and addition tables for the nine elements of $\mathbf{Z}_3(\alpha)$, written in the order $0, 1, 2, \alpha, 2\alpha, 1 + \alpha, 1 + 2\alpha, 2 + \alpha$, and $2 + 2\alpha$.

35.8 Mark each of the following true or false.

___ a) The number π is transcendental over $\mathbf{Q}$.

___ b) $\mathbf{C}$ is a simple extension of $\mathbf{R}$.

___ c) Every element of a field F is algebraic over F.

___ d) $\mathbf{R}$ is an extension field of $\mathbf{Q}$.

___ e) $\mathbf{Q}$ is an extension field of $\mathbf{Z}_2$.

___ f) Let $\alpha \in \mathbf{C}$ be algebraic over $\mathbf{Q}$ of degree n. If $f(\alpha) = 0$ for nonzero $f(x) \in \mathbf{Q}[x]$ then (degree $f(x)) \geq n$.

___ g) Let $\alpha \in \mathbf{C}$ be algebraic over $\mathbf{Q}$ of degree n. If $f(\alpha) = 0$ for nonzero $f(x) \in \mathbf{R}[x]$ then (degree $f(x)) \geq n$.

___ h) Every nonconstant polynomial in $F[x]$ has a zero in some extension field of F.

___ i) Every nonconstant polynomial in $F[x]$ has a zero in every extension field of F.

___ j) If x is an indeterminate, $\mathbf{Q}[\pi] \simeq \mathbf{Q}[x]$.

35.9 Let E be an extension field of a field F and let $\alpha \in E$ be algebraic over F. The polynomial irr(α, F) is sometimes referred to as the **minimal polynomial for α over F**. Why is this designation appropriate?

35.10 a) Show that there exists an irreducible polynomial of degree 3 in $\mathbf{Z}_3[x]$.

b) Show from (a) that there exists a finite field of twenty-seven elements. [*Hint:* Use Exercise 35.6.]

35.11 Consider the prime field $\mathbf{Z}_p$ of characteristic $p \neq 0$.

a) Show that, for $p \neq 2$, not every element in $\mathbf{Z}_p$ is a square of an element of $\mathbf{Z}_p$. [*Hint:* $1^2 = (p-1)^2 = 1$ in $\mathbf{Z}_p$. Deduce the desired conclusion *by counting*.]

b) Using part (a), show that there exist finite fields of p^2 elements for every prime p in $\mathbf{Z}^+$.

35.12 We have stated without proof that π and e are transcendental over $\mathbf{Q}$.

a) Find a subfield F of $\mathbf{R}$ such that π is algebraic of degree 3 over F.

b) Find a subfield E of $\mathbf{R}$ such that e^2 is algebraic of degree 5 over E.

35.13 Let E be an extension field of a field F and let $\alpha \in E$ be transcendental over F. Show that every element of $F(\alpha)$ that is not in F is also transcendental over F.

35.14 a) Show that $x^3 + x^2 + 1$ is irreducible over $\mathbf{Z}_2$.

b) Let α be a zero of $x^3 + x^2 + 1$ in an extension field of $\mathbf{Z}_2$. Show that $x^3 + x^2 + 1$ factors into three linear factors in $(\mathbf{Z}_2(\alpha))[x]$ by actually finding this factorization. [*Hint:* Every element of $\mathbf{Z}_2(\alpha)$ is of the form

$$a_0 + a_1\alpha + a_2\alpha^2 \qquad \text{for} \quad a_i = 0, 1.$$

Divide $x^3 + x^2 + 1$ by $x - \alpha$ by long division. Show that the quotient also has a zero in $\mathbf{Z}_2(\alpha)$ by simply trying the eight possible elements. Then complete the factorization.]

35.15 Show that $\{a + b\sqrt[3]{2} + c(\sqrt[3]{2})^2 \,|\, a, b, c \in \mathbf{Q}\}$ is a subfield of $\mathbf{R}$ by using the ideas of this section, rather than by a formal verification of the field axioms. [*Hint:* Use Theorem 35.4.]

35.16 Let E be an extension field of $\mathbf{Z}_2$ and let $\alpha \in E$ be algebraic of degree 3 over $\mathbf{Z}_2$. Classify the groups $\langle \mathbf{Z}_2(\alpha), + \rangle$ and $\langle (\mathbf{Z}_2(\alpha))^*, \cdot \rangle$ according to the Fundamental Theorem of Finitely Generated Abelian Groups. As usual, $(\mathbf{Z}_2(\alpha))^*$ is the set of nonzero elements of $\mathbf{Z}_2(\alpha)$.

35.17 Following the idea of Exercise 35.10 show that there exists a field of eight elements; of sixteen elements; of twenty-five elements.

35.18 Let F be a finite field of characteristic p. Show that every element of F is algebraic over the prime field $\mathbf{Z}_p \le F$. [*Hint:* Let F^* be the set of nonzero elements of F. Apply group theory to the group $\langle F^*, \cdot \rangle$ to show that every $\alpha \in F^*$ is a zero of some polynomial in $\mathbf{Z}_p[x]$ of the form $x^n - 1$.]

35.19 Use Exercises 35.6 and 35.18 to show that every finite field is of prime-power order, that is, it has a prime power number of elements.

thirty-six

Vector Spaces

36.1 DEFINITION AND ELEMENTARY PROPERTIES

The topic of vector spaces is the cornerstone of linear algebra. Since linear algebra is not the subject for study in this text, our treatment of vector spaces will be brief, designed to develop only the concepts of linear independence and dimension that we need for our field theory.

You are probably familiar with the terms *vector* and *scalar* from your course in calculus. Here we allow scalars to be elements of any field, not just the real numbers, and develop the theory by axioms just as for the other algebraic structures we have studied. The properties appearing in these axioms should look familiar to you.

Definition Let F be a field. A ***vector space over*** F (or ***F-vector space***) consists of an abelian group V under addition together with an operation of scalar multiplication of each element of V by each element of F on the left, such that for all $a, b \in F$ and $\alpha, \beta \in V$ the following conditions are satisfied:

$\mathcal{V}_1$. $a\alpha \in V$.
$\mathcal{V}_2$. $a(b\alpha) = (ab)\alpha$.
$\mathcal{V}_3$. $(a + b)\alpha = (a\alpha) + (b\alpha)$.
$\mathcal{V}_4$. $a(\alpha + \beta) = (a\alpha) + (a\beta)$.
$\mathcal{V}_5$. $1\alpha = \alpha$.

The elements of V are **vectors** and the elements of F are **scalars.** When only one field F is under discussion, we drop the reference to F and refer to a *vector space.*

Note that multiplication for a vector space is not a binary operation on one set in the sense we defined it in Chapter 1. It is rather a rule that associates an element $a\alpha$ of V with each ordered pair (a, α), consisting of an element a of F and an element α of V. This can be viewed as a *function* mapping $F \times V$ into V. The *nice* way to define a binary operation on a set S is similarly to say that it is a function from $S \times S$ into S, but we wished to be more naive in Chapter 1. Both the additive identity for V, the 0-vector, and the additive identity for F, the 0-scalar, will be denoted by 0.

Example 36.1 Consider the abelian group $\langle \mathbf{R}^n, + \rangle = \mathbf{R} \times \mathbf{R} \times \cdots \times \mathbf{R}$ for n factors, which consists of ordered n-tuples under addition by components. Define scalar multiplication for scalars in $\mathbf{R}$ by

$$r\alpha = (ra_1, \ldots, ra_n)$$

for $r \in \mathbf{R}$ and $\alpha = (a_1, \ldots, a_n) \in \mathbf{R}^n$. With these operations, $\mathbf{R}^n$ becomes a vector space over $\mathbf{R}$. The axioms for a vector space are easily checked. In particular, for $n = 2$, you will have no trouble convincing yourself that $\mathbf{R}^2 = \mathbf{R} \times \mathbf{R}$ as a vector space over $\mathbf{R}$ can be viewed as all "vectors whose starting points are the origin of the Euclidean plane" in the sense often studied in calculus courses. ∎

Example 36.2 For any field F, $F[x]$ can be viewed as a vector space over F, where addition of vectors is ordinary addition of polynomials in $F[x]$ and scalar multiplication of an element of $F[x]$ by an element of F is ordinary multiplication in $F[x]$. The axioms $\mathcal{V}_1$ through $\mathcal{V}_5$ for a vector space then follow immediately from the axioms for the integral domain $F[x]$. ∎

Example 36.3 Let E be an extension field of a field F. Then E can be regarded as a vector space over F, where addition of vectors is the usual addition in E and scalar multiplication is the usual field multiplication in E with $a \in F$ and $\alpha \in E$. The axioms follow at once from the field axioms for E. Here our field of scalars is actually a subset of our space of vectors. *It is this example that is the important one for us.* ∎

We are assuming nothing about vector spaces from previous work and shall prove everything we need from the definition, even though you may be familiar with the results from your calculus course.

Theorem 36.1 *If V is a vector space over F, then $0\alpha = 0$, $a0 = 0$ and $(-a)\alpha = a(-\alpha) = -(a\alpha)$ for all $a \in F$ and $\alpha \in V$.*

Proof. The equation $0\alpha = 0$ is to be read "(0-scalar)α = 0-vector," and likewise, $a0 = 0$ is to be read "a(0-vector) = 0-vector." The proofs here are very similar to those in Theorem 23.1 for a ring and again depend heavily on the distributive laws $\mathscr{V}_3$ and $\mathscr{V}_4$. Now

$$(0\alpha) = (0 + 0)\alpha = (0\alpha) + (0\alpha)$$

is an equation in the abelian group $\langle V, + \rangle$, so by the group cancellation law, $0 = 0\alpha$. Likewise, from

$$a0 = a(0 + 0) = a0 + a0,$$

we conclude that $a0 = 0$. Then

$$0 = 0\alpha = (a + (-a))\alpha = a\alpha + (-a)\alpha,$$

so $(-a)\alpha = -(a\alpha)$. Likewise, from

$$0 = a0 = a(\alpha + (-\alpha)) = a\alpha + a(-\alpha),$$

we conclude that $a(-\alpha) = -(a\alpha)$ also. ∎

36.2 LINEAR INDEPENDENCE AND BASES

Definition Let V be a vector space over F. The vectors in a subset $S = \{\alpha_i \mid i \in I\}$ of V *span* (or *generate*) V if for every $\beta \in V$, we have

$$\beta = a_1\alpha_{i_1} + a_2\alpha_{i_2} + \cdots + a_n\alpha_{i_n}$$

for some $a_j \in F$ and $\alpha_{i_j} \in S$, $j = 1, \ldots, n$. A vector $\sum_{j=1}^{n} a_j\alpha_{i_j}$ is a *linear combination of the α_{i_j}*.

Example 36.4 In the vector space $\mathbf{R}^n$ over $\mathbf{R}$ of Example 36.1, the vectors

$$(1, 0, \ldots, 0), (0, 1, \ldots, 0), \ldots, (0, 0, \ldots, 1)$$

clearly span $\mathbf{R}^n$, for

$$(a_1, a_2, \ldots, a_n) = a_1(1, 0, \ldots, 0) + a_2(0, 1, \ldots, 0) + \cdots + a_n(0, 0, \ldots, 1).$$

Also, the monomials x^m for $m \geq 0$ span $F[x]$ over F, the vector space of Example 36.2. ∎

Example 36.5   Let F be a field and E an extension field of F. Let $\alpha \in E$ be algebraic over F. Then $F(\alpha)$ is a vector space over F and by Theorem 35.4, it is spanned by the vectors in $\{1, \alpha, \ldots, \alpha^{n-1}\}$, where $n = \deg(\alpha, F)$. *This is the important example for us.* ∎

Definition   A vector space V over a field F is **finite dimensional** if there is a finite subset of V whose vectors span V.

Example 36.6   Example 36.4 shows that $\mathbf{R}^n$ is finite-dimensional. The vector space $F[x]$ over F is *not* finite-dimensional, since polynomials of arbitrarily large degree could obviously not be linear combinations of elements of any *finite* set of polynomials. ∎

Example 36.7   If $F \leq E$ and $\alpha \in E$ is algebraic over the field F, Example 36.5 shows that $F(\alpha)$ is a finite-dimensional vector space over F. *This is the most important example for us.* ∎

The next definition contains the most important idea in this section.

Definition   The vectors in a subset $S = \{\alpha_i \mid i \in I\}$ of a vector space V over a field F are **linearly independent over F** if $\sum_{j=1}^{n} a_j \alpha_{i_j} = 0$ implies that $a_j = 0$ for $j = 1, \ldots, n$. If the vectors are not linearly independent over F, they are **linearly dependent over F**.

Thus the vectors in $\{\alpha_i \mid i \in I\}$ are linearly independent over F if the only way the 0-vector can be expressed as a linear combination of the vectors α_i is to have all scalar coefficients equal to 0. If the vectors are linearly dependent over F, then there exist $a_j \in F$ for $j = 1, \ldots, n$ such that $\sum_{j=1}^{n} a_j \alpha_{i_j} = 0$, where not all $a_j = 0$.

Example 36.8   It is clear that the vectors of the set of vectors spanning the space $\mathbf{R}^n$ that are given in Example 36.4 are linearly independent over $\mathbf{R}$. Likewise, the vectors in $\{x^m \mid m \geq 0\}$ are linearly independent vectors of $F[x]$ over F. Note that $(1, -1)$, $(2, 1)$, and $(-3, 2)$ are linearly dependent in $\mathbf{R}^2$ over $\mathbf{R}$, since

$$7(1, -1) + (2, 1) + 3(-3, 2) = (0, 0) = 0. \quad ∎$$

Example 36.9   Let E be an extension field of a field F, and let $\alpha \in E$ be algebraic over F. If $\deg(\alpha, F) = n$, then by Theorem 35.4, every element of $F(\alpha)$ can be *uniquely* expressed in the form

$$b_0 + b_1\alpha + \cdots + b_{n-1}\alpha^{n-1}$$

for $b_i \in F$. In particular, $0 = 0 + 0\alpha + \cdots + 0\alpha^{n-1}$ must be a *unique* such expression for 0. Thus the elements $1, \alpha, \ldots, \alpha^{n-1}$ are linearly independent vectors in $F(\alpha)$ over the field F. They also span $F(\alpha)$, so by the next definition, $1, \alpha, \ldots, \alpha^{n-1}$ form a *basis* for $F(\alpha)$ over F. *This is the important example for us.* In fact, this is the reason we are doing this material on vector spaces. ■

Definition   If V is a vector space over a field F, the vectors in a subset $B = \{\beta_i | i \in I\}$ of V form a **basis for V over F** if they span V and are linearly independent.

36.3 DIMENSION

The only other results we wish to prove about vector spaces are that every finite-dimensional vector space has a basis, and that any two bases of a finite-dimensional vector space have the same number of elements. Both these facts are true without the assumption that the vector space is finite-dimensional, but the proofs require more knowledge of set theory than we are assuming, and the finite-dimensional case is all we need. First we give an easy lemma.

Lemma 36.1   *Let V be a vector space over a field F, and let $\alpha \in V$. If α is a linear combination of vectors β_i for $i = 1, \ldots, m$ and each β_i is a linear combination of vectors γ_j for $j = 1, \ldots, n$, then α is a linear combination of the γ_j.*

Proof.   Let $\alpha = \sum_{i=1}^{m} a_i \beta_i$, and let $\beta_i = \sum_{i=1}^{n} b_{ij} \gamma_j$, where a_i and b_{ij} are in F. Then

$$\alpha = \sum_{i=1}^{m} a_i \left(\sum_{j=1}^{n} b_{ij} \gamma_j \right) = \left(\sum_{j=1}^{n} \sum_{i=1}^{m} a_i b_{ij} \gamma_j \right),$$

and $(\sum_{i=1}^{m} a_i b_{ij}) \in F$. ■

Theorem 36.2   *In a finite-dimensional vector space, every finite set of vectors spanning the space contains a subset that is a basis.*

Proof.   Let V be finite-dimensional over F, and let vectors $\alpha_1, \ldots, \alpha_n$ in V span V. Let us list the α_i in a row. Examine each α_i in succession, starting at the left with $i = 1$, and discard the first α_j that is some linear combination of the preceding α_i for $i < j$. Then continue, starting with the following α_{j+1}, and discard the next α_k that is some linear combination of its remaining predecessors, and so on. When we reach α_n after a finite number of steps, those α_i remaining in our list are such that none is a linear combination of the preceding α_i in this reduced list. Lemma 36.1 shows that any vector that is a linear combination of the original collection of α_i is still a linear combination of our reduced, and possibly smaller, set in which no α_i is a linear combination of its predecessors. Thus the vectors in the reduced set of α_i again span V.

For the reduced set, suppose that

$$a_1\alpha_{i_1} + \cdots + a_r\alpha_{i_r} = 0$$

for $i_1 < i_2 < \cdots < i_r$ and that some $a_j \neq 0$. We may assume from Theorem 36.1 that $a_r \neq 0$, or we could drop $a_r\alpha_{i_r}$ from the left side of the equation. Then, using Theorem 36.1 again, we obtain

$$\alpha_{i_r} = \left(-\frac{a_1}{a_r}\right)\alpha_{i_1} + \cdots + \left(-\frac{a_{r-1}}{a_r}\right)\alpha_{i_{r-1}},$$

which shows that α_{i_r} is a linear combination of its predecessors, contradicting our construction. Thus the vectors α_i in the reduced set both span V and are linearly independent, so they form a basis for V over F. ∎

Corollary *A finite-dimensional vector space has a finite basis.*

Proof. By definition, a finite-dimensional vector space has a finite set of vectors that span the space. Theorem 36.2 completes the proof. ∎

The next theorem is the culmination of our work on vector spaces.

Theorem 36.3 *Let $S = \{\alpha_1, \ldots, \alpha_r\}$ be a finite set of linearly independent vectors of a finite-dimensional vector space V over a field F. Then S can be enlarged to a basis for V over F. Furthermore, if $B = \{\beta_1, \ldots, \beta_n\}$ is any basis for V over F, then $r \leq n$.*

Proof. By the corollary of Theorem 36.2, there is a basis $B = \{\beta_1, \ldots, \beta_n\}$ for V over F. Consider the finite sequence of vectors

$$\alpha_1, \ldots, \alpha_r, \beta_1, \ldots, \beta_n.$$

These vectors span V, since B is a basis. Following the technique, used in Theorem 36.2, of discarding in turn each vector that is a linear combination of its remaining predecessors, working from left to right, we arrive at a basis for V. Clearly, no α_i is cast out, since the α_i are linearly independent. Thus S can be enlarged to a basis for V over F.

For the second part of the conclusion, consider the sequence

$$\alpha_1, \beta_1, \ldots, \beta_n.$$

These vectors are not linearly independent over F, because α_1 is a linear combination

$$\alpha_1 = b_1\beta_1 + \cdots + b_n\beta_n,$$

since the β_i form a basis. Thus

$$\alpha_1 + (-b_1)\beta_1 + \cdots + (-b_n)\beta_n = 0.$$

The vectors in the sequence do span V, and if we form a basis by the technique of working from left to right and casting out in turn each vector that is a linear combination of its remaining predecessors, at least one β_i must be cast out, giving a basis

$$\{\alpha_1, \beta_1^{(1)}, \ldots, \beta_m^{(1)}\},$$

where $m \leq n - 1$. Applying the same technique to the sequence of vectors

$$\alpha_1, \alpha_2, \beta_1^{(1)}, \ldots, \beta_m^{(1)},$$

we arrive at a new basis

$$\{\alpha_1, \alpha_2, \beta_1^{(2)}, \ldots, \beta_s^{(2)}\},$$

with $s \leq n - 2$. Continuing, we arrive finally at a basis

$$\{\alpha_1, \ldots, \alpha_r, \beta_1^{(r)}, \ldots, \beta_t^{(r)}\},$$

where $0 \leq t \leq n - r$. Thus $r \leq n$. ∎

Corollary   *Any two bases of a finite-dimensional vector space V over F have the same number of elements.*

Proof.   Let $B = \{\beta_1, \ldots, \beta_n\}$ and $B' = \{\beta'_1, \ldots, \beta'_m\}$ be two bases. Then by Theorem 36.3, regarding B as an independent set of vectors and B' as a basis, we see that $n \leq m$. A symmetric argument gives $m \leq n$, so $m = n$. ∎

Definition   If V is a finite-dimensional vector space over a field F, the number of elements in a basis (independent of the choice of basis, by Theorem 36.3) is the ***dimension of V over F***.

Example 36.10   Let E be an extension field of a field F, and let $\alpha \in E$. Example 36.9 shows that if α is algebraic over F and $\deg(\alpha, F) = n$, then the dimension of $F(\alpha)$ as a vector space over F is n. *This is the important example for us.* ∎

36.4 AN APPLICATION TO
FIELD THEORY

We collect the results of field theory contained in Examples 36.3, 36.5, 36.7, 36.9, and 36.10, and incorporate them into one theorem. The last sentence of this theorem gives an additional elegant application of these vector space ideas to field theory.

Theorem 36.4   *Let E be an extension field of F, and let $\alpha \in E$ be algebraic over F. If $\deg(\alpha, F) = n$, then $F(\alpha)$ is an n-dimensional vector space over F with basis $\{1, \alpha, \ldots, \alpha^{n-1}\}$. Furthermore, every element β of $F(\alpha)$ is algebraic over F, and $\deg(\beta, F) \leq \deg(\alpha, F)$.*

Proof.   We have shown everything in the preceding examples except the *very important result* stated in the last sentence of the above theorem. Let $\beta \in F(\alpha)$, where α is algebraic over F of degree n. Consider the elements

$$1, \beta, \beta^2, \ldots, \beta^n.$$

These can't be $n + 1$ distinct elements of $F(\alpha)$ that are linearly independent over F, for by Theorem 36.3 any basis of $F(\alpha)$ over F would have to contain at least as many elements as are in any set of linearly independent vectors over F. However, the basis $\{1, \alpha, \ldots, \alpha^{n-1}\}$ has just n elements. If $\beta^i = \beta^j$, then $\beta^i - \beta^j = 0$, so in any case there exist $b_i \in F$ such that

$$b_0 + b_1\beta + b_2\beta^2 + \cdots + b_n\beta^n = 0,$$

where not all $b_i = 0$. Then $f(x) = b_n x^n + \cdots + b_1 x + b_0$ is a nonzero element of $F[x]$ such that $f(\beta) = 0$. Therefore, β is algebraic over F and $\deg(\beta, F)$ is at most n. ∎

Exercises

36.1 Find three bases for $\mathbf{R}^2$ over $\mathbf{R}$, no two of which have a vector in common.

36.2 Determine whether the given set of vectors is a basis for $\mathbf{R}^3$ over $\mathbf{R}$.

a) $\{(1, 1, 0), (1, 0, 1), (0, 1, 1)\}$

b) $\{(-1, 1, 2), (2, -3, 1), (10, -14, 0)\}$

36.3 According to Theorem 36.4, the element $1 + \alpha$ of $\mathbf{Z}_2(\alpha)$ of Example 35.9 is algebraic over $\mathbf{Z}_2$. Find the irreducible polynomial for $1 + \alpha$ in $\mathbf{Z}_2[x]$.

36.4 Give a basis for each of the following vector spaces over the indicated fields.

a) $\mathbf{Q}(\sqrt{2})$ over $\mathbf{Q}$ b) $\mathbf{R}(\sqrt{2})$ over $\mathbf{R}$

c) $\mathbf{Q}(\sqrt[3]{2})$ over $\mathbf{Q}$ d) $\mathbf{C}$ over $\mathbf{R}$

e) $\mathbf{Q}(i)$ over $\mathbf{Q}$ f) $\mathbf{Q}(\sqrt[4]{2})$ over $\mathbf{Q}$

†**36.5** Prove that if V is a finite-dimensional vector space over a field F, then a subset $\{\beta_i\}$ of V is a basis for V over F if and only if every vector in V can be expressed *uniquely* as a linear combination of the β_i.

36.6 Mark each of the following true or false.

___ a) The sum of two vectors is a vector

___ b) The sum of two scalars is a vector.

___ c) The product of two scalars is a scalar.

— d) The product of a scalar and a vector is a vector.

— e) Every vector space has a finite basis.

— f) The vectors in a basis are linearly dependent.

— g) The 0-vector may be part of a basis.

— h) If $F \leq E$ and $\alpha \in E$ is algebraic over the field F, then α^2 is algebraic over F.

— i) If $F \leq E$ and $\alpha \in E$ is algebraic over the field F, then $\alpha + \alpha^2$ is algebraic over F.

— j) Every vector space has a basis.

The exercises that follow deal with the further study of vector spaces. In many cases, we ask you to define for vector spaces some concept that is analogous to one we have studied for other algebraic structures. These exercises should improve your ability to recognize parallel and related situations in algebra. Any of these exercises may assume that you have knowledge of concepts defined in the preceding exercises.

36.7 Let V be a vector space over a field F.

a) Define a *subspace of the vector space V over F*.

b) Prove that an intersection of subspaces of V is again a subspace of V over F.

36.8 Let V be a vector space over a field F, and let $S = \{\alpha_i | i \in I\}$ be a collection of vectors in V.

a) Using Exercise 36.7(b), define the *subspace of V generated by S*.

b) Prove that the vectors in the subspace of V generated by S are precisely the (finite) linear combinations of vectors in S. (Compare with Theorem 9.1.)

36.9 Let $V_1, \ldots, V_n$ be vector spaces over the same field F. Define the *direct sum* $V_1 \oplus \cdots \oplus V_n$ of the vector spaces V_i for $i = 1, \ldots, n$, and show that the direct sum is again a vector space over F.

36.10 Generalize Example 36.1 to obtain the vector space F^n of ordered n-tuples of elements of F over the field F, for any field F. What is a basis for F^n?

36.11 Let F be any field. Consider the "system of m simultaneous linear equations in n unknowns"

$$a_{11}X_1 + a_{12}X_2 + \cdots + a_{1n}X_n = b_1,$$
$$a_{21}X_1 + a_{22}X_2 + \cdots + a_{2n}X_n = b_2,$$
$$\vdots$$
$$a_{m1}X_1 + a_{m2}X_2 + \cdots + a_{mn}X_n = b_m,$$

where $a_{ij}, b_i \in F$.

a) Show that the "system has a solution" if and only if the vector $\beta = (b_1, \ldots, b_m)$ of F^m lies in the subspace of F^m generated by the vectors $\alpha_j = (a_{1j}, \ldots, a_{mj})$. (This result is trivial to prove, being practically the definition of a solution, but should really be regarded as the *fundamental existence theorem for a simultaneous solution of a system of linear equations*.)

b) From (a), show that if $n = m$ and $\{\alpha_j | j = 1, \ldots, n\}$ is a basis for F^n, then the system always has a unique solution.

36.12 Define an *isomorphism of a vector space V over a field F with a vector space V' over the same field F*.

36.13 Prove that every finite-dimensional vector space V of dimension n over a field F is isomorphic to the vector space F^n of Exercise 36.10.

36.14 Let V and V' be vector spaces over the same field F. A function $\phi: V \to V'$ is a **linear transformation of V into V'** if the following conditions are satisfied for all $\alpha, \beta \in V$ and $a \in F$:

$$(\alpha + \beta)\phi = \alpha\phi + \beta\phi.$$
$$(a\alpha)\phi = a(\alpha\phi).$$

a) If $\{\beta_i \mid i \in I\}$ is a basis for V over F, show that a linear transformation $\phi: V \to V'$ is completely determined by the vectors $\beta_i \phi \in V'$.

b) Let $\{\beta_i \mid i \in I\}$ be a basis for V, and let $\{\beta'_i \mid i \in I\}$ be any set of vectors, not necessarily distinct, of V'. Show that there exists exactly one linear transformation $\phi: V \to V'$ such that $\beta_i \phi = \beta'_i$.

36.15 Let V and V' be vector spaces over the same field F, and let $\phi: V \to V'$ be a linear transformation.

a) To what concept that we have studied for the algebraic structures of groups and rings does the concept of a *linear transformation* correspond?

b) Define the *kernel* (or *null space*) *of* ϕ, and show that it is a subspace of V.

c) Describe when ϕ is an isomorphism of V with V'.

36.16 Let V be a vector space over a field F, and let S be a subspace of V. Define the *quotient space* V/S, and show that it is a vector space over F.

36.17 Let V and V' be vector spaces over the same field F, and let V be finite dimensional over F. Let $\dim(V)$ be the dimension of the vector space V over F. Let $\phi: V \to V'$ be a linear transformation.

a) Show that $V\phi$ is a subspace of V'.

b) Show that $\dim(V\phi) = \dim(V) - \dim(\text{kernel } \phi)$. [*Hint*: Choose a convenient basis for V, using Theorem 36.3. For example, enlarge a basis for (kernel ϕ) to a basis for V.]

*__36.18__ Let S and T be subspaces of a vector space V over a field F.

a) Define the *join $S \vee T$ of S and T*, and show that $S \vee T$ is also a subspace of V over F. (Compare with Section 8.2. $S \vee T$ is usually denoted by $S + T$ in the literature.)

b) Describe the elements in $S \vee T$ in terms of those in S and those in T.

*__36.19__ Let V be a finite-dimensional vector space over a field F, and let $\dim(V)$ be the dimension of V over F. Show that if S and T are subspaces of V over F, then

$$\dim(S \vee T) = \dim(S) + \dim(T) - \dim(S \cap T).$$

[*Hint*: The dimension of a space is the number of elements in a basis. Use Theorem 36.3 to choose convenient bases, so that the proof becomes easy. First choose any basis $\{\alpha_i\}$ for $S \cap T$, and then augment it with vectors β_j so that $\{\alpha_i, \beta_j\}$ is a basis for S. Repeat the process so that $\{\alpha_i, \gamma_k\}$ is a basis for T. Show that $\{\alpha_i, \beta_j, \gamma_k\}$ is a basis for $S \vee T$.]

Further Algebraic Structures

This is an expository section designed to give you an idea of some additional important algebraic structures and their relation to the structures we have studied.

*37.1 GROUPS WITH OPERATORS

Definition A *group with operators* consists of a group G and a set $\mathcal{O}$, the *set of operators*, together with an operation of external multiplication of each element of G by each element of $\mathcal{O}$ on the right such that for all $\alpha, \beta \in G$ and $a \in \mathcal{O}$, the following conditions are satisfied:

1. $(\alpha a) \in G$.
2. $(\alpha\beta)a = (\alpha a)(\beta a)$.

We shall somewhat incorrectly speak of the $\mathcal{O}$-*group* G.

Note how we have followed closely the form of our definition of a vector space from Chapter 36. The operation of external multiplication is really a function $\phi: G \times \mathcal{O} \to G$, where $(\alpha, a)\phi$ is denoted by αa for $\alpha \in G$ and $a \in \mathcal{O}$. Of course, whether you write the operator a on the right of α, as we did here, or on the left is a matter of preference and convenience.

While we did not require any structure on the set $\mathcal{O}$ in our definition, it frequently happens that $\mathcal{O}$ has some natural algebraic structure. Let us give a few examples.

Example 37.1 Every abelian group G gives rise to a natural group with operators. Let us write the group operation of G multiplicatively. Letting $\mathcal{O} = \mathbf{Z}$, for $\alpha \in G$ and $n \in \mathbf{Z}$, define $\alpha n = \alpha^n$. Since G is abelian, we have

$$(\alpha\beta)n = (\alpha\beta)^n = \alpha^n\beta^n = (\alpha n)(\beta n).$$

Thus every abelian group G can be regarded as a $\mathbf{Z}$-group. Here $\mathbf{Z}$ has a natural ring structure. ∎

Example 37.2 If V is a vector space over a field F, V can be regarded in a natural way as a (left) F-group. Here F has a field structure. ∎

Consider an $\mathcal{O}$-group G. For a *fixed* $a \in \mathcal{O}$, the mapping $\rho_a: G \to G$ defined by $\alpha\rho_a = \alpha a$ for $\alpha \in G$ is a homomorphism of G into G, since $(\alpha\beta)a = (\alpha a)(\beta a)$. This suggests our next example.

Example 37.3 Let G be any group, and let $\mathcal{O}$ be any set of homomorphisms of G into itself (such homomorphisms are *endomorphisms of G*). For $\alpha, \beta \in G$ and $\phi \in \mathcal{O}$, the property $(\alpha\beta)\phi = (\alpha\phi)(\beta\phi)$ for the endomorphism ϕ shows that G can be naturally viewed as an $\mathcal{O}$-group. ∎

A substantial chunk of our abelian group theory could have been done for $\mathcal{O}$-groups. Starting back with subgroups, an **admissible subgroup,** or $\mathcal{O}$ **-subgroup** of an $\mathcal{O}$-group G, is a subgroup H of $\langle G, \cdot \rangle$ such that $\alpha a \in H$ for all $\alpha \in H$ and $a \in \mathcal{O}$, that is, such that H is closed under external multiplication by elements of $\mathcal{O}$. If one is constantly working with $\mathcal{O}$-groups, one simply drops the term *admissible* and speaks of a subgroup of the $\mathcal{O}$-group G, always meaning an $\mathcal{O}$-subgroup. A couple more examples will show how elegantly these ideas relate to our past work.

Example 37.4 Let G be any group, and let $\mathcal{I}$ be the set of all inner automorphisms of G. As in Example 37.3, G becomes an $\mathcal{I}$-group in a natural way. An (admissible) subgroup H of G then must have the property that $ai_g = g^{-1}ag$ is in H for all $a \in H$ and all $g \in G$. Thus the $\mathcal{I}$-subgroups H of the $\mathcal{I}$-group G are essentially the normal subgroups of G. ∎

Example 37.5 Let R be a ring. The additive group $\langle R, + \rangle$ of R can be regarded as an R-group, where for a group element a in $\langle R, + \rangle$ and $r \in R$, one defines ar by the ring multiplication. The right distributive law in R gives $(a + b)r = (ar) + (br)$, which is exactly the condition that $\langle R, + \rangle$ be an

R-group. An R-subgroup N of the R-group $\langle R, + \rangle$ then must be a subgroup of $\langle R, + \rangle$ satisfying $ar \in N$ for all $a \in N$ and $r \in R$. Thus the R-subgroups are essentially the right ideals of R. If R is a commutative ring, then the R-subgroups are essentially the ideals of R. ∎

One can form the factor group of an $\mathcal{O}$-group G modulo a normal $\mathcal{O}$-subgroup; the factor group also becomes an $\mathcal{O}$-group in a natural way when one defines external multiplication on cosets by using representatives. One has the concept of an $\mathcal{O}$-homomorphism from one $\mathcal{O}$-group into another. We shall ask you to try to come up with the proper definitions for these ideas in the exercises.

Finally, we state without proof the Jordan-Hölder theorem for an $\mathcal{O}$-group. The definitions of subnormal series and composition series are analogous to our definitions of Part I; you need only require that every subgroup be an $\mathcal{O}$-subgroup.

Theorem 37.1 (Jordan-Hölder) *Any two composition series of an $\mathcal{O}$-group G are isomorphic.*

Taking $\mathcal{O} = \{\iota\}$, where ι is the identity map of a group G onto itself, we recover the Jordan-Hölder theorem of Chapter 14 for composition series. Taking $\mathcal{O} = \mathcal{I}$, the set of inner automorphisms of G, we recover the Jordan-Holder theorem of Chapter 14 for a principal series. And finally, but most impressively, taking the Jordan-Hölder theorem for the F-group V, where V is a finite-dimensional vector space over F, we recover the invariance of dimension of V, for a simple F-factor group of the F-group V will be a vector space (an F-group) of dimension 1 over F.

*37.2 MODULES

Definition   Let R be a ring. A (*left*) R-*module* consists of an abelian group M together with an operation of external multiplication of each element of M by each element of R on the left such that for all $\alpha, \beta \in M$ and r, $s \in R$, the following conditions are satisfied:

1. $(r\alpha) \in M$.
2. $r(\alpha + \beta) = r\alpha + r\beta$.
3. $(r + s)\alpha = r\alpha + s\alpha$.
4. $(rs)\alpha = r(s\alpha)$.

We shall somewhat incorrectly speak of the R-*module* M.

An R-module is very much like a vector space except that the scalars need only form a ring. If R is a ring with unity and $1\alpha = \alpha$ for all $\alpha \in M$, then M is a **unitary R-module.**

Example 37.6 Every abelian group G can be regarded as a **Z**-module if we define $n\alpha = \alpha^n$ for $\alpha \in G$ and $n \in \mathbf{Z}$. We have used multiplicative notation for the operation in G. The axioms for a module are easily verified. ∎

Example 37.7 For an ideal N in R, $\langle N, + \rangle$ can be viewed as an R-module, where for $\alpha \in N$ and $r \in R$, $r\alpha$ is the ordinary ring multiplication of r and α, both viewed as elements of the ring R. ∎

One can speak of submodules, quotient modules, and R-homomorphisms from one R-module into another, all by the natural definitions. One can also take a direct sum of R-modules, arriving again at an R-module.

Definition An R-module M is **cyclic** if there exists $\alpha \in M$ such that $M = \{r\alpha \,|\, r \in R\}$.

Thus a cyclic R-module is generated by a single element. The idea of a *set of generators for an R-module* is the natural generalization of the idea of a spanning set of vectors of a vector space. A nice result for us to state without proof and then to illustrate for some special cases, in light of our past work, is the following theorem.

Theorem 37.2 *If R is a* PID, *then every finitely generated R-module is isomorphic to a direct sum of cyclic R-modules.*

Taking $R = \mathbf{Z}$ and referring to Example 37.6, we see from the theorem that every finitely generated abelian group is isomorphic to a direct sum of cyclic groups. This is a large part of the Fundamental Theorem of Finitely Generated Abelian Groups. For $R = F$, where F is a field, applying Theorem 37.2 to a finite-dimensional vector space V over F, we see that V is isomorphic to a direct sum of vector spaces of dimension 1 over F.

*37.3 ALGEBRAS

Definition An **algebra** consists of a vector space V over a field F, together with a binary operation of multiplication on the set V of vectors, such that for all $a \in F$ and $\alpha, \beta, \gamma \in V$, the following conditions are satisfied:

1. $(a\alpha)\beta = a(\alpha\beta) = \alpha(a\beta)$.
2. $(\alpha + \beta)\gamma = \alpha\gamma + \beta\gamma$.
3. $\alpha(\beta + \gamma) = \alpha\beta + \alpha\gamma$.

We shall somewhat incorrectly speak of an **algebra V over F**. Also, V is an **associative algebra over F** if, in addition to the preceding three conditions,

4. $(\alpha\beta)\gamma = \alpha(\beta\gamma)$ for all $\alpha, \beta, \gamma \in V$.

Example 37.8   If E is an extension field of a field F, then $V = E$ can be viewed as an associative algebra over F, where addition and multiplication of elements of V are field addition and multiplication in E, and scalar multiplication by elements of F is again field multiplication in E. ■

Example 37.9   For any group G and field F, the *group algebra* $F(G)$ defined in Section 25.3 is an associative algebra over F. ■

Definition   An algebra V over a field F is a ***division algebra over*** F if V has a unity for multiplication and contains a multiplicative inverse of each nonzero element. (Note that associativity of multiplication is *not* assumed.)

Example 37.10   An extension field E of a field F can be viewed as an associative division algebra over F. Also the quaternions $\mathcal{Q}$ of Section 25.4 form an associative division algebra over the real numbers. ■

We conclude with a statement, for your information, of some famous results regarding division algebras over the real numbers.

Theorem 37.3   *The real numbers, the complex numbers, and the quaternions are the only (up to isomorphism) associative division algebras over the real numbers (Frobenius, 1878). The only additional division algebra over the real numbers is the Cayley algebra, which is a vector space of dimension 8 over* **R** *(Bott and Milnor, 1957).*

Exercises

*37.1 Our definition of an $\mathcal{O}$-group did *not* start

An $\mathcal{O}$-group $\langle \ , \ , \ldots, \ \rangle$ is $\ldots$

in a fashion similar to our definitions of a group and a ring. If an $\mathcal{O}$-group were defined in this way, what would $\langle \ , \ , \ldots, \ \rangle$ be? Be sure that you get *all* the sets and *all* the operations involved.

*37.2 Repeat Exercise 37.1 for an R-module.

*37.3 Repeat Exercise 37.1 for an algebra.

*37.4 Show that an intersection of admissible subgroups of an $\mathcal{O}$-group G is again an admissible subgroup of G.

*37.5 Let G be any group, and let $\mathcal{A}$ be the set of *all* automorphisms of G. By Example 37.3, G can be regarded as an $\mathcal{A}$-group. An $\mathcal{A}$-subgroup of G is a **characteristic subgroup** of G.

Every subgroup of every abelian group is a normal subgroup, that is, an $\mathscr{I}$-subgroup, but give an example showing that not every subgroup of every abelian group is a characteristic subgroup.

*37.6 Show that the factor group of an $\mathcal{O}$-group G modulo an admissible normal subgroup can again be considered an $\mathcal{O}$-group in a natural way. That is, define the external multiplication on the cosets, show it is well defined, and check the axioms for an $\mathcal{O}$-group.

*37.7 Define the concept of an $\mathcal{O}$-homomorphism of an $\mathcal{O}$-group G into an $\mathcal{O}$-group G'. Show that the kernel is an $\mathcal{O}$-subgroup of G.

*37.8 Prove that for a left R-module M,

$$0\alpha = 0, \qquad r0 = 0,$$

and

$$(-r)\alpha = -(r\alpha) = r(-\alpha)$$

for every $\alpha \in M$ and $r \in R$.

*37.9 Let M be a left R-module, and let $\alpha \in M$. Show that $L_\alpha = \{a \in R \,|\, a\alpha = 0\}$ is a left ideal of R.

*37.10 Define a *submodule of a (left) R-module M* and a *quotient module of a (left) R-module M modulo a submodule N*.

*37.11 Define an *R-homomorphism of a (left) R-module M into a (left) R-module M'*.

*37.12 Show that the ring $\langle M_n(F), +, \cdot \rangle$ of Section 25.1 becomes an algebra of dimension n^2 over F, if one defines scalar multiplication by $b(a_{ij}) = (ba_{ij})$ for $(a_{ij}) \in M_n(F)$ and $b \in F$.

*37.13 Let V be an algebra of finite dimension and with a basis

$$B = \{\beta_i \,|\, i = 1, \ldots, n\}$$

over a field F. Show that the multiplication of vectors in V is completely determined by the n^2 products $\beta_r\beta_s$ for each ordered pair (β_r, β_s) of basis vectors from B.

*37.14 Let V be an algebra of finite dimension and with a basis

$$B = \{\beta_i \,|\, i = 1, \ldots, n\}$$

over a field F. Show that V is an associative algebra over F if and only if

$$\beta_r(\beta_s\beta_t) = (\beta_r\beta_s)\beta_t$$

for each of the n^3 ordered triples $(\beta_r, \beta_s, \beta_t)$ of basis vectors from B.

*37.15 Let V be a finite-dimensional vector space over a field F with a basis $B = \{\beta_i \,|\, i = 1, \ldots, n\}$ over F. Let $\{c_{rst} \,|\, r, s, t = 1, \ldots, n\}$ be any collection of n^3 scalars in F. Show that there exists exactly one binary operation of multiplication on V such that V is an algebra over F under this multiplication and such that

$$\beta_r\beta_s = \sum_t c_{rst}\beta_t$$

for every ordered pair (β_r, β_s) of basis vectors from B. The scalars c_{rst} are the **structure constants of the algebra.**

thirty-eight

Algebraic Extensions

38.1 FINITE EXTENSIONS

In Theorem 36.4 we saw that if E is an extension field of a field F and $\alpha \in E$ is algebraic over F, then every element of $F(\alpha)$ is algebraic over F. In studying zeros of polynomials in $F[x]$, we shall be interested almost exclusively in extensions of F containing only elements algebraic over F.

Definition An extension field E of a field F is an *algebraic extension of F* if every element in E is algebraic over F.

Definition If an extension field E of a field F is of finite dimension n as a vector space over F, then E is a *finite extension of degree n over F*. We shall let $[E : F]$ be the degree n of E over F.

We shall often use the fact that if E is a finite extension of F, then $[E : F] = 1$ if and only if $E = F$. We need only observe that by Theorem 36.3, $\{1\}$ can always be enlarged to a basis for E over F. Then $[E : F] = 1$ implies that $E = F(1) = F$. The converse is obvious.

Let us repeat the argument of Theorem 36.4 to show that a finite extension E of a field F must be an algebraic extension of F.

342

Theorem 38.1 *A finite extension field E of a field F is an algebraic extension of F.*

Proof. We must show that for $\alpha \in E$, α is algebraic over F. By Theorem 36.3, if $[E : F] = n$, then

$$1, \alpha, \ldots, \alpha^n$$

cannot be linearly independent elements, so there exist $a_i \in F$ such that

$$a_n \alpha^n + \cdots + a_1 \alpha + a_0 = 0,$$

and not all $a_i = 0$. Then $f(x) = a_n x^n + \cdots + a_1 x + a_0$ is a nonzero polynomial in $F[x]$, and $f(\alpha) = 0$. Therefore, α is algebraic over F. ∎

We cannot overemphasize the importance of our next theorem. It plays a role in field theory analogous to the role of the theorem of Lagrange in group theory. While its proof follows easily from our brief work with vector spaces, it is a tool of incredible power. We shall later be using the theorem constantly in our Galois theory arguments. Also, an elegant application of it in the starred chapter that follows shows the impossibility of performing certain geometric constructions with a straightedge and a compass. *Never underestimate a theorem that counts something.* ∎

Theorem 38.2 *If E is a finite extension field of a field F, and K is a finite extension field of E, then K is a finite extension of F, and*

$$[K : F] = [K : E][E : F].$$

Proof. Let $\{\alpha_i \,|\, i = 1, \ldots, n\}$ be a basis for E as a vector space over F, and let $\{\beta_j \,|\, j = 1, \ldots m\}$ be a basis for K as a vector space over E. The theorem will be proved if we can show that the mn elements $\alpha_i \beta_j$ form a basis for K, viewed as a vector space over F.

Let γ be any element of K. Since the β_j form a basis for K over E, we have

$$\gamma = \sum_{j=1}^{m} b_j \beta_j$$

for $b_j \in E$. Since the α_i form a basis for E over F, we have

$$b_j = \sum_{i=1}^{n} a_{ij} \alpha_i$$

for $a_{ij} \in F$. Then

$$\gamma = \sum_{j=1}^{m} \left(\sum_{i=1}^{n} a_{ij} \alpha_i \right) \beta_j = \sum_{i,j} a_{ij} (\alpha_i \beta_j),$$

so the mn vectors $\alpha_i \beta_j$ span K over F.

It remains for us to show that the mn elements $\alpha_i\beta_j$ are independent over F. Suppose that $\sum_{i,j} c_{ij}(\alpha_i\beta_j) = 0$, with $c_{ij} \in F$. Then

$$\sum_{j=1}^{m} \left(\sum_{i=1}^{n} c_{ij}\alpha_i \right) \beta_j = 0,$$

and $(\sum_{i=1}^{n} c_{ij}\alpha_i) \in E$. Since the elements β_j are independent over E, we must have

$$\sum_{i=1}^{n} c_{ij}\alpha_i = 0$$

for all j. But now the α_i are independent over F, so $\sum_{i=1}^{n} c_{ij}\alpha_i = 0$ implies that $c_{ij} = 0$ for all i and j. Thus the $\alpha_i\beta_j$ not only span K over F but also are independent over F. Thus they form a basis for K over F. ∎

Note that we proved this theorem by actually exhibiting a basis. It is worth remembering that if $\{\alpha_i \,|\, i = 1, \ldots, n\}$ is a basis for E over F and $\{\beta_j \,|\, j = 1, \ldots, m\}$ is a basis for K over E, for fields $F \le E \le K$, then the set $\{\alpha_i\beta_j\}$ of mn products is a basis for K over F. Figure 38.1 gives a diagram for this situation. We shall illustrate this further in a moment.

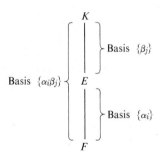

Fig. 38.1

Corollary 1 *If F_i is a field for $i = 1, \ldots, r$ and F_{i+1} is a finite extension of F_i, then F_r is a finite extension of F_1, and*

$$[F_r : F_1] = [F_r : F_{r-1}][F_{r-1} : F_{r-2}] \cdots [F_2 : F_1].$$

Proof. Proof of Corollary 1 is an obvious extension of Theorem 38.2, by induction. ∎

Corollary 2 *If E is an extension field of F, $\alpha \in E$ is algebraic over F, and $\beta \in F(\alpha)$, then $\deg(\beta, F)$ divides $\deg(\alpha, F)$.*

Proof. By Theorem 36.4, $\deg(\alpha, F) = [F(\alpha) : F]$ and $\deg(\beta, F) = [F(\beta) : F]$. We have $F \le F(\beta) \le F(\alpha)$, so by Theorem 38.2 $[F(\beta) : F]$ divides $[F(\alpha) : F]$. ∎

The following example illustrates a type of argument one often makes using Theorem 38.2 or its corollaries.

Example 38.1 By Corollary 2 of Theorem 38.2, there is no element of $Q(\sqrt{2})$ that is a zero of $x^3 - 2$. Note that $\deg(\sqrt{2}, Q) = 2$, while a zero of $x^3 - 2$ is of degree 3 over Q, but 3 does not divide 2. ∎

Let E be an extension field of a field F, and let α_1, α_2 be elements of E, not necessarily algebraic over F. By definition, $F(\alpha_1)$ is the smallest extension field of F in E that contains α_1. Similarly, $(F(\alpha_1))(\alpha_2)$ can be characterized as the smallest extension field of F in E containing both α_1 and α_2. We could equally well have started with α_2, so clearly $(F(\alpha_1))(\alpha_2) = (F(\alpha_2))(\alpha_1)$. We denote this field by $F(\alpha_1, \alpha_2)$. Similarly, for $\alpha_i \in E$, $F(\alpha_1, \dots, \alpha_n)$ is the smallest extension field of F containing all the α_i for $i = 1, \dots, n$. We obtain the field $F(\alpha_1, \dots, \alpha_n)$ from the field F by **adjoining to F the elements** α_i in E. You will have no difficulty in checking that, analogous to an intersection of subgroups of a group, an intersection of subfields of a field E is again a subfield of E. It is obvious that $F(\alpha_1, \dots, \alpha_n)$ is the intersection of all subfields of E containing F and all the α_i for $i = 1, \dots, n$.

Example 38.2 Consider $Q(\sqrt{2})$. Theorem 36.4 shows that $\{1, \sqrt{2}\}$ is a basis for $Q(\sqrt{2})$ over Q. By computing, we find $\text{irr}(\sqrt{2} + \sqrt{3}, Q) = x^4 - 10x^2 + 1$, so $[Q(\sqrt{2} + \sqrt{3}) : Q] = 4$. Thus $(\sqrt{2} + \sqrt{3}) \notin Q(\sqrt{2})$, so $\sqrt{3} \notin Q(\sqrt{2})$. Consequently, $\{1, \sqrt{3}\}$ is a basis for $Q(\sqrt{2}, \sqrt{3}) = (Q(\sqrt{2}))(\sqrt{3})$ over $Q(\sqrt{2})$. The proof of Theorem 38.2 (see the comment following the theorem) then shows that $\{1, \sqrt{2}, \sqrt{3}, \sqrt{6}\}$ is a basis for $Q(\sqrt{2}, \sqrt{3})$ over Q. ∎

Example 38.3 Let $2^{1/3}$ be the real cube root of 2 and $2^{1/2}$ be the positive square root of 2. Then, as we saw in Example 38.1, $2^{1/3} \notin Q(2^{1/2})$. Thus $[Q(2^{1/2}, 2^{1/3}) : Q(2^{1/2})] = 3$. Then $\{1, 2^{1/2}\}$ is a basis for $Q(2^{1/2})$ over Q, and $\{1, 2^{1/3}, 2^{2/3}\}$ is a basis for $Q(2^{1/2}, 2^{1/3})$ over $Q(2^{1/2})$. Furthermore, by Theorem 38.2 (see the comment following the theorem),

$$\{1, 2^{1/2}, 2^{1/3}, 2^{5/6}, 2^{2/3}, 2^{7/6}\}$$

is a basis for $Q(2^{1/2}, 2^{1/3})$ over Q. Clearly, since $2^{7/6} = (2)2^{1/6}$, we have $2^{1/6} \in Q(2^{1/2}, 2^{1/3})$. Now $2^{1/6}$ is a zero of $x^6 - 2$, which is irreducible over Q, by Eisenstein's criterion, with $p = 2$. Thus

$$Q \le Q(2^{1/6}) \le Q(2^{1/2}, 2^{1/3})$$

and by Theorem 38.2,

$$6 = [Q(2^{1/2}, 2^{1/3}) : Q] = [Q(2^{1/2}, 2^{1/3}) : Q(2^{1/6})][Q(2^{1/6}) : Q]$$
$$= [Q(2^{1/2}, 2^{1/3}) : Q(2^{1/6})](6).$$

Therefore, we must have

$$[\mathbf{Q}(2^{1/2}, 2^{1/3}) : \mathbf{Q}(2^{1/6})] = 1,$$

so $\mathbf{Q}(2^{1/2}, 2^{1/3}) = \mathbf{Q}(2^{1/6})$, by the comment preceding Theorem 38.1. ∎

Example 38.3 shows that it is possible for an extension $F(\alpha_1, \ldots, \alpha_n)$ of a field F to be actually a simple extension, even though $n > 1$.

Let us characterize extensions of F of the form $F(\alpha_1, \ldots, \alpha_n)$ in the case that all the α_i are algebraic over F.

Theorem 38.3 *Let E be an algebraic extension of a field F. Then there exist a finite number of elements $\alpha_1, \ldots, \alpha_n$ in E such that $E = F(\alpha_1, \ldots, \alpha_n)$ if and only if E is a finite-dimensional vector space over F, that is, if and only if E is a finite extension of F.*

Proof. Suppose that $E = F(\alpha_1, \ldots, \alpha_n)$. Since E is an algebraic extension of F, each α_i is algebraic over F, so clearly each α_i is algebraic over every extension field of F in E. Thus $F(\alpha_1)$ is algebraic over F, and in general, $F(\alpha_1, \ldots, \alpha_j)$ is algebraic over $F(\alpha_1, \ldots, \alpha_{j-1})$ for $j = 2, \ldots, n$. Corollary 1 of Theorem 38.2 applied to the sequence of finite extensions

$$F, F(\alpha_1), F(\alpha_1, \alpha_2), \ldots, F(\alpha_1, \ldots, \alpha_n) = E$$

then shows that E is a finite extension of F.

Conversely, suppose that E is a finite algebraic extension of F. If $[E:F] = 1$, then $E = F(1) = F$, and we are done. If $E \neq F$, let $\alpha_1 \in E$, where $\alpha_1 \notin F$. Then $[F(\alpha_1):F] > 1$. If $F(\alpha_1) = E$, we are done; if not, let $\alpha_2 \in E$, where $\alpha_2 \notin F(\alpha_1)$. Continuing this process, we see from Theorem 38.2 that since $[E:F]$ is finite, we must arrive at α_n such that

$$F(\alpha_1, \ldots, \alpha_n) = E. \quad \blacksquare$$

38.2 ALGEBRAICALLY CLOSED FIELDS AND ALGEBRAIC CLOSURES

We have not yet observed that if E is an extension of a field F and $\alpha, \beta \in E$ are algebraic over F, then so are $\alpha + \beta$, $\alpha\beta$, $\alpha - \beta$, and α/β, if $\beta \neq 0$. This follows easily from Theorem 38.3 and is also included in the following theorem.

Theorem 38.4 *Let E be an extension field of F. Then*

$$\bar{F}_E = \{\alpha \in E \,|\, \alpha \text{ is algebraic over } F\}$$

*is a subfield of E, the **algebraic closure of F in E**.*

Proof. Let α, $\beta \in \bar{F}_E$. Then Theorem 38.3 shows that $F(\alpha, \beta)$ is a finite extension of F, and by Theorem 38.1 every element of $F(\alpha, \beta)$ is algebraic over F, that is, $F(\alpha, \beta) \subseteq \bar{F}_E$. Thus $\bar{F}_E$ contains $\alpha + \beta$, $\alpha\beta$, $\alpha - \beta$, and also α/β for $\beta \neq 0$, so $\bar{F}_E$ is a subfield of E. ∎

Corollary   *The set of all algebraic numbers forms a field.*

Proof.   Proof of this corollary is immediate from Theorem 38.4, for the set of all algebraic numbers is the algebraic closure of **Q** in **C**. ∎

It is well known that the complex numbers have the property that every nonconstant polynomial in $\mathbf{C}[x]$ has a zero in **C**. This is known as the *Fundamental Theorem of Algebra.* An analytic proof of this theorem is given later in this chapter under a starred heading. We now give a definition generalizing this important concept to other fields.

Definition   A field F is ***algebraically closed*** if every nonconstant polynomial in $F[x]$ has a zero in F.

The next theorem shows that the concept of a field being algebraically closed can also be defined in terms of factorization of polynomials over the field.

Theorem 38.5   *A field F is algebrically closed if and only if every nonconstant polynomial in $F[x]$ factors in $F[x]$ into linear factors.*

Proof.   Let F be algebraically closed, and let $f(x)$ be a nonconstant polynomial in $F[x]$. Then $f(x)$ has a zero $a \in F$. By Corollary 1 of Theorem 31.1, $x - a$ is a factor of $f(x)$, so $f(x) = (x - a)g(x)$. Then if $g(x)$ is nonconstant, it has a zero $b \in F$, and $f(x) = (x - a)(x - b)h(x)$. Continuing, we get a factorization of $f(x)$ in $F[x]$ into linear factors.

Conversely, suppose that every nonconstant polynomial of $F[x]$ has a factorization into linear factors. If $ax - b$ is a linear factor of $f(x)$, then b/a is a zero of $f(x)$. Thus F is algebraically closed. ∎

Corollary   *An algebraically closed field F has no proper algebraic extensions, that is, no algebraic extensions E with $F < E$.*

Proof.   Let E be an algebraic extension of F, so $F \leq E$. Then if $\alpha \in E$, we have $\mathrm{irr}(\alpha, F) = x - \alpha$, by Theorem 38.5, since F is algebraically closed. Thus $\alpha \in F$, and we must have $F = E$. ∎

In starred Section 38.3 we shall show that just as there exists an algebraically closed extension **C** of the real numbers **R**, for any field F there similarly exists an algebraic extension $\bar{F}$ of F, with the property that $\bar{F}$ is algebraically closed. Chapter 41 will show that such an extension $\bar{F}$ is unique, up to isomorphism of course. Naively, to find $\bar{F}$ you proceed as follows. If not every polynomial $f(x)$ in $F[x]$ has a zero, then adjoin a zero α of such an $f(x)$ to F, thus obtaining the field $F(\alpha)$. *Theorem 35.1, Kronecker'. theorem, is strongly used here, of course.* If $F(\alpha)$ is still not algebraically closed, then continue the process further. The trouble is that, contrary to the situation for the algebraic closure **C** of **R**, you may have to do this a (possibly large) infinite number of times. It can easily be shown (see Exercises 38.1. and 38.16) that $\bar{\mathbf{Q}}$ is isomorphic to the field of all algebraic numbers, and that one cannot obtain $\bar{\mathbf{Q}}$ from **Q** by adjoining a finite number of algebraic numbers. We shall have to first discuss some set-theoretic machinery Zorn's lemma, in order to be able to handle such a situation. This machinery is a bit complex, so we are putting it under a starred heading. However, the existence theorem for $\bar{F}$ is very important, and we state it here so that you will realize that you should know this fact. There is nothing wrong with your assuming its validity.

> **Theorem 38.5** *Every field F has an **algebraic closure**, that is, an algebraic extension $\bar{F}$ that is algebraically closed.*

*38.3 THE EXISTENCE OF AN ALGEBRAIC CLOSURE

We shall prove that every field has an algebraic extension that is algebraically closed. We feel that you should have the opportunity to see some proof involving the *Axiom of Choice* by the time you finish a course in algebra This is a natural place for such a proof. We shall use an equivalent form Zorn's lemma, of the Axiom of Choice. To state Zorn's lemma, we have to give a set-theoretic definition.

> **Definition** A *partial ordering of a set* S is given by a relation $\leq$ defined for certain ordered pairs of elements of S such that the following conditions are satisfied:
>
> 1. $a \leq a$ for all $a \in S$ (*reflexive law*).
> 2. If $a \leq b$ and $b \leq a$, then $a = b$ (*antisymmetric law*).
> 3. If $a \leq b$ and $b \leq c$, then $a \leq c$ (*transitive law*).

In a *partially* ordered set, not every two elements need be **comparable** that is, for $a, b \in S$, you need not have either $a \leq b$ or $b \leq a$. As usual, $a <$ denotes $a \leq b$ but $a \neq b$.

A subset T of a partially ordered set S is a **chain** if every two elements and b in T are comparable, that is, either $a \leq b$ or $b \leq a$ (or both). An element $u \in S$ is an **upper bound** for a subset A of a partially ordered set S if $a \leq$ for all $a \in A$. Finally, an element m of a partially ordered set S is **maxima** if there is no $s \in S$ such that $m < s$.

Example 38.4 The collection of all subsets of a set forms a partially ordered set under the relation $\leq$ given by $\subseteq$. For example, if the whole set is **R**, we have $\mathbf{Z} \subseteq \mathbf{Q}$. Note, however, that for **Z** and $\mathbf{Q}^+$, neither $\mathbf{Z} \subseteq \mathbf{Q}^+$ nor $\mathbf{Q}^+ \subseteq \mathbf{Z}$.

> **Zorn's Lemma** *If S is a partially ordered set such that every chain in S has an upper bound in S, then S has at least one maximal element.*

There is no question of *proving* Zorn's lemma. The lemma is equivalent to the Axiom of Choice. Thus we are really taking Zorn's lemma here as an *axiom* for our set theory. We refer you to the literature for a statement of the Axiom of Choice and a proof of its equivalence to Zorn's lemma.

Zorn's lemma is often useful when you want to show the existence of a largest or maximal structure of some kind. If a field F has an algebraic extension $\bar{F}$ that is algebraically closed, then $\bar{F}$ will certainly be a maximal algebraic extension of F, for since $\bar{F}$ is algebraically closed, it can have no proper algebraic extensions.

The idea of our proof of Theorem 38.6 is very simple. Given a field F, we shall first describe a class of algebraic extensions of F that is so large that it must contain (up to isomorphism) any conceivable algebraic extension of F. We then define a partial ordering, the ordinary subfield ordering, on this class, and show that the hypotheses of Zorn's lemma are satisfied. By Zorn's lemma, there will exist a maximal algebraic extension $\bar{F}$ of F in this class. We shall then argue that this $\bar{F}$ can have no proper algebraic extensions, so it must be algebraically closed.

Our proof differs a bit from the one found in many texts. We like it because it uses no algebra other than that of Theorems 35.1 and 38.2. Thus it throws into sharp relief the tremendous strength of both Kronecker's theorem and Zorn's lemma. The proof looks long, but only because we are writing out every little step in hideous detail. To the professional mathematician, the construction of the proof from the information in the preceding paragraph is a routine matter. This proof was suggested to the author during his graduate student days by a fellow graduate student, Norman Shapiro, who also had a strong preference for it.

We are now ready to carry out our proof of Theorem 38.6, which we restate here.

> **Theorem 38.6** *Every field F has an algebraic closure $\bar{F}$.*

Proof. It can be shown in set theory that given any set, there exists a set with *strictly more* elements. Suppose we form a set

$$A = \{\omega_{f_i} \mid f \in F[x]; i = 0, \ldots, (\text{degree } f)\}$$

that has an element for every possible zero of any $f(x) \in F[x]$. Let Ω be a set with strictly more elements than A. By forming $\Omega \cup F$ if necessary, we can assume $F \subset \Omega$. Consider all possible fields that are algebraic extensions of F and that, as sets, consist of elements of Ω. One such algebraic extension is F itself. If E is any extension field of F, and if $\gamma \in E$ is a zero of $f(x) \in F[x]$ for $\gamma \notin F$ and $\deg(\gamma, F) = n$, then renaming γ by ω for $\omega \in \Omega$ and $\omega \notin F$, and renaming elements $a_0 + a_1\gamma + \cdots + a_{n-1}\gamma^{n-1}$ of $F(\gamma)$ by distinct elements of Ω as the a_i range over F, we can consider our renamed $F(\gamma)$ to be an algebraic extension field $F(\omega)$ of F, with $F(\omega) \subset \Omega$ and $f(\omega) = 0$. The set Ω has enough elements to form $F(\omega)$, since Ω has more than enough elements to provide n different zeros for each element of each degree n in any subset of $F[x]$.

All algebraic extension fields E_j of F, with $E_j \subseteq \Omega$, form a set

$$S = \{E_j \mid j \in J\}$$

that is partially ordered under our usual subfield inclusion $\leq$. One element of S is F itself. The preceding paragraph shows that if F is far away from being algebraically closed, there will be many fields E_j in S.

Let $T = \{E_{j_k}\}$ be a chain in S, and let $W = \bigcup_k E_{j_k}$. We now make W into a field. Let $\alpha, \beta \in W$. Then there exist $E_{j_1}, E_{j_2} \in S$, with $\alpha \in E_{j_1}$ and $\beta \in E_{j_2}$. Since T is a chain, one of the fields E_{j_1} and E_{j_2} is a subfield of the other, say $E_{j_1} \leq E_{j_2}$. Then $\alpha, \beta \in E_{j_2}$, and we use the field operations of E_{j_2} to *define* the sum of α and β in W as $(\alpha + \beta) \in E_{j_2}$ and, likewise, the product as $(\alpha\beta) \in E_{j_2}$. These operations are well defined in W; they are independent of our choice of E_{j_2}, since if $\alpha, \beta \in E_{j_3}$ also, for E_{j_3} in T, then one of the fields E_{j_2} and E_{j_3} is a subfield of the other, since T is a chain. Thus we have operations of addition and multiplication defined on W.

All the field axioms for W under these operations now follow from the fact that these operations were defined in terms of addition and multiplication in fields. Thus, for example, $1 \in F$ serves as multiplicative identity in W, since for $\alpha \in W$, if $1, \alpha \in E_{j_1}$, then we have $1\alpha = \alpha$ in E_{j_1}, so $1\alpha = \alpha$ in W, by definition of multiplication in W. Also, as further illustration, to check the distributive laws, let $\alpha, \beta, \gamma \in W$. Since T is a chain, we can find one field in T containing all three elements α, β, and γ, and in this field the distributive laws for α, β, and γ hold. Thus they hold in W. Therefore, we can view W as a field, and by construction, $E_{j_k} \leq W$ for every $E_{j_k} \in T$.

If we can show that W is algebraic over F, then $W \in S$ will be an upper bound for T. But if $\alpha \in W$, then $\alpha \in E_{j_1}$ for some E_{j_1} in T, so α is algebraic over F. Hence W is an algebraic extension of F and is an upper bound for T.

The hypotheses of Zorn's lemma are thus fulfilled, so there is a maximal element $\bar{F}$ of S. We claim that $\bar{F}$ is algebraically closed. Let $f(x) \in \bar{F}[x]$, where $f(x) \notin \bar{F}$. Suppose that $f(x)$ has no zero in $\bar{F}$. Since Ω has many more elements than $\bar{F}$ has, we can take $\omega \in \Omega$, where $\omega \notin \bar{F}$, and form a field $\bar{F}(\omega) \subseteq \Omega$, with ω a zero of $f(x)$, as we saw in the first paragraph of this proof.

Let β be in $\bar{F}(\omega)$. Then by Theorem 36.4, β is a zero of a polynomial

$$g(x) = \alpha_0 + \alpha_1 x + \cdots + \alpha_n x^n$$

in $\bar{F}[x]$, with $\alpha_i \in \bar{F}$, and hence α_i algebraic over F. Then by Theorem 38.3, $F(\alpha_0, \ldots, \alpha_n)$ is a finite extension of F, and since β is algebraic over $F(\alpha_0, \ldots, \alpha_n)$, we also see that $F(\alpha_0, \ldots, \alpha_n, \beta)$ is a finite extension over $F(\alpha_0, \ldots, \alpha_n)$. Theorem 38.2 then shows that $F(\alpha_0, \ldots, \alpha_n, \beta)$ is a finite extension of F, so by Theorem 38.1, β is algebraic over F. Hence $\bar{F}(\omega) \in S$ and $\bar{F} < \bar{F}(\omega)$, which contradicts the choice of $\bar{F}$ as maximal in S. Thus $f(x)$ must have had a zero in $\bar{F}$, so $\bar{F}$ is algebraically closed. ∎

The mechanics of the preceding proof are largely routine to the professional mathematician. Since it may be the first proof that you have ever seen using Zorn's lemma, we wrote the proof out in simply hideous detail. However, the construction and arguments involved should properly be regarded as easy and routine.

It is well known that **C** is an algebraically closed field. Although there are more algebraic proofs of this fact, we give an analytic proof as the one most accessible to the student who has had a course in functions of a complex variable.

Theorem 38.7 (Fundamental Theorem of Algebra) *The field **C** of complex numbers is an algebraically closed field.*

Proof. Let the polynomial $f(z) \in \mathbf{C}[z]$ have no zero in **C**. Then $1/f(z)$ gives an entire function, that is, $1/f$ is analytic everywhere. Also if $f \notin \mathbf{C}$, $\lim_{|c| \to \infty} |f(c)| = \infty$, so $\lim_{|c| \to \infty} |1/f(c)| = 0$. Thus $1/f$ must be bounded in the plane. Hence by Liouville's theorem of complex function theory, $1/f$ is constant, and thus f is constant. Therefore, a nonconstant polynomial in $\mathbf{C}[z]$ must have a zero in **C**, so **C** is algebraically closed. ∎

Exercises

38.1 Find the degree and a basis for each of the given field extensions.
a) $\mathbf{Q}(\sqrt{2})$ over $\mathbf{Q}$
b) $\mathbf{Q}(\sqrt{2}, \sqrt{3})$ over $\mathbf{Q}$
c) $\mathbf{Q}(\sqrt{2}, \sqrt{3}, \sqrt{5})$ over $\mathbf{Q}$
d) $\mathbf{Q}(\sqrt[3]{2}, \sqrt{3})$ over $\mathbf{Q}$
e) $\mathbf{Q}(\sqrt{2}, \sqrt[3]{2})$ over $\mathbf{Q}$

38.2 Find the degree of each of the following field extensions. Be prepared to justify your answers.
a) $\mathbf{Q}(\sqrt{2} + \sqrt{3})$ over $\mathbf{Q}$
b) $\mathbf{Q}(\sqrt{2}, \sqrt{3})$ over $\mathbf{Q}$
c) $\mathbf{Q}(\sqrt{2}, \sqrt[3]{5})$ over $\mathbf{Q}$
d) $\mathbf{Q}(\sqrt[3]{2}, \sqrt[3]{6}, \sqrt[3]{24})$ over $\mathbf{Q}$
e) $\mathbf{Q}(\sqrt{2}, \sqrt{6})$ over $\mathbf{Q}(\sqrt{3})$
f) $\mathbf{Q}(\sqrt{2} + \sqrt{3})$ over $\mathbf{Q}(\sqrt{3})$

g) $\mathbf{Q}(\sqrt{2}, \sqrt{3})$ over $\mathbf{Q}(\sqrt{2} + \sqrt{3})$

h) $\mathbf{Q}(\sqrt{2}, \sqrt{6} + \sqrt{10})$ over $\mathbf{Q}(\sqrt{3} + \sqrt{5})$

38.3 Find a basis for each of the field extensions given in (e) through (h) of Exercise 38.2.

38.4 Show by an example that for a proper extension field E of a field F, the algebraic closure of F in E need not be algebraically closed.

38.5 Let $(a + bi) \in \mathbf{C}$ for $a, b \in \mathbf{R}$ with $b \neq 0$. Show that $\mathbf{C} = \mathbf{R}(a + bi)$.

†**38.6** Show that if E is a finite extension of a field F and $[E : F]$ is a prime number, then E is a simple extension of F and, indeed, $E = F(\alpha)$ for every $\alpha \in E$ not in F.

38.7 Mark each of the following true or false.

___ a) Every finite extension of a field is an algebraic extension.

___ b) Every algebraic extension of a field is a finite extension.

___ c) The top field of a finite tower of finite extensions of fields is a finite extension of the bottom field.

___ d) $\mathbf{R}$ is algebraically closed.

___ e) $\mathbf{Q}$ is its own algebraic closure in $\mathbf{R}$, that is, $\mathbf{Q}$ is **algebraically closed in R.**

___ f) $\mathbf{C}$ is algebraically closed in $\mathbf{C}(x)$, where x is an indeterminate.

___ g) $\mathbf{C}(x)$ is algebraically closed, where x is an indeterminate.

___ h) The field $\mathbf{C}(x)$ has no algebraic closure, since $\mathbf{C}$ already contains all algebraic numbers.

___ i) An algebraically closed field must be of characteristic 0.

___ j) If E is an algebraically closed extension field of F, then E is an algebraic extension of F.

38.8 Prove that $x^2 - 3$ is irreducible over $\mathbf{Q}(\sqrt[3]{2})$.

38.9 What degree field extensions can we obtain by successively adjoining to a field F a square root of an element of F not a square in F, then a square root of some non-square in this new field, and so on? Argue from this that a zero of $x^{14} - 3x^2 + 12$ over $\mathbf{Q}$ can never be expressed as a rational function of square roots, square roots of rational functions of square roots, and so on of elements of $\mathbf{Q}$.

38.10 Let E be a finite extension field of F. Let D be an integral domain such that $F \subseteq D \subseteq E$. Show that D is a field.

38.11 Prove in detail that $\mathbf{Q}(\sqrt{3} + \sqrt{7}) = \mathbf{Q}(\sqrt{3}, \sqrt{7})$.

38.12 Generalizing Exercise 38.11, show that if $\sqrt{a} + \sqrt{b} \neq 0$, then $\mathbf{Q}(\sqrt{a} + \sqrt{b}) = \mathbf{Q}(\sqrt{a}, \sqrt{b})$ for all a and b in $\mathbf{Q}$. [*Hint*: Compute $(\sqrt{a} + \sqrt{b})^2$.]

38.13 Let E be a finite extension of a field F, and let $p(x) \in F[x]$ be irreducible over F and have degree that is not a divisor of $[E : F]$. Show that $p(x)$ has no zeros in E.

38.14 Let E be an extension field of F. Let $\alpha \in E$ be algebraic of odd degree over F. Show that α^2 is algebraic of odd degree over F, and $F(\alpha) = F(\alpha^2)$.

38.15 Show that if F, E, and K are fields with $F \leq E \leq K$, then K is algebraic over F if and only if E is algebraic over F, and K is algebraic over E. (You must *not* assume the extensions are finite.)

38.16 Let E be an extension field of a field F. Prove that every $\alpha \in E$ which is not in the algebraic closure $\bar{F}_E$ of F in E is transcendental over $\bar{F}_E$.

38.17 Let E be an algebraically closed extension field of a field F. Show that the algebraic closure $\bar{F}_E$ of F in E is algebraically closed. (Applying this exercise to $\mathbf{C}$ and $\mathbf{Q}$, we see that the field of all algebraic numbers is an algebraically closed field.)

38.18 Show that if E is an algebraic extension of a field F and contains all zeros in $\bar{F}$ of every $f(x) \in F[x]$, then E is an algebraically closed field.

38.19 Show that no finite field of odd characteristic is algebraically closed. (Actually, no finite field of characteristic 2 is algebraically closed either.) [*Hint:* By counting, show that for such a finite field F, some polynomial $x^2 - a$, for some $a \in F$, has no zero in F. See Exercise 35.11.]

38.20 Prove that, as asserted in the text, the algebraic closure of $\mathbf{Q}$ in $\mathbf{C}$ is not a finite extension of $\mathbf{Q}$.

38.21 Argue that every finite extension field of $\mathbf{R}$ either is $\mathbf{R}$ itself or is isomorphic to $\mathbf{C}$.

38.22 Use Zorn's lemma to show that every proper ideal of a ring R with unity is contained in some maximal ideal.

Geometric Constructions

In this chapter we digress briefly to give an application demonstrating the power of Theorem 38.2. For a more detailed study of geometric constructions, you are referred to Courant and Robbins [43, Chapter III].

We are interested in what types of figures can be constructed with a compass and a straightedge in the sense of classical Euclidean plane geometry. You no doubt remember being told that "it is impossible to trisect the angle." We shall discuss this and other classic questions.

*39.1 CONSTRUCTIBLE NUMBERS

Let us imagine that we are given only a single line segment that we shall define to be *one unit* in length. A real number α is **constructible** if one can construct a line segment of length $|\alpha|$ in a finite number of steps from this given segment of unit length by using a straightedge and a compass. Recall that using a straightedge and a compass it is possible, among other things, to erect a perpendicular to a given line at a known point on the line and to find a line passing through a given point and parallel to a given line. Our first result is the following theorem.

Theorem 39.1 *If α and β are constructible real numbers, then so are $\alpha + \beta, \alpha - \beta, \alpha\beta$, and α/β, if $\beta \neq 0$.*

Proof. We are given that α and β are constructible, so there are line segments of lengths $|\alpha|$ and $|\beta|$ available to us. For $\alpha, \beta > 0$, extend a line segment of length α with the straightedge. Start at one end of the original segment of length α, and lay off on the extension the length β with the compass. This constructs a line segment of length $\alpha + \beta$; $\alpha - \beta$ is similarly constructible (see Fig. 39.1). If α and β are not both positive, an obvious breakdown into cases according to their signs shows that $\alpha + \beta$ and $\alpha - \beta$ are still constructible.

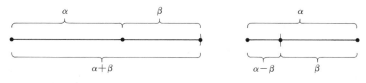

Fig. 39.1

The construction of $\alpha\beta$ is indicated in Fig. 39.2. We shall let $\overline{OA}$ be the line segment from the point O to the point A, and shall let $|\overline{OA}|$ be the length of this line segment. If $\overline{OA}$ is of length $|\alpha|$, find a line l through O not containing $\overline{OA}$. Then find the points P and B on l such that $\overline{OP}$ is of length 1 and $\overline{OB}$ is of length $|\beta|$. Draw $\overline{PA}$ and construct l' through B, parallel to $\overline{PA}$ and intersecting $\overline{OA}$ extended at Q. By similar triangles, we have

$$\frac{1}{|\alpha|} = \frac{|\beta|}{|\overline{OQ}|},$$

so $\overline{OQ}$ is of length $|\alpha\beta|$.

Finally, Fig. 39.3 shows that α/β is constructible, if $\beta \neq 0$. Let $\overline{OA}$ be of length $|\alpha|$, and find l through O not containing $\overline{OA}$. Then find B and P on l

Fig. 39.2

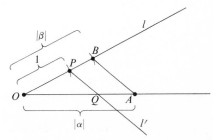

Fig. 39.3

such that $\overline{OB}$ is of length $|\beta|$ and $\overline{OP}$ is of length 1. Draw $\overline{BA}$ and construct l' through P, parallel to $\overline{BA}$, and intersecting $\overline{OA}$ at Q. Again by similar triangles, we have

$$\frac{|\overline{OQ}|}{1} = \frac{|\alpha|}{|\beta|},$$

so $\overline{OQ}$ is of length $|\alpha/\beta|$. ∎

Corollary *The set of all constructible real numbers forms a subfield F of the field of real numbers.*

Proof. Proof of this corollary is immediate from Theorem 39.1. ∎

Clearly, the field F of all constructible real numbers contains $\mathbf{Q}$, the field of rational numbers, since $\mathbf{Q}$ is the smallest subfield of $\mathbf{R}$.

From now on, we proceed analytically. We can construct any rational number. Regarding our given segment

$$0 \underline{\hspace{3cm}} 1$$

of length 1 as the basic unit on an x-axis, we can locate any point (q_1, q_2) in the plane with both coordinates rational. Any further point in the plane that we can locate by using a compass and a straightedge can be found in one of the following three ways:

1. as an intersection of two lines, each of which passes through two known points having rational coordinates,

2. as an intersection of a line that passes through two points having rational coordinates and a circle whose center has rational coordinates and the square of whose radius is rational,

3. as an intersection of two circles whose centers have rational coordinates and the squares of whose radii are rational.

Equations of lines and circles of the type discussed in 1, 2, and 3 are of the form

$$ax + by + c = 0$$

and

$$x^2 + y^2 + dx + ey + f = 0,$$

where $a, b, c, d, e,$ and f are all in $\mathbf{Q}$. Since in case 3 the intersection of two circles with equations

$$x^2 + y^2 + d_1 x + e_1 y + f_1 = 0$$

and

$$x^2 + y^2 + d_2 x + e_2 y + f_2 = 0$$

is the same as the intersection of the first circle, having equation

$$x^2 + y^2 + d_1 x + e_1 y + f_1 = 0,$$

and the line (the common chord), having equation

$$(d_1 - d_2)x + (e_1 - e_2)y + f_1 - f_2 = 0,$$

it is clear that case 3 can be reduced to case 2. For case 1, a simultaneous solution of two linear equations with rational coefficients can only lead to rational values of x and y, giving us no new points. However, finding a simultaneous solution of a linear equation with rational coefficients and a quadratic equation with rational coefficients, as in case 2, leads, upon substitution, to a quadratic equation. Such an equation, when solved by the quadratic formula, may have solutions involving square roots of numbers that are not squares in $\mathbf{Q}$.

In the preceding argument, nothing was really used involving $\mathbf{Q}$ except field axioms. If H is the smallest field containing those real numbers constructed so far, the argument shows that the "next new number" constructed lies in a field $H(\sqrt{\alpha})$ for some $\alpha \in H$, where $\alpha > 0$. We have proved half of our next theorem.

Theorem 39.2   *The field F of constructible real numbers consists precisely of all real numbers that we can obtain from $\mathbf{Q}$ by taking square roots of positive numbers a finite number of times and applying a finite number of field operations.*

Proof.   We have shown that F can contain no numbers except those we obtain from $\mathbf{Q}$ by taking a finite number of square roots of positive numbers and applying a finite number of field operations. However, if $\alpha > 0$ is constructible, then Fig. 39.4 shows that $\sqrt{\alpha}$ is constructible. Let $\overline{OA}$ have length α, and find P on $\overline{OA}$ extended so that $\overline{OP}$ has length 1. Find the midpoint of $\overline{PA}$ and draw a semicircle with $\overline{PA}$ as diameter. Erect a perpendicular to

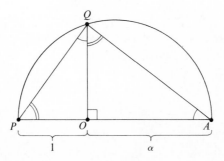

Fig. 39.4

$\overline{PA}$ at O, intersecting the semicircle at Q. Then the triangles OPQ and OQA are similar, so

$$\frac{|\overline{OQ}|}{|\overline{OA}|} = \frac{|\overline{OP}|}{|\overline{OQ}|},$$

and $|\overline{OQ}|^2 = 1\alpha = \alpha$. Thus $\overline{OQ}$ is of length $\sqrt{\alpha}$. Therefore square roots of constructible numbers are constructible.

Theorem 39.1 showed that field operations are possible by construction. ∎

Corollary   *If γ is constructible and $\gamma \notin \mathbf{Q}$, then there is a finite sequence of real numbers $\alpha_1, \ldots, \alpha_n = \gamma$ such that $\mathbf{Q}(\alpha_1, \ldots, \alpha_i)$ is an extension of $\mathbf{Q}(\alpha_1, \ldots, \alpha_{i-1})$ of degree 2. In particular, $[\mathbf{Q}(\gamma):\mathbf{Q}] = 2^r$ for some integer $r \geq 0$.*

Proof.   The existence of the α_i is immediate from Theorem 39.2. Then

$$2^n = [\mathbf{Q}(\alpha_1, \ldots, \alpha_n):\mathbf{Q}]$$
$$= [\mathbf{Q}(\alpha_1, \ldots, \alpha_n):\mathbf{Q}(\gamma)][\mathbf{Q}(\gamma):\mathbf{Q}],$$

by Theorem 38.2, which completes the proof. ∎

We hope you realize that the preceding ideas, although rather messy to write down in detail, are very simple.

*39.2 THE IMPOSSIBILITY OF CERTAIN CONSTRUCTIONS

We now can show the impossibility of certain geometric constructions.

Theorem 39.3   *"Doubling the cube is impossible," that is, given a side of a cube, it is not always possible to construct with a straightedge and a compass the side of a cube that has double the volume of the original cube.*

Proof.   Let the given cube have a side of length 1, and hence a volume of 1. The cube being sought would have to have a volume of 2, and hence a side of length $\sqrt[3]{2}$. But $\sqrt[3]{2}$ is a zero of irreducible $x^3 - 2$ over $\mathbf{Q}$, so

$$[\mathbf{Q}(\sqrt[3]{2}):\mathbf{Q}] = 3.$$

The corollary of Theorem 39.2 shows that to double this cube of volume 1, we would need to have $3 = 2^r$ for some integer r. Clearly, no such r exists. ∎

Theorem 39.4   *"Squaring the circle is impossible," that is, given a circle, it is not always possible to construct with a straightedge and a compass a square having area equal to the area of the given circle.*

Proof.   Let the given circle have a radius of 1, and hence an area of π. We would need to construct a square of side $\sqrt{\pi}$. But π is transcendental over **Q**, so $\sqrt{\pi}$ is transcendental over **Q** also. ∎

Theorem 39.5   *"Trisecting the angle is impossible," that is, there exists an angle that cannot be trisected with a straightedge and a compass.*

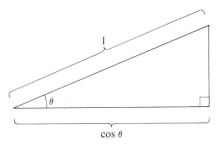

Fig. 39.5

Proof.   Figure 39.5 indicates that the angle θ can be constructed if and only if a segment of length $|\cos \theta|$ can be constructed. Now 60° is a constructible angle, and we shall show that it cannot be trisected. Note that

$$\begin{aligned}
\cos 3\theta &= \cos (2\theta + \theta) \\
&= \cos 2\theta \cos \theta - \sin 2\theta \sin \theta \\
&= (2 \cos^2 \theta - 1) \cos \theta - 2 \sin \theta \cos \theta \sin \theta \\
&= (2 \cos^2 \theta - 1) \cos \theta - 2 \cos \theta(1 - \cos^2 \theta) \\
&= 4 \cos^3 \theta - 3 \cos \theta.
\end{aligned}$$

Let $\theta = 20°$, so that $\cos 3\theta = \frac{1}{2}$, and let $\alpha = \cos 20°$. From our identity $4 \cos^3 \theta - 3 \cos \theta = \cos 3\theta$, we see that

$$4\alpha^3 - 3\alpha = \tfrac{1}{2}.$$

Thus α is a zero of $8x^3 - 6x - 1$. This polynomial is irreducible in **Q**$[x]$, since, by Theorem 31.3, it is enough to show that it does not factor in **Z**$[x]$. But a factorization in **Z**$[x]$ would entail a linear factor of the form $(8x \pm 1)$, $(4x \pm 1)$, $(2x \pm 1)$, or $(x \pm 1)$. One can easily check that none of the numbers $\pm\frac{1}{8}$, $\pm\frac{1}{4}$, $\pm\frac{1}{2}$, and ± 1 is a zero of $8x^3 - 6x - 1$. Thus

$$[\mathbf{Q}(\alpha) : \mathbf{Q}] = 3,$$

so by the corollary of Theorem 39.2, α is not constructible. Hence 60° cannot be trisected. ∎

Note that the regular n-gon is constructible for $n \geq 3$ if and only if the angle $2\pi/n$ is constructible, which is the case if and only if a line segment of length $\cos(2\pi/n)$ is constructible. We shall return to the study of the constructibility of certain regular n-gons in Chapter 48.

Exercises

*39.1 Using Theorem 39.5, show that the regular 9-gon is not constructible.

*39.2 Show *algebraically* that it is possible to construct an angle of 30°.

*39.3 Referring to Fig. 39.6, where $\overline{AQ}$ bisects angle OAP, show that the regular 10-gon is constructible (and therefore that the regular pentagon is also). [*Hint:* Triangle OAP is similar to triangle APQ. Show algebraically that r is constructible.]

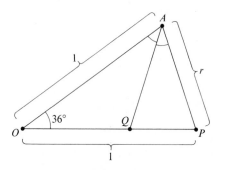

Fig. 39.6

*39.4 Using the results of Exercise 39.3 where needed, show that the following are true.

a) The regular 20-gon is constructible.

b) The regular 30-gon is constructible.

c) The angle 72° can be trisected.

d) The regular 15-gon can be constructed.

*39.5 Mark each of the following true or false.

— a) It is impossible to double any cube of constructible edge by compass and straightedge constructions.

— b) It is impossible to double every cube of constructible edge by compass and straightedge constructions.

— c) It is impossible to square any circle of constructible radius by straightedge and compass constructions.

— d) No constructible angle can be trisected by straightedge and compass constructions.

— e) Every constructible number is of degree 2^r over $\mathbf{Q}$ for some integer $r \geq 0$.

— f) We have shown that every real number of degree 2^r over **Q** for some integer $r \geq 0$ is constructible.

— g) The fact that **Z** is a UFD was used strongly at the conclusion of Theorems 39.3 and 39.5.

— h) Counting arguments are exceedingly powerful mathematical tools.

— i) One can find any constructible number in a finite number of steps by starting with a given segment of unit length and using a straightedge and a compass.

— j) One can find the totality of all constructible numbers in a finite number of steps by starting with a given segment of unit length and using a straightedge and a compass.

forty

Automorphisms of Fields

40.1 THE BASIC ISOMORPHISMS OF ALGEBRAIC FIELD THEORY

Let F be a field, and let $\bar{F}$ be an algebraic closure of F, that is, an algebraic extension of F that is algebraically closed. Such a field $\bar{F}$ exists, by Theorem 38.6. Our selection of a particular $\bar{F}$ is not critical, since, as we shall show in Chapter 41, any two algebraic closures of F are isomorphic under a map leaving F fixed. *From now on in our work, we shall assume that all algebraic extensions and all elements algebraic over a field F under consideration are contained in one fixed algebraic closure $\bar{F}$ of F.*

Remember that we are engaged in the study of zeros of polynomials. In the terminology of Chapter 38, studying zeros of polynomials in $F[x]$ amounts to studying the structure of algebraic extensions of F and of elements algebraic over F. We shall show that if E is an algebraic extension of F with $\alpha, \beta \in E$, then α and β have the same algebraic properties if and only if $\mathrm{irr}(\alpha, F) = \mathrm{irr}(\beta, F)$. We shall phrase this fact in terms of mappings, as we have been doing all along in field theory. We achieve this by showing the existence of an isomorphism $\psi_{\alpha, \beta}$ of $F(\alpha)$ onto $F(\beta)$ that maps each ele-

362

ment of F onto itself and maps α and β, in the case that $\mathrm{irr}(\alpha, F) = \mathrm{irr}(\beta, F)$. The next theorem exhibits this isomorphism $\psi_{\alpha,\beta}$. These isomorphisms will become our fundamental tools for the study of algebraic extensions; they supplant the *evaluation homomorphisms* ϕ_α of Chapter 30, which make their last contribution in defining these isomorphisms. For this reason, we shall refer to the isomorphism $\psi_{\alpha,\beta}$ as a *basic isomorphism of algebraic field theory*. Before stating and proving this theorem, let us introduce some more terminology.

Definition   Let E be an algebraic extension of a field F. Two elements $\alpha, \beta \in E$ are **conjugate over** F if $\mathrm{irr}(\alpha, F) = \mathrm{irr}(\beta, F)$, that is, if α and β are zeros of the same irreducible polynomial over F.

Example 40.1   The concept of conjugate elements just defined conforms with the classic idea of *conjugate complex numbers* if we understand that by conjugate complex numbers we mean numbers that are *conjugate over* **R**. If $a, b \in \mathbf{R}$ and $b \neq 0$, the conjugate complex numbers $a + bi$ and $a - bi$ are both zeros of $x^2 - 2ax + a^2 + b^2$, which is irreducible in $\mathbf{R}[x]$. ∎

Theorem 40.1 (*The Basic Isomorphisms of Algebraic Field Theory*)   Let F be a field, and let α and β be algebraic over F with $\deg(\alpha, F) = n$. The map $\psi_{\alpha,\beta} : F(\alpha) \to F(\beta)$ defined by

$$(c_0 + c_1\alpha + \cdots + c_{n-1}\alpha^{n-1})\psi_{\alpha,\beta} = c_0 + c_1\beta + \cdots + c_{n-1}\beta^{n-1}$$

for $c_i \in F$ is an isomorphism of $F(\alpha)$ onto $F(\beta)$ if and only if α and β are conjugate over F.

Proof.   Suppose that $\psi_{\alpha,\beta} : F(\alpha) \to F(\beta)$ as defined in the statement of the theorem is an isomorphism. Let $\mathrm{irr}(\alpha, F) = a_0 + a_1 x + \cdots + a_n x^n$. Then $a_0 + a_1\alpha + \cdots + a_n\alpha^n = 0$, so

$$(a_0 + a_1\alpha + \cdots + a_n\alpha^n)\psi_{\alpha,\beta} = a_0 + a_1\beta + \cdots + a_n\beta^n = 0.$$

By the last assertion in the statement of Theorem 35.3, this implies that $\mathrm{irr}(\beta, F)$ divides $\mathrm{irr}(\alpha, F)$. A similar argument using the isomorphism $(\psi_{\alpha,\beta})^{-1} = \psi_{\beta,\alpha}$ shows that $\mathrm{irr}(\alpha, F)$ divides $\mathrm{irr}(\beta, F)$. Therefore, since both polynomials are monic, $\mathrm{irr}(\alpha, F) = \mathrm{irr}(\beta, F)$, so α and β are conjugate over F.

Conversely, suppose $\mathrm{irr}(\alpha, F) = \mathrm{irr}(\beta, F) = p(x)$. Then the evaluation homomorphisms $\phi_\alpha : F[x] \to F(\alpha)$ and $\phi_\beta : F[x] \to F(\beta)$ both have the same kernel $\langle p(x) \rangle$. By Theorem 29.3, corresponding to $\phi_\alpha : F[x] \to F(\alpha)$, there is a natural isomorphism ψ_α mapping $F[x]/\langle p(x) \rangle$ onto $(F[x])\phi_\alpha = F(\alpha)$. Similarly, ϕ_β gives rise to an isomorphism ψ_β mapping $F[x]/\langle p(x) \rangle$ onto $F(\beta)$. Let $\psi_{\alpha,\beta} = (\psi_\alpha)^{-1}\psi_\beta$. These mappings are diagrammed in Fig. 40.1, where the dashed lines indicate corresponding elements under the mappings. As the composition of two isomorphisms, $\psi_{\alpha,\beta}$ is again an isomorphism and

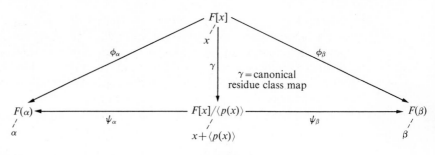

Fig. 40.1

maps $F(\alpha)$ onto $F(\beta)$. Also, for $(c_0 + c_1\alpha + \cdots + c_{n-1}\alpha^{n-1}) \in F(\alpha)$, we have

$$(c_0 + c_1\alpha + \cdots + c_{n-1}\alpha^{n-1})\psi_{\alpha,\beta}$$
$$= (c_0 + c_1\alpha + \cdots + c_{n-1}\alpha^{n-1})(\psi_\alpha^{-1}\psi_\beta)$$
$$= [(c_0 + c_1 x + \cdots + c_{n-1}x^{n-1}) + \langle p(x)\rangle]\psi_\beta$$
$$= c_0 + c_1\beta + \cdots + c_{n-1}\beta^{n-1}.$$

Thus $\psi_{\alpha,\beta}$ is the map defined in the statement of the theorem. ∎

The following corollary of Theorem 40.1 is the cornerstone of our proof of the important Isomorphism Extension Theorem of Chapter 41 and of most of the rest of our work.

Corollary 1 *Let α be algebraic over a field F. Every isomorphism ψ mapping $F(\alpha)$ into $\bar{F}$ such that $a\psi = a$ for $a \in F$ maps α onto a conjugate β of α over F. Conversely, for each conjugate β of α over F, there exists exactly one isomorphism $\psi_{\alpha,\beta}$ of $F(\alpha)$ into $\bar{F}$ mapping α onto β and mapping each $a \in F$ onto itself.*

Proof. Let ψ be an isomorphism mapping $F(\alpha)$ into $\bar{F}$ such that $a\psi = a$ for $a \in F$. Let $\mathrm{irr}(\alpha, F) = a_0 + a_1 x + \cdots + a_n x^n$. Then

$$a_0 + a_1\alpha + \cdots + a_n\alpha^n = 0,$$

so

$$0 = (a_0 + a_1\alpha + \cdots + a_n\alpha^n)\psi = a_0 + a_1(\alpha\psi) + \cdots + a_n(\alpha\psi)^n,$$

and $\beta = \alpha\psi$ is a conjugate of α.

Conversely, for each conjugate β of α over F, the isomorphism $\psi_{\alpha,\beta}$ of Theorem 40.1 is an isomorphism with the desired properties. That $\psi_{\alpha,\beta}$ is the only such isomorphism follows from the fact that an isomorphism of $F(\alpha)$ is completely determined by its values on elements of F and its value on α. ∎

As a second corollary of Theorem 40.1, we can prove a result with which you are probably already familiar.

Corollary 2 Let $f(x) \in \mathbf{R}[x]$. If $f(a + bi) = 0$ for $(a + bi) \in \mathbf{C}$, where $a, b \in \mathbf{R}$, then $f(a - bi) = 0$ also. Loosely, complex zeros of polynomials with real coefficients occur in conjugate pairs.

Proof. We have seen that $\mathbf{C} = \mathbf{R}(i)$, and, of course, $\mathbf{C} = \mathbf{R}(-i)$ also. Now

$$\text{irr}(i, \mathbf{R}) = \text{irr}(-i, \mathbf{R}) = x^2 + 1,$$

so i and $-i$ are conjugate over $\mathbf{R}$. By Theorem 40.1, the map $\psi_{i,-i} : \mathbf{C} \to \mathbf{C}$ given by $(a + bi)\psi_{i,-i} = a - bi$ is an isomorphism. Thus, if for $a_i \in \mathbf{R}$,

$$f(a + bi) = a_0 + a_1(a + bi) + \cdots + a_n(a + bi)^n = 0,$$

then

$$0 = (f(a + bi))\psi_{i,-i} = a_0 + a_1(a - bi) + \cdots + a_n(a - bi)^n$$
$$= f(a - bi),$$

that is, $f(a - bi) = 0$ also. ∎

Example 40.2 Consider $\mathbf{Q}(\sqrt{2})$ over $\mathbf{Q}$. The zeros of $\text{irr}(\sqrt{2}, \mathbf{Q}) = x^2 - 2$ are $\sqrt{2}$ and $-\sqrt{2}$, so $\sqrt{2}$ and $-\sqrt{2}$ are conjugate over $\mathbf{Q}$. According to Theorem 40.1, the map $\psi_{\sqrt{2},-\sqrt{2}} : \mathbf{Q}(\sqrt{2}) \to \mathbf{Q}(\sqrt{2})$ defined by

$$(a + b\sqrt{2})\psi_{\sqrt{2},-\sqrt{2}} = a - b\sqrt{2}$$

is an isomorphism of $\mathbf{Q}(\sqrt{2})$ onto itself. ∎

40.2 AUTOMORPHISMS AND FIXED FIELDS

As illustrated in the preceding corollary and example, a field may have a nontrivial isomorphism onto itself. *Such maps will be of utmost importance in the work that follows.*

Definition An isomorphism of a field onto itself is an ***automorphism of the field.***

Definition If σ is an isomorphism of a field E into some field, then an element a of E is **left fixed by** σ, if $a\sigma = a$. A collection S of isomorphisms of E **leaves a subfield F of E fixed** if each $a \in F$ is left fixed by every $\sigma \in S$. If $\{\sigma\}$ leaves F fixed, then σ **leaves F fixed.**

It is our purpose to study the structure of an algebraic extension E of a field F by studying the automorphisms of E that leave fixed each element of F. We shall presently show that these automorphisms form a group in a natural way. We can then apply the results of Part I concerning group structure to get information about the structure of our field extension. Thus much of our preceding work is now being brought together. The next three theorems are very easy to prove, but the ideas contained in them form the foundation for everything that follows. These theorems are therefore of great importance to us. They really amount to observations, rather than theorems; it is the *ideas* contained in them that are important. A big step in mathematics does not always consist of proving a *hard* theorem, but may consist of noticing how certain known mathematics may relate to new situations. Here we are bringing group theory into our study of zeros of polynomials. Be sure that you understand the concepts involved. Unlikely as it may seem, they are the key to the solution of our *final goal* in this text.

> *Final goal (to be more precisely stated later): To show that not all zeros of every quintic (degree 5) polynomial $f(x)$ can be expressed in terms of radicals starting with elements in the field of coefficients of $f(x)$.*

If $\{\sigma_i \mid i \in I\}$ is a collection of automorphisms of a field E, the elements of E about which $\{\sigma_i \mid i \in I\}$ gives the least information are those $a \in E$ left fixed by every σ_i for $i \in I$. This first of our three theorems contains almost all that can be said about these fixed elements of E.

Theorem 40.2   *Let $\{\sigma_i \mid i \in I\}$ be a collection of automorphisms of a field E. Then the set $E_{\{\sigma_i\}}$ of all $a \in E$ left fixed by every σ_i for $i \in I$ forms a subfield of E.*

Proof.   If $a\sigma_i = a$ and $b\sigma_i = b$ for all $i \in I$, then

$$(a \pm b)\sigma_i = a\sigma_i \pm b\sigma_i = a \pm b$$

and

$$(ab)\sigma_i = (a\sigma_i)(b\sigma_i) = ab$$

for all $i \in I$. Also, if $b \neq 0$, then

$$(a/b)\sigma_i = (a\sigma_i)/(b\sigma_i) = a/b$$

for all $i \in I$. Since the σ_i are automorphisms, we have

$$0\sigma_i = 0 \quad \text{and} \quad 1\sigma_i = 1$$

for all $i \in I$. Hence $0, 1 \in E_{\{\sigma_i\}}$. Thus $E_{\{\sigma_i\}}$ is a subfield of E. ∎

Definition   The field $E_{\{\sigma_i\}}$ of Theorem 40.2 is the **fixed field of** $\{\sigma_i \mid i \in I\}$. For a single automorphism σ, we shall refer to $E_{\{\sigma\}}$ as the **fixed field of** σ.

Example 40.3   Consider the automorphism $\psi_{\sqrt{2}, -\sqrt{2}}$ of $Q(\sqrt{2})$ given in Example 40.2. For $a, b \in Q$, we have

$$(a + b\sqrt{2})\psi_{\sqrt{2}, -\sqrt{2}} = a - b\sqrt{2}$$

and $a - b\sqrt{2} = a + b\sqrt{2}$ if and only if $b = 0$. Thus the fixed field of $\psi_{\sqrt{2},-\sqrt{2}}$ is $\mathbf{Q}$. ∎

Note that an automorphism of a field E is in particular a one-to-one mapping of E onto E, that is, a *permutation of E*. If σ and τ are automorphisms of E, then the permutation $\sigma\tau$ is again an automorphism of E, since, in general, a compositum of homomorphisms is again a homomorphism. This is how group theory makes its entrance.

Theorem 40.3   *The set of all automorphisms of a field E is a group under function composition.*

Proof.   Multiplication of automorphisms of E is defined by function composition, and is thus associative (it is *permutation multiplication*). The identity permutation $\iota : E \to E$ given by $\alpha\iota = \alpha$ for $\alpha \in E$ is obviously an automorphism of E. If σ is an automorphism, then the permutation σ^{-1} is also obviously an automorphism. Thus all automorphisms of E form a subgroup of S_E, the group of all permutations of E given by Theorem 4.1. ∎

Theorem 40.4   *Let E be a field, and let F be a subfield of E. Then the set $G(E/F)$ of all automorphisms of E leaving F fixed forms a subgroup of the group of all automorphisms of E. Furthermore, $F \leq E_{G(E/F)}$.*

Proof.   For $\sigma, \tau \in G(E/F)$ and $a \in F$, we have

$$a(\sigma\tau) = (a\sigma)\tau = a\tau = a,$$

so $\sigma\tau \in G(E/F)$. Clearly, the identity automorphism ι is in $G(E/F)$. Also, if $a\sigma = a$ for $a \in F$, then $a = a\sigma^{-1}$, so $\sigma \in G(E/F)$ implies that $\sigma^{-1} \in G(E/F)$. Thus $G(E/F)$ is a subgroup of the group of all automorphisms of E.

Since every element of F is left fixed by every element of $G(E/F)$, it follows immediately that the field $E_{G(E/F)}$ of *all* elements of E left fixed by $G(E/F)$ contains F. ∎

Definition   The group $G(E/F)$ of the preceding theorem is the ***group of automorphisms of E leaving F fixed***, or, more briefly, the ***group of E over F***.

Do not think of E/F as denoting a quotient space of some sort, but rather as meaning that E is an extension field of the field F.

The ideas contained in the preceding three theorems are illustrated in the following example. We urge you to study this example carefully.

Example 40.4   Consider the field $\mathbf{Q}(\sqrt{2}, \sqrt{3})$. If we view $\mathbf{Q}(\sqrt{2}, \sqrt{3})$ as $(\mathbf{Q}(\sqrt{3}))(\sqrt{2})$, the basic isomorphism $\psi_{\sqrt{2},-\sqrt{2}}$ of Theorem 40.1 defined by

$$(a + b\sqrt{2})\psi_{\sqrt{2},-\sqrt{2}} = a - b\sqrt{2}$$

for $a, b \in \mathbf{Q}(\sqrt{3})$ is an automorphism of $\mathbf{Q}(\sqrt{2}, \sqrt{3})$, having $\mathbf{Q}(\sqrt{3})$ as fixed field. Similarly, we have the automorphism $\psi_{\sqrt{3}, -\sqrt{3}}$ of $\mathbf{Q}(\sqrt{2}, \sqrt{3})$, having $\mathbf{Q}(\sqrt{2})$ as fixed field. Since the product of two automorphisms is an automorphism, we can consider $\psi_{\sqrt{2}, -\sqrt{2}} \psi_{\sqrt{3}, -\sqrt{3}}$, which **moves** both $\sqrt{2}$ and $\sqrt{3}$, that is, leaves neither number fixed. Let

$$\iota = \text{the identity automorphism,}$$
$$\sigma_1 = \psi_{\sqrt{2}, -\sqrt{2}},$$
$$\sigma_2 = \psi_{\sqrt{3}, -\sqrt{3}}, \text{ and}$$
$$\sigma_3 = \psi_{\sqrt{2}, -\sqrt{2}} \psi_{\sqrt{3}, -\sqrt{3}}.$$

The group of all automorphisms of $\mathbf{Q}(\sqrt{2}, \sqrt{3})$ has a fixed field, by Theorem 40.2. This fixed field must contain $\mathbf{Q}$, since every automorphism of a field leaves 1 and hence the prime subfield fixed. A basis for $\mathbf{Q}(\sqrt{2}, \sqrt{3})$ over $\mathbf{Q}$ is $\{1, \sqrt{2}, \sqrt{3}, \sqrt{6}\}$. Since $\sqrt{2}\sigma_1 = -\sqrt{2}$, $\sqrt{6}\sigma_1 = -\sqrt{6}$ and $\sqrt{3}\sigma_2 = -\sqrt{3}$, we see that $\mathbf{Q}$ is exactly the fixed field of $\{\iota, \sigma_1, \sigma_2, \sigma_3\}$. It is easily seen that $G = \{\iota, \sigma_1, \sigma_2, \sigma_3\}$ is a group under automorphism multiplication (function composition). The group table for G is given in Table 40.1. For example,

$$\sigma_1 \sigma_3 = \psi_{\sqrt{2}, -\sqrt{2}}(\psi_{\sqrt{2}, -\sqrt{2}} \psi_{\sqrt{3}, -\sqrt{3}}) = \psi_{\sqrt{3} - \sqrt{3}} = \sigma_2.$$

Table 40.1

	ι	σ_1	σ_2	σ_3
ι	ι	σ_1	σ_2	σ_3
σ_1	σ_1	ι	σ_3	σ_2
σ_2	σ_2	σ_3	ι	σ_1
σ_3	σ_3	σ_2	σ_1	ι

The group G is isomorphic to the Klein 4-group. We can show that G is the full group $G(\mathbf{Q}(\sqrt{2}, \sqrt{3})/\mathbf{Q})$, for every automorphism τ of $\mathbf{Q}(\sqrt{2}, \sqrt{3})$ maps $\sqrt{2}$ onto either $\pm\sqrt{2}$, by Corollary 1 of Theorem 40.1. Similarly, τ maps $\sqrt{3}$ onto either $\pm\sqrt{3}$. But since $\{1, \sqrt{2}, \sqrt{3}, \sqrt{2}\sqrt{3}\}$ is a basis for $\mathbf{Q}(\sqrt{2}, \sqrt{3})$ over $\mathbf{Q}$, an automorphism of $\mathbf{Q}(\sqrt{2}, \sqrt{3})$ leaving $\mathbf{Q}$ fixed is determined by its values on $\sqrt{2}$ and $\sqrt{3}$. Clearly, $\iota, \sigma_1, \sigma_2$, and σ_3 then give all possible combinations of values on $\sqrt{2}$ and $\sqrt{3}$, and hence are all possible automorphisms of $\mathbf{Q}(\sqrt{2}, \sqrt{3})$.

Note that $G(\mathbf{Q}(\sqrt{2}, \sqrt{3})/\mathbf{Q})$ has order 4, and $[\mathbf{Q}(\sqrt{2}, \sqrt{3}):\mathbf{Q}] = 4$. *This is no accident*, but rather an instance of a quite general situation, as we shall see later. ■

40.3 THE FROBENIUS AUTOMORPHISM

Let F be a finite field. We shall show later that the group of all automorphisms of F is cyclic. Now a cyclic group has by definition a generating element, and it may have several generating elements. For an abstract cyclic group there is no way of distinguishing any one generator as being more important than any other. However, for the cyclic group of all automorphisms of a finite field there is a canonical (natural) generator, the *Frobenius automorphism* (classically, the *Frobenius substitution*). This fact is of considerable importance in some advanced work in algebra. The next theorem exhibits this Frobenius automorphism.

Theorem 40.5   *Let F be a finite field of characteristic p. Then the map $\sigma_p : F \to F$ defined by $a\sigma_p = a^p$ for $a \in F$ is an automorphism, the **Frobenius automorphism**, of F. Also, $F_{\{\sigma_p\}} \simeq \mathbf{Z}_p$.*

Proof.   Let $a, b \in F$. Applying the binomial theorem to $(a + b)^p$, we have

$$(a + b)^p = a^p + (p \cdot 1)a^{p-1}b + \left(\frac{p(p-1)}{2} \cdot 1\right)a^{p-2}b^2$$

$$+ \cdots + (p \cdot 1)ab^{p-1} + b^p$$

$$= a^p + 0a^{p-1}b + 0a^{p-2}b^2 + \cdots + 0ab^{p-1} + b^p$$

$$= a^p + b^p.$$

Thus we have

$$(a + b)\sigma_p = (a + b)^p = a^p + b^p = a\sigma_p + b\sigma_p.$$

Of course,

$$(ab)\sigma_p = (ab)^p = a^p b^p = (a\sigma_p)(b\sigma_p),$$

so σ_p is at least a homomorphism. If $a\sigma_p = 0$, then $a^p = 0$, and $a = 0$, so the kernel of σ_p is $\{0\}$, and σ_p is an isomorphic mapping. Finally, since F is finite, σ_p is onto, by counting. Thus σ_p is an automorphism of F.

The prime field $\mathbf{Z}_p$ must be contained (up to isomorphism) in F, since F is of characteristic p. For $c \in \mathbf{Z}_p$, we have $c\sigma_p = c^p = c$, by Fermat's theorem (see the corollary of Theorem 24.6). Thus the polynomial $x^p - x$ has p zeros in F, namely the elements of $\mathbf{Z}_p$. By Corollary 2 of Theorem 31.1, a polynomial of degree n over a field can have at most n zeros in the field. Since the elements fixed under σ_p are precisely the zeros in F of $x^p - x$, we see that

$$\mathbf{Z}_p = F_{\{\sigma_p\}}. \quad \blacksquare$$

Freshmen in college still sometimes make the error of saying that $(a + b)^n = a^n + b^n$. Here we see that this *freshman exponentiation* $(a + b)^p = a^p + b^p$ with exponent p is actually valid in a field F of characteristic p.

Exercises

40.1 Find all conjugates of each of the given numbers over the given fields.

a) $\sqrt{2}$ over $\mathbf{Q}$

b) $\sqrt{2}$ over $\mathbf{R}$

c) $3 + \sqrt{2}$ over $\mathbf{Q}$

d) $\sqrt{2} - \sqrt{3}$ over $\mathbf{Q}$

e) $\sqrt{2} + i$ over $\mathbf{Q}$

f) $\sqrt{2} + i$ over $\mathbf{R}$

g) $\sqrt{1 + \sqrt{2}}$ over $\mathbf{Q}$

h) $\sqrt{1 + \sqrt{2}}$ over $\mathbf{Q}(\sqrt{2})$

40.2 Consider the field $E = \mathbf{Q}(\sqrt{2}, \sqrt{3}, \sqrt{5})$. In the notation of Theorem 40.1, we have the following basic isomorphisms (which are here automorphisms of E):

$$\psi_{\sqrt{2}, -\sqrt{2}} : (\mathbf{Q}(\sqrt{3}, \sqrt{5}))(\sqrt{2}) \to (\mathbf{Q}(\sqrt{3}, \sqrt{5}))(-\sqrt{2}),$$
$$\psi_{\sqrt{3}, -\sqrt{3}} : (\mathbf{Q}(\sqrt{2}, \sqrt{5}))(\sqrt{3}) \to (\mathbf{Q}(\sqrt{2}, \sqrt{5}))(-\sqrt{3}),$$
$$\psi_{\sqrt{5}, -\sqrt{5}} : (\mathbf{Q}(\sqrt{2}, \sqrt{3}))(\sqrt{5}) \to (\mathbf{Q}(\sqrt{2}, \sqrt{3}))(-\sqrt{5}).$$

For shorter notation, let $\tau_2 = \psi_{\sqrt{2}, -\sqrt{2}}$, $\tau_3 = \psi_{\sqrt{3}, -\sqrt{3}}$, and $\tau_5 = \psi_{\sqrt{5}, -\sqrt{5}}$. Compute each of the following.

a) $\sqrt{3}\tau_2$

b) $(\sqrt{2} + \sqrt{5})\tau_2$

c) $(\sqrt{2} + 3\sqrt{5})(\tau_2\tau_3)$

d) $\dfrac{\sqrt{2} - 3\sqrt{5}}{2\sqrt{3} - \sqrt{2}}(\tau_3\tau_5)$

e) $(\sqrt{2} + \sqrt{45})(\tau_2\tau_3\tau_5{}^2)$

f) $[(\sqrt{2} - \sqrt{3})\tau_5 + \sqrt{30}(\tau_5\tau_2)]\tau_3$

40.3 The fields $\mathbf{Q}(\sqrt{2})$ and $\mathbf{Q}(3 + \sqrt{2})$ are the same, of course. Let $\alpha = 3 + \sqrt{2}$.

a) Find a conjugate $\beta \neq \alpha$ of α over $\mathbf{Q}$.

b) Referring to (a), compare the basic automorphism $\psi_{\sqrt{2}, -\sqrt{2}}$ of $\mathbf{Q}(\sqrt{2})$ with the basic automorphism $\psi_{\alpha, \beta}$.

40.4 Referring to Example 40.4, find the following fixed fields in $E = \mathbf{Q}(\sqrt{2}, \sqrt{3})$.

a) $E_{\{\sigma_1, \sigma_3\}}$

b) $E_{\{\sigma_3\}}$

c) $E_{\{\sigma_2, \sigma_3\}}$

40.5 Referring to Exercise 40.2, find the fixed field of each of the following automorphisms or sets of automorphisms of E.

a) τ_3

b) $\tau_3{}^2$

c) $\{\tau_2, \tau_3\}$

d) $\tau_2\tau_5$

e) $\tau_2\tau_3\tau_5$

f) $\{\tau_2, \tau_3, \tau_5\}$

†**40.6** Let α be algebraic of degree n over F. Show from Corollary 1 of Theorem 40.1 that there are at most n different isomorphisms of $F(\alpha)$ into $\bar{F}$.

40.7 Mark each of the following true or false.

— a) For all α, $\beta \in E$, there is always an automorphism of E mapping α onto β.

— b) For α, β algebraic over a field F, there is always an isomorphism of $F(\alpha)$ onto $F(\beta)$.

— c) For α, β algebraic and conjugate over a field F, there is always an isomorphism of $F(\alpha)$ onto $F(\beta)$.

— d) Every automorphism of every field E leaves fixed every element of the prime subfield of E.

— e) Every automorphism of every field E leaves fixed an infinite number of elements of E.

— f) Every automorphism of every field E leaves fixed at least two elements of E.

— g) Every automorphism of every field E of characteristic 0 leaves fixed an infinite number of elements of E.

— h) All automorphisms of a field E form a group under function composition.

— i) The set of all elements of a field E left fixed by a single automorphism of E forms a subfield of E.

— j) For fields $F \leq E \leq K$, $G(K/E) \leq G(K/F)$.

40.8 Refer to Exercise 40.2 for the following:

a) Show that each of the automorphisms τ_2, τ_3, and τ_5 is of order 2 in $G(E/\mathbf{Q})$. (Remember what is meant by the *order* of an element of a group.)

b) Find the subgroup H of $G(E/\mathbf{Q})$ generated by the elements τ_2, τ_3, and τ_5, and give the group table. [*Hint:* There are eight elements.]

c) Just as was done in Example 40.4, argue that the group H of (b) is the full group $G(E/\mathbf{Q})$.

40.9 Describe the value of the Frobenius automorphism σ_2 on each element of the finite field of four elements given in Example 35.9. Find the fixed field of σ_2.

40.10 Describe the value of the Frobenius automorphism σ_3 on each element of the finite field of nine elements given in Exercise 35.7. Find the fixed field of σ_3.

40.11 Let F be a field of characteristic $p \neq 0$. Give an example to show that the map $\sigma_p : F \to F$ given by $a\sigma_p = a^p$ for $a \in F$ need not be an automorphism in the case that F is infinite. What may go wrong?

40.12 Let $F(\alpha_1, \ldots, \alpha_n)$ be an extension field of F. Show that any automorphism σ of $F(\alpha_1, \ldots, \alpha_n)$ leaving F fixed is completely determined by the n values $\alpha_i\sigma$.

40.13 Let E be an algebraic extension of a field F, and let σ be an automorphism of E leaving F fixed. Let $\alpha \in E$. Show that σ induces a permutation of the set of all zeros of $\mathrm{irr}(\alpha, F)$ that are in E.

40.14 Let E be an algebraic extension of a field F. Let $S = \{\sigma_i \mid i \in I\}$ be a collection of automorphisms of E such that every σ_i leaves each element of F fixed. Show that if S generates the subgroup H of $G(E/F)$, then $E_S = E_H$.

40.15 We saw in the corollary of Theorem 31.4 that the cyclotomic polynomial

$$\Phi_p(x) = \frac{x^p - 1}{x - 1} = x^{p-1} + x^{p-2} + \cdots + x + 1$$

is irreducible over $\mathbf{Q}$ for every prime p. Let ζ be a zero of $\Phi_p(x)$, and consider the field $\mathbf{Q}(\zeta)$.

a) Show that $\zeta, \zeta^2, \ldots, \zeta^{p-1}$ are distinct zeros of $\Phi_p(x)$, and conclude that they are all the zeros of $\Phi_p(x)$.

b) Deduce from Corollary 1 of Theorem 40.1 and part (a) of this exercise that $G(\mathbf{Q}(\zeta)/\mathbf{Q})$ is abelian of order $p - 1$.

c) Show that the fixed field of $G(\mathbf{Q}(\zeta)/\mathbf{Q})$ is $\mathbf{Q}$. [*Hint*: Show that

$$\{\zeta, \zeta^2, \ldots, \zeta^{p-1}\}$$

is a basis for $\mathbf{Q}(\zeta)$ over $\mathbf{Q}$, and consider which linear combinations of $\zeta, \zeta^2, \ldots, \zeta^{p-1}$ are left fixed by all elements of $G(\mathbf{Q}(\zeta)/\mathbf{Q})$.]

40.16 Theorem 40.1 described basic isomorphisms for the case where α and β were conjugate algebraic elements over F. Is there a similar isomorphism of $F(\alpha)$ with $F(\beta)$ in the case that α and β are both transcendental over F?

40.17 Let F be a field, and let x be an indeterminate over F. Determine all automorphisms of $F(x)$ leaving F fixed, by describing their values on x.

40.18 Prove the following sequence of theorems.

a) An automorphism of a field E carries elements that are squares of elements in E onto elements that are squares of elements of E.

b) An automorphism of the field $\mathbf{R}$ of real numbers carries positive numbers onto positive numbers.

c) If σ is an automorphism of $\mathbf{R}$ and $a < b$, where $a, b \in \mathbf{R}$, then $a\sigma < b\sigma$.

d) An automorphism of $\mathbf{R}$ is completely determined by its values on elements of $\mathbf{Q}$.

e) The only automorphism of $\mathbf{R}$ is the identity automorphism.

forty-one

The Isomorphism Extension Theorem

41.1 THE EXTENSION THEOREM

Let us continue studying automorphisms of fields. In this chapter and the next, we shall be concerned with both the existence and the number of automorphisms of a field E.

Suppose that E is an algebraic extension of F and that we want to find some automorphisms of E. We know from Theorem 40.1 that if $\alpha, \beta \in E$ are conjugate over F, then there is an isomorphism $\psi_{\alpha,\beta}$ of $F(\alpha)$ onto $F(\beta)$. Of course, $\alpha, \beta \in E$ implies both $F(\alpha) \leq E$ and $F(\beta) \leq E$. It is natural to wonder whether the domain of definition of $\psi_{\alpha,\beta}$ can be enlarged from $F(\alpha)$ to a bigger field, perhaps all of E, and whether this might perhaps lead to an automorphism of E. A mapping diagram of this situation is shown in Fig. 41.1. Rather than speak of "enlarging the domain of definition of $\psi_{\alpha,\beta}$," it is customary to speak of **"extending the map $\psi_{\alpha,\beta}$ to a map τ,"** which is a mapping of all of E.

Remember that we are always assuming that all algebraic extensions of F under consideration are contained in a fixed algebraic closure $\bar{F}$ of F. The Isomorphism Extension Theorem shows that the mapping $\psi_{\alpha,\beta}$ can indeed

Fig. 41.1

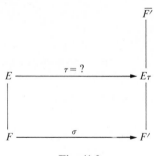

Fig. 41.2

always be extended to an *isomorphism* of E into $\bar{F}$. Whether this extension gives an *automorphism* of E, that is, maps E onto itself, is a question we shall study in Chapter 42. Thus this extension theorem, used in conjunction with our basic isomorphisms $\psi_{\alpha,\beta}$, will guarantee the existence of lots of *isomorphic mappings*, at least, for many fields. This theorem may well be the only extension theorem you have encountered. Such theorems are very important in mathematics, particularly in algebraic and topological situations.

Let us take a more general look at this situation. Suppose that E is an algebraic extension of a field F and that we have an isomorphism σ of F onto a field F'. Let $\bar{F}'$ be an algebraic closure of F'. We would like to extend σ to an isomorphism τ of E into $\bar{F}'$. This situation is shown in Fig. 41.2. Naively, we pick $\alpha \in E$ but not in F, and try to extend σ to $F(\alpha)$. If

$$p(x) = \mathrm{irr}(\alpha, F) = a_0 + a_1 x + \cdots + a_n x^n,$$

let β be a zero in $\bar{F}'$ of

$$q(x) = a_0 \sigma + (a_1 \sigma)x + \cdots + (a_n \sigma)x^n.$$

Here $q(x) \in F'[x]$. Since σ is an isomorphism, it is clear that $q(x)$ is irreducible in $F'[x]$. It is also pretty clear that $F(\alpha)$ can be mapped isomorphically onto $F''(\beta)$ by a map extending σ and mapping α onto β. (This isn't quite Theorem 40.1, but it is close to it; a few elements have been renamed by the isomorphism σ.) If $F(\alpha) = E$, we are done. If $F(\alpha) \neq E$, we have to find another element in E not in $F(\alpha)$ and continue the process. It is a situation very much like that in the construction of an algebraic closure $\bar{F}$ of a field F. Again the trouble is that, in general, where E is not a finite extension, the process may have to be repeated a (possibly large) infinite number of times, so we need Zorn's lemma to handle it. For this reason, we give the general proof of Theorem 41.1 under a starred heading at the end of this chapter.

Theorem 41.1 (Isomorphism Extension Theorem)   *Let E be an algebraic extension of a field F. Let σ be an isomorphism of F onto a field F'. Let*

$\bar{F}'$ *be an algebraic closure of* F'. *Then* σ *can be extended to an isomorphism* τ *of* E *into* $\bar{F}'$ *such that* $a\tau = a\sigma$ *for all* $a \in F$.

We give as a corollary the existence of an extension of one of our basic isomorphisms $\psi_{\alpha,\beta}$, as discussed at the start of this section.

Corollary 1 *If* $E \leq \bar{F}$ *is an algebraic extension of* F *and* $\alpha, \beta \in E$ *are conjugate over* F, *then the basic isomorphism* $\psi_{\alpha,\beta} : F(\alpha) \to F(\beta)$, *given by Theorem 40.1, can be extended to an isomorphism of* E *into* $\bar{F}$.

Proof. Proof of this corollary is immediate from Theorem 41.1 if in the statement of the theorem we replace F by $F(\alpha)$, F' by $F(\beta)$, and $\bar{F}'$ by $\bar{F}$. ∎

As another corollary, we can show, as we promised earlier, that an algebraic closure of F is unique, up to an isomorphism leaving F fixed.

Corollary 2 *Let* $\bar{F}$ *and* $\bar{F}'$ *be two algebraic closures of* $\bar{F}$. *Then* $\bar{F}$ *is isomorphic to* $\bar{F}'$ *under an isomorphism leaving each element of* F *fixed.*

Proof. By Theorem 41.1, the identity isomorphism of F onto F can be extended to an isomorphism τ mapping $\bar{F}$ into $\bar{F}'$ that leaves F fixed (see Fig. 41.3). We need only show τ is onto $\bar{F}'$. But by Theorem 41.1, the map $\tau^{-1} : \bar{F}\tau \to \bar{F}$ can be extended to an isomorphism of $\bar{F}'$ into $\bar{F}$. Since τ^{-1} is already onto $\bar{F}$, we must have $\bar{F}\tau = \bar{F}'$. ∎

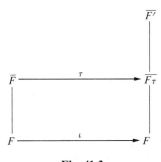

Fig. 41.3

41.2 THE INDEX OF A FIELD EXTENSION

Having discussed the question of *existence*, we turn now to the question of *how many*. For a *finite* extension E of a field F, we would like to count how many isomorphisms there are of E into $\bar{F}$ that leave F fixed. We shall show that there are only a finite number of such isomorphisms. Since every automorphism in $G(E/F)$ is such an isomorphism, a count of these isomorphisms will include all these automorphisms. Example 40.4 showed that $G(Q(\sqrt{2}, \sqrt{3})/Q)$ has four elements, and that $4 = [Q(\sqrt{2}, \sqrt{3}):Q]$. While such an equality is not always true, it is true in a very important case. The next theorem takes the first big step in proving this. We state the theorem in more general terms than we shall need, but it does not make the proof any harder.

Theorem 41.2 *Let E be a finite extension of a field F. Let σ be an isomorphism of F onto a field F', and let $\overline{F'}$ be an algebraic closure of F'. Then the number of extensions of σ to an isomorphism τ of E into $\overline{F'}$ is finite, and independent of F', $\overline{F'}$, and σ. That is, this number of extensions is completely determined by the two fields E and F; it is intrinsic to them.*

Proof. The diagram in Fig. 41.4 may help you to follow the construction that we are about to make. This diagram is constructed in the following way.

Fig. 41.4

Consider two isomorphisms

$$\sigma_1 : F \xrightarrow{\text{onto}} F'_1, \qquad \sigma_2 : F \xrightarrow{\text{onto}} F'_2,$$

where $\overline{F'_1}$ and $\overline{F'_2}$ are algebraic closures of F'_1 and F'_2, respectively. Now $\sigma_1^{-1}\sigma_2$ is an isomorphism of F'_1 onto F'_2. Then by Theorem 41.1 and its second corollary, there is an isomorphism

$$\lambda : \overline{F'_1} \xrightarrow{\text{onto}} \overline{F'_2}$$

extending this isomorphism $\sigma_1^{-1}\sigma_2 : F' \xrightarrow{\text{onto}} F'_2$. Referring to Fig. 41.4, corresponding to each $\tau_1 : E \to \overline{F'_1}$ that extends σ_1, one obtains an isomorphism $\tau_2 : E \to \overline{F'_2}$, by starting at E and going first to the left, then up, and then to the right. Written algebraically,

$$\alpha\tau_2 = \alpha\tau_1\lambda$$

for $\alpha \in E$. Clearly τ_2 extends σ_2. The fact that we could have *started* with τ_2 and recovered τ_1 by defining

$$\alpha\tau_1 = \alpha\tau_2\lambda^{-1},$$

that is, by chasing the other way around the diagram, shows that the correspondence between $\tau_1 : E \to F_1'$ and $\tau_2 : E \to F_2'$ is one to one. In view of this one-to-one correspondence, the number of τ extending σ is independent of F', $\bar{F}'$, and σ.

That the number of mappings extending σ is finite follows from the fact that since E is a finite extension of F, $E = F(\alpha_1, \ldots, \alpha_n)$ for some $\alpha_1, \ldots, \alpha_n$ in E, by Theorem 38.3. There are only a finite number of possible candidates for the images $\alpha_i\tau$ in F', for if

$$\text{irr } (\alpha_i, F) = a_{i0} + a_{i1}x + \cdots + a_{im_i}x^{m_i},$$

where $a_{ik} \in F$, then $\alpha_i\tau$ must be one of the zeros in $\bar{F}'$ of

$$\left[a_{i0}\sigma + (a_{i1}\sigma)x + \cdots + (a_{im_i}\sigma)x^{m_i} \right] \in F'[x]. \quad \blacksquare$$

Definition   Let E be a finite extension of a field F. The number of isomorphisms of E into $\bar{F}$ leaving F fixed is the **index** $\{E : F\}$ **of** E **over** F.

Corollary   *If $F \le E \le K$, where K is a finite extension field of the field F, then $\{K : F\} = \{K : E\} \{E : F\}$.*

Proof.   It follows from Theorem 41.2 that each of the $\{E : F\}$ isomorphisms $\tau_i : E \to \bar{F}$ leaving F fixed has $\{K : E\}$ extensions to an isomorphism of K into $\bar{F}$. $\blacksquare$

The preceding corollary was really the main thing we were after. Note that it counts something. *Never underestimate a result that counts something,* even if it is only called a "corollary."

We shall show in Chapter 43 that unless F is an infinite field of characteristic $p \ne 0$, we always have $[E : F] = \{E : F\}$, for every finite extension field E of F. For the case $E = F(\alpha)$, the $\{F(\alpha) : F\}$ extensions of the identity map $\iota : F \to F$ to maps of $F(\alpha)$ into $\bar{F}$ are given by the basic isomorphisms $\psi_{\alpha,\beta}$, for each conjugate β in $\bar{F}$ of α over F. Thus if irr(α, F) has n *distinct* zeros in $\bar{F}$, we have $\{E : F\} = n$. We shall show later that unless F is infinite and of characteristic $p \ne 0$, the number of distinct zeros of irr(α, F) is deg$(\alpha, F) = [F(\alpha) : F]$.

Example 41.1   Consider $E = \mathbf{Q}(\sqrt{2}, \sqrt{3})$ over $\mathbf{Q}$, as in Example 40.4. Our work in Example 40.4 shows that $\{E : \mathbf{Q}\} = [E : \mathbf{Q}] = 4$. Also, $\{E : \mathbf{Q}(\sqrt{2})\} = 2$, and $\{\mathbf{Q}(\sqrt{2}) : \mathbf{Q}\} = 2$, so

$$4 = \{E : \mathbf{Q}\} = \{E : \mathbf{Q}(\sqrt{2})\} \{\mathbf{Q}(\sqrt{2}) : \mathbf{Q}\} = (2)(2).$$

This illustrates the corollary of Theorem 41.2. $\blacksquare$

*41.3 PROOF OF THE EXTENSION THEOREM

We restate the extension theorem.

Theorem 41.1 (Isomorphism Extension Theorem)   *Let E be an algebraic extension of a field F. Let σ be an isomorphism of F onto a field F'. Let $\overline{F'}$ be an algebraic closure of F'. Then σ can be extended to an isomorphism τ of E into $\overline{F'}$ such that $a\tau = a\sigma$ for $a \in F$.*

Proof.   Consider all pairs (L, λ), where L is a field such that $F \leq L \leq E$ and λ is an isomorphism of L into $\overline{F'}$ such that $a\lambda = a\sigma$ for $a \in F$. The set S of such pairs (L, λ) is nonempty, since (F, σ) is such a pair. Define a partial ordering on S by $(L_1, \lambda_1) \leq (L_2, \lambda_2)$, if $L_1 \leq L_2$ and $a\lambda_2 = a\lambda_1$ for $a \in L_1$. It is easily checked that this relation $\leq$ does give a partial ordering of S.

Let $T = \{(H_i, \lambda_i) \mid i \in I\}$ be a chain of S. We claim that $H = \bigcup_{i \in I} H_i$ is a subfield of E. Let $a, b \in H$, where $a \in H_1$ and $b \in H_2$; then either $H_1 \leq H_2$ or $H_2 \leq H_1$, since T is a chain. If, say, $H_1 \leq H_2$, then $a, b \in H_2$, so $a \pm b$, ab, and a/b for $b \neq 0$ are all in H_2 and hence in H. Since for each $i \in I$, $F \subseteq H_i \subseteq E$, we have $F \subseteq H \subseteq E$. Thus H is a subfield of E.

Define $\lambda : H \to \overline{F'}$ as follows. Let $c \in H$. Then $c \in H_i$ for some $i \in I$, and let

$$c\lambda = c\lambda_i.$$

The map λ is well defined, because if $c \in H_1$ and $c \in H_2$, then either $(H_1, \lambda_1) \leq (H_2, \lambda_2)$ or $(H_2, \lambda_2) \leq (H_1, \lambda_1)$, since T is a chain. In either case, $c\lambda_1 = c\lambda_2$. We claim that λ is an isomorphism of H into $\overline{F'}$. If $a, b \in H$, then there is an H_i such that $a, b \in H_i$, and

$$(a + b)\lambda = (a + b)\lambda_i = a\lambda_i + b\lambda_i = a\lambda + b\lambda.$$

Similarly,

$$(ab)\lambda = (ab)\lambda_i = (a\lambda_i)(b\lambda_i) = (a\lambda)(b\lambda).$$

If $a\lambda = 0$, then $a \in H_i$ for some i implies that $a\lambda_i = 0$, so $a = 0$. Therefore, λ is an isomorphism. Thus $(H, \lambda) \in S$, and it is clear from our definitions of H and λ that (H, λ) is an upper bound for T.

We have shown that every chain of S has an upper bound in S, so the hypotheses of Zorn's lemma are satisfied. Hence there exists a maximal element (K, τ) of S. Let $K\tau = K'$, where $K' \leq \overline{F'}$. Now if $K \neq E$, let $\alpha \in E$ but $\alpha \notin K$. Now α is algebraic over F, so α is algebraic over K. Also, let $p(x) = \mathrm{irr}(\alpha, K)$. Let ψ_α be the canonical isomorphism

$$\psi_\alpha : K[x]/\langle p(x) \rangle \to K(\alpha),$$

corresponding to the basic homomorphism $\phi_\alpha: K[x] \to K(\alpha)$. If

$$p(x) = a_0 + a_1 x + \cdots + a_n x^n,$$

consider

$$q(x) = a_0\tau + (a_1\tau)x + \cdots + (a_n\tau)x^n$$

in $K'[x]$. Obviously, since τ is an isomorphism, $q(x)$ is irreducible in $K'[x]$. Since $K' \leq \bar{F}'$, there is a zero α' of $q(x)$ in $\bar{F}'$. Let

$$\psi_{\alpha'}: K'[x]/\langle q(x)\rangle \to K'(\alpha')$$

be the isomorphism analogous to ψ_α. Finally, let

$$\bar{\tau}: K[x]/\langle p(x)\rangle \to K'[x]/\langle q(x)\rangle$$

be the obvious isomorphism extending τ on K and mapping $x + \langle p(x)\rangle$ onto $x + \langle q(x)\rangle$. (See Fig. 41.5.) Then the composition of maps

$$(\psi_\alpha)^{-1}\bar{\tau}\psi_{\alpha'}: K(\alpha) \to K'(\alpha')$$

is an isomorphism of $K(\alpha)$ into $\bar{F}'$. Clearly, $(K, \tau) < (K(\alpha), (\psi_\alpha)^{-1}\bar{\tau}\psi_{\alpha'})$, which contradicts that (K, τ) is maximal. Therefore we must have had $K = E$. ∎

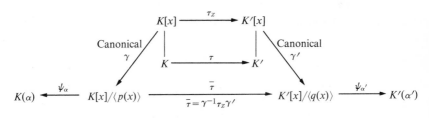

Fig. 41.5

Unfortunately, you probably consider this last proof to be hopelessly complicated. To professional algebraists, this construction is so standard that they usually just write down something like "the proof follows by Zorn's lemma" as soon as they have defined the partial ordering $\leq$ on the set S. We wrote out the proof in hideous detail, since it may be only the second proof you have seen involving Zorn's lemma.

Exercises

41.1 Let $E = \mathbf{Q}(\sqrt{2}, \sqrt{3}, \sqrt{5})$. For each isomorphic mapping of a subfield of E given below, give all extensions of the mapping to a isomorphic mapping of E into $\bar{\mathbf{Q}}$. Describe the extensions by giving values on the generating set $\{\sqrt{2}, \sqrt{3}, \sqrt{5}\}$ for E over $\mathbf{Q}$.

a) $\iota: \mathbf{Q}(\sqrt{2}, \sqrt{15}) \to \mathbf{Q}(\sqrt{2}, \sqrt{15})$, where ι is the identity map

b) $\sigma: \mathbf{Q}(\sqrt{2}, \sqrt{15}) \to \mathbf{Q}(\sqrt{2}, \sqrt{15})$ where $\sqrt{2}\sigma = \sqrt{2}$ and $\sqrt{15}\sigma = -\sqrt{15}$

c) $\psi_{\sqrt{30}, -\sqrt{30}}: \mathbf{Q}(\sqrt{30}) \to \mathbf{Q}(\sqrt{30})$

41.2 It is a fact, which you can verify by cubing, that the zeros of $x^3 - 2$ in $\mathbf{Q}$ are

$$\alpha_1 = \sqrt[3]{2}, \qquad \alpha_2 = \sqrt[3]{2}\,\frac{-1 + i\sqrt{3}}{2}, \qquad \text{and} \qquad \alpha_3 = \sqrt[3]{2}\,\frac{-1 - i\sqrt{3}}{2},$$

where $\sqrt[3]{2}$, as usual, is the real cube root of 2.

a) Describe all extensions of the identity map of $\mathbf{Q}$ to an isomorphism mapping $\mathbf{Q}(\sqrt[3]{2})$ into $\overline{\mathbf{Q}}$.

b) Describe all extensions of the identity map of $\mathbf{Q}$ to an isomorphism mapping $\mathbf{Q}(\sqrt[3]{2}, \sqrt{3})$ into $\overline{\mathbf{Q}}$.

c) Describe all extensions of the automorphism $\psi_{\sqrt{3}, -\sqrt{3}}$ of $\mathbf{Q}(\sqrt{3})$ to an isomorphism mapping $\mathbf{Q}(i, \sqrt{3}, \sqrt[3]{2})$ into $\overline{\mathbf{Q}}$.

41.3 Let σ be the automorphism of $\mathbf{Q}(\pi)$ that maps π onto $-\pi$.

a) Describe the fixed field of σ.

b) Describe all extensions of σ to an isomorphism mapping the field $\mathbf{Q}(\sqrt{\pi})$ into $\overline{\mathbf{Q}(\pi)}$.

†**41.4** Let K be an algebraically closed field. Show that every isomorphism σ of K into itself such that K is algebraic over $K\sigma$ is an automorphism of K, that is, is an onto map. [*Hint:* Apply Theorem 41.1 to σ^{-1}.]

41.5 Mark each of the following true or false.

— a) Let $F(\alpha)$ be any simple extension of a field F. Then every isomorphism of F into $\overline{F}$ has an extension to an isomorphism of $F(\alpha)$ into $\overline{F}$.

— b) Let $F(\alpha)$ be any simple algebraic extension of a field F. Then every isomorphism of F into $\overline{F}$ has an extension to an isomorphism of $F(\alpha)$ into $\overline{F}$.

— c) An isomorphism of F into $\overline{F}$ has the same number of extensions to each simple algebraic extension of F.

— d) Algebraic closures of isomorphic fields are always isomorphic.

— e) Algebraic closures of fields that are not isomorphic are never isomorphic.

— f) Any algebraic closure of $\mathbf{Q}(\sqrt{2})$ is isomorphic to any algebraic closure of $\mathbf{Q}(\sqrt{17})$.

— g) The index of a finite extension E over a field F is finite.

— h) The index behaves multiplicatively with respect to finite towers of finite extensions of fields.

— i) Our remarks prior to the statement of Theorem 41.1 in Section 41.1 essentially constitute a proof of this theorem for a finite extension E over F.

— j) Corollary 2 of Theorem 41.1 shows that $\mathbf{C}$ is isomorphic to $\overline{\mathbf{Q}}$.

41.6 Let E be an algebraic extension of a field F. Show that every isomorphism of E into $\overline{F}$ leaving F fixed can be extended to an automorphism of $\overline{F}$.

41.7 Prove that if E is an algebraic extension of a field F, then two algebraic closures $\overline{F}$ and $\overline{E}$ of F and E, respectively, are isomorphic.

41.8 Prove that the algebraic closure of $\mathbf{Q}(\sqrt{\pi})$ in $\mathbf{C}$ is isomorphic to any algebraic closure of $\bar{\mathbf{Q}}(x)$, where $\bar{\mathbf{Q}}$ is the field of algebraic numbers and x is an indeterminate.

41.9 Prove that if E is a finite extension of a field F, then $\{E:F\} \leq [E:F]$. [*Hint:* The remarks preceding Example 41.1 essentially showed this for a simple algebraic extension $F(\alpha)$ of F. Use the fact that a finite extension is a tower of simple extensions, together with the multiplicative properties of the index and degree.]

Splitting
Fields

We are going to be interested chiefly in *automorphisms* of a field E, rather than mere isomorphic mappings of E. You see, it is the *automorphisms* of a field that form a group. We wonder whether for some extension field E of a field F, *every* isomorphic mapping of E into $\bar{F}$ leaving F fixed is actually an automorphism of E.

Suppose E is an algebraic extension of a field F. If $\alpha \in E$ and $\beta \in \bar{F}$ is a conjugate of α over F, then there is a basic isomorphism

$$\psi_{\alpha,\beta}: F(\alpha) \to F(\beta).$$

By Corollary 1 of Theorem 41.1, $\psi_{\alpha,\beta}$ can be extended to an isomorphic mapping of E into $\bar{F}$. Now if $\beta \notin E$, such an isomorphic mapping of E can't be an automorphism of E. *Thus, if an algebraic extension E of a field F is such that all its isomorphic mappings into $\bar{F}$ leaving F fixed are actually automorphisms of E, then for every $\alpha \in E$, all conjugates of α over F must be in E also.* This observation seemed to come very easily. We point out that we used a lot of power, namely the existence of the basic isomorphisms and the Isomorphism Extension Theorem.

These ideas suggest the formulation of the following definition.

Definition   Let F be a field with algebraic closure $\bar{F}$. Let $\{f_i(x)|i \in I\}$ be a collection of polynomials in $F[x]$. A field $E \leq \bar{F}$ is the ***splitting field*** ***of*** $\{f_i(x)|i \in I\}$ ***over*** F if E is the smallest subfield of $\bar{F}$ containing F and all the zeros in $\bar{F}$ of each of the $f_i(x)$ for $i \in I$. A field $K \leq \bar{F}$ is a ***splitting field over*** F if it is the splitting field of some set of polynomials in $F[x]$.

For one polynomial $f(x) \in F[x]$, we shall often refer to the splitting field of $\{f(x)\}$ over F as the **splitting field of** $f(x)$ **over** F. It is clear that the splitting field of $\{f_i(x)|i \in I\}$ over F in $\bar{F}$ is the intersection of all subfields of $\bar{F}$ containing F and all zeros in $\bar{F}$ of each $f_i(x)$ for $i \in I$. Thus such a splitting field surely does exist.

We now show that splitting fields over F are precisely those fields $E \leq \bar{F}$ with the property that all isomorphic mappings of E into $\bar{F}$ leaving F fixed are automorphisms of E. This will be a corollary of the next theorem. *Once more, we are characterizing a concept in terms of mappings.* Remember, we are always assuming that all algebraic extensions of a field F under consideration are in one fixed algebraic closure $\bar{F}$ of F.

Theorem 42.1   *A field E, where $F \leq E \leq \bar{F}$, is a splitting field over F if and only if every automorphism of $\bar{F}$ leaving F fixed maps E onto itself and thus induces an automorphism of E leaving F fixed.*

Proof.   Let E be a splitting field over F in $\bar{F}$ of $\{f_i(x)|i \in I\}$, and let σ be an automorphism of $\bar{F}$ leaving F fixed. Let $\{\alpha_j|j \in J\}$ be the collection of all zeros in $\bar{F}$ of all the $f_i(x)$ for $i \in I$. Now our previous work shows that for a fixed α_j, $F(\alpha_j)$ has as elements all expressions of the form

$$g(\alpha_j) = a_0 + a_1\alpha_j + \cdots + a_{n_j-1}\alpha_j^{n_j-1},$$

where n_j is the degree of $\mathrm{irr}(\alpha_j, F)$ and $a_k \in F$. Consider the set S of all *finite* sums of *finite* products of elements of the form $g(\alpha_j)$ for all $j \in J$. The set S is a subset of E obviously closed under addition and multiplication and containing $0, 1$ and the additive inverse of each element. Since each element of S is in some $F(\alpha_{j_1}, \ldots, \alpha_{j_r}) \subseteq S$, we see that S also contains the multiplicative inverse of each nonzero element. Thus S is a subfield of E containing all α_j for $j \in J$. By definition of the splitting field E of $\{f_i(x)|i \in I\}$, we see that we must have $S = E$. All this work was just to show that $\{\alpha_j|j \in J\}$ *generates* E over F, in the sense of taking *finite* sums and *finite* products. Knowing this, we see immediately that the value of σ on any element of E is completely determined by the values $\alpha_j\sigma$. But by Corollary 1 of Theorem 40.1, $\alpha_j\sigma$ must also be a zero of $\mathrm{irr}(\alpha_j, F)$. By Theorem 35.3, $\mathrm{irr}(\alpha_j, F)$

divides the $f_i(x)$ for which $f_i(\alpha_j) = 0$, so $\alpha_j\sigma \in E$ also. Thus σ maps E into E isomorphically. However, the same is true of the automorphism σ^{-1} of $\bar{F}$. Since for $\beta \in E$,

$$\beta = (\beta\sigma^{-1})\sigma,$$

we see that σ maps E onto E, and thus induces an automorphism of E.

Suppose, conversely, that every automorphism of $\bar{F}$ leaving F fixed induces an automorphism of E. Let $g(x)$ be an *irreducible* polynomial in $F[x]$ having a zero α in E. If β is any zero of $g(x)$ in $\bar{F}$, then by Theorem 40.1 there is a basic isomorphism $\psi_{\alpha,\beta}$ of $F(\alpha)$ onto $F(\beta)$ leaving F fixed. By Theorem 41.1, $\psi_{\alpha,\beta}$ can be extended to an isomorphism τ of $\bar{F}$ into $\bar{F}$. But then

$$\tau^{-1}:\bar{F}\tau \to \bar{F}$$

can be extended to an isomorphism mapping $\bar{F}$ into $\bar{F}$. Since the image of τ^{-1} is already all of $\bar{F}$, we see that τ must have been onto $\bar{F}$, so τ is an automorphism of $\bar{F}$ leaving F fixed. Then by assumption, τ induces an automorphism of E, so $\alpha\tau = \beta$ is in E. We have shown that if $g(x)$ is an irreducible polynomial in $F[x]$ having one zero in E, then all zeros of $g(x)$ in $\bar{F}$ are in E. Hence if $\{g_k(x)\}$ is the set of *all* irreducible polynomials in $F[x]$ having a zero in E, then E is the splitting field of $\{g_k(x)\}$. ∎

Definition Let E be an extension field of a field F. A polynomial $f(x) \in F[x]$ **splits in** E if it factors into a product of linear factors in $E[x]$.

Corollary 1 *If $E \leq \bar{F}$ is a splitting field over F, then every irreducible polynomial in $F[x]$ having a zero in E splits in E.*

Proof. If E is a splitting field over F in $\bar{F}$, then every automorphism of $\bar{F}$ induces an automorphism of E. The second half of the proof of Theorem 42.1 showed precisely that E is also the splitting field over F of the set $\{g_k(x)\}$ of *all* irreducible polynomials in $F[x]$ having a zero in E. Thus an irreducible polynomial $f(x)$ of $F[x]$ having a zero in E has all its zeros in $\bar{F}$ in E. Therefore, its factorization into linear factors in $\bar{F}[x]$, given by Theorem 38.5 actually takes place in $E[x]$, so $f(x)$ splits in E. ∎

Corollary 2 *If $E \leq \bar{F}$ is a splitting field over F, then every isomorphic mapping of E into $\bar{F}$ leaving F fixed is actually an automorphism of E. In particular, if E is a splitting field of finite degree over F, then*

$$\{E:F\} = |G(E/F)|.$$

Proof. Every isomorphism σ mapping E into $\bar{F}$ leaving F fixed can be extended to an automorphism τ of $\bar{F}$, by Theorem 41.1, together with the onto argument of the second half of the proof of Theorem 42.1. If E is a splitting field over F, then by Theorem 42.1 τ restricted to E, that is σ, is an automorphism of E. Thus for a splitting field E over F, every isomorphic mapping of E into $\bar{F}$ leaving F fixed is an automorphism of E.

The equation $\{E:F\} = |G(E/F)|$ then follows immediately for a splitting field E of finite degree over F, since $\{E:F\}$ was defined as the number of different isomorphic mappings of E into $\bar{F}$ leaving F fixed. ∎

Example 42.1 Obviously, $\mathbf{Q}(\sqrt{2}, \sqrt{3})$ is the splitting field of

$$\{x^2 - 2, x^2 - 3\}$$

over $\mathbf{Q}$. Example 40.4 showed that the mappings ι, σ_1, σ_2, and σ_3 are all the automorphisms of $\mathbf{Q}(\sqrt{2}, \sqrt{3})$ leaving $\mathbf{Q}$ fixed. (Actually, since every automorphism of a field must leave the prime subfield fixed, we see that these are the only automorphisms of $\mathbf{Q}(\sqrt{2}, \sqrt{3})$.) Then

$$\{\mathbf{Q}(\sqrt{2}, \sqrt{3}):\mathbf{Q}\} = |G(\mathbf{Q}(\sqrt{2}, \sqrt{3})/\mathbf{Q})| = 4,$$

illustrating Corollary 2. ∎

We wish to determine conditions under which

$$|G(E/F)| = \{E:F\} = [E:F]$$

for finite extensions E of F. This is our next topic. We shall show in the following section that this equation always holds when E is a splitting field over a field F of characteristic 0 or when F is a finite field. This equation need not be true when F is an infinite field of characteristic $p \neq 0$.

Example 42.2 Let $\sqrt[3]{2}$ be the real cube root of 2, as usual. Now $x^3 - 2$ does not split in $\mathbf{Q}(\sqrt[3]{2})$, for $\mathbf{Q}(\sqrt[3]{2}) < \mathbf{R}$, and only one zero of $x^3 - 2$ is real. Thus $x^3 - 2$ factors in $(\mathbf{Q}(\sqrt[3]{2}))[x]$ into a linear factor $x - \sqrt[3]{2}$ and an irreducible quadratic factor. The splitting field E of $x^3 - 2$ over $\mathbf{Q}$ is therefore of degree 2 over $\mathbf{Q}(\sqrt[3]{2})$. Then

$$[E:\mathbf{Q}] = [E:\mathbf{Q}(\sqrt[3]{2})][\mathbf{Q}(\sqrt[3]{2}):\mathbf{Q}] = (2)(3) = 6.$$

We have shown that the splitting field over $\mathbf{Q}$ of $x^3 - 2$ is of degree 6 over $\mathbf{Q}$.

You can verify by cubing that

$$\sqrt[3]{2}\,\frac{-1 + i\sqrt{3}}{2} \quad \text{and} \quad \sqrt[3]{2}\,\frac{-1 - i\sqrt{3}}{2}$$

are the other zeros of $x^3 - 2$ in $\mathbf{C}$. Thus the splitting field E of $x^3 - 2$ over $\mathbf{Q}$ is $\mathbf{Q}(\sqrt[3]{2}, i\sqrt{3})$. (This is *not* the same field as $\mathbf{Q}(\sqrt[3]{2}, i, \sqrt{3})$, which is of degree 12 over $\mathbf{Q}$.) Further study of this interesting example is left to the exercises (see Exercises 42.3, 42.8, 42.12, and 42.14). ∎

Exercises

42.1 For each of the given polynomials in $Q[x]$, find the degree over Q of the splitting field over Q of the polynomial.

a) $x^2 + 3$ b) $x^4 - 1$ c) $(x^2 - 2)(x^2 - 3)$

d) $x^3 - 3$ e) $x^3 - 1$ f) $(x^2 - 2)(x^3 - 2)$

42.2 Let $f(x)$ be a polynomial in $F[x]$ of degree n. Let $E \leq \bar{F}$ be the splitting field of $f(x)$ over F in $\bar{F}$. What bounds can be put on $[E:F]$?

42.3 Refer to Example 42.2 to answer the following questions.

a) What is the order of $G(Q(\sqrt[3]{2})/Q)$?

b) What is the order of $G(Q(\sqrt[3]{2}, i\sqrt{3})/Q)$?

c) What is the order of $G(Q(\sqrt[3]{2}, i\sqrt{3})/Q(\sqrt[3]{2}))$?

42.4 Let α be a zero of $x^3 + x^2 + 1$ over Z_2. Show that $x^3 + x^2 + 1$ splits in $Z_2(\alpha)$. [*Hint:* There are eight elements in $Z_2(\alpha)$. Exhibit two more zeros of $x^3 + x^2 + 1$, in addition to α, among these eight elements.]

†**42.5** Show that if a finite extension E of a field F is a splitting field over F, then E is a splitting field of one polynomial in $F[x]$.

42.6 Mark each of the following true or false.

— a) Let α, $\beta \in E$, where $E \leq \bar{F}$ is a splitting field over F. Then there exists an automorphism of E leaving F fixed and mapping α onto β if and only if irr$(\alpha, F) =$ irr(β, F).

— b) R is a splitting field over Q.

— c) R is a splitting field over R.

— d) C is a splitting field over R.

— e) $Q(i)$ is a splitting field over Q.

— f) $Q(\pi)$ is a splitting field over $Q(\pi^2)$.

— g) For every splitting field E over F, where $E \leq \bar{F}$, every isomorphic mapping of E is an automorphism of E.

— h) For every splitting field E over F, where $E \leq \bar{F}$, every isomorphism mapping E into $\bar{F}$ is an automorphism of E.

— i) For every splitting field E over F, where $E \leq \bar{F}$, every isomorphism mapping E into $\bar{F}$ and leaving F fixed is an automorphism of E.

— j) Every algebraic closure $\bar{F}$ of a field F is a splitting field over F.

42.7 Show by an example that Corollary 1 of Theorem 42.1 is no longer true if the word *irreducible* is deleted.

42.8 a) Is $|G(E/F)|$ multiplicative for finite towers of finite extensions, that is, is

$$|G(K/F)| = |G(K/E)||G(E/F)| \qquad \text{for} \quad F \leq E \leq K \leq \bar{F}?$$

Why? [*Hint:* Use Exercise 42.3.]

b) Is $|G(E/F)|$ multiplicative for finite towers of finite extensions, each of which is a splitting field over the bottom field? Why?

42.9 Show that if $[E:F] = 2$, then E is a splitting field over F.

42.10 Show that for $F \leq E \leq \bar{F}$, E is a splitting field over F if and only if E contains all conjugates over F in $\bar{F}$ of each of its elements.

42.11 Show that $\mathbf{Q}(\sqrt[3]{2})$ has only the identity automorphism.

42.12 Referring to Example 42.2, show that

$$G(\mathbf{Q}(\sqrt[3]{2}, i\sqrt{3})/\mathbf{Q}(i\sqrt{3})) \simeq \langle \mathbf{Z}_3, + \rangle.$$

42.13 a) Show that an automorphism of a splitting field E over F of a polynomial $f(x) \in F[x]$ permutes the zeros of $f(x)$ in E.

b) Show that an automorphism of a splitting field E over F of a polynomial $f(x) \in F[x]$ is completely determined by the permutation of the zeros of $f(x)$ in E given in (a).

c) Show that if E is a splitting field over F of a polynomial $f(x) \in F[x]$, then $G(E/F)$ can be viewed in a natural way as a certain group of permutations.

42.14 Let E be the splitting field of $x^3 - 2$ over $\mathbf{Q}$, as in Example 42.2.

a) What is the order of $G(E/\mathbf{Q})$? [*Hint:* Use Corollary 2 of Theorem 42.2 and the corollary of Theorem 41.1 applied to the tower $\mathbf{Q} \leq \mathbf{Q}(i\sqrt{3}) \leq E$.]

b) Show that $G(E/\mathbf{Q}) \simeq S_3$, the symmetric group on three letters. [*Hint:* Use Exercise 42.13, together with (a).]

42.15 Show that for a prime p, the splitting field over $\mathbf{Q}$ of $x^p - 1$ is of degree $p - 1$ over $\mathbf{Q}$. [*Hint:* Refer to the corollary of Theorem 31.4.]

42.16 Let $\bar{F}$ and $\bar{F'}$ be two algebraic closures of a field F, and let $f(x) \in F[x]$. Show that the splitting field E over F of $f(x)$ in $\bar{F}$ is isomorphic to the splitting field E' over F of $f(x)$ in $\bar{F'}$. [*Hint:* Use Theorem 41.1.]

forty-three

Separable Extensions

43.1 MULTIPLICITY OF ZEROS
OF A POLYNOMIAL

Remember that we are now always assuming that all algebraic extensions of a field F under consideration are contained in one fixed algebraic closure $\bar{F}$ of F.

Our next aim is to determine, for a finite extension E of F, under what conditions $\{E:F\} = [E:F]$. The key to answering this question is to consider the multiplicity of zeros of polynomials.

Definition Let $f(x) \in F[x]$. An element α of $\bar{F}$ such that $f(\alpha) = 0$ is a *zero of $f(x)$ of multiplicity* v if v is the greatest integer such that $(x - \alpha)^v$ is a factor of $f(x)$ in $\bar{F}[x]$.

The next theorem shows that the multiplicities of the zeros of one given *irreducible* polynomial over a field are all the same. The ease with which we can prove this theorem is a further indication of the power of our basic isomorphisms and of our whole approach to the study of zeros of polynomials by means of mappings.

388

Theorem 43.1   *Let $f(x)$ be irreducible in $F[x]$. Then all zeros of $f(x)$ in $\bar{F}$ have the same multiplicity.*

Proof.   Let α and β be zeros of $f(x)$ in $\bar{F}$. Then by Theorem 40.1 there is a basic isomorphism $\psi_{\alpha,\beta}: F(\alpha) \xrightarrow{\text{onto}} F(\beta)$. By Corollary 1 of Theorem 41.1, $\psi_{\alpha,\beta}$ can be extended to an isomorphism $\tau: \bar{F} \to \bar{F}$. Then τ induces a natural isomorphism $\tau_x: \bar{F}[x] \to \bar{F}[x]$, with $x\tau_x = x$. Now τ_x leaves $f(x)$ fixed, since $f(x) \in F[x]$ and $\psi_{\alpha,\beta}$ leaves F fixed. However,

$$((x - \alpha)^{\nu})\tau_x = (x - \beta)^{\nu},$$

which shows that the multiplicity of β in $f(x)$ is greater than or equal to the multiplicity of α. A symmetric argument gives the reverse inequality, so the multiplicity of α equals that of β. ∎

Corollary   *If $f(x)$ is irreducible in $F[x]$, then $f(x)$ has a factorization in $\bar{F}[x]$ of the form*

$$a \prod_{i} (x - a_i)^{\nu},$$

where the α_i are the distinct zeros of $f(x)$ in $\bar{F}$ and $a \in F$.

Proof.   Proof of the corollary is immediate from Theorem 43.1. ∎

At this point, we should probably show by an example that the phenomenon of a zero of multiplicity greater than 1 of an irreducible polynomial can occur. We shall show later in this chapter that it can only occur for a polynomial over an infinite field of characteristic $p \neq 0$.

Example 43.1   Let $E = \mathbf{Z}_p(y)$, where y is an indeterminate. Let $t = y^p$, and let F be the subfield $\mathbf{Z}_p(t)$ of E. (See Fig. 43.1.) Now $E = F(y)$ is algebraic over F, for y is a zero of $(x^p - t) \in F[x]$. By Theorem 35.3, irr(y, F) must divide $x^p - t$ in $F[x]$. [Actually, irr$(y, F) = x^p - t$. We leave a proof of this to the exercises (see Exercise 43.4).] Since $F(y)$ is clearly not equal to F, we

$$E = \mathbf{Z}_p(y) = F(y)$$

$$F = \mathbf{Z}_p(t) = \mathbf{Z}_p(y^p)$$

$$\mathbf{Z}_p$$

Fig. 43.1

must have the degree of $\mathrm{irr}(y, F) \geq 2$. But note that

$$x^p - t = x^p - y^p = (x - y)^p,$$

since E has characteristic p (see Theorem 40.5 and the following comment). Thus y is a zero of $\mathrm{irr}(y, F)$ of multiplicity > 1. Actually, $x^p - t = \mathrm{irr}(y, F)$, so the multiplicity of y is p. ■

From here on we rely heavily on Theorem 41.2 and its corollary. Theorem 40.1 and its corollary show that for a simple algebraic extension $F(\alpha)$ of F there is one extension of the identity isomorphism $\imath$ mapping F into F for every distinct zero of $\mathrm{irr}(\alpha, F)$ and that these are the only extensions of $\imath$. Thus $\{F(\alpha):F\}$ *is the number of distinct zeros of* $\mathrm{irr}(\alpha, F)$.

In view of our work with the theorem of Lagrange and Theorem 38.2, you should recognize the potential of a theorem like this next one.

Theorem 43.2   *If E is a finite extension of F, then $\{E:F\}$ divides $[E:F]$.*

Proof.   By Theorem 38.3, if E is finite over F, then $E = F(\alpha_1, \ldots, \alpha_n)$, where $\alpha_i \in \bar{F}$. Let $\mathrm{irr}(\alpha_i, F(\alpha_1, \ldots, \alpha_{i-1}))$ have α_i as one of n_i distinct zeros that are all of a common multiplicity v_i, by Theorem 43.1. Then

$$[F(\alpha_1, \ldots, \alpha_i) : F(\alpha_1, \ldots, \alpha_{i-1})] = n_i v_i$$
$$= \{F(\alpha_1, \ldots, \alpha_i) : F(\alpha_1, \ldots, \alpha_{i-1})\} v_i.$$

By Theorem 38.2 and the corollary of Theorem 41.2,

$$[E:F] = \prod_i n_i v_i,$$

and

$$\{E:F\} = \prod_i n_i.$$

Therefore, $\{E:F\}$ divides $[E:F]$. ■

43.2 SEPARABLE EXTENSIONS

Definition   A finite extension E of F is a **separable extension of F** if $\{E:F\} = [E:F]$. An element α of $\bar{F}$ is **separable over F** if $F(\alpha)$ is a separable extension of F. An irreducible polynomial $f(x) \in F[x]$ is **separable over F** if every zero of $f(x)$ in $\bar{F}$ is separable over F.

To make things a little easier for the student, we have restricted our definition of a separable extension of a field F to *finite* extensions E of F. For the corresponding definition for infinite extensions, see Exercise 43.6.

We know that $\{F(\alpha):F\}$ is the number of distinct zeros of irr(α, F). Also, the multiplicity of α in irr(α, F) is the same as the multiplicity of each conjugate of α over F, by Theorem 43.1. *Thus α is separable over F if and only if* irr(α, F) *has all zeros of multiplicity* 1. This tells us at once that *an irreducible polynomial $f(x) \in F[x]$ is separable over F if and only if $f(x)$ has all zeros of multiplicity* 1.

Theorem 43.3 *If K is a finite extension of E and E is a finite extension of F, that is, $F \leq E \leq K$, then K is separable over F if and only if K is separable over E and E is separable over F.*

Proof. Now

$$[K:F] = [K:E][E:F],$$

and

$$\{K:F\} = \{K:E\}\{E:F\}.$$

Then if K is separable over F, so that $[K:F] = \{K:F\}$, we must have $[K:E] = \{K:E\}$ and $[E:F] = \{E:F\}$, since in each case the index divides the degree, by Theorem 43.2. Thus, if K is separable over F, then K is separable over E and E is separable over F.

The converse is equally easy, for $[K:E] = \{K:E\}$ and $[E:F] = \{E:F\}$ implies that

$$[K:F] = [K:E][E:F] = \{K:E\}\{E:F\} = \{K:F\}. \quad \blacksquare$$

Theorem 43.3 can be extended in the obvious way, by induction, to any finite tower of finite extensions. The top field is a separable extension of the bottom one if and only if each field is a separable extension of the one immediately under it.

Corollary *If E is a finite extension of F, then E is separable over F if and only if each α in E is separable over F.*

Proof. Suppose that E is separable over F, and let $\alpha \in E$. Then

$$F \leq F(\alpha) \leq E,$$

and Theorem 43.3 shows that $F(\alpha)$ is separable over F.

Suppose, conversely, that every $\alpha \in E$ is separable over F. Since E is finite over F, there exist $\alpha_1, \ldots, \alpha_n$ such that

$$F < F(\alpha_1) < F(\alpha_1, \alpha_2) < \cdots < E = F(\alpha_1, \ldots, \alpha_n).$$

Now, clearly, since α_i is separable over F, α_i is separable over $F(\alpha_1, \ldots, \alpha_{i-1})$, since

$$q(x) = \mathrm{irr}(\alpha_i, F(\alpha_1, \ldots, \alpha_{i-1}))$$

divides irr(α_i, F), so that α_i is a zero of $q(x)$ of multiplicity 1. Thus $F(\alpha_1, \ldots, \alpha_i)$ is separable over $F(\alpha_1, \ldots, \alpha_{i-1})$, so E is separable over F, by Theorem 43.3, extended by induction. ∎

43.3 PERFECT FIELDS

We now turn to the task of proving that α can fail to be separable over F only if F is an infinite field of characteristic $p \neq 0$. One method is to introduce formal derivatives of polynomials. While this is an elegant technique, and also a useful one, we shall, for the sake of brevity, use the following lemma instead. Formal derivatives are developed in Exercises 43.8 through 43.15.

Lemma 43.1 *Let $\bar{F}$ be an algebraic closure of F, and let*

$$f(x) = x^n + a_{n-1}x^{n-1} + \cdots + a_1 x + a_0$$

be any monic polynomial in $\bar{F}[x]$. If $(f(x))^m \in F[x]$ and $m \cdot 1 \neq 0$ in F, then $f(x) \in F[x]$, that is, all $a_i \in F$.

Proof. We must show that $a_i \in F$, and we proceed, by induction on r, to show that $a_{n-r} \in F$. For $r = 1$,

$$(f(x))^m = x^{mn} + (m \cdot 1)a_{n-1}x^{mn-1} + \cdots + a_0{}^m.$$

Since $(f(x))^m \in F[x]$, we have, in particular,

$$(m \cdot 1)a_{n-1} \in F.$$

Thus $a_{n-1} \in F$, since $m \cdot 1 \neq 0$ in F.

As induction hypothesis, suppose that $a_{n-r} \in F$ for $r = 1, 2, \ldots, k$. Then the coefficient of $x^{mn-(k+1)}$ in $(f(x))^m$ is of the form

$$(m \cdot 1)a_{n-(k+1)} + g_{k+1}(a_{n-1}, a_{n-2}, \ldots, a_{n-k}),$$

where $g_{k+1}(a_{n-1}, a_{n-2}, \ldots, a_{n-k})$ is a formal polynomial in $a_{n-1}, a_{n-2}, \ldots,$ a_{n-k}. By the induction assumption, $g_{k+1}(a_{n-1}, a_{n-2}, \ldots, a_{n-k}) \in F$, so again $a_{n-(k+1)} \in F$, since $m \cdot 1 \neq 0$ in F. ∎

We are now in a position to handle fields F of characteristic zero and to show that for a finite extension E of F, we have $\{E:F\} = [E:F]$. By definition, this amounts to proving that every finite extension of a field of characteristic zero is a separable extension. First, we give a definition.

Definition A field is *perfect* if every finite extension is a separable extension.

Theorem 43.4 *Every field of characteristic zero is perfect.*

Proof. Let E be a finite extension of a field F of characteristic zero, and let $\alpha \in E$. Then $f(x) = \operatorname{irr}(\alpha, F)$ factors in $\bar{F}[x]$ into $\prod_i(x - \alpha_i)^v$, where the α_i are the distinct zeros of $\operatorname{irr}(\alpha, F)$, and say, $\alpha = \alpha_1$. Thus

$$f(x) = \left(\prod_i (x - \alpha_i)\right)^v,$$

and since $v \cdot 1 \neq 0$ for a field F of characteristic 0, we must have

$$\left(\prod_i (x - \alpha_i)\right) \in F[x],$$

by Lemma 43.1. Since $f(x)$ is irreducible and of minimal degree in $F[x]$ having α as a zero, we then see that $v = 1$. Therefore, α is separable over F for all $\alpha \in E$. By the corollary of Theorem 43.3, this means that E is a separable extension of F. ∎

Lemma 43.1 will also get us through for the case of a finite field, although the proof is a bit harder. Don't feel that you can't understand the meaning of a theorem just because you get lost in the proof.

Theorem 43.5 *Every finite field is perfect.*

Proof. Let F be a finite field of characteristic p, and let E be a finite extension of F. Let $\alpha \in E$. We need to show that α is separable over F. Now $f(x) = \operatorname{irr}(\alpha, F)$ factors in $\bar{F}$ into $\prod_i(x - \alpha_i)^v$, where the α_i are the distinct zeros of $f(x)$, and, say, $\alpha = \alpha_1$. Let $v = p^t e$, where p does not divide e. Then

$$f(x) = \prod_i (x - \alpha_i)^v = \left(\prod_i (x - \alpha_i)^{p^t}\right)^e$$

is in $F[x]$, and by Lemma 43.1, $\prod_i(x - \alpha_i)^{p^t}$ is in $F[x]$, since $e \cdot 1 \neq 0$ in F. Since $f(x) = \operatorname{irr}(\alpha, F)$ is of minimal degree over F having α as a zero, we must have $e = 1$.

Theorem 40.5 and the remark following it show then that

$$f(x) = \prod_i (x - \alpha_i)^{p^t} = \prod_i (x^{p^t} - \alpha_i^{p^t}).$$

Thus, if we regard $f(x)$ as $g(x^{p^t})$, we must have $g(x) \in F[x]$. Now $g(x)$ is separable over F with distinct zeros $\alpha_i^{p^t}$. Consider $F(\alpha_1^{p^t}) = F(\alpha^{p^t})$. Then $F(\alpha^{p^t})$ is separable over F. Since $x^{p^t} - \alpha^{p^t} = (x - \alpha)^{p^t}$, we see that α is the only zero of $x^{p^t} - \alpha^{p^t}$ in $\bar{F}$. As a finite-dimensional vector space over a finite field F, $F(\alpha^{p^t})$ must be again a finite field. Hence the map

$$\sigma_p : F(\alpha^{p^t}) \to F(\alpha^{p^t})$$

given by $a\sigma_p = a^p$ for $a \in F(\alpha^{p^t})$ is an automorphism of $F(\alpha^{p^t})$, by Theorem 40.5. Consequently, $(\sigma_p)^t$ is also an automorphism of $F(\alpha^{p^t})$, and

$$a((\sigma_p)^t) = a^{p^t}.$$

Since an automorphism of $F(\alpha^{p^t})$ is an onto map, there is $\beta \in F(\alpha^{p^t})$ such that $\beta((\sigma_p)^t) = \alpha^{p^t}$. But then $\beta^{p^t} = \alpha^{p^t}$, and we saw that α was the only zero of $x^{p^t} - \alpha^{p^t}$, so we must have $\beta = \alpha$. Since $\beta \in F(\alpha^{p^t})$, we have $F(\alpha) = F(\alpha^{p^t})$. Since $F(\alpha^{p^t})$ was separable over F, we now see that $F(\alpha)$ is separable over F. Therefore, α is separable over F and $t = 0$.

We have shown that for $\alpha \in E$, α is separable over F. Then by the corollary of Theorem 43.3, E is a separable extension of F. ∎

We have completed our aim, which was to show that fields of characteristic 0 and finite fields have only separable finite extensions, that is, these fields are perfect. *For finite extensions E of such perfect fields F, we have then $[E:F] = \{E:F\}$.*

*43.4 THE PRIMITIVE ELEMENT THEOREM

The following interesting theorem will be of use to us later under another starred heading.

Theorem 43.6 (Primitive Element Theorem) *Let E be a finite separable extension of an infinite field F. Then there exists $\alpha \in E$ such that $E = F(\alpha)$. (Such an element α is a **primitive element**.) That is, a finite separable extension of an infinite field is a simple extension.*

Proof. We prove this in the case that $E = F(\beta, \gamma)$. The induction argument is obvious. Let irr(β, F) have distinct zeros $\beta = \beta_1, \ldots, \beta_n$, and let irr$(\gamma, F)$ have distinct zeros $\gamma = \gamma_1, \ldots, \gamma_m$ in $\bar{F}$, where all zeros are of multiplicity 1, since E is a separable extension of F. Since F is infinite, we can find $a \in F$ such that

$$a \neq (\beta_i - \beta)/(\gamma - \gamma_j)$$

for all i and j, with $j \neq 1$. That is, $a(\gamma - \gamma_j) \neq \beta_i - \beta$. Letting $\alpha = \beta + a\gamma$, we have $\alpha = \beta + a\gamma \neq \beta_i + a\gamma_j$, so

$$\alpha - a\gamma_j \neq \beta_i$$

for all i and all $j \neq 1$. Let $f(x) = $ irr(β, F), and consider

$$h(x) = f(\alpha - ax) \in (F(\alpha))[x].$$

Now $h(\gamma) = f(\beta) = 0$. However, $h(\gamma_j) \neq 0$ for $j \neq 1$ by construction, since the β_i were the only zeros of $f(x)$. Hence $h(x)$ and $g(x) = \text{irr}(\gamma, F)$ have a common factor in $(F(\alpha))[x]$, namely $\text{irr}(\gamma, F(\alpha))$, which must be linear, since γ is the only common zero of $g(x)$ and $h(x)$. Thus $\gamma \in F(\alpha)$, and therefore $\beta = \alpha - a\gamma$ is in $F(\alpha)$. Hence $F(\beta, \gamma) = F(\alpha)$. $\blacksquare$

Corollary   *A finite extension of a field of characteristic zero is a simple extension.*

Proof.   Proof of this corollary follows at once from Theorem 43.6 and the fact that every field of characteristic 0 is infinite and perfect. $\blacksquare$

Actually, a finite extension of a finite field F is also a simple extension of F. This will be shown in Chapter 45. Thus, once again, the only possible "bad case" is an extension of an infinite field of characteristic $p \neq 0$. In summary, *a finite separable extension of a field F is a simple extension of F.*

Exercises

43.1 Give an example of an $f(x) \in \mathbf{Q}[x]$ that has no zeros in $\mathbf{Q}$ but whose zeros in $\mathbf{C}$ are all of multiplicity 2. Explain how this is consistent with Theorem 43.4, which shows that $\mathbf{Q}$ is perfect.

†**43.2** Show that if α, $\beta \in \bar{F}$ are both separable over F, then $\alpha \pm \beta$, $\alpha\beta$, and α/β, if $\beta \neq 0$, are all separable over F. [*Hint:* Use Theorem 43.3 and its corollary.]

43.3 Mark each of the following true or false.

— a) Every finite extension of every field F is separable over F.

— b) Every finite extension of every finite field F is separable over F.

— c) Every field of characteristic 0 is perfect.

— d) Every polynomial of degree n over every field F always has n distinct zeros in $\bar{F}$.

— e) Every polynomial of degree n over every perfect field F always has n distinct zeros in $\bar{F}$.

— f) Every irreducible polynomial of degree n over every perfect field F always has n distinct zeros in $\bar{F}$.

— g) Every algebraically closed field is perfect.

— h) Every field F has an algebraic extension E that is perfect.

— i) If E is a finite separable splitting field extension of F, then $|G(E/F)| = [E:F]$.

— j) If E is a finite splitting field extension of F, then $|G(E/F)|$ divides $[E:F]$.

43.4 Show that $\{1, y, \ldots, y^{n-1}\}$ is a basis for $\mathbf{Z}_p(y)$ over $\mathbf{Z}_p(y^p)$, where y is an indeterminate. Referring to Example 43.1, conclude by a degree argument that $x^p - t$ is irreducible over $\mathbf{Z}_p(t)$, where $t = y^p$.

43.5 Prove that if E is an algebraic extension of a perfect field F, then E is perfect.

43.6 A (possibly infinite) algebraic extension E of a field F is a **separable extension of** F if for every $\alpha \in E$, $F(\alpha)$ is a separable extension of F, in the sense defined in the text. Show that if E is a (possibly infinite) separable extension of F and K is a (possibly infinite) separable extension of E, then K is a separable extension of F.

43.7 Let E be an algebraic extension of a field F. Show that the set of all elements in E that are separable over F forms a subfield of E, the **separable closure of** F **in** E. [*Hint:* Use Exercise 43.2.]

Exercises 43.8 through 43.15 introduce formal derivatives in $F[x]$.

43.8 Let F be any field, and let $f(x) = a_0 + a_1 x + \cdots + a_i x^i + \cdots + a_n x^n$ be in $F[x]$. The **derivative** $f'(x)$ **of** $f(x)$ is the polynomial

$$f'(x) = a_1 + \cdots + (i \cdot 1)a_i x^{i-1} + \cdots + (n \cdot 1)a_n x^{n-1},$$

where $i \cdot 1$ has its usual meaning for $i \in \mathbf{Z}^+$ and $1 \in F$. *These are formal derivatives; no "limits" are involved here.*

a) Prove that the map $D : F[x] \to F[x]$ given by $D(f(x)) = f'(x)$ is a homomorphism of $\langle F[x], + \rangle$.

b) Find the kernel of D in the case that F is of characteristic 0.

c) Find the kernel of D in the case that F is of characteristic $p \neq 0$.

43.9 Continuing the ideas of Exercise 43.8, show that:

a) $D(af(x)) = aD(f(x))$ for all $f(x) \in F[x]$ and $a \in F$.

b) $D(f(x)g(x)) = f(x)g'(x) + f'(x)g(x)$ for all $f(x), g(x) \in F[x]$. [*Hint:* Use part (a) of this exercise and the preceding exercise and proceed by induction on the degree of $f(x)g(x)$.]

c) $D((f(x))^m) = (m \cdot 1)f(x)^{m-1}f'(x)$ for all $f(x) \in F[x]$. [*Hint:* Use (b).]

43.10 Let $f(x) \in F[x]$, and let $\alpha \in \bar{F}$ be a zero of $f(x)$ of multiplicity v. Show that $v > 1$ if and only if α is also a zero of $f'(x)$. [*Hint:* Apply (b) and (c) of Exercise 43.9 to the factorization $f(x) = (x - \alpha)^v g(x)$ of $f(x)$ in $\bar{F}[x]$.]

43.11 Show from Exercise 43.10 that every irreducible polynomial over a field F of characteristic 0 is separable. [*Hint:* Use the fact that $\mathrm{irr}(\alpha, F)$ is the *minimal* polynomial for α over F.]

43.12 Show from Exercise 43.10 that an irreducible polynomial $q(x)$ over a field F of characteristic $p \neq 0$ is not separable if and only if each exponent of each term of $q(x)$ is divisible by p.

43.13 Generalize Exercise 43.10, showing that $f(x) \in F[x]$ has no zero of multiplicity > 1 if and only if $f(x)$ and $f'(x)$ have no common nonconstant factor in $\bar{F}[x]$.

*43.14 Working a bit harder than in Exercise 43.13, show that $f(x) \in F[x]$ has no zero of multiplicity > 1 if and only if $f(x)$ and $f'(x)$ have no common nonconstant factor in $F[x]$. [*Hint:* Use Theorem 33.3 to show that if 1 is a gcd of $f(x)$ and $f'(x)$ in $F[x]$, it is a gcd of these polynomials in $\bar{F}[x]$ also.]

*43.15 Describe a feasible computational procedure for determining whether $f(x) \in F[x]$ has a zero of multiplicity > 1, without actually finding the zeros of $f(x)$. [*Hint:* Use Exercise 43.14.]

*43.16 Find $\alpha \in \mathbf{Q}(\sqrt{2}, \sqrt{3})$ such that $\mathbf{Q}(\sqrt{2}, \sqrt{3}) = \mathbf{Q}(\alpha)$. Verify by direct computation that $\sqrt{2}$ and $\sqrt{3}$ can indeed be expressed as formal polynomials in your α, with coefficients in $\mathbf{Q}$.

*43.17 Observe where the hypothesis that F was infinite was used in Theorem 43.6. Show that if F is finite with s elements, and if β and γ are algebraic over F and of degrees n and m, respectively, then there exists α such that $F(\beta, \gamma) = F(\alpha)$, provided $s > mn$. (This result will be superseded by the work in Chapter 45).

*forty-four

Totally Inseparable Extensions

*44.1 TOTALLY INSEPARABLE EXTENSIONS

We develop our theory of totally inseparable extensions in a fashion parallel to our development of separable extensions.

Definition A finite extension E of a field F is a *totally inseparable extension of* F if $\{E:F\} = 1 < [E:F]$. An element α of $\bar{F}$ is *totally inseparable over* F if $F(\alpha)$ is totally inseparable over F.

We know that $\{F(\alpha):F\}$ is the number of distinct zeros of $\mathrm{irr}(\alpha, F)$. *Thus α is totally inseparable over F if and only if $\mathrm{irr}(\alpha, F)$ has only one zero that is of multiplicity >1.*

Example 44.1 With reference to Example 43.1, it is clear that $\mathbf{Z}_p(y)$ is totally inseparable over $\mathbf{Z}_p(y^p)$, where y is an indeterminate. ∎

Theorem 44.1 (Counterpart of Theorem 43.3) *If K is a finite extension of E, E is a finite extension of F, and $F < E < K$, then K is totally insep-*

398

arable over F if and only if K is totally inseparable over E and E is totally inseparable over F.

Proof. Since $F < E < K$, we have $[K:E] > 1$ and $[E:F] > 1$. Suppose K is totally inseparable over F. Then $\{K:F\} = 1$, and

$$\{K:F\} = \{K:E\}\{E:F\},$$

so we must have

$$\{K:E\} = 1 < [K:E] \qquad \text{and} \qquad \{E:F\} = 1 < [E:F].$$

Thus K is totally inseparable over E, and E is totally inseparable over F.

Conversely, if K is totally inseparable over E and E is totally inseparable over F, then

$$\{K:F\} = \{K:E\}\{E:F\} = (1)(1) = 1,$$

and $[K:F] < 1$. Thus K is totally inseparable over F. ∎

Theorem 44.1 can be extended in the obvious way, by induction, to any finite proper tower of finite extensions. The top field is a totally inseparable extension of the bottom one if and only if each field is a totally inseparable extension of the one immediately under it.

Corollary (Counterpart of the Corollary of Theorem 43.3) *If E is a finite extension of F, then E is totally inseparable over F if and only if each α in E, α ∉ F, is totally inseparable over F.*

Proof. Suppose that E is totally inseparable over F, and let $\alpha \in E$, with $\alpha \notin F$. Then

$$F < F(\alpha) \leq E.$$

If $F(\alpha) = E$, we are done, by the definition of α totally inseparable over F. If $F < F(\alpha) < E$, then Theorem 44.1 shows that since E is totally inseparable over F, $F(\alpha)$ is totally inseparable over F.

Conversely, suppose that for every $\alpha \in E$, with $\alpha \notin F$, α is totally inseparable over F. Since E is finite over F, there exist $\alpha_1, \ldots, \alpha_n$ such that

$$F < F(\alpha_1) < F(\alpha_1, \alpha_2) < \cdots < E = F(\alpha_1, \ldots, \alpha_n).$$

Now since α_i is totally inseparable over F, α_i is totally inseparable over $F(\alpha_1, \ldots, \alpha_{i-1})$, for $q(x) = \text{irr}(\alpha_i, F(\alpha_1, \ldots, \alpha_{i-1}))$ divides $\text{irr}(\alpha_i, F)$ so that α_i is the only zero of $q(x)$ and is of multiplicity > 1. Thus $F(\alpha_1, \ldots, \alpha_i)$ is totally inseparable over $F(\alpha_1, \ldots, \alpha_{i-1})$, and E is totally inseparable over F, by Theorem 44.1, extended by induction. ∎

Obviously, thus far we have so closely paralleled our work in Chapter 43 that we could easily have handled these ideas together.

*44.2 SEPARABLE CLOSURES·

We now come to our main reason for including this material.

> **Theorem 44.2** Let F have characteristic $p \neq 0$, and let E be a finite
> extension of F. Then $\alpha \in E$, $\alpha \notin F$, is totally inseparable over F if and only
> if there is some integer $t \geq 1$ such that $\alpha^{p^t} \in F$. Furthermore, there is a
> unique extension K of F, with $F \leq K \leq E$, such that K is separable over
> F, and either $E = K$ or E is totally inseparable over K.

Proof. Let $\alpha \in E$, $\alpha \notin F$, be totally inseparable over F. Then $\mathrm{irr}(\alpha, F)$ has
just one zero α of multiplicity >1, and, as shown in the proof of Theorem
43.5, $\mathrm{irr}(\alpha, F)$ must be of the form

$$x^{p^t} - \alpha^{p^t}.$$

Hence $\alpha^{p^t} \in F$ for some $t \geq 1$.

Conversely, if $\alpha^{p^t} \in F$ for some $t \geq 1$, where $\alpha \in E$ and $\alpha \notin F$, then

$$x^{p^t} - \alpha^{p^t} = (x - \alpha)^{p^t},$$

and $(x^{p^t} - \alpha^{p^t}) \in F[x]$, showing that $\mathrm{irr}(\alpha, F)$ divides $(x - \alpha)^{p^t}$. Thus $\mathrm{irr}(\alpha, F)$
has α as its only zero and this zero is of multiplicity >1, so α is totally insep-
arable over F.

For the second part of the theorem, let $E = F(\alpha_1, \ldots, \alpha_n)$. Then if

$$\mathrm{irr}(\alpha_i, F) = \prod_j (x^{p^{t_i}} - \alpha_{ij}{}^{p^{t_i}}),$$

with $\alpha_{i1} = \alpha_i$, let $\beta_{ij} = \alpha_{ij}{}^{p^{t_i}}$. We have $F(\beta_{11}, \beta_{21}, \ldots, \beta_{n1}) \leq E$, and β_{i1} is a
zero of

$$f_i(x) = \prod_j (x - \beta_{ij}),$$

where $f_i(x) \in F[x]$. Now since raising to the power p is an isomorphism σ_p
of E into E, raising to the power p^t is the isomorphic mapping $(\sigma_p)^t$ of E
into E. Thus since the α_{ij} are all distinct for a fixed i, so are the β_{ij} for a
fixed i. Therefore, β_{ij} is separable over F, because it is a zero of a polynomial
$f_i(x)$ in $F[x]$ with zeros of multiplicity 1. Then

$$K = F(\beta_{11}, \beta_{21}, \ldots, \beta_{n1})$$

is separable over F, by the proof of the corollary of Theorem 43.3. If all
$p^{t_i} = 1$, then $K = E$. If some $p^{t_i} \neq 1$, then $K \neq E$, and $\alpha_i{}^{p^{t_i}} = \beta_{i1}$ is in K,
showing that each $\alpha_i \notin K$ is totally inseparable over K, by the first part of
this theorem. Hence $E = K(\alpha_1, \ldots, \alpha_n)$ is totally inseparable over K, by the
proof of the corollary of Theorem 44.1.

It follows from the corollaries of Theorems 43.3 and 44.1 that the field K of Theorem 44.2 consists of all elements α in E that are separable over F. Thus K is unique. ∎

Definition The unique field K of Theorem 44.2 is the **separable closure of F in E**.

The preceding theorem shows the precise structure of totally inseparable extensions of a field of characteristic p. Such an extension can be obtained by repeatedly adjoining pth roots of elements (which are not already pth powers) to obtain larger and larger fields.

We remark that Theorem 44.2 is true for infinite algebraic extensions E of F. The proof of the first assertion of the theorem is valid for the case of infinite extensions also. For the second part, since $\alpha \pm \beta$, $\alpha\beta$, and α/β, for $\beta \neq 0$, are all contained in the field $F(\alpha, \beta)$, all elements of E separable over F form a subfield K of E, the **separable closure of F in E**. It follows that an $\alpha \in E$, $\alpha \notin K$, is totally inseparable over K, since α and all coefficients of irr(α, K) are in a finite extension of F, and then Theorem 44.2 can be applied.

Exercises

*__44.1__ Let y and z be indeterminates, and let $u = y^{12}$ and $v = z^{18}$. Describe the separable closure of $\mathbf{Z}_3(u, v)$ in $\mathbf{Z}_3(y, z)$.

*__44.2__ Let y and z be indeterminates, and let $u = y^{12}$ and $v = y^2 z^{18}$. Describe the separable closure of $\mathbf{Z}_3(u, v)$ in $\mathbf{Z}_3(y, z)$.

*__44.3__ Show that if E is an algebraic extension of a field F, then the set of all elements of E totally inseparable over F forms a subfield of E, the **totally inseparable closure of F in E**.

*__44.4__ Referring to Exercise 44.1, describe the totally inseparable closure (see Exercise 44.3) of $\mathbf{Z}_3(u, v)$ in $\mathbf{Z}_3(y, z)$.

*__44.5__ Referring to Exercise 44.2, describe the totally inseparable closure of $\mathbf{Z}_3(u, v)$ in $\mathbf{Z}_3(y, z)$.

*__44.6__ Mark each of the following true or false.

___ a) No proper algebraic extension of an infinite field of characteristic $p \neq 0$ is ever a separable extension.

___ b) If $F(\alpha)$ is totally inseparable over F of characteristic $p \neq 0$, then $\alpha^{p^t} \in F$ for some $t > 0$.

___ c) For an indeterminate y, $\mathbf{Z}_5(y)$ is separable over $\mathbf{Z}_5(y^5)$.

___ d) For an indeterminate y, $\mathbf{Z}_5(y)$ is separable over $\mathbf{Z}_5(y^{10})$.

___ e) For an indeterminate y, $\mathbf{Z}_5(y)$ is totally inseparable over $\mathbf{Z}_5(y^{10})$.

___ f) If F is a field and α is algebraic over F, then α is either separable or totally inseparable over F.

   — g) If E is an algebraic extension of a field F, then F has a separable closure in E.

   — h) If E is an algebraic extension of a field F, then E is totally inseparable over the separable closure of F in E.

   — i) If E is an algebraic extension of a field F and E is not a separable extension of F, then E is totally inseparable over the separable closure of F in E.

   — j) If α is totally inseparable over F, then α is the only zero of irr(α, F).

*44.7 Show that a field F of characteristic $p \neq 0$ is perfect if and only if $F^p = F$, that is, every element of F is a pth power of some element of F.

*44.8 Let E be a finite extension of a field F of characteristic p. In the notation of Exercise 44.7, show that $E^p = E$ if and only if $F^p = F$. [*Hint:* The map $\sigma_p : E \to E$ defined by $\alpha\sigma_p = \alpha^p$ for $\alpha \in E$ is an isomorphism. Consider the diagram in Fig. 44.1, and make degree arguments.]

Fig. 44.1

forty-five

Finite Fields

It is the purpose of this chapter to determine the structure of all finite fields. We shall show that for every prime p and positive integer n, there is exactly one finite field (up to isomorphism) of order p^n. This field $\mathrm{GF}(p^n)$ is usually referred to as the **Galois field of order** p^n. We shall be using quite a bit of our material on cyclic groups. The proofs are simple and elegant.

45.1 THE STRUCTURE OF A FINITE FIELD

It is easy to see that all finite fields must have prime-power order.

> **Theorem 45.1** *Let E be a finite extension of degree n over a finite field F. If F has q element, then E has q^n elements.*

Proof. Let $\{\alpha_1, \ldots, \alpha_n\}$ be a basis for E as a vector space over F. Then every $\beta \in E$ can be *uniquely* written in the form

$$\beta = b_1\alpha_1 + \cdots + b_n\alpha_n$$

for $b_i \in F$. Since each b_i may be any of the q elements of F, the total number of such distinct linear combinations of the α_i is q^n. ∎

Corollary   *If E is a finite field of characteristic p, then E contains exactly p^n elements for some positive integer n.*

Proof.   Every finite field E is a finite extension of a prime field isomorphic to the field Z_p, where p is the characteristic of E. The corollary follows at once from Theorem 45.1. ∎

We now turn to the study of the multiplicative structure of a finite field. The following theorem will show us how any finite field can be formed from the prime subfield.

Theorem 45.2   *A finite field E of p^n elements is the splitting field of $x^{p^n} - x$ over its prime subfield Z_p (up to isomorphism).*

Proof.   Let E be a finite field with p^n elements, where p is the characteristic of E. The set E^* of nonzero elements of E forms a multiplicative group of order $p^n - 1$ under the field multiplication. For $\alpha \in E^*$, the order of α in this group divides the order $p^n - 1$ of the group. Thus for $\alpha \in E^*$, we have $\alpha^{p^n - 1} = 1$, so $\alpha^{p^n} = \alpha$. Therefore, every element in E is a zero of $x^{p^n} - x$. Since $x^{p^n} - x$ can have at most p^n zeros, we see that E is the splitting field of $x^{p^n} - x$ over Z_p. ∎

Definition   An element α of a field is an ***nth root of unity*** if $\alpha^n = 1$. It is a ***primitive nth root of unity*** if $\alpha^n = 1$ and $\alpha^m \neq 1$ for $0 < m < n$.

Thus the nonzero elements of a finite field of p^n elements are all $(p^n - 1)$th roots of unity.

Let F be *any field*, and let U_n be the set of all nth roots of unity in F. It is easy for us to see that U_n is a group under field multiplication. If $a^n = 1$ and $b^n = 1$, then

$$(ab)^n = a^n b^n = 1,$$

so multiplication is closed on U_n. Checking the group axioms is equally trivial. We claim that U_n is a cyclic group. In fact, the following more general result is true.

Theorem 45.3   *If G is a finite multiplicative subgroup of the multiplicative group $\langle F^*, \cdot \rangle$ of nonzero elements of a field F, then G is cyclic.*

Proof.   By Theorem 9.3, as a finite abelian group, G is isomorphic to a direct product $Z_{m_1} \times Z_{m_2} \times \cdots \times Z_{m_r}$ of cyclic groups, where m_i divides m_{i+1}. Let us think of each of the Z_{m_i} as a cyclic group of order m_i in *multiplicative* notation. Then for $a_i \in Z_{m_i}$, $a_i^{m_i} = 1$, so $a_i^{m_r} = 1$, since m_i divides m_r. Thus for all $\alpha \in G$, we have $\alpha^{m_r} = 1$, so every element of G is a zero of $x^{m_r} - 1$.

But G has $\prod_{i=1}^{r} m_i$ elements, while $x^{m_r} - 1$ can have at most m_r zeros in a field. Therefore, we must have $r = 1$, so G is cyclic. ∎

Corollary 1 *The multiplicative group of all nonzero elements of a finite field under field multiplication is cyclic.*

Proof. Proof of Corollary 1 is immediate from Theorem 45.3. ∎

Corollary 2 *A finite extension E of a finite field F is a simple extension of F.*

Proof. Let α be a generator for the cyclic group E^* of nonzero elements of E. Then, obviously, $E = F(\alpha)$. ∎

Example 45.1 Consider the finite field $\mathbf{Z}_{11}$. By Corollary 1 of Theorem 45.3, $\langle \mathbf{Z}_{11}^*, \cdot \rangle$ is cyclic. Let us try to find a generator of $\mathbf{Z}_{11}^*$ by brute force and ignorance. We start by trying 2. Since $|\mathbf{Z}_{11}^*| = 10$, 2 must be an element of $\mathbf{Z}_{11}^*$ of order dividing 10, that is, either 2, 5, or 10. Now

$$2^2 = 4, \qquad 2^4 = 4^2 = 5, \qquad \text{and} \qquad 2^5 = (2)(5) = 10 = -1.$$

Thus neither 2^2 nor 2^5 is 1, but, of course, $2^{10} = 1$, so 2 is a generator of $\mathbf{Z}_{11}^*$, that is, 2 is a primitive 10th root of unity in $\mathbf{Z}_{11}$. We were lucky.

By the theory of cyclic groups, all the generators of $\mathbf{Z}_{11}^*$, that is, all the primitive 10th roots of unity in $\mathbf{Z}_{11}$, are then of the form 2^n, where n is relatively prime to 10. These elements are

$$2^1 = 2, \qquad 2^3 = 8,$$
$$2^7 = 7, \qquad 2^9 = 6.$$

The primitive 5th roots of unity in $\mathbf{Z}_{11}$ are of the form 2^m, where the gcd of m and 10 is 2, that is,

$$2^2 = 4, \qquad 2^4 = 5,$$
$$2^6 = 9, \qquad 2^8 = 3.$$

The primitive square root of unity in $\mathbf{Z}_{11}$ is $2^5 = 10 = -1$. ∎

45.2 THE EXISTENCE OF $GF(p^n)$

We turn now to the question of the existence of a finite field of order p^r for every prime power p^r, $r > 0$. We need the following lemma.

Lemma 45.1 *If F is a finite field of characteristic p, then $x^{p^n} - x$ has p^n distinct zeros in the splitting field $K \leq \bar{F}$ of $x^{p^n} - x$ over F.*

Proof. Let F be a finite field of characteristic p, and let K be the splitting field in $\bar{F}$ of the polynomial $x^{p^n} - x$ over F. We are to show that $x^{p^n} - x$

has p^n distinct zeros in K. Since we have not taken time to introduce an algebraic theory of derivatives, this elegant technique is not available to us, so we proceed by brute force. Obviously, 0 is a zero of $x^{p^n} - x$ of multiplicity 1. Suppose $\alpha \neq 0$ is a zero of $x^{p^n} - x$, and hence is a zero of $f(x) = x^{p^n - 1} - 1$. Then $x - \alpha$ is a factor of $f(x)$ in $K[x]$, and by long division, we find that

$$\frac{f(x)}{(x - \alpha)} = g(x)$$
$$= x^{p^n - 2} + \alpha x^{p^n - 3} + \alpha^2 x^{p^n - 4} + \cdots + \alpha^{p^n - 3} x + \alpha^{p^n - 2}.$$

Now $g(x)$ has $p^n - 1$ summands, and in $g(\alpha)$, each summand is

$$\alpha^{p^n - 2} = \frac{\alpha^{p^n - 1}}{\alpha} = \frac{1}{\alpha}.$$

Thus

$$g(\alpha) = [(p^n - 1) \cdot 1] \frac{1}{\alpha} = -\frac{1}{\alpha}.$$

since we are in a field of characteristic p. Therefore, $g(\alpha) \neq 0$, so α is a zero of $f(x)$ of multiplicity 1. ∎

Theorem 45.4 *A finite field* $\mathrm{GF}(p^n)$ *of* p^n *elements exists for every prime power* p^n.

Proof. Let $K \leq \overline{\mathbf{Z}_p}$ be the splitting field of $x^{p^n} - x$ over $\mathbf{Z}_p$, and let F be the subset of K consisting of all zeros of $x^{p^n} - x$ in K. Then for $\alpha, \beta \in F$, the equations

$$(\alpha \pm \beta)^{p^n} = \alpha^{p^n} \pm \beta^{p^n} = \alpha \pm \beta$$

and

$$(\alpha\beta)^{p^n} = \alpha^{p^n}\beta^{p^n} = \alpha\beta$$

show that F is closed under addition, subtraction, and multiplication. Clearly, 0 and 1 are zeros of $x^{p^n} - x$. For $\alpha \neq 0$, $\alpha^{p^n} = \alpha$ implies that $(1/\alpha)^{p^n} = 1/\alpha$. Thus F is a subfield of K containing $\mathbf{Z}_p$. Since K is the smallest extension of $\mathbf{Z}_p$ containing the zeros of $x^{p^n} - x$, we see that we must have $K = F$. Therefore, K is the desired field of p^n elements, since Lemma 45.1 showed that $x^{p^n} - x$ has p^n distinct zeros in $\overline{\mathbf{Z}_p}$. ∎

Corollary *If* F *is any finite field, then for every positive integer* n, *there is an irreducible polynomial* $F[x]$ *of degree* n.

Proof. Let F have $q = p^r$ elements, where p is the characteristic of F. By Theorem 45.4, there is a field $K \leq \overline{F}$ containing $\mathbf{Z}_p$ (up to isomorphism) and consisting precisely of the zeros of $x^{p^{rn}} - x$. Every element of F is a zero of $x^{p^r} - x$, by Theorem 45.2. Now $p^{rs} = p^r p^{r(s-1)}$. Applying this equation repeatedly to the exponents and using the fact that for $\alpha \in F$ we have $\alpha^{p^r} = \alpha$, we see that for $\alpha \in F$,

$$\alpha^{p^{rn}} = \alpha^{p^{r(n-1)}} = \alpha^{p^{r(n-2)}} = \cdots = \alpha^{p^r} = \alpha.$$

Thus $F \le K$. Then Theorem 45.1 shows that we must have $[K:F] = n$. We have seen that K is simple over F in Corollary 2 of Theorem 45.3, so $K = F(\beta)$ for some $\beta \in K$. Therefore, irr(β, F) must be of degree n. ∎

Exercises

45.1 Find all generators of each of the following cyclic groups. (Recall the theory of cyclic groups.)

a) $\langle \mathbf{Z}_7^*, \cdot \rangle$ b) $\langle \mathbf{Z}_{17}^*, \cdot \rangle$ c) $\langle \mathbf{Z}_{23}^*, \cdot \rangle$

45.2 (Recall the theory of cyclic groups.)

a) Find the number of primitive 8th roots of unity in GF(9).

b) Find the number of primitive 18th roots of unity in GF(19).

c) Find the number of primitive 15th roots of unity in GF(31).

d) Find the number of primitive 10th roots of unity in GF(23).

45.3 Let $\overline{\mathbf{Z}}_2$ be an algebraic closure of $\mathbf{Z}_2$, and let α, $\beta \in \overline{\mathbf{Z}}_2$ be zeros of $x^3 + x^2 + 1$ and $x^3 + x + 1$, respectively. Using the results of this section, show that $\mathbf{Z}_2(\alpha) = \mathbf{Z}_2(\beta)$.

†**45.4** Show that every irreducible polynomial in $\mathbf{Z}_p[x]$ is a divisor of $x^{p^n} - x$ for some n.

45.5 Mark each of the following true or false.

  — a) The nonzero elements of every finite field form a cyclic group under multiplication.

  — b) The elements of every finite field form a cyclic group under addition.

  — c) The zeros in $\mathbf{C}$ of $(x^{28} - 1) \in \mathbf{Q}[x]$ form a cyclic group under multiplication.

  — d) There exists a finite field of 60 elements.

  — e) There exists a finite field of 125 elements.

  — f) There exists a finite field of 36 elements.

  — g) The complex number i is a primitive 4th root of unity.

  — h) There exists an irreducible polynomial of degree 58 in $\mathbf{Z}_2[x]$.

  — i) The nonzero elements of $\mathbf{Q}$ form a cyclic group $\mathbf{Q}^*$ under field multiplication.

  — j) If F is a finite field, then every isomorphism mapping F into an algebraic closure $\overline{F}$ of F is an automorphism of F.

45.6 Let F be a finite field of p^n elements containing the prime subfield $\mathbf{Z}_p$. Show that if $\alpha \in F$ is a generator of the cyclic group $\langle F^*, \cdot \rangle$ of nonzero elements of F, then $\deg(\alpha, \mathbf{Z}_p) = n$.

45.7 Show that a finite field of p^n elements has exactly one subfield of p^m elements for each divisor m of n.

45.8 Show that $x^{p^n} - x$ is the product of all monic irreducible polynomials in $\mathbf{Z}_p[x]$ of a degree d dividing n.

45.9 Let p be an odd prime.

a) Show that for $a \in \mathbf{Z}$, where $a \not\equiv 0 \pmod{p}$, the congruence $x^2 \equiv a \pmod{p}$ has a solution in $\mathbf{Z}$ if and only if $a^{(p-1)/2} \equiv 1 \pmod{p}$. [*Hint:* Formulate an equivalent statement in the finite field $\mathbf{Z}_p$, and use the theory of cyclic groups.]

b) Using part (a), determine whether or not the polynomial $x^2 - 6$ is irreducible in $\mathbf{Z}_{17}[x]$.

45.10 Show that two finite fields of the same order are isomorphic.

45.11 Use Exercise 43.10 to show that $x^{p^n} - x$ has no zeros of multiplicity >1 in $\overline{\mathbf{Z}_p}$. (See the proof of Lemma 45.1.)

45.12 Let E be a finite field of order p^n.

a) Show that the Frobenius automorphism σ_p has order n.

b) Deduce from (a) that $G(E/\mathbf{Z}_p)$ is cyclic of order n with generator σ_p. [*Hint:* Remember that

$$|G(E/F)| = \{E : F\} = [E : F]$$

for a finite separable splitting field extension E over F.]

forty-six

Galois
Theory

46.1 RÉSUMÉ

This chapter is perhaps the climax in elegance of the subject matter of the entire text. The Galois theory gives a beautiful interplay of group and field theory. Starting with Chapter 40, our work has been aimed at this goal. We shall start by recalling for you the main results we have developed that you should have well in mind.

1. Let $F \leq E \leq \bar{F}$, $\alpha \in E$, and let β be a conjugate of α over F, that is, $\text{irr}(\alpha, F)$ has β as a zero also. Then there is an isomorphism $\psi_{\alpha,\beta}$ mapping $F(\alpha)$ onto $F(\beta)$ that leaves F fixed and maps α onto β.

2. If $F \leq E \leq \bar{F}$ and $\alpha \in E$, then an automorphism σ of $\bar{F}$ that leaves F fixed *must* map α onto some conjugate of α over F.

3. If $F \leq E$, the collection of all automorphisms of E leaving F fixed forms a group $G(E/F)$. For any subset S of $G(E/F)$, the set of all elements of E left fixed by all elements of S is a field E_S. Also, $F \leq E_{G(E/F)}$.

409

4. A field E, $F \leq E \leq \bar{F}$, is a splitting field over F if and only if every isomorphism of E into $\bar{F}$ leaving F fixed is an automorphism of E. If E is a finite extension and a splitting field over F, then $|G(E/F)| = \{E : F\}$.

5. If E is a finite extension of F, then $\{E : F\}$ divides $[E : F]$. If E is also separable over F, then $\{E : F\} = [E : F]$. Also, E is separable over F if and only if irr(α, F) has all zeros of multiplicity 1 for every $\alpha \in E$.

6. If E is a finite extension of F and is a separable splitting field over F, then $|G(E/F)| = \{E : F\} = [E : F]$.

46.2 NORMAL EXTENSIONS

We are going to be interested in finite extensions K of F such that every isomorphism of K leaving F fixed is an automorphism of K and such that

$$[K : F] = \{K : F\}.$$

In view of résumé numbers 4 and 5, these are the finite extensions of F that are separable splitting fields over F.

Definition   A finite extension K of F is a **finite normal extension of** F if K is a separable splitting field over F.

Suppose that K is a finite normal extension of F, where $K \leq \bar{F}$, as usual. Then by result 4, every automorphism of $\bar{F}$ leaving F fixed induces an automorphism of K. As before, we let $G(K/F)$ be the group of all automorphisms of K leaving F fixed. After one more result, we shall be ready to illustrate the main theorem.

Theorem 46.1   *Let K be a finite normal extension of F, and let E be an extension of F, where $F \leq E \leq K \leq \bar{F}$. Then K is a finite normal extension of E, and $G(K/E)$ is precisely the subgroup of $G(K/F)$ consisting of all those automorphisms that leave E fixed. Moreover, two automorphisms σ and τ in $G(K/F)$ induce the same isomorphism of E into $\bar{F}$ if and only if they are in the same right coset of $G(K/E)$ in $G(K/F)$.*

Proof.   If K is the splitting field of a set $\{f_i(x) | i \in I\}$ of polynomials in $F[x]$, then clearly K is the splitting field over E of this same set of polynomials viewed as elements of $E[x]$. Theorem 43.3 shows that K is separable over E, since K is separable over F. Thus K is a normal extension of E. This establishes our first contention.

Clearly, every element of $G(K/E)$ is an automorphism of K leaving F fixed, since it even leaves the possibly larger field E fixed. Thus $G(K/E)$ can be viewed as a subset of $G(K/F)$. Since $G(K/E)$ is a group under function composition also, we see that $G(K/E) \leq G(K/F)$.

Finally, for σ and τ in $G(K/F)$, σ and τ are in the same right coset of $G(K/E)$ if and only if $\sigma\tau^{-1} \in G(K/E)$ or if and only if $\sigma = \mu\tau$ for $\mu \in G(K/E)$. But if $\sigma = \mu\tau$ for $\mu \in G(K/E)$, then for $\alpha \in E$, we have

$$\alpha\sigma = \alpha(\mu\tau) = (\alpha\mu)\tau = \alpha\tau,$$

since $\alpha\mu = \alpha$ for $\alpha \in E$. Conversely, if $\alpha\sigma = \alpha\tau$ for all $\alpha \in E$, then

$$\alpha(\sigma\tau^{-1}) = \alpha$$

for all $\alpha \in E$, so $\sigma\tau^{-1}$ leaves E fixed, and $\mu = \sigma\tau^{-1}$ is thus in $G(K/E)$. ∎

The preceding theorem shows that there is a one-to-one correspondence between right cosets of $G(K/E)$ in $G(K/F)$ and isomorphisms of E leaving F fixed. Note that we can't say that these right cosets correspond to *automorphisms* of E over F, since E may not be a splitting field over F. Of course, if E is a *normal* extension of F, then these isomorphisms would be automorphisms of E over F. You might guess that this will happen if and only if $G(K/E)$ is a *normal* subgroup of $G(K/F)$. and this is indeed the case. That is, the two different uses of the word *normal* are really closely related. Thus if E is a normal extension of F, then the right cosets of $G(K/E)$ in $G(K/F)$ can be viewed as elements of the *factor group* $G(K/F)/G(K/E)$, which is then a group of automorphisms acting on E and leaving F fixed. We shall show that this factor group is isomorphic to $G(E/F)$.

46.3 THE MAIN THEOREM

The Main Theorem of Galois Theory states that for a finite normal extension K of a field F, there is a one-to-one correspondence between the subgroups of $G(K/F)$ and the intermediate fields E, where $F \leq E \leq K$. *This correspondence associates with each intermediate field E the subgroup $G(K/E)$. Of course, we can also go the other way and start with a subgroup H of $G(K/F)$ and associate with H its fixed field K_H.* We shall illustrate this with an easy example, then state the theorem and discuss its proof.

Example 46.1 Let $K = \mathbf{Q}(\sqrt{2}, \sqrt{3})$. Now K is a normal extension of $\mathbf{Q}$, and Example 40.4 showed that there are four automorphisms of K leaving $\mathbf{Q}$ fixed. We recall them by giving their values on the basis $\{1, \sqrt{2}, \sqrt{3}, \sqrt{6}\}$ for K over $\mathbf{Q}$.

    ι: The identity map.
    σ_1: Maps $\sqrt{2}$ onto $-\sqrt{2}$, $\sqrt{6}$ onto $-\sqrt{6}$, and leaves the others fixed.
    σ_2: Maps $\sqrt{3}$ onto $-\sqrt{3}$, $\sqrt{6}$ onto $-\sqrt{6}$, and leaves the others fixed.
    σ_3: Maps $\sqrt{2}$ onto $-\sqrt{2}$, $\sqrt{3}$ onto $-\sqrt{3}$, and leaves the others fixed.

We saw that $\{\iota, \sigma_1, \sigma_2, \sigma_3\}$ is isomorphic to the Klein 4-group. The complete list of subgroups, with each subgroup paired off with the corresponding intermediate field that it leaves fixed, is as follows:

$$\{\iota, \sigma_1, \sigma_2, \sigma_3\} \leftrightarrow \mathbf{Q},$$
$$\{\iota, \sigma_1\} \leftrightarrow \mathbf{Q}(\sqrt{3}),$$
$$\{\iota, \sigma_2\} \leftrightarrow \mathbf{Q}(\sqrt{2}),$$
$$\{\iota, \sigma_3\} \leftrightarrow \mathbf{Q}(\sqrt{6}),$$
$$\{\iota\} \leftrightarrow \mathbf{Q}(\sqrt{2}, \sqrt{3}).$$

All subgroups of the abelian group $\{\iota, \sigma_1, \sigma_2, \sigma_3\}$ are normal subgroups, and clearly all the intermediate fields are normal extensions of $\mathbf{Q}$. Isn't that elegant?

Note that if one subgroup is contained in another, then the larger of the two subgroups corresponds to the smaller of the two corresponding fixed fields. The reason for this is clear. The larger the subgroup, that is, the more auto-morphisms, the smaller the fixed field, that is, the fewer elements left fixed. In Fig. 46.1 we give the corresponding lattice diagrams for the subgroups and intermediate fields. Note again that the groups near the top correspond to the fields near the bottom. That is, one lattice looks like the other inverted or turned upside down. Since here each lattice actually looks like itself turned upside down, this is not a good example for us to use to illustrate this lattice

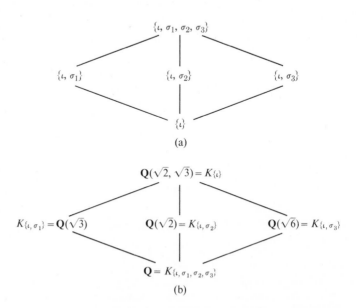

(a)

(b)

Fig. 46.1 (a) Group lattice diagram. (b) Field lattice diagram.

inversion principle. If you will turn ahead to Fig. 47.2, you will see diagrams in which the lattices do not look like their own inversions. ∎

Definition   If K is a finite normal extension of a field F, the $G(K/F)$ is the *Galois group of K over F.*

We shall now state the main theorem, then give another example, and finally, under a starred heading, complete the proof of the main theorem.

Theorem 46.2 (Main Theorem of Galois Theory)   *Let K be a finite normal extension of a field F, with Galois group $G(K/F)$. For a field E, where $F \le E \le K$, let $E\lambda$ be the subgroup of $G(K/F)$ leaving E fixed. Then λ is a one-to-one map of the set of all such intermediate fields E onto the set of all subgroups of $G(K/F)$. The following properties hold for λ:*

1. $E\lambda = G(K/E)$.
2. $E = K_{G(K/E)} = K_{E\lambda}$.
3. *For* $H \le G(K/F)$, $K_H\lambda = H$.
4. $[K:E] = |E\lambda|$; $[E:F] = \{G(K/F):E\lambda\}$, *the number of cosets of $E\lambda$ in $G(K/F)$.*
5. *E is a normal extension of F if and only if $E\lambda$ is a normal subgroup of $G(K/F)$. When $E\lambda$ is a normal subgroup of $G(K/F)$, then*

$$G(E/F) \simeq G(K/F)/G(K/E).$$

6. *The lattice of subgroups of $G(K/F)$ is the inverted lattice of intermediate fields of K over F.*

Observations on the Proof.   We have really already proved a substantial part of this theorem. Let us see just how much we have left to prove.

Property 1 is just the definition of λ found in the statement of the theorem. For property 2, Theorem 40.4 shows that

$$E \le K_{G(K/E)}.$$

Let $\alpha \in K$, where $\alpha \notin E$. Since K is a normal extension of E, by using a basic isomorphism and the Isomorphism Extension Theorem, we can find an automorphism of K leaving E fixed and mapping α onto a different zero of irr(α, F). This implies that

$$K_{G(K/E)} \le E,$$

so $E = K_{G(K/E)}$. This disposes of property 2 and also tells us that λ is one to one, for if $E_1\lambda = E_2\lambda$, then by property 2, we have

$$E_1 = K_{E_1\lambda} = K_{E_2\lambda} = E_2.$$

Now property 3 is going to be our main job. This amounts exactly to showing that λ is an onto map. Of course, for $H \le G(K/F)$, we have $H \le K_H\lambda$, for H surely is included in the set of all automorphisms leaving K_H fixed. Here we will be using strongly our property $[K : E] = \{K : E\}$.

Property 4 is clear from $[K : E] = \{K : E\}$, $[E : F] = \{E : F\}$, and the last statement in Theorem 46.1.

We shall have to show that the two senses of the word *normal* correspond for property 5.

We have already disposed of property 6 in Example 46.1.

Thus only properties 3 and 5 remain to be proved.

The Main Theorem of Galois Theory is a strong tool in the study of zeros of polynomials. If $f(x) \in F[x]$ is such that every irreducible factor of $f(x)$ is separable over F, then the splitting field K of $f(x)$ over F is a normal extension of F. The Galois group $G(K/F)$ is the **group of the polynomial** $f(x)$ **over** F. The structure of this group may give considerable information regarding the zeros of $f(x)$. This will be strikingly illustrated in Chapter 49 when we achieve our *final goal*.

46.4 GALOIS GROUPS
OVER FINITE FIELDS

Let K be a finite extension of a *finite field* F. We have seen that K is a separable extension of F (a finite field is perfect). Suppose that the order of F is p^r and $[K : F] = n$, so the order of K is p^{rn}. Then we have seen that K is the splitting field of $x^{p^{rn}} - x$ over F. Hence K is a normal extension of F.

Now one automorphism of K that leaves F fixed is σ_{p^r}, where for $\alpha \in K$, $\alpha\sigma_{p^r} = \alpha^{p^r}$. Note that $\alpha(\sigma_{p^r})^i = \alpha^{p^{ri}}$. Since a polynomial of degree p^{ri} can have at most p^{ri} zeros in a field, we see that the smallest power of σ_{p^r} that could possibly leave all p^{rn} elements of K fixed is the nth power. That is, the order of the element σ_{p^r} in $G(K/F)$ is at least n. Therefore, since $|G(K/F)| = [K : F] = n$, it must be that $G(K/F)$ is cyclic and generated by σ_{p^r}. We summarize these arguments in a theorem.

> **Theorem 46.3** Let K be a finite extension of degree n of a finite field F of p^r elements. Then $G(K/F)$ is cyclic of order n, and is generated by σ_{p^r}, where for $\alpha \in K$, $\alpha\sigma_{p^r} = \alpha^{p^r}$.

We use this theorem to give another illustration of the Main Theorem of Galois Theory.

Example 46.2 Let $F = Z_p$, and let $K = \mathrm{GF}(p^{12})$, so $[K : F] = 12$. Then $G(K/F)$ is isomorphic to the cyclic group $\langle Z_{12}, + \rangle$. The lattice diagram for the subgroups and for the intermediate fields is given in Fig. 46.2. Again, each lattice is not only the inversion of the other, but unfortunately, also

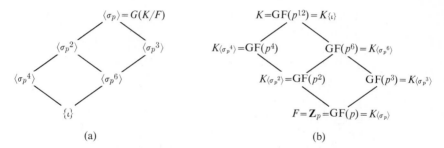

Fig. 46.2 (a) Group lattice diagram. (b) Field lattice diagram.

looks like the inversion of itself. Examples where the lattices do not look like their own inversions are given in the next (starred) section. We describe the cyclic subgroups of $G(K/F) = \langle \sigma_p \rangle$ by giving generators, e.g.,

$$\langle \sigma_p{}^4 \rangle = \{\iota, \sigma_p{}^4, \sigma_p{}^8\}. \quad \blacksquare$$

*46.5 PROOF OF THE MAIN THEOREM COMPLETED

We saw that properties 3 and 5 are all that remain to be proved in the Main Theorem of Galois Theory.

Proof. Turning to property 3, we must show that for $H \leq G(K/F)$, $K_H \lambda = H$. We know that $H \leq K_H \lambda \leq G(K/F)$. Thus what we really must show is that it is impossible to have H a *proper* subgroup of $K_H \lambda$. We shall suppose that

$$H < K_H \lambda,$$

and shall derive a contradiction. If K_H is infinite, then as a finite separable extension of an infinite field, $K = K_H(\alpha)$ for some $\alpha \in K$, by Theorem 43.6. On the other hand, if K_H is finite, then we still have $K = K_H(\alpha)$ for some $\alpha \in K$, by Corollary 2 of Theorem 45.3. Let

$$n = [K : K_H] = \{K : K_H\} = |G(K/K_H)|.$$

Then $H < G(K/K_H)$ implies that $|H| < |G(K/K_H)| = n$. Thus we would have to have $|H| < [K : K_H] = n$. Let the elements of H be $\sigma_1, \ldots, \sigma_{|H|}$, and consider the polynomial

$$f(x) = \prod_{i=1}^{|H|} (x - \alpha\sigma_i).$$

Then $f(x)$ is of degree $|H| < n$. Now the coefficients of each power of x in $f(x)$ are *symmetric* expressions in the $\alpha\sigma_i$. For example, the coefficient of $x^{|H|-1}$ is $-\alpha\sigma_1 - \alpha\sigma_2 - \cdots - \alpha\sigma_{|H|}$. Thus these coefficients are invariant under each isomorphism $\sigma_i \in H$, since if $\sigma \in H$, then

$$\sigma_1\sigma, \ldots, \sigma_{|H|}\sigma$$

is again the sequence $\sigma_1, \ldots, \sigma_{|H|}$, except for order, H being a group. Hence $f(x)$ has coefficients in K_H, and since some σ_i is ι, we see that some $\alpha\sigma_i$ is α, so $f(\alpha) = 0$. Therefore, we would have

$$\deg(\alpha, K_H) \le |H| < n = [K : K_H] = [K_H(\alpha) : K_H].$$

This is impossible. Thus we have proved property 3.

We turn to property 5. Every extension E of F, $F \le E \le K$, is separable over F, by Theorem 43.3. Thus E is normal over F if and only if E is a splitting field over F. By the Isomorphism Extension Theorem, every isomorphism of E into $\bar{F}$ leaving F fixed can be extended to an *automorphism* of K, since K is *normal* over F. Thus the automorphisms of $G(K/F)$ induce all possible isomorphisms of E into $\bar{F}$ leaving F fixed. By Theorem 42.1, this shows that E is a splitting field over F, and hence normal over F, if and only if for all $\sigma \in G(K/F)$ and $\alpha \in E$,

$$(\alpha\sigma) \in E.$$

By property 2, E is the fixed field of $G(K/E)$, so $(\alpha\sigma) \in E$ if and only if for all $\tau \in G(K/E)$,

$$(\alpha\sigma)\tau = \alpha\sigma.$$

This in turn holds if and only if

$$\alpha(\sigma\tau\sigma^{-1}) = \alpha$$

for all $\alpha \in E$, $\sigma \in G(K/F)$, and $\tau \in G(K/E)$. But this means that for all $\sigma \in G(K/F)$ and $\tau \in G(K/E)$, $\sigma\tau\sigma^{-1}$ leaves every element of E fixed, that is,

$$(\sigma\tau\sigma^{-1}) \in G(K/E).$$

This is precisely the condition that $G(K/E)$ be a normal subgroup of $G(K/F)$.

It remains for us to show that when E is a normal extension of F, $G(E/F) \simeq G(K/F)/G(K/E)$. For $\sigma \in G(K/F)$, let σ_E be the *automorphism* of E induced by σ (we are assuming that E is a *normal* extension of F). Thus $\sigma_E \in G(E/F)$. The map $\phi : G(K/F) \to G(E/F)$ given by

$$\sigma\phi = \sigma_E$$

for $\sigma \in G(K/F)$ is obviously a homomorphism. By the Isomorphism Extension Theorem, every automorphism of E leaving F fixed can be extended to some automorphism of K, that is, it is τ_E for some $\tau \in G(K/F)$. Thus ϕ is onto $G(E/F)$. The kernel of ϕ is clearly $G(K/E)$. Therefore, by the Fundamental Isomorphism Theorem, $G(E/F) \simeq G(K/F)/G(K/E)$. Furthermore, this isomorphism is a natural one. ∎

Exercises

46.1 The field $K = \mathbf{Q}(\sqrt{2}, \sqrt{3}, \sqrt{5})$ is a finite normal extension of $\mathbf{Q}$. Fill in the blanks below. The notation is that of Theorem 46.2.

a) $[K:\mathbf{Q}] = $ _____.

b) $|G(K/\mathbf{Q})| = $ _____.

c) $|\mathbf{Q}\lambda| = $ _____.

d) $|(\mathbf{Q}(\sqrt{2}, \sqrt{3}))\lambda| = $ _____.

e) $|(\mathbf{Q}(\sqrt{6}))\lambda| = $ _____.

f) $|(\mathbf{Q}(\sqrt{30}))\lambda| = $ _____.

g) $|(\mathbf{Q}(\sqrt{2} + \sqrt{6}))\lambda| = $ _____.

h) $|K\lambda| = $ _____.

46.2 Describe the group of the polynomial $(x^4 - 1) \in \mathbf{Q}[x]$ over $\mathbf{Q}$.

46.3 Give the order and describe a generator of the group $G(GF(729)/GF(9))$.

46.4 Give an example of two finite normal extensions K_1 and K_2 of the same field F such that K_1 and K_2 are not isomorphic fields but $G(K_1/F) \simeq G(K_2/F)$.

46.5 Let K be the splitting field of $x^3 - 2$ over $\mathbf{Q}$. (Refer to Example 42.2.)

a) Describe the six elements of $G(K/\mathbf{Q})$ by giving their values on $\sqrt[3]{2}$ and $i\sqrt{3}$. (By Example 42.2, $K = \mathbf{Q}(\sqrt[3]{2}, i\sqrt{3})$.)

b) To what group we have seen before is $G(K/\mathbf{Q})$ isomorphic?

c) Using the notation given in the answer to (a) in the back of the text, give the lattice diagrams for the subfields of K and for the subgroups of $G(K/\mathbf{Q})$, indicating corresponding intermediate fields and subgroups, as we did in Fig. 46.1.

†**46.6** A finite normal extension K of field F is **abelian over** F if $G(K/F)$ is an abelian group. Show that if K is abelian over F and E is a normal extension of F, where $F \leq E \leq K$, then K is abelian over E and E is abelian over F.

46.7 Mark each of the following true or false.

___ a) Two different subgroups of a Galois group may have the same fixed field.

___ b) In the notation of Theorem 46.2, if $F \leq E < L \leq K$, then $E\lambda < L\lambda$.

___ c) If K is a finite normal extension of F, then K is a normal extension of E, where $F \leq E \leq K$.

___ d) If two finite normal extensions E and L of a field F have isomorphic Galois groups, then $[E:F] = [L:F]$.

___ e) If E is a finite normal extension of F and H is a normal subgroup of $G(E/F)$, then E_H is a normal extension of F.

___ f) If E is any finite normal simple extension of a field F, then the Galois group $G(E/F)$ is a simple group.

___ g) No Galois group is simple.

___ h) The Galois group of a finite extension of a finite field is abelian.

— i) An extension E of degree 2 over a field F is always a normal extension of F.

— j) An extension E of degree 2 over a field F is always a normal extension of F if the characteristic of F is not 2.

46.8 Let K be a finite normal extension of a field F. Prove that for every $\alpha \in K$, the **norm of α over** F, given by

$$N_{K/F}(\alpha) = \prod_{\sigma \in G(K/F)} \alpha\sigma,$$

and the **trace of** α **over** F, given by

$$Tr_{K/F}(\alpha) = \sum_{\sigma \in G(K/F)} \alpha\sigma,$$

are elements of F.

46.9 Consider $K = \mathbf{Q}(\sqrt{2}, \sqrt{3})$. Referring to Exercise 46.8, compute each of the following (see Example 46.1).

a) $N_{K/\mathbf{Q}}(\sqrt{2})$

b) $N_{K/\mathbf{Q}}(\sqrt{2} + \sqrt{3})$

c) $N_{K/\mathbf{Q}}(\sqrt{6})$

d) $N_{K/\mathbf{Q}}(2)$

e) $Tr_{K/\mathbf{Q}}(\sqrt{2})$

f) $Tr_{K/\mathbf{Q}}(\sqrt{2} + \sqrt{3})$

g) $Tr_{K/\mathbf{Q}}(\sqrt{6})$

h) $Tr_{K/\mathbf{Q}}(2)$

46.10 Let K be a normal extension of F, and let $K = F(\alpha)$. Let

$$\mathrm{irr}(\alpha, F) = x^n + a_{n-1}x^{n-1} + \cdots + a_1 x + a_0.$$

Referring to Exercise 46.8, show that

a) $N_{K/F}(\alpha) = (-1)^n a_0$,

b) $Tr_{K/F}(\alpha) = -a_{n-1}$.

46.11 Describe the group of the polynomial $(x^4 - 5x^2 + 6) \in \mathbf{Q}[x]$ over $\mathbf{Q}$.

46.12 Describe the group of the polynomial $(x^3 - 1) \in \mathbf{Q}[x]$ over $\mathbf{Q}$.

46.13 Let $f(x) \in F[x]$ be a polynomial of degree n such that each irreducible factor is separable over F. Show that the order of the group of $f(x)$ over F divides $n!$.

46.14 Let $f(x) \in F[x]$ be a polynomial such that every irreducible factor of $f(x)$ is a separable polynomial over F. Show that the group of $f(x)$ over F can be viewed in a natural way as a group of permutations of the zeros of $f(x)$ in $\bar{F}$.

46.15 Let F be a field and let ζ be a primitive nth root of unity in $\bar{F}$, where the characteristic of F is either 0 or does not divide n.

a) Show that $F(\zeta)$ is a normal extension of F.

b) Show that $G(F(\zeta)/F)$ is abelian. [*Hint*: Every $\sigma \in G(F(\zeta)/F)$ maps ζ onto some ζ^r and is completely determined by this value r.]

46.16 A finite normal extension K of a field F is **cyclic over** F if $G(K/F)$ is a cyclic group.

a) Show that if K is cyclic over F and E is a normal extension of F, where $F \leq E \leq K$, then E is cyclic over F and K is cyclic over E.

b) Show that if K is cyclic over F, then there exists exactly one field E, $F \leq E \leq K$, of degree d over F for each divisor d of $[K : F]$.

46.17 Let K be a finite normal extension of F.

a) For $\alpha \in K$, show that

$$f(x) = \prod_{\sigma \in G(K/F)} (x - \alpha\sigma)$$

is in $F[x]$.

b) Referring to (a), show that $f(x)$ is a power of $\mathrm{irr}(\alpha, F)$, and $f(x) = \mathrm{irr}(\alpha, F)$ if and only if $E = F(\alpha)$.

46.18 Let K be a finite normal extension of a field F, and let E and L be extensions of F contained in K, as shown in Fig. 46.3. Describe $G\{K/(E \vee L)\}$ in terms of $G(K/E)$ and $G(K/L)$.

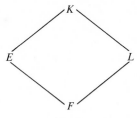

Fig. 46.3

*46.19 With reference to the situation in Exercise 46.18, describe $G\{K/(E \cap L)\}$ in terms of $G(K/E)$ and $G(K/L)$.

*forty-seven

Illustrations of Galois Theory

---

*47.1 SYMMETRIC FUNCTIONS

Let F be a field, and let $y_1, \ldots, y_n$ be n indeterminates. There are some obvious automorphisms of $F(y_1, \ldots, y_n)$ leaving F fixed, namely those defined by permutations of $\{y_1, \ldots, y_n\}$. To be less clear but more explicit, let σ be a permutation of $\{1, \ldots, n\}$, that is, $\sigma \in S_n$. Then σ gives rise to a natural map $\bar{\sigma}: F(y_1, \ldots, y_n) \to F(y_1, \ldots, y_n)$ given by

$$\frac{f(y_1, \ldots, y_n)}{g(y_1, \ldots, y_n)} \bar{\sigma} = \frac{f(y_{1\sigma}, \ldots, y_{n\sigma})}{g(y_{1\sigma}, \ldots, y_{n\sigma})}$$

for $f(y_1, \ldots, y_n)$, $g(y_1, \ldots, y_n) \in F[y_1, \ldots, y_n]$, with $g(y_1, \ldots, y_n) \neq 0$. It is immediate that $\bar{\sigma}$ is an automorphism of $F(y_1, \ldots, y_n)$ leaving F fixed. The elements of $F(y_1, \ldots, y_n)$ left fixed by *all* $\bar{\sigma}$, for all $\sigma \in S_n$, are those rational functions that are *symmetric* in the indeterminates $y_1, \ldots, y_n$.

Definition An element of $F(y_1, \ldots, y_n)$ is a **symmetric function in** $y_1, \ldots, y_n$ **over** F, if it is left fixed by all permutations of $y_1, \ldots, y_n$ in the sense just explained.

420

Let $\bar{S}_n$ be the group of all the automorphisms $\bar{\sigma}$ for $\sigma \in S_n$. Obviously, $\bar{S}_n$ is naturally isomorphic to S_n. Let K be the subfield of $F(y_1, \ldots, y_n)$, which is the fixed field of $\bar{S}_n$. Consider the polynomial

$$f(x) = \prod_{i=1}^{n} (x - y_i);$$

this polynomial $f(x) \in (F(y_1, \ldots, y_n))[x]$ is a **general polynomial of degree** n. Let $\bar{\sigma}_x$ be the extension of $\bar{\sigma}$, in the natural way, to $(F(y_1, \ldots, y_n))[x]$, where $x\bar{\sigma}_x = x$. Now clearly $f(x)$ is left fixed by each map $\bar{\sigma}_x$ for $\sigma \in S_n$, that is,

$$\prod_{i=1}^{n} (x - y_i) = \prod_{i=1}^{n} (x - y_{i\sigma}).$$

Thus the coefficients of $f(x)$ are in K; they are symmetric functions in the $y_1, \ldots, y_n$. As illustration, note that the constant term of $f(x)$ is

$$(-1)^n y_1 y_2 \cdots y_n,$$

the coefficient of x^{n-1} is $-(y_1 + y_2 + \cdots + y_n)$, and so on. These are obviously symmetric functions in $y_1, \ldots, y_n$.

Definition The ith **elementary symmetric function in** $y_1, \ldots, y_n$ is $s_i = (-1)^i a_i$, where a_i is the coefficient of x^{n-i} in the general polynomial $\prod_{i=1}^{n} (x - y_i)$.

Thus the first elementary symmetric function in $y_1, \ldots, y_n$ is

$$s_1 = y_1 + y_2 + \cdots + y_n,$$

the second is $s_2 = y_1 y_2 + y_1 y_3 + \cdots + y_{n-1} y_n$, and so on, and the nth is $s_n = y_1 y_2 \cdots y_n$.

Consider the field $E = F(s_1, \ldots, s_n)$. Of course, $E \le K$, where K is the field of all symmetric functions in $y_1, \ldots, y_n$ over F. But $F(y_1, \ldots, y_n)$ is a finite normal extension of E, namely the splitting field of

$$f(x) = \prod_{i=1}^{n} (x - y_i)$$

over E. Since the degree of $f(x)$ is n, we have at once

$$[F(y_1, \ldots, y_n) : E] \le n!$$

(see Exercise 42.2). However, since K is the fixed field of $\bar{S}_n$ and

$$|\bar{S}_n| = |S_n| = n!,$$

we have also

$$n! \leq \{F(y_1, \ldots, y_n): K\} \leq [F(y_1, \ldots, y_n): K].$$

Therefore,

$$n! \leq [F(y_1, \ldots, y_n): K] \leq [F(y_1, \ldots, y_n): E] \leq n!,$$

so

$$K = E.$$

The full Galois group of $F(y_1, \ldots, y_n)$ over E is therefore $\overline{S}_n$. The fact that $K = E$ shows that every symmetric function can be expressed as a rational function of the elementary symmetric functions $s_1, \ldots, s_n$. We summarize these results in a theorem.

> **Theorem 47.1** Let $s_1, \ldots, s_n$ be the elementary symmetric functions in the indeterminates $y_1, \ldots, y_n$. Then every symmetric function of $y_1, \ldots, y_n$ over F is a rational function of the elementary symmetric functions. Also, $F(y_1, \ldots, y_n)$ is a finite normal extension of degree $n!$ of $F(s_1, \ldots, s_n)$, and the Galois group of this extension is naturally isomorphic to S_n.

In view of Cayley's theorem, it can be deduced from Theorem 47.1 that any finite group can occur as a Galois group (up to isomorphism). (See Exercise 47.13.)

*47.2 EXAMPLES

Let us give our promised example of a finite normal extension having a Galois group whose lattice of subgroups does not look like its own inversion.

Example 47.1 Consider the splitting field in C of $x^4 - 2$ over Q. Now $x^4 - 2$ is irreducible over Q, by Eisenstein's criterion, with $p = 2$. Let $\alpha = \sqrt[4]{2}$ be the real positive zero of $x^4 - 2$. Then the four zeros of $x^4 - 2$ in C are obviously α, $-\alpha$, $i\alpha$, and $-i\alpha$, where i is the usual zero of $x^2 + 1$ in C. The splitting field K of $x^4 - 2$ over Q thus contains $(i\alpha)/\alpha = i$. Since α is a real number, $Q(\alpha) < R$, so $Q(\alpha) \neq K$. However, since $Q(\alpha, i)$ contains all zeros of $x^4 - 2$, we see that $Q(\alpha, i) = K$. Letting $E = Q(\alpha)$, we have the diagram in Fig. 47.1.

Now $\{1, \alpha, \alpha^2, \alpha^3\}$ is a basis for E over Q, and $\{1, i\}$ is a basis for K over E. Thus

$$\{1, \alpha, \alpha^2, \alpha^3, i, i\alpha, i\alpha^2, i\alpha^3\}$$

is a basis for K over Q. Since $[K:Q] = 8$, we must have $|G(K/Q)| = 8$, so we need to find eight automorphisms of K leaving Q fixed. We know

$K = \mathbf{Q}(\alpha, i)$

$E = \mathbf{Q}(\alpha)$

$\mathbf{Q}$

Fig. 47.1

that any such automorphism σ is completely determined by its values on elements of the basis $\{1, \alpha, \alpha^2, \alpha^3, i, i\alpha, i\alpha^2, i\alpha^3\}$, and these values are in turn determined by $\alpha\sigma$ and $i\sigma$. But $\alpha\sigma$ must always be a conjugate of α over $\mathbf{Q}$, that is, one of the four zeros of $\mathrm{irr}(\alpha, \mathbf{Q}) = x^4 - 2$. Likewise, $i\sigma$ must be a zero of $\mathrm{irr}(i, \mathbf{Q}) = x^2 + 1$. Thus the four possibilities for $\alpha\sigma$, combined with the two possibilities for $i\sigma$, must give all eight automorphisms. We describe these in Table 47.1.

Table 47.1

	ρ_0	ρ_1	ρ_2	ρ_3	μ_1	δ_1	μ_2	δ_2
$\alpha \rightarrow$	α	$i\alpha$	$-\alpha$	$-i\alpha$	α	$i\alpha$	$-\alpha$	$-i\alpha$
$i \rightarrow$	i	i	i	i	$-i$	$-i$	$-i$	$-i$

For example, $\alpha\rho_3 = -i\alpha$ and $i\rho_3 = i$, while ρ_0 is the identity automorphism. Now

$$\alpha(\rho_1\mu_1) = (\alpha\rho_1)\mu_1 = (i\alpha)\mu_1 = (i\mu_1)(\alpha\mu_1) = -i\alpha,$$

and, similarly,

$$i(\rho_1\mu_1) = -i,$$

so $\rho_1\mu_1 = \delta_2$. A similar computation shows that

$$\alpha(\mu_1\rho_1) = i\alpha \qquad \text{and} \qquad i(\mu_1\rho_1) = -i.$$

Thus $\mu_1\rho_1 = \delta_1$, so $\rho_1\mu_1 \neq \mu_1\rho_1$ and $G(K/\mathbf{Q})$ is not abelian. Therefore, $G(K/\mathbf{Q})$ must be isomorphic to one of the two nonabelian groups of order 8 described in Example 22.5. Computing from Table 47.1, we see that ρ_1 is of order 4, μ_1 is of order 2, $\{\rho_1, \mu_1\}$ generates $G(K/\mathbf{Q})$, and $\mu_1\rho_1 = \rho_1{}^3\mu_1 = \delta_1$. Thus $G(K/\mathbf{Q})$ is isomorphic to the group G_1 of Example 22.5, the *octic group*. We chose our notation for the elements of $G(K/\mathbf{Q})$ so that its group table would coincide with the table for the octic group in Table 4.2. The lattice

(a)

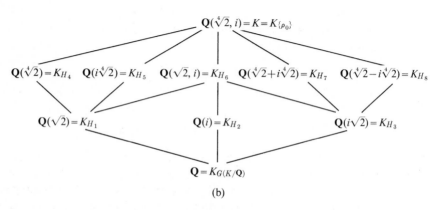

(b)

Fig. 47.2 (a) Group lattice diagram. (b) Field lattice diagram.

of subgroups H_i of $G(K/\mathbf{Q})$ is that given in Fig. 4.7. We repeat it here in Fig. 47.2 and also give the corresponding lattice of intermediate fields between $\mathbf{Q}$ and K. This finally illustrates nicely that one lattice is the inversion of the other.

The determination of the fixed fields K_{H_i} sometimes requires a bit of ingenuity. Let's illustrate. Finding K_{H_2} is easy, for we merely have to find an extension of $\mathbf{Q}$ of degree 2 left fixed by $\{\rho_0, \rho_1, \rho_2, \rho_3\}$. Since all ρ_j leave i fixed, $\mathbf{Q}(i)$ is clearly the field we are after. To find K_{H_4}, we have to find an extension of $\mathbf{Q}$ of degree 4 left fixed by ρ_0 and μ_1. Since μ_1 leaves α fixed and α is a zero of $\mathrm{irr}(\alpha, \mathbf{Q}) = x^4 - 2$, we see that $\mathbf{Q}(\alpha)$ is of degree 4 over $\mathbf{Q}$ and is left fixed by $\{\rho_0, \mu_1\}$. *By Galois theory, it is the only such field.* Here we are using strongly the one-to-one correspondence given by the Galois theory. If we find one field that fits the bill, it is the one we are after. Finding K_{H_7} requires more ingenuity. Since $H_7 = \{\rho_0, \delta_1\}$ is a group, for any $\beta \in K$ we see that $\beta\rho_0 + \beta\delta_1$ is left fixed by ρ_0 and δ_1. Taking $\beta = \alpha$, we see that

$\alpha\rho_0 + \alpha\delta_1 = \alpha + i\alpha$ is left fixed by H_7. We can check and see that ρ_0 and δ_1 are the only automorphisms leaving $\alpha + i\alpha$ fixed. Thus by the one-to-one correspondence, we must have

$$\mathbf{Q}(\alpha + i\alpha) = \mathbf{Q}(\sqrt[4]{2} + i\sqrt[4]{2}) = K_{H_7}.$$

Suppose one wished to find $\mathrm{irr}(\alpha + i\alpha, \mathbf{Q})$. If $\gamma = \alpha + i\alpha$, then for every conjugate of γ over $\mathbf{Q}$, there exists an automorphism of K mapping γ into that conjugate. Thus we need only compute the various different values $\gamma\sigma$ for $\sigma \in G(K/\mathbf{Q})$ to find the other zeros of $\mathrm{irr}(\gamma, \mathbf{Q})$. By Theorem 46.1, elements σ of $G(K/\mathbf{Q})$ giving these different values can be found by taking a set of representatives of the right cosets of $G(K/\mathbf{Q}(\gamma)) = \{\rho_0, \delta_1\}$ in $G(K/\mathbf{Q})$. A set of representatives for these right cosets is

$$\{\rho_0, \rho_1, \rho_2, \rho_3\}.$$

The conjugates of $\gamma = \alpha + i\alpha$ are thus $\alpha + i\alpha$, $i\alpha - \alpha$, $-\alpha - i\alpha$, and $-i\alpha + \alpha$. Hence

$$\begin{aligned}
\mathrm{irr}(\gamma, \mathbf{Q}) &= [(x - (\alpha + i\alpha))(x - (i\alpha - \alpha))] \\
&\quad \cdot [(x - (-\alpha - i\alpha))(x - (-i\alpha + \alpha))] \\
&= (x^2 - 2i\alpha x - 2\alpha^2)(x^2 + 2i\alpha x - 2\alpha^2) \\
&= x^4 + 4\alpha^4 = x^4 + 8. \quad \blacksquare
\end{aligned}$$

We have seen examples in which the splitting field of a quartic (4th degree) polynomial over a field F is an extension of F of degree 8 (Example 47.1) and of degree 24 (Theorem 47.1, with $n = 4$). The degree of an extension of a field F that is a splitting field of a quartic over F must clearly always divide $4! = 24$. Of course, the splitting field of $(x - 2)^4$ over $\mathbf{Q}$ is $\mathbf{Q}$, an extension of degree 1, and the splitting field of $(x^2 - 2)^2$ over $\mathbf{Q}$ is $\mathbf{Q}(\sqrt{2})$, an extension of degree 2. Our last example will give an extension of degree 4 for the splitting field of a quartic.

Example 47.2 Consider the splitting field of $x^4 + 1$ over $\mathbf{Q}$. By Theorem 31.3, one can show that $x^4 + 1$ is irreducible over $\mathbf{Q}$, by arguing that it does not factor in $\mathbf{Z}[x]$. This is easy to show (see Exercise 47.1). One can easily verify, by trial, that the zeros of $x^4 + 1$ are $(1 \pm i)/\sqrt{2}$ and $(-1 \pm i)/\sqrt{2}$. A computation shows that if

$$\alpha = \frac{1 + i}{\sqrt{2}},$$

then

$$\alpha^3 = \frac{-1 + i}{\sqrt{2}}, \qquad \alpha^5 = \frac{-1 - i}{\sqrt{2}}, \qquad \text{and} \qquad \alpha^7 = \frac{1 - i}{\sqrt{2}}.$$

(These facts that are pulled out of thin air are covered in the first few weeks of a course in functions of a complex variable or may be derived from the work in the next chapter.) Thus the splitting field K of $x^4 + 1$ over $\mathbf{Q}$ is $\mathbf{Q}(\alpha)$, and $[K : \mathbf{Q}] = 4$. Let us compute $G(K/\mathbf{Q})$ and give the group and field lattice diagrams. Since there exist automorphisms of K mapping α onto each conjugate of α, and since an automorphism σ of $\mathbf{Q}(\alpha)$ is completely determined by $\alpha\sigma$, we see that the four elements of $G(K/\mathbf{Q})$ are defined by Table 47.2. Since

$$\alpha(\sigma_j\sigma_k) = \alpha^j\sigma_k = (\alpha\sigma_k)^j = (\alpha^k)^j = \alpha^{jk}$$

Table 47.2

	σ_1	σ_3	σ_5	σ_7
$\alpha \to$	α	α^3	α^5	α^7

and $\alpha^8 = 1$, we see that $G(K/\mathbf{Q})$ is isomorphic to the group $\{1, 3, 5, 7\}$ under multiplication modulo 8. This is the group G_8 of Theorem 24.7. Since $\sigma_j^2 = \sigma_1$, the identity, for all j, $G(K/\mathbf{Q})$ must be isomorphic to the Klein 4-group. The lattice diagrams are given in Fig. 47.3.

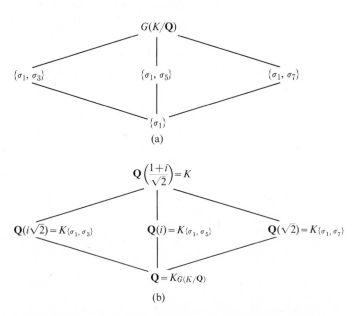

Fig. 47.3 (a) Group lattice diagram. (b) Field lattice diagram.

To find $K_{\{\sigma_1,\sigma_3\}}$, it is only necessary to find an element of K not in $\mathbf{Q}$ left fixed by $\{\sigma_1, \sigma_3\}$, since $[K_{\{\sigma_1,\sigma_3\}} : \mathbf{Q}] = 2$. Clearly, $\alpha\sigma_1 + \alpha\sigma_3$ is left fixed by both σ_1 and σ_3, since $\{\sigma_1, \sigma_3\}$ is a group. We have

$$\alpha\sigma_1 + \alpha\sigma_3 = \alpha + \alpha^3 = i\sqrt{2}.$$

Similarly,

$$\alpha\sigma_1 + \alpha\sigma_7 = \alpha + \alpha^7 = \sqrt{2}$$

is left fixed by $\{\sigma_1, \sigma_7\}$. This technique is of no use in finding $E_{\{\sigma_1,\sigma_5\}}$, for

$$\alpha\sigma_1 + \alpha\sigma_5 = \alpha + \alpha^5 = 0,$$

and $0 \in \mathbf{Q}$. But by a similar argument, $(\alpha\sigma_1)(\alpha\sigma_5)$ is left fixed by both σ_1 and σ_5, and

$$(\alpha\sigma_1)(\alpha\sigma_5) = \alpha\alpha^5 = -i.$$

Thus $\mathbf{Q}(-i) = \mathbf{Q}(i)$ is the field we are after. ∎

Exercises

*47.1 Show that $x^4 + 1$ is irreducible in $\mathbf{Q}[x]$, as we asserted in Example 47.2.

*47.2 Verify that the intermediate fields given in the field lattice diagram in Figure 47.3 are correct. (Some are verified in the text. Verify the rest.)

*47.3 For each field in the field lattice diagram in Fig. 47.2, find a primitive element generating the field over $\mathbf{Q}$ (see Theorem 43.6) and give its irreducible polynomial over $\mathbf{Q}$.

*47.4 Let ζ be a primitive 5th root of unity in $\mathbf{C}$.

a) Show that $\mathbf{Q}(\zeta)$ is the splitting field of $x^5 - 1$ over $\mathbf{Q}$.

b) Show that every automorphism of $K = \mathbf{Q}(\zeta)$ maps ζ onto some power ζ^r of ζ.

c) Using (b), describe the elements of $G(K/\mathbf{Q})$.

d) Give the group and field lattice diagrams for $\mathbf{Q}(\zeta)$ over $\mathbf{Q}$, computing the intermediate fields as we did in Examples 47.1 and 47.2.

*47.5 Describe the group of the polynomial $(x^5 - 2) \in (\mathbf{Q}(\zeta))[x]$ over $\mathbf{Q}(\zeta)$, where ζ is a primitive 5th root of unity.

*47.6 Repeat Exercise 47.4 for ζ a primitive 7th root of unity in $\mathbf{C}$.

*47.7 In the easiest way possible, describe the group of the polynomial

$$(x^8 - 1) \in \mathbf{Q}[x]$$

over $\mathbf{Q}$.

*47.8 Find the splitting field K in $\mathbf{C}$ of the polynomial $(x^4 - 4x^2 - 1) \in \mathbf{Q}[x]$. Compute the group of the polynomial over $\mathbf{Q}$ and exhibit the correspondence between the subgroups of $G(K/\mathbf{Q})$ and the intermediate fields. In other words, do the complete job.

*47.9 Express each of the following symmetric functions in y_1, y_2, y_3 over $\mathbf{Q}$ as a rational function of the elementary symmetric functions s_1, s_2, s_3.

a) $y_1^2 + y_2^2 + y_3^2$

b) $(y_1 - y_2)^2(y_1 - y_3)^2(y_2 - y_3)^2$

c) $\dfrac{y_1}{y_2} + \dfrac{y_2}{y_1} + \dfrac{y_1}{y_3} + \dfrac{y_3}{y_1} + \dfrac{y_2}{y_3} + \dfrac{y_3}{y_2}$

*47.10 Let $\alpha_1, \alpha_2, \alpha_3$ be the zeros in $\mathbf{C}$ of the polynomial

$$(x^3 - 4x^2 + 6x - 2) \in \mathbf{Q}[x].$$

Find the polynomial having as zeros precisely the following.

a) $\alpha_1 + \alpha_2 + \alpha_3$

b) $\alpha_1^2, \alpha_2^2, \alpha_3^2$

c) $(\alpha_1 - \alpha_2)^2, (\alpha_1 - \alpha_3)^2, (\alpha_2 - \alpha_3)^2$

*47.11 Let $f(x) \in F[x]$ be a monic polynomial of degree n having all its irreducible factors separable over F. Let $K \le \bar{F}$ be the splitting field of $f(x)$ over F, and suppose that $f(x)$ factors in $K[x]$ into

$$\prod_{i=1}^{n} (x - \alpha_i).$$

Let

$$\Delta(f) = \prod_{i<j} (\alpha_i - \alpha_j);$$

the product $(\Delta(f))^2$ is the **discriminant of** $f(x)$.

a) Show that $\Delta(f) = 0$ if and only if $f(x)$ has as a factor the square of some irreducible polynomial in $F[x]$.

b) Show that $(\Delta(f))^2 \in F$.

c) $G(K/F)$ may be viewed as a subgroup of $\bar{S}_n$, where $\bar{S}_n$ is the group of all permutations of $\{\alpha_i \mid i = 1, \ldots, n\}$. Show that $G(K/F)$, when viewed in this fashion, is a subgroup of $\bar{A}_n$, the group formed by all even permutations of $\{\alpha_i \mid i = 1, \ldots, n\}$, if and only if $\Delta(f) \in F$.

*47.12 An element of $\mathbf{C}$ is an **algebraic integer** if it is a zero of some *monic* polynomial in $\mathbf{Z}[x]$. Show that the set of all algebraic integers forms a subring of $\mathbf{C}$.

*47.13 Show that every finite group is isomorphic to some Galois group $G(K/F)$ for some finite normal extension K of some field F.

forty-eight

Cyclotomic Extensions

*48.1 THE GALOIS GROUP OF A CYCLOTOMIC EXTENSION

This section deals with extension fields of a field F obtained by adjoining to F some roots of unity. The case of a finite field F was covered in Chapter 45, so we shall be primarily concerned with the case where F is infinite.

Definition The splitting field of $x^n - 1$ over F is the nth *cyclotomic extension of* F.

Suppose that F is any field, and consider $(x^n - 1) \in F[x]$. By long division, as in the proof of Lemma 45.1, one sees that if α is a zero of $x^n - 1$ and $g(x) = (x^n - 1)/(x - \alpha)$, then $g(\alpha) = (n \cdot 1)(1/\alpha) \neq 0$, provided that the characteristic of F does not divide n. Therefore, under this condition, the splitting field of $x^n - 1$ is a separable and thus a normal extension of F.

Assume from now on that this is the case, and let K be the splitting field of $x^n - 1$ over F. Then $x^n - 1$ has n distinct zeros in K, and by Theorem 45.3 these form a cyclic group of order n under the field multiplication. We

429

saw in the corollary of Theorem 6.4 that a cyclic group of order n has $\varphi(n)$ generators, where φ is the Euler phi-function introduced prior to Theorem 24.8. For our situation here, these $\varphi(n)$ generators are exactly the primitive nth roots of unity.

Definition The polynomial

$$\Phi_n(x) = \prod_{i=1}^{\varphi(n)} (x - \alpha_i),$$

where the α_i are the primitive nth roots of unity in $\overline{F}$, is the **nth cyclotomic polynomial over F.**

Since an automorphism of the Galois group $G(K/F)$ must permute the primitive nth roots of unity, we see that $\Phi_n(x)$ is left fixed under every element of $G(K/F)$ regarded as extended in the natural way to $K[x]$. Thus $\Phi_n(x) \in F[x]$. In particular, for $F = \mathbf{Q}$, $\Phi_n(x) \in \mathbf{Q}[x]$, and $\Phi_n(x)$ is a divisor of $x^n - 1$. Thus over $\mathbf{Q}$, we must actually have $\Phi_n(x) \in \mathbf{Z}[x]$, by Theorem 31.3. We have seen that $\Phi_p(x)$ is irreducible over $\mathbf{Q}$, in the corollary of Theorem 31.4. While $\Phi_n(x)$ need not be irreducible in the case of the fields $\mathbf{Z}_p$, it can be shown that over $\mathbf{Q}$, $\Phi_n(x)$ is irreducible.

Let us now limit our discussion to characteristic 0, in particular to subfields of the complex numbers. Let i be the usual complex zero of $x^2 + 1$. Using trigonometric identities, the student can easily check formally that

$$(\cos \theta_1 + i \sin \theta_1)(\cos \theta_2 + i \sin \theta_2) = \cos (\theta_1 + \theta_2) + i \sin (\theta_1 + \theta_2).$$

It is then immediate by induction that

$$(\cos \theta + i \sin \theta)^n = \cos n\theta + i \sin n\theta.$$

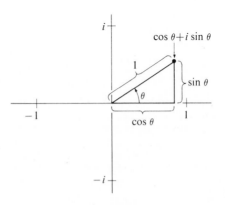

Fig. 48.1

In particular, if $\theta = 2\pi/n$, we have

$$\left(\cos\frac{2\pi}{n} + i\sin\frac{2\pi}{n}\right)^n = \cos 2\pi + i\sin 2\pi = 1,$$

so $\cos(2\pi/n) + i\sin(2\pi/n)$ is an nth root of unity. Figure 48.1 may help you to visualize all this. It is quite obvious from this figure that the least integer m such that $(\cos(2\pi/n) + i\sin(2\pi/n))^m = 1$ is n. *Thus* $\cos(2\pi/n) + i\sin(2\pi/n)$ *is a primitive* nth *root of unity, a zero of*

$$\Phi_n(x) \in \mathbf{Q}[x].$$

Example 48.1   A primitive 8th root of unity in $\mathbf{C}$ is

$$\zeta = \cos\frac{2\pi}{8} + i\sin\frac{2\pi}{8}$$

$$= \cos\frac{\pi}{4} + i\sin\frac{\pi}{4}$$

$$= \frac{1}{\sqrt{2}} + i\frac{1}{\sqrt{2}} = \frac{1+i}{\sqrt{2}}.$$

By the theory of cyclic groups, in particular by the corollary of Theorem 6.4, all the primitive 8th roots of unity in $\mathbf{Q}$ are ζ, ζ^3, ζ^5, and ζ^7, so

$$\Phi_8(x) = (x - \zeta)(x - \zeta^3)(x - \zeta^5)(x - \zeta^7).$$

The student may compute directly from this expression that $\Phi_8(x) = x^4 + 1$ (see Exercise 48.1). Compare this with Example 47.2. ∎

Let us still restrict our work to $F = \mathbf{Q}$, and let us assume, without proof, that $\Phi_n(x)$ is irreducible over $\mathbf{Q}$. Let

$$\zeta = \cos\frac{2\pi}{n} + i\sin\frac{2\pi}{n},$$

so that ζ is a primitive nth root of unity. Note that ζ is a generator of the cyclic multiplicative group of order n consisting of *all* nth roots of unity. All the primitive nth roots of unity, that is, all the generators of this group, are of the form ζ^m for $1 \le m < n$ and m relatively prime to n. The field $\mathbf{Q}(\zeta)$ is the whole splitting field of $x^n - 1$ over $\mathbf{Q}$. Let $K = \mathbf{Q}(\zeta)$. If ζ^m is another primitive nth root of unity, then since ζ and ζ^m are conjugate over $\mathbf{Q}$, there is an automorphism τ_m in $G(K/\mathbf{Q})$ mapping ζ onto ζ^m. Let τ_r be the similar automorphism in $G(K/\mathbf{Q})$ corresponding to the primitive nth root of unity ζ^r. Then

$$\zeta(\tau_r\tau_m) = (\zeta^r)\tau_m = (\zeta\tau_m)^r = (\zeta^m)^r = \zeta^{rm}.$$

This shows that the Galois group $G(K/Q)$ is isomorphic to the group G_n of Theorem 24.7 consisting of elements of Z_n relatively prime to n under multiplication modulo n. This group has $\varphi(n)$ elements and is, of course, abelian.

This material is easy. Special cases have appeared several times in the text and exercises. For example, α of Example 47.2 is a primitive 8th root of unity, and we made arguments in that example identical to those given here. We summarize these results in a theorem.

> **Theorem 48.1** The Galois group of the nth cyclotomic extension of Q has $\varphi(n)$ elements and is isomorphic to the group consisting of the positive integers less than n and relatively prime to n under multiplication modulo n.

Example 48.2 Example 47.2 illustrates this theorem, for it is easy to see that the splitting field of $x^4 + 1$ is the same as the splitting field of $x^8 - 1$ over Q. This follows from the fact that $\Phi_8(x) = x^4 + 1$ (see Example 48.1 and Exercise 48.1). ∎

> **Corollary** The Galois group of the pth cyclotomic extension of Q for a prime p is cyclic of order $p - 1$.

Proof. By Theorem 48.1, the Galois group of the pth cyclotomic extension of Q has $\varphi(p) = p - 1$ elements, and is isomorphic to the group of positive integers less than p and relatively prime to p under multiplication modulo p. This is exactly the multiplicative group $\langle Z_p^*, \cdot \rangle$ of nonzero elements of the field Z_p under field multiplication. By Corollary 1 of Theorem 45.3, this group is cyclic. ∎

*48.2 CONSTRUCTIBLE POLYGONS

We conclude with an application determining which regular n-gons are constructible with a compass and a straightedge. We saw in Chapter 39 that the regular n-gon is constructible if and only if $\cos(2\pi/n)$ is a constructible real number. Now let

$$\zeta = \cos\frac{2\pi}{n} + i\sin\frac{2\pi}{n}.$$

Then

$$\frac{1}{\zeta} = \cos\frac{2\pi}{n} - i\sin\frac{2\pi}{n},$$

for

$$\left(\cos\frac{2\pi}{n} + i\sin\frac{2\pi}{n}\right)\left(\cos\frac{2\pi}{n} - i\sin\frac{2\pi}{n}\right) = \cos^2\frac{2\pi}{n} + \sin^2\frac{2\pi}{n} = 1.$$

But then

$$\zeta + \frac{1}{\zeta} = 2 \cos \frac{2\pi}{n}.$$

Thus the corollary of Theorem 39.2 shows that the regular n-gon is constructible only if $\zeta + 1/\zeta$ generates an extension of $\mathbf{Q}$ of degree a power of 2.

If K is the splitting field of $x^n - 1$ over $\mathbf{Q}$, then $[K:\mathbf{Q}] = \varphi(n)$, by Theorem 48.1. If $\sigma \in G(K/\mathbf{Q})$ and $\zeta\sigma = \zeta^r$, then

$$\left(\zeta + \frac{1}{\zeta} \right)\sigma = \zeta^r + \frac{1}{\zeta^r}$$

$$= \left(\cos \frac{2\pi r}{n} + i \sin \frac{2\pi r}{n} \right) + \left(\cos \frac{2\pi r}{n} - i \sin \frac{2\pi r}{n} \right)$$

$$= 2 \cos \frac{2\pi r}{n}.$$

But for $1 < r < n$, we have $2 \cos (2\pi r/n) = 2 \cos (2\pi/n)$ only in the case that $r = n - 1$. Thus the only elements of $G(K/\mathbf{Q})$ carrying $\zeta + 1/\zeta$ onto itself are the identity automorphism and the automorphism τ, with $\zeta\tau = \zeta^{n-1} = 1/\zeta$. This shows that the subgroup of $G(K/\mathbf{Q})$ leaving $\mathbf{Q}(\zeta + 1/\zeta)$ fixed is of order 2, so by Galois theory,

$$\left[\mathbf{Q}\left(\zeta + \frac{1}{\zeta} \right) : \mathbf{Q} \right] = \frac{\varphi(n)}{2}.$$

Hence the regular n-gon is constructible only if $\varphi(n)/2$, and therefore also $\varphi(n)$, is a power of 2.

It can be shown by elementary arguments in number theory that if

$$n = 2^v p_1^{s_1} \cdots p_t^{s_t},$$

where the p_i are the distinct odd primes dividing n, then

$$\varphi(n) = 2^{v-1} p_1^{s_1-1} \cdots p_t^{s_t-1}(p_1 - 1) \cdots (p_t - 1). \tag{48.1}$$

If $\varphi(n)$ is to be a power of 2, then every odd prime dividing n must appear only to the first power and must be one more than a power of 2. Thus we must have each

$$p_i = 2^m + 1$$

for some m. Since -1 is a zero of $x^q + 1$ for q an odd prime, $x + 1$ divides $x^q + 1$ for q an odd prime. Thus, if $m = qu$, where q is an odd prime, then $2^m + 1 = (2^u)^q + 1$ is divisible by $2^u + 1$. Therefore, for $p_i = 2^m + 1$ to be prime, it must be that m is divisible by 2 only, so p_i has to have the form

$$p_i = 2^{(2^k)} + 1,$$

a **Fermat prime.** Fermat conjectured that these numbers $2^{(2^k)} + 1$ were prime for all nonnegative integers k. Euler showed that while $k = 0, 1, 2, 3,$ and 4 give the primes 3, 5, 17, 257, and 65537, for $k = 5$, one finds that $2^{(2^5)} + 1$ is divisible by 641. It has been shown that for $5 \le k \le 16$, all the numbers $2^{(2^k)} + 1$ are composite. The case $k = 17$ is unsolved as this text goes to press. It is unknown whether the number of Fermat primes is finite or infinite.

We have thus shown that the only regular n-gons that might be constructible are those where the odd primes dividing n are Fermat primes whose squares do not divide n. In particular, the only regular p-gons that might be constructible for p a prime greater than 2 are those where p is a Fermat prime.

Example 48.3   The regular 7-gon is not constructible, since 7 is not a Fermat prime. Similarly, the regular 18-gon is not constructible, for while 3 is a Fermat prime, its square divides 18. ∎

It is a fact that we now demonstrate that all these regular n-gons that are candidates for being constructible are indeed actually constructible. Let ζ again be the primitive nth root of unity $\cos(2\pi/n) + i \sin(2\pi/n)$. We saw above that

$$2 \cos \frac{2\pi}{n} = \zeta + \frac{1}{\zeta},$$

and that

$$\left[\mathbf{Q}\left(\zeta + \frac{1}{\zeta} \right) : \mathbf{Q} \right] = \frac{\varphi(n)}{2}.$$

Suppose now that $\varphi(n)$ is a power 2^s of 2. Let E be $\mathbf{Q}(\zeta + 1/\zeta)$. We saw above that $\mathbf{Q}(\zeta + 1/\zeta)$ is the subfield of $K = \mathbf{Q}(\zeta)$ left fixed by $H_1 = \{\iota, \tau\}$, where ι is the identity element of $G(K/\mathbf{Q})$ and $\zeta\tau = 1/\zeta$. By Sylow theory, there exist additional subgroups H_j of order 2^j of $G(\mathbf{Q}(\zeta)/\mathbf{Q})$ for $j = 0, 2, 3, \ldots, s$ such that

$$\{\iota\} = H_0 < H_1 < \cdots < H_s = G(\mathbf{Q}(\zeta)/\mathbf{Q}).$$

By Galois theory,

$$\mathbf{Q} = K_{H_s} < K_{H_{s-1}} < \cdots < K_{H_1} = \mathbf{Q}\left(\zeta + \frac{1}{\zeta} \right),$$

and $[K_{H_{j-1}} : K_{H_j}] = 2$. Note that $(\zeta + 1/\zeta) \in \mathbf{R}$, so $\mathbf{Q}(\zeta + 1/\zeta) < \mathbf{R}$. If $K_{H_{j-1}} = K_{H_j}(\alpha_j)$, then α_j is a zero of some $(a_j x^2 + b_j x + c_j) \in K_{H_j}[x]$. By the familiar "quadratic formula," we have

$$K_{H_{j-1}} = K_{H_j}(\sqrt{b_j^2 - 4a_j c_j}).$$

Since we saw in Chapter 39 that construction of square roots of positive constructible numbers can be achieved by a straightedge and a compass, we see that every element in $\mathbf{Q}(\zeta + 1/\zeta)$, in particular $\cos(2\pi/n)$, is constructible. Hence the regular n-gons where $\varphi(n)$ is a power of 2 are constructible.

We summarize our work under this heading in a theorem.

Theorem 48.2 *The regular n-gon is constructible with a compass and a straightedge if and only if all the odd primes dividing n are Fermat primes whose squares do not divide n.*

Example 48.4 The regular 60-gon is constructible, since $60 = (2^2)(3)(5)$ and 3 and 5 are both Fermat primes.

Exercises

*48.1 Referring to Example 48.1, complete the indicated computation, showing that $\Phi_8(x) = x^4 + 1$. [*Suggestion:* Compute the product in terms of ζ, and then use the fact that $\zeta^8 = 1$ and $\zeta^4 = -1$ to simplify the coefficients.]

*48.2 Classify the group of the polynomial $(x^{20} - 1) \in \mathbf{Q}[x]$ over $\mathbf{Q}$ according to the Fundamental Theorem of Finitely Generated Abelian Groups. [*Hint:* Use Theorem 48.1.]

*48.3 Using the formula for $\varphi(n)$ in terms of the factorization of n, as given in Eq. (48.1) of Section 48.2, compute the following.

a) $\varphi(60)$ b) $\varphi(1000)$ c) $\varphi(8100)$

*48.4 Give the first thirty values of $n \geq 3$ for which the regular n-gon is constructible with a straightedge and a compass.

†*48.5 Show that if F is a field of characteristic not dividing n, then

$$x^n - 1 = \prod_{d \mid n} \Phi_d(x)$$

in $F[x]$, where the product is over all divisors d of n.

*48.6 Find the cyclotomic polynomial $\Phi_n(x)$ over $\mathbf{Q}$ for $n = 1, 2, 3, 4, 5$, and 6. [*Hint:* Use Exercise 48.5.]

*48.7 Mark each of the following true or false.

___ a) $\Phi_n(x)$ is irreducible over every field of characteristic 0.

___ b) Every zero in $\mathbf{C}$ of $\Phi_n(x)$ is a primitive nth root of unity.

___ c) The group of $\Phi_n(x) \in \mathbf{Q}[x]$ over $\mathbf{Q}$ has order n.

___ d) The group of $\Phi_n(x) \in \mathbf{Q}[x]$ over $\mathbf{Q}$ is abelian.

___ e) The Galois group of the splitting field of $\Phi_n(x)$ over $\mathbf{Q}$ has order $\varphi(n)$.

___ f) The regular 25-gon is constructible with a straightedge and a compass.

___ g) The regular 17-gon is constructible with a straightedge and a compass.

___ h) For a prime p, the regular p-gon is constructible if and only if p is a Fermat prime.

___ i) All integers of the form $2^{(2^k)} + 1$ for nonnegative integers k are Fermat primes.

___ j) All Fermat primes are numbers of the form $2^{(2^k)} + 1$ for nonnegative integers k.

*48.8 Find the smallest angle of integral degree, that is, $1°, 2°, 3°$, and so on, constructible with a straightedge and a compass. [*Hint:* Constructing a $1°$ angle amounts to constructing the regular 360-gon, and so on.]

*48.9 Let K be the splitting field of $x^{12} - 1$ over $\mathbf{Q}$.

a) Find $[K:\mathbf{Q}]$.

b) Show that for $\sigma \in G(K/\mathbf{Q})$, σ^2 is the identity automorphism. Classify $G(K/\mathbf{Q})$ according to the Fundamental Theorem of Finitely Generated Abelian Groups.

*48.10 Find $\Phi_3(x)$ over $\mathbf{Z}_2$. Find $\Phi_8(x)$ over $\mathbf{Z}_3$.

*48.11 How many elements are there in the splitting field of $x^6 - 1$ over $\mathbf{Z}_3$?

*48.12 Find $\Phi_{12}(x)$ in $\mathbf{Q}[x]$. [*Hint:* Use Exercises 48.5 and 48.6.]

*48.13 Show that in $\mathbf{Q}[x]$, $\Phi_{2n}(x) = \Phi_n(-x)$ for odd integers $n > 1$. [*Hint:* Use Exercise 48.5 and the factorization $x^{2n} - 1 = -(x^n - 1)((-x)^n - 1)$. Proceed by induction.]

*48.14 Let $n, m \in \mathbf{Z}^+$ be relatively prime. Show that the splitting field in $\mathbf{C}$ of $x^{nm} - 1$ over $\mathbf{Q}$ is the same as the splitting field in $\mathbf{C}$ of $(x^n - 1)(x^m - 1)$ over $\mathbf{Q}$.

*48.15 Let $n, m \in \mathbf{Z}^+$ be relatively prime. Show that the group of $(x^{nm} - 1) \in \mathbf{Q}[x]$ over $\mathbf{Q}$ is isomorphic to the direct product of the groups of $(x^n - 1) \in \mathbf{Q}[x]$ and of $(x^m - 1) \in \mathbf{Q}[x]$ over $\mathbf{Q}$. [*Hint:* Using Galois theory, show that the groups of $x^m - 1$ and $x^n - 1$ can both be regarded as subgroups of the group of $x^{nm} - 1$. Then use Theorem 8.6.]

forty-nine

Insolvability of the Quintic

49.1 THE PROBLEM

You are familiar with the fact that a quadratic polynomial $f(x) = ax^2 + bx + c$, $a \neq 0$, with real coefficients has $(-b \pm \sqrt{b^2 - 4ac})/2a$ as zeros in **C**. Actually, this is true for $f(x) \in F[x]$, where F is any field of characteristic $\neq 2$ and the zeros are in $\bar{F}$. You are asked to show this in Exercise 49.1. Thus, for example, $(x^2 + 2x + 3) \in \mathbf{Q}[x]$ has its zeros in $\mathbf{Q}(\sqrt{-2})$. One wonders whether the zeros of a cubic polynomial over **Q** can also always be expressed in terms of radicals. The answer is yes, and, indeed, even the zeros of a polynomial of degree 4 over **Q** can be expressed in terms of radicals. After mathematicians had tried for years to find the "radical formula" for zeros of a 5th degree polynomial, it was a triumph when Abel proved that a quintic need not be solvable by radicals. Our first job will be to describe precisely what this means. You will be delighted to see what a large amount of the algebra we have developed is used in the forthcoming discussion.

49.2 EXTENSIONS BY RADICALS

Definition   An extension K of a field F is an ***extension of F by radicals***
if there are elements $\alpha_1, \ldots, \alpha_r \in K$ and positive integers $n_1, \ldots, n_r$
such that $K = F(\alpha_1, \ldots, \alpha_r)$, $\alpha_1^{n_1} \in F$ and $\alpha_i^{n_i} \in F(\alpha_1, \ldots, \alpha_{i-1})$ for $1 <$
$i \le r$. A polynomial $f(x) \in F[x]$ is *solvable by radicals over* F if the
splitting field E of $f(x)$ over F is contained in an extension of F by radicals.

A polynomial $f(x) \in F[x]$ is thus solvable by radicals over F if we can
obtain every zero of $f(x)$ by using a finite sequence of the operations of
addition, subtraction, multiplication, division, and taking n_ith roots, starting with elements of F. Now to say that the quintic is not solvable in the
classic case, that is, characteristic 0, is not to say that no quintic is solvable,
as the following example shows.

Example 49.1   The polynomial $x^5 - 1$ is solvable by radicals over **Q**. The
splitting field K of $x^5 - 1$ is generated over **Q** by a primitive 5th root ζ of
unity. Then $\zeta^5 = 1$, and $K = \mathbf{Q}(\zeta)$. Similarly, $x^5 - 2$ is solvable by radicals over **Q**, for its splitting field over **Q** is generated by $\sqrt[5]{2}$ and ζ, where
$\sqrt[5]{2}$ is the real zero of $x^5 - 2$. ∎

To say that the quintic is insolvable in the classic case means that there
exists *some* polynomial of degree 5 with real coefficients that is not solvable
by radicals. We shall show this. *We assume throughout this chapter that all
fields mentioned have characteristic 0.*

The outline of the argument is very easy to give, and it is worthwhile
to try to remember it.

1. *We shall show that a polynomial $f(x) \in F[x]$ is solvable by radicals over
   F (if and) only if its splitting field E over F has a solvable Galois group.*
   Recall that a solvable group is one having a composition series with
   abelian quotients. While this theorem goes both ways, we shall not
   prove the "if" part.

2. *We shall show that there is a subfield F of the real numbers and a polynomial $f(x) \in F[x]$ of degree 5 with a splitting field E over F such that
   $G(E/F) \simeq S_5$, the symmetric group on 5 letters.* Recall that a composition
   series for S_5 is $\{i\} < A_5 < S_5$. Since A_5 is not abelian, we will be done.

The following lemma does most of our work for step 1.

Lemma 49.1   *Let F be a field of characteristic 0, and let $a \in F$. If K is
the splitting field of $x^n - a$ over F, then $G(K/F)$ is a solvable group.*

Proof.   Suppose first that F contains all the nth roots of unity. By Theorem
45.3 and the comments preceding it, the nth roots of unity form a cyclic
subgroup of $\langle F^*, \cdot \rangle$. Let ζ be a generator of the subgroup. (Actually, the

generators are exactly the *primitive* nth roots of unity.) Then the nth roots of unity are

$$1, \zeta, \zeta^2, \ldots, \zeta^{n-1}.$$

If $\beta \in \bar{F}$ is a zero of $(x^n - a) \in F[x]$, then all zeros of $x^n - a$ are

$$\beta, \zeta\beta, \zeta^2\beta, \ldots, \zeta^{n-1}\beta.$$

Since $K = F(\beta)$, an automorphism σ in $G(K/F)$ is determined by the value $\beta\sigma$ of the automorphism σ on β. If $\beta\sigma = \zeta^i\beta$ and $\beta\tau = \zeta^j\beta$, where $\tau \in G(K/F)$, then

$$\beta(\sigma\tau) = (\beta\sigma)\tau = (\zeta^i\beta)\tau = \zeta^i(\beta\tau) = \zeta^i\zeta^j\beta,$$

since $\zeta^i \in F$. Similarly,

$$\beta(\tau\sigma) = \zeta^j\zeta^i\beta.$$

Thus $\sigma\tau = \tau\sigma$, and $G(K/F)$ is abelian and therefore solvable.

Now suppose that F does not contain a primitive nth root of unity. Let ζ be a generator of the cyclic group of nth roots of unity under multiplication in $\bar{F}$. Let β again be a zero of $x^n - a$. Since β and $\zeta\beta$ are both in the splitting field K of $x^n - a$, $\zeta = (\zeta\beta)/\beta$ is in K. Let $F' = F(\zeta)$, so we have $F < F' \le K$. Now F' is a normal extension of F, since F' is the splitting field of $x^n - 1$. Since $F' = F(\zeta)$, an automorphism η in $G(F'/F)$ is determined by $\zeta\eta$, and we must have $\zeta\eta = \zeta^i$ for some i, since all zeros of $x^n - 1$ are powers of ζ. If $\zeta\mu = \zeta^j$ for $\mu \in G(F'/F)$, then

$$\zeta(\eta\mu) = (\zeta\eta)\mu = \zeta^i\mu = (\zeta\mu)^i = (\zeta^j)^i = \zeta^{ij},$$

and, similarly,

$$\zeta(\mu\eta) = \zeta^{ij}.$$

Thus $G(F'/F)$ is abelian. By the Main Theorem of Galois Theory,

$$\{\iota\} \le G(K/F') \le G(K/F)$$

is a normal series and hence a subnormal series of groups. The first part of the proof shows that $G(K/F')$ is abelian, and Galois theory tells us that $G(K/F)/G(K/F')$ is isomorphic to $G(F'/F)$, which is abelian. It is easy to see that if a group has a subnormal series of subgroups with abelian quotient groups, then any refinement of this series also has abelian quotient groups. (See Exercise 49.6.) Thus a composition series of $G(K/F)$ must have abelian quotient groups, so $G(K/F)$ is solvable. ∎

The following theorem shows that if K is a normal extension of F by radicals, then $G(K/F)$ is solvable. Exercise 49.8 then shows that if $f(x) \in F[x]$ is solvable by radicals and has splitting field E, then $G(E/F)$ is solvable. This will complete part 1 of our program.

Theorem 49.1   *If K is a normal extension by radicals of a field F of characteristic 0, then $G(K/F)$ is solvable.*

Proof. We know that there exist $\alpha_1, \ldots, \alpha_r \in K$ and positive integers $n_1, \ldots, n_r$ such that $K = F(\alpha_1, \ldots, \alpha_r)$, $\alpha_1^{n_1} \in F$ and $\alpha_i^{n_i} \in F(\alpha_1, \ldots, \alpha_{i-1})$ for $1 < i \le r$. Let $K_0 = F$, and let K_i be the splitting field of $x^{n_i} - \alpha_i^{n_i}$ over K_{i-1}. Then $K \le K_r$, and Lemma 49.1 shows that $G(K_i/K_{i-1})$ is solvable. Since $G(K_i/K_{i-1}) \simeq G(K_r/K_{i-1})/G(K_r/K_i)$, the normal series

$$\{\iota\} \le G(K_r/K_{r-1}) \le G(K_r/K_{r-2}) \le \cdots \le G(K_r/K_0) = G(K_r/F)$$

has solvable quotient groups. Exercise 49.7 shows that this series then has a refinement to a composition series with abelian quotients, so $G(K_r/F)$ is solvable. Since $G(K/F) \simeq G(K_r/F)/G(K_r/K)$, Exercise 15.16 shows that $G(K/F)$ is solvable. ∎

49.3 THE INSOLVABILITY OF THE QUINTIC

It remains for us to show that there is a subfield F of the real numbers and a polynomial $f(x) \in F[x]$ of degree 5 such that the splitting field E of $f(x)$ over F has a Galois group isomorphic to S_5.

Let $y_1 \in \mathbf{R}$ be transcendental over $\mathbf{Q}$, $y_2 \in \mathbf{R}$ be transcendental over $\mathbf{Q}(y_1)$, and so on, until you get $y_5 \in \mathbf{R}$ transcendental over $\mathbf{Q}(y_1, \ldots, y_4)$. It can easily be shown by a counting argument that such transcendental real numbers exist. Transcendentals found in this fashion are **independent transcendental elements over Q**. Let $E = \mathbf{Q}(y_1, \ldots, y_5)$, and let

$$f(x) = \prod_{i=1}^{5}(x - y_i).$$

Thus $f(x) \in E[x]$. Now the coefficients of $f(x)$ are, except possibly for sign, among the so-called *elementary symmetric functions* in the y_i, namely

$$s_1 = y_1 + y_2 + \cdots + y_5,$$
$$s_2 = y_1 y_2 + y_1 y_3 + y_1 y_4 + y_1 y_5 + y_2 y_3$$
$$\quad + y_2 y_4 + y_2 y_5 + y_3 y_4 + y_3 y_5 + y_4 y_5,$$
$$\vdots$$
$$s_5 = y_1 y_2 y_3 y_4 y_5.$$

The coefficient of x^i in $f(x)$ is $\pm s_{5-i}$. Let $F = \mathbf{Q}(s_1, s_2, \ldots, s_5)$; then $f(x) \in F[x]$ (see Fig. 49.1). Clearly, E is the splitting field over F of $f(x)$. Since the y_i behave as indeterminates over $\mathbf{Q}$, for each $\sigma \in S_5$, the symmetric

Fig. 49.1

group on five letters, σ induces an automorphism $\bar{\sigma}$ of E defined by $a\bar{\sigma} = a$ for $a \in \mathbf{Q}$ and $y_i\bar{\sigma} = y_{i\sigma}$. Since $\prod_{i=1}^{5}(x - y_i)$ is the same polynomial as $\prod_{i=1}^{5}(x - y_{i\sigma})$, we have

$$s_i\bar{\sigma} = s_i$$

for each i, so $\bar{\sigma}$ leaves F fixed, and hence $\bar{\sigma} \in G(E/F)$. Now S_5 has order $5!$, so

$$|G(E/F)| \geq 5!.$$

Since the splitting field of a polynomial of degree 5 over F has degree at most $5!$ over F, we see that

$$|G(E/F)| \leq 5!.$$

Thus $|G(E/F)| = 5!$, and the automorphisms $\bar{\sigma}$ make up the full Galois group $G(E/F)$. Therefore, $G(E/F) \simeq S_5$, so $G(E/F)$ is not solvable. By the comments preceding Theorem 49.1, $f(x)$ is not solvable by radicals over F. We summarize this in a theorem.

Theorem 49.2 Let $y_1, \ldots, y_5$ be independent transcendental real numbers over $\mathbf{Q}$. The polynomial

$$f(x) = \prod_{i=1}^{5}(x - y_i)$$

is not solvable by radicals over $F = \mathbf{Q}(s_1, \ldots, s_5)$, where s_i is the ith elementary symmetric function in $y_1, \ldots, y_5$.

It is evident that a generalization of these arguments shows that (*final goal*) a polynomial of degree n need not be solvable by radicals for $n \geq 5$.

In conclusion, we comment that there exist polynomials of degree 5 in $\mathbf{Q}[x]$ that are not solvable by radicals over $\mathbf{Q}$. A demonstration of this is left to the exercises (see Exercise 49.9).

Exercises

49.1 Let F be a field, and let $f(x) = ax^2 + bx + c$ be in $F[x]$, where $a \neq 0$. Show that if the characteristic of F is not 2, the splitting field of $f(x)$ over F is $F(\sqrt{b^2 - 4ac})$. [*Hint*: Complete the square, just as in your high school work, to derive the "quadratic formula."]

49.2 Can the splitting field K of $x^2 + x + 1$ over $\mathbf{Z}_2$ be obtained by adjoining a square root to $\mathbf{Z}_2$ of an element in $\mathbf{Z}_2$? Is K an extension of $\mathbf{Z}_2$ by radicals?

49.3 Show that if F is a field of characteristic different from 2 and

$$f(x) = ax^4 + bx^2 + c,$$

where $a \neq 0$, then $f(x)$ is solvable by radicals over F.

49.4 Is every polynomial in $F[x]$ of the form $ax^8 + bx^6 + cx^4 + dx^2 + e$, where $a \neq 0$, solvable by radicals over F, if F is of characteristic 0? Why?

49.5 Mark each of the following true or false.

—   a) Let F be a field of characteristic 0. A polynomial in $F[x]$ is solvable by radicals if and only if its splitting field in $\bar{F}$ is contained in an extension of F by radicals.

—   b) Let F be a field of characteristic 0. A polynomial in $F[x]$ is solvable by radicals if and only if its splitting field in $\bar{F}$ has a solvable Galois group over F.

—   c) The splitting field of $x^{17} - 5$ over $\mathbf{Q}$ has a solvable Galois group.

—   d) The numbers π and $\sqrt{\pi}$ are independent transcendental numbers over $\mathbf{Q}$.

—   e) The Galois group of a finite extension of a finite field is solvable.

—   f) No quintic polynomial is solvable by radicals over any field.

—   g) Every 4th degree polynomial over a field of characteristic 0 is solvable by radicals.

—   h) The zeros of a cubic polynomial over a field F of characteristic 0 can always be attained by means of a finite sequence of operations of addition, subtraction, multiplication, division, and taking square roots starting with elements in F.

—   i) The zeros of a cubic polynomial over a field F of characteristic 0 can never be attained by means of a finite sequence of operations of addition, subtraction, multiplication, division, and taking square roots, starting with elements in F.

—   j) The theory of normal series of groups plays an important role in applications of Galois theory.

*49.6 Show that for a finite group, every refinement of a subnormal series with abelian quotients also has abelian quotients, thus completing the proof of Lemma 49.1. [*Hint*: Use Theorem 15.3.]

*49.7 Show that for a finite group, a subnormal series with solvable quotient groups can be refined to a composition series with abelian quotients, thus completing the proof of Theorem 49.1. [*Hint*: Use Theorem 15.3.]

*49.8 Let K be a normal extension by radicals of a field F of characteristic 0, and let E be a normal extension of F, $F \leq E \leq K$. Show that $G(E/F)$ is solvable. [*Hint*: Use Galois theory, Theorem 49.1, and Exercise 15.16.]

*49.9 This exercise exhibits a polynomial of degree 5 in $Q[x]$ that is not solvable by radicals over Q.

a) Show that if a subgroup H of S_5 contains a cycle of length 5 and a transposition τ, then $H = S_5$. [*Hint:* Show that H contains every transposition of S_5, and apply the corollary of Theorem 5.1. See Exercise 9.15.]

b) Show that if $f(x)$ is an irreducible polynomial in $Q[x]$ of degree 5 having exactly two complex and three real zeros in C, then the group of $f(x)$ over Q is S_5. [*Hint:* Use Sylow theory to show that the group has an element of order 5. Use the fact that $f(x)$ has exactly two complex zeros to show that the group has an element of order 2. Then apply (a).]

c) The polynomial $f(x) = 2x^5 - 5x^4 + 5$ is irreducible in $Q[x]$, by the Eisenstein criterion, with $p = 5$. Use the techniques of calculus to find relative maxima and minima and to "graph the polynomial function f" well enough to see that $f(x)$ must have exactly three real zeros in C. Conclude from (b) and Theorem 49.1 that $f(x)$ is not solvable by radicals over Q.

appendix

Mathematical Induction

Sometimes we want to prove that a statement about positive integers is true for all positive integers or perhaps for some finite or infinite sequence of consecutive integers. Such proofs are accomplished using mathematical induction. The validity of the method rests on an axiom of the positive integers.

Induction Axiom *Let S be a subset of* $\mathbf{Z}^+$ *satisfying*

$1 \in S$ *and,*
if $k \in S$, *then* $(k + 1) \in S$.

Then $S = \mathbf{Z}^+$.

From this axiom we obtain immediately the method of mathematical induction.

Mathematical Induction *Let P(n) be a statement concerning the positive integer n. Suppose that*

P(1) is true and,
if P(k) is true, then P(k + 1) is true.

Then P(n) is true for all $n \in \mathbf{Z}^+$.

444

Most of the time, you want to show that $P(n)$ holds for all $n \in \mathbf{Z}^+$. If you wish only to show that it holds for $r, r + 1, r + 2, \ldots, s - 1, s$, then you show that $P(r)$ is true and that $P(k)$ implies $P(k + 1)$ for $r \le k \le s - 1$. Note that r may be any integer in $\mathbf{Z}$, positive, negative, or zero.

Example A1 Let us prove the formula

$$1 + 2 + \cdots + n = \frac{n(n + 1)}{2} \tag{A1}$$

for the sum of the arithmetic progression, using mathematical induction.

We let $P(n)$ be the statement that Formula (A1) is true. For $n = 1$, we obtain

$$\frac{n(n + 1)}{2} = \frac{1(2)}{2} = 1,$$

so $P(1)$ is true.

Suppose $P(k)$ is true, so

$$1 + 2 + \cdots + k = \frac{k(k + 1)}{2}.$$

Then

$$1 + 2 + \cdots + (k + 1) = (1 + 2 + \cdots + k) + (k + 1)$$

$$= \frac{k(k + 1)}{2} + (k + 1) = \frac{k^2 + k + 2k + 2}{2}$$

$$= \frac{k^2 + 3k + 2}{2} = \frac{(k + 1)(k + 2)}{2}$$

so $P(k + 1)$ holds. Thus Formula (A1) is true for all $n \in \mathbf{Z}^+$. ∎

Example A2 Let us show that a set of n elements has 2^n subsets in all, for $n \in \{0, 1, 2, 3, \ldots\} = \mathbf{Z}^+ \bigcup \{0\}$.

This time we start our induction with $n = 0$. We let S be a finite set such that $|S| = n$, and we wish to show

$$P(n): S \text{ has } 2^n \text{ subsets.} \tag{A2}$$

If $|S| = 0$, then $S = \varnothing$ and has only one subset, namely, $\varnothing$. Since $2^0 = 1$, we see that $P(0)$ is true.

Suppose $P(k)$ is true, and let S have $k + 1$ elements. Let one element of S be c. Then $S - \{c\}$ has k elements, and hence 2^k subsets. Now every subset of S either contains c or does not contain c. Those not containing c are subsets of $S - \{c\}$, so there are 2^k of them by the induction hypothesis. Each subset containing c consists of one of the 2^k subsets not containing c with c adjoined.

There are 2^k such subsets also. The total number of subsets of S is then

$$2^k + 2^k = 2^k(2) = 2^{k+1},$$

so $P(k + 1)$ is true. Thus $P(n)$ is true for $n = 0$ and $n \in \mathbf{Z}^+$. ∎

Example A3 Let $x \in \mathbf{R}$, $x > -1$, $x \neq 0$. Let us show that $(1 + x)^n > 1 + nx$ for every $n \geq 2$, $n \in \mathbf{Z}^+$.

We let $P(n)$ be the statement

$$(1 + x)^n > 1 + nx. \tag{A3}$$

(Note that $P(1)$ is false.) Then $P(2)$ is the statement $(1 + x)^2 > 1 + 2x$. Now $(1 + x)^2 = 1 + 2x + x^2$, and $x^2 > 0$ since $x \neq 0$. Thus $(1 + x)^2 > 1 + 2x$, so $P(2)$ is true.

Suppose $P(k)$ is true, so

$$(1 + x)^k > 1 + kx. \tag{A4}$$

Now $1 + x > 0$ since $x > -1$. Multiplying both sides of Eq. (A4) by $1 + x$, we obtain

$$(1 + x)^{k+1} > (1 + kx)(1 + x) = 1 + (k + 1)x + kx^2.$$

Since $kx^2 \geq 0$, we see that $P(k + 1)$ is true. Thus $P(n)$ is true for $n \geq 2$, $n \in \mathbf{Z}^+$. ∎

In conclusion, we mention that you may find occasion to use *complete induction*, where the statement

if $P(k)$ is true, then $P(k + 1)$ is true

is replaced by the statement

if $P(m)$ is true for $1 \leq m \leq k$, then $P(k + 1)$ is true.

Again, you are trying to show that $P(k + 1)$ is true, knowing $P(k)$ is true. But if you have reached the stage in induction where $P(k)$ has been proved, then you know $P(m)$ is true for $1 \leq m \leq k$, so the strengthened hypothesis in the second statement is permissible.

Exercises

A1 Show that $1^2 + 2^2 + 3^2 + \cdots + n^2 = \dfrac{n(n + 1)(2n + 1)}{6}$ for $n \in \mathbf{Z}^+$.

A2 Show that $1^3 + 2^3 + 3^3 + \cdots + n^3 = \dfrac{n^2(n + 1)^2}{4}$ for $n \in \mathbf{Z}^+$.

A3 Show that $1 + 3 + 5 + \cdots + (2n - 1) = n^2$ for $n \in \mathbf{Z}^+$.

A4 Show that $\dfrac{1}{1 \cdot 2} + \dfrac{1}{2 \cdot 3} + \dfrac{1}{3 \cdot 4} + \cdots + \dfrac{1}{n(n + 1)} = \dfrac{n}{n + 1}$ for $n \in \mathbf{Z}^+$.

A5 Prove by induction that if $a, r \in \mathbf{R}$ and $r \ne -1$, then $a + ar + ar^2 + \cdots + ar^n = a(1 - r^{n+1})/(1 - r)$ for $n \in \mathbf{Z}^+$.

A6 Find the flaw in this argument.

We show by induction that any two positive integers are equal. We use induction on the maximum of the two numbers. Let $P(n)$ be the statement that two positive integers with maximum value n are equal.

Since the only two positive integers with maximum value 1 are 1 and 1, we see that $P(1)$ is true.

Suppose $P(k)$ is true, and let r and s be positive integers with maximum value $k + 1$. Then the maximum of $r - 1$ and $s - 1$ is k, so $r - 1 = s - 1$ by induction hypothesis. Therefore $r = s$. Thus $P(k + 1)$ is true, so $P(n)$ is true for all $n \in \mathbf{Z}^+$.

A7 Criticize this argument.

Let us show that every positive integer has some interesting property. Let $P(n)$ be the statement that n has an interesting property. We use complete induction.

Of course $P(1)$ is true, since 1 is the multiplicative identity, which is surely an interesting property of 1.

Suppose $P(m)$ is true for $1 \le m \le k$. If $P(k + 1)$ were not true, then $k + 1$ would be the smallest integer without an interesting property, which would, in itself, be an interesting property of $k + 1$. So $P(k + 1)$ must be true. Thus $P(n)$ is true for all $n \in \mathbf{Z}^+$.

A8 We have never been able really to see any flaw in (a). Try your luck with it and then answer (b).

a) A murderer is sentenced to be executed. He asks the judge not to let him know the day of the execution. The judge says, "I sentence you to be executed at 10 A.M. some day of this coming January, but I promise that you will not be aware you are being executed that day until they come to get you at 8 A.M." The criminal goes to his cell and proceeds to prove he can't be executed in January as follows:

Let $P(n)$ be the statement that I can't be executed on January $(31 - n)$. I want to prove $P(n)$ for $0 \le n \le 30$. Now I can't be executed on January 31, for since that is the last day of the month and I am to be executed that month, I would know that was the day before 8 A.M., contrary to the judge's sentence. Thus $P(0)$ is true. Suppose $P(m)$ is true for $0 \le m \le k$ where $k \le 29$. That is, suppose I can't be executed on January $(31 - k)$ through January 31. Then January $(31 - k - 1)$ must be the last possible day for execution, and I would be aware that was the day before 8 A.M., contrary to the judge's sentence. Thus I can't be executed on January $(31 - (k + 1))$, so $P(k + 1)$ is true. Therefore I can't be executed in January.

(Of course, the criminal was executed on January 17.)

b) An instructor teaches a class five days a week, Monday through Friday. She tells her class that she will give one more quiz on one day during the final week of classes, but that the students will not know for sure the quiz will be that day until they come to the classroom. What is the last day of the week she can give the quiz to satisfy these conditions?

Bibliography

Classic Works

1. N. Bourbaki, *Éléments de Mathématique*, Book II of Part I, *Algèbre*. Paris: Hermann, 1942–58.
2. N. Jacobson, *Lectures in Abstract Algebra*. Princeton, N.J.: Van Nostrand, vols. I, 1951, II, 1953, and III, 1964.
3. O. Schreier and E. Sperner, *Introduction to Modern Algebra and Matrix Theory* (English translation), second edition. New York: Chelsea, 1959.
4. B. L. van der Waerden, *Modern Algebra* (English translation). New York: Ungar, vols. I, 1949, and II, 1950.

General Algebra Texts

5. A. A. Albert, *Fundamental Concepts of Higher Algebra*. Chicago: University of Chicago Press, 1956.
6. G. Birkhoff and S. Mac Lane, *A Survey of Modern Algebra*, third edition. New York: Macmillan, 1965.
7. R. A. Dean, *Elements of Abstract Algebra*. New York: Wiley, 1966.
8. I. N. Herstein, *Topics in Algebra*. New York: Blaisdell, 1964.
9. T. W. Hungerford, *Algebra*. New York: Holt, Rinehart and Winston, 1974.
10. R. E. Johnson, *University Algebra*. Englewood Cliffs, N.J.: Prentice-Hall, 1966.
11. S. Lang, *Algebra*. Reading, Mass.: Addison-Wesley, 1965.
12. N. H. McCoy, *Introduction to Modern Algebra*. Boston: Allyn and Bacon, 1960.
13. G. D. Mostow, J. H. Sampson, and J. Meyer, *Fundamental Structures of Algebra*. New York: McGraw-Hill, 1963.

14. W. W. Sawyer, *A Concrete Approach to Abstract Algebra*. San Francisco: Freeman, 1959.

15. S. Warner, *Modern Algebra*. Englewood Cliffs, N.J.: Prentice-Hall, vols. I and II, 1965.

Group Theory

16. W. Burnside, *Theory of Groups of Finite Order*, second edition. New York: Dover, 1955.

17. H. S. M. Coxeter and W. O. Moser, *Generators and Relations for Discrete Groups*, second edition. Berlin: Springer, 1965.

18. M. Hall, Jr., *The Theory of Groups*. New York: Macmillan, 1959.

19. A. G. Kurosh, *The Theory of Groups* (English translation). New York: Chelsea, vols. I, 1955, and II, 1956.

20. W. Ledermann, *Introduction to the Theory of Finite Groups*, fourth revised edition. New York: Interscience, 1961.

21. J. G. Thompson and W. Feit, "Solvability of Groups of Odd Order." *Pac. J. Math.*, **13** (1963), 775–1029.

22. M. A. Rabin, "Recursive Unsolvability of Group Theoretic Problems." *Ann. Math.*, **67** (1958), 172–194.

Ring Theory

23. E. Artin, C. J. Nesbitt, and R. M. Thrall, *Rings with Minimum Condition*. Ann Arbor: University of Michigan Press, 1944.

24. N. H. McCoy, *Rings and Ideals* (Carus Monograph No. 8). Buffalo: The Mathematical Association of America; LaSalle, Illinois: Open Court, 1948.

25. N. H. McCoy, *The Theory of Rings*. New York: Macmillan, 1964.

Field Theory

26. E. Artin, *Galois Theory* (Notre Dame Mathematical Lecture No. 2), second edition. Notre Dame, Indiana: University of Notre Dame Press, 1944.

27. O. Zariski and P. Samuel, *Commutative Algebra*. Princeton, N.J.: Van Nostrand, vol. I, 1958.

Number Theory

28. G. H. Hardy and E. M. Wright, *An Introduction to the Theory of Numbers*, fourth edition. Oxford: Clarendon Press, 1960.

29. S. Lang, *Algebraic Numbers*. Reading, Mass.: Addison-Wesley, 1964.

30. W. J. LeVeque, *Elementary Theory of Numbers*. Reading, Mass.: Addison-Wesley, 1962.

31. W. J. LeVeque, *Topics in Number Theory*. Reading, Mass.: Addison-Wesley, 2 vols., 1956.

32. T. Nagell, *Introduction to Number Theory*. New York: Wiley, 1951.

33. I. Nivin and H. S. Zuckerman, *An Introduction to the Theory of Numbers*. New York: Wiley, 1960.

34. H. Pollard, *The Theory of Algebraic Numbers* (Carus Monograph No. 9). Buffalo: The Mathematical Association of America; New York: Wiley, 1950.

35. D. Shanks, *Solved and Unsolved Problems in Number Theory*. Washington: Spartan Books, vol. I, 1962.

36. B. M. Stewart, *Theory of Numbers*, second edition. New York: Macmillan, 1964.

37. J. V. Uspensky and M. H. Heaslet, *Elementary Number Theory*. New York: McGraw-Hill, 1939.

38. E. Weiss, *Algebraic Number Theory*. New York: McGraw-Hill, 1963.

Homological Algebra

39. J. P. Jans, *Rings and Homology*. New York: Holt, 1964.

40. S. Mac Lane, *Homology*. Berlin: Springer, 1963.

Other References

41. A. A. Albert (editor), *Studies in Modern Algebra* (MAA Studies in Mathematics, vol. 2). Buffalo: The Mathematical Association of America; Englewood Cliffs, N.J.: Prentice-Hall, 1963.

42. E. Artin, *Geometric Algebra*. New York: Interscience, 1957.

43. R. Courant and R. Robbins, *What Is Mathematics?* Oxford: Oxford University Press, 1941.

44. H. S. M. Coxeter, *Introduction to Geometry*. New York: Wiley, 1961.

45. R. H. Crowell and R. H. Fox, *Introduction to Knot Theory*. New York: Ginn, 1963.

Answers and Comments

Chapter 0

0.1 $\{-\sqrt{3}, \sqrt{3}\}$

0.3 $\{1, -1, 2, -2, 3, -3, 4, -4, 5, -5, 6, -6, 10, -10, 12, -12, 15, -15, 20, -20, 30, -30, 60, -60\}$

0.5 Not a set (not well defined)

0.7 The set $\varnothing$

0.9 The set $\mathbf{Q}$

0.11 Not an equivalence relation

0.13 An equivalence relation; $\bar{0} = \{0\}, \bar{a} = \{a, -a\}$ for each nonzero $a \in \mathbf{R}$

0.15 An equivalence relation; $\bar{1} = \{1, 2, \ldots, 9\}$, $\overline{10} = \{10, 11, \ldots, 99\}$, $\overline{100} = \{100, 101, \ldots, 999\}$, and in general $\overline{10^n} = \{10^n, 10^n + 1, \ldots, 10^{n+1} - 1\}$

0.17 An equivalence relation;
$\bar{1} = \{1, 3, 5, 7, \ldots\} = \{2(n-1) + 1 \,|\, n \in \mathbf{Z}^+\}$
$\bar{2} = \{2, 4, 6, 8, \ldots\} = \{2n \,|\, n \in \mathbf{Z}^+\}$

0.19 (We don't want to spoil your fun with this one.)

0.21 2

0.23 15

Chapter 1

1.1 a) e, b, a b) a, a. You can't say. c) a, c. $*$ is not associative.
d) No, $b * e \neq e * b$.

1.3 d, c, c, d

1.7   F T F F F T T T T F

1.9   1. 16. 19,683. $n^{(n^2)}$.

1.11   a) Two binary operations $*$ and $*'$ on the same set S give **algebraic structures of the same type**, if each $x \in S$ has a counterpart $x' \in S$ such that the correspondence $x \leftrightarrow x'$ is one to one and such that $(a * b)' = a' *' b'$ for all $a, b \in S$. (This may be expressed in other ways.)

   b) 10

Chapter 2

2.1   a) No. $\mathscr{G}_3$ fails.   b) No. $\mathscr{G}_1$ fails.   c) Yes   d) No. $\mathscr{G}_3$ fails at $a = 0$.   e) Yes
   f) Yes

2.3   *Partial answer*: $(a * b' * c)' = c' * b * a'$

2.7

	e	a	b
e	e	a	b
a	a	e	e
b	b	e	e

(Other answers are possible.)

2.9   c) $-\frac{1}{3}$

Chapter 3

3.1   a) Yes   b) No   c) Yes   d) Yes   e) Yes   f) No

3.3   a) $0, 25, 50, -25, -50$ b) $1, \frac{1}{2}, 2, 4, \frac{1}{4}$
   c) $1, \pi, \pi^2, 1/\pi, 1/\pi^2$ (Other answers are possible.)

3.5   a)

	0	1	2	3	4	5
0	0	1	2	3	4	5
1	1	2	3	4	5	0
2	2	3	4	5	0	1
3	3	4	5	0	1	2
4	4	5	0	1	2	3
5	5	0	1	2	3	4

   b) $\langle 2 \rangle = \langle 4 \rangle = \{0, 2, 4\}$
      $\langle 3 \rangle = \{0, 3\}$
      $\langle 1 \rangle = \langle 5 \rangle = \mathbf{Z}_6$
   c) 1 and 5

3.7   T F T F F T F F T F

   d) A group $\{e\}$ of one element has only one (improper) subgroup.

3.19   The Klein 4-group V is an example.

Chapter 4

4.1 a) $\begin{pmatrix} 1 & 2 & 3 & 4 & 5 & 6 \\ 1 & 2 & 3 & 6 & 5 & 4 \end{pmatrix}$ b) $\begin{pmatrix} 1 & 2 & 3 & 4 & 5 & 6 \\ 2 & 4 & 1 & 5 & 6 & 3 \end{pmatrix}$

c) $\begin{pmatrix} 1 & 2 & 3 & 4 & 5 & 6 \\ 3 & 4 & 1 & 6 & 2 & 5 \end{pmatrix}$ d) $\begin{pmatrix} 1 & 2 & 3 & 4 & 5 & 6 \\ 5 & 1 & 6 & 2 & 4 & 3 \end{pmatrix}$ e) $\begin{pmatrix} 1 & 2 & 3 & 4 & 5 & 6 \\ 2 & 6 & 1 & 5 & 4 & 3 \end{pmatrix}$

4.3 a) $\langle \rho_1 \rangle = \langle \rho_2 \rangle = \{\rho_0, \rho_1, \rho_2\}$
$\langle \mu_1 \rangle = \{\rho_0, \mu_1\}$

b)

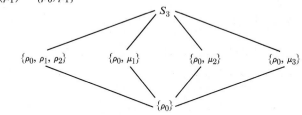

4.7 S_3 is an example. See Exercise 4.3(b).

4.9 $|D_n| = 2n$

4.13 a) Yes   b) No (not closed)   c) No (no inverse)   d) Yes

Chapter 5

5.1 a) $\begin{pmatrix} 1 & 2 & 3 & 4 & 5 & 6 & 7 & 8 \\ 4 & 5 & 3 & 7 & 1 & 6 & 8 & 2 \end{pmatrix}$ b) $\begin{pmatrix} 1 & 2 & 3 & 4 & 5 & 6 & 7 & 8 \\ 3 & 7 & 2 & 8 & 5 & 4 & 1 & 6 \end{pmatrix}$

c) $\begin{pmatrix} 1 & 2 & 3 & 4 & 5 & 6 & 7 & 8 \\ 5 & 8 & 3 & 2 & 7 & 6 & 1 & 4 \end{pmatrix}$

5.5 a) 4   b) A cycle of length n has order n.

c) σ has order 6; τ has order 4.   d) (a) 6 (b) 6 (c) 8

e) The order of a permutation expressed as a product of disjoint cycles is the least common multiple of the lengths of the cycles.

Chapter 6

6.1 2, 4, 4, 16

6.3 a) 6   b) 7   c) 4   d) 8   e) An infinite number

6.5 Z_6: 1, 2, 3, 6
Z_8: 1, 2, 4, 8
Z_{12}: 1, 2, 3, 4, 6, 12
Z_{60}: 1, 2, 3, 4, 5, 6, 10, 12, 15, 20, 30, 60
Z_{17}: 1, 17

6.7 S_3 gives a counterexample.

6.9 Consider $(xax^{-1})^2$.

6.11 $(p - 1)(q - 1)$

Chapter 7

7.1 1) Z_4 is cyclic while V is not.

2) Z_4 has only two elements that are solutions of $x + x = 0$, while V has four solutions of the corresponding equation $x^2 = e$.

7.5 T T F T F T T T F T

7.15 *Partial answer: $a\psi = a - 1$*

Chapter 8

8.1

Element	Order	Element	Order
(0, 0)	1	(0, 2)	2
(1, 0)	2	(1, 2)	2
(0, 1)	4	(0, 3)	4
(1, 1)	4	(1, 3)	4

The group is not cyclic.

8.5 $Z_{20} \times Z_3, Z_{15} \times Z_4, Z_{12} \times Z_5, Z_5 \times Z_3 \times Z_4$

8.7 T T F T F F F F F T

8.9 *Partial answer:* There are seven of them.

8.13 S_3

8.15 If a group G is the internal direct product of abelian subgroups, then G is abelian.

8.19 $HK = \{\rho_0, \rho_2, \mu_1, \mu_2\}; H \vee K = S_3$

Chapter 9

9.1 For 720: $Z_{16} \times Z_9 \times Z_5 \simeq Z_{720},$

$Z_2 \times Z_8 \times Z_9 \times Z_5 \simeq Z_2 \times Z_{360},$

$Z_4 \times Z_4 \times Z_9 \times Z_5 \simeq Z_4 \times Z_{180},$

$Z_2 \times Z_2 \times Z_4 \times Z_9 \times Z_5 \simeq Z_2 \times Z_2 \times Z_{180},$

$Z_2 \times Z_2 \times Z_2 \times Z_2 \times Z_9 \times Z_5 \simeq Z_2 \times Z_2 \times Z_2 \times Z_{90},$

$Z_{16} \times Z_3 \times Z_3 \times Z_5 \simeq Z_3 \times Z_{240},$

$Z_2 \times Z_8 \times Z_3 \times Z_3 \times Z_5 \simeq Z_6 \times Z_{120},$

$Z_4 \times Z_4 \times Z_3 \times Z_3 \times Z_5 \simeq Z_{12} \times Z_{60},$

$Z_2 \times Z_2 \times Z_4 \times Z_3 \times Z_3 \times Z_5 \simeq Z_2 \times Z_6 \times Z_{60},$

$Z_2 \times Z_2 \times Z_2 \times Z_2 \times Z_3 \times Z_3 \times Z_5 \simeq Z_2 \times Z_2 \times Z_6 \times Z_{30},$

For 1089: $Z_9 \times Z_{121} \simeq Z_{1089},$

$Z_3 \times Z_3 \times Z_{121} \simeq Z_3 \times Z_{363},$

$Z_9 \times Z_{11} \times Z_{11} \simeq Z_{11} \times Z_{99},$

$Z_3 \times Z_3 \times Z_{11} \times Z_{11} \simeq Z_{33} \times Z_{33}$

9.3 $\{2, 3\}$ generates Z_{12}. $\{4, 6\}$ generates $\langle 2 \rangle$. $\{8, 6, 10\}$ generates $\langle 2 \rangle$.

9.7 There are three of order 24. There are two of order 25. There are six of order (24)(25).

9.9 49

9.13 The numbers are the same.

9.17 A *subgroup* of S_n may not be able to be generated by two elements.

9.19 c) $\phi : G \to \mathbf{Z} \times \mathbf{Z}$ where $(a, b)\phi = \left(a, \dfrac{b - a}{10} \right)$.

Chapter 10

10.5 *Partial answer:* The group has order 48.

10.7 c) 432   d) $|H| = 6; H \simeq S_3$.

Chapter 11

11.1 a) 18   b) 8   c) 6

11.3 *Partial answer:* The left cosets are

$$e = (0, 0) + \langle (1, 2) \rangle = \{(0, 0), (1, 2)\},$$
$$a = (1, 0) + \langle (1, 2) \rangle = \{(1, 0), (0, 2)\},$$
$$b = (0, 1) + \langle (1, 2) \rangle = \{(0, 1), (1, 3)\},$$
$$c = (1, 1) + \langle (1, 2) \rangle = \{(1, 1), (0, 3)\}.$$

The induced operation on the left cosets is well defined, and they do form a group. This group of cosets is isomorphic to $\mathbf{Z}_4$.

Chapter 12

12.1 a) When working with a factor group G/H, one would let a and b be elements of G, not elements of G/H. The student probably does not understand what elements of G/H look like and can write nothing sensible concerning them.

   b) We must show that G/H is abelian. Let aH and bH be two elements of G/H.

12.3 a) 3   b) 4   c) 8   d) 4   e) 6

12.5 a) $\mathbf{Z}$ b) $\mathbf{Z}$ c) $\mathbf{Z} \times \mathbf{Z}$

12.7 $\{\rho_0, \mu_1\}, \{\rho_0, \mu_2\}$ and $\{\rho_0, \mu_3\}$

12.9 T T T T T F T F T F

   f) $\mathbf{Z}$ is torsion free but $\mathbf{Z}/n\mathbf{Z} \simeq \mathbf{Z}_n$, a torsion group.

   h) For $n > 2$, S_n is nonabelian, but $S_n/A_n \simeq \mathbf{Z}_2$, and $\mathbf{Z}_2$ is abelian.

   j) $n\mathbf{R} = \mathbf{R}$, so $\mathbf{R}/n\mathbf{R}$ is of order 1.

12.23 $G' = \{\rho_0, \rho_2\}$

Chapter 13

13.1 a) Yes. (Image ϕ) $= \mathbf{Z}$. (Kernel ϕ) $= \{0\}$.

   b) No. $2 = (\frac{3}{2})\phi + (\frac{3}{2})\phi \neq 3\phi = 3$.

   c) Yes. (Image ϕ) $= \mathbf{R}^+$. (Kernel ϕ) $= \{1, -1\}$.

   d) Yes. (Image ϕ) $= \mathbf{Z}_2$. (Kernel ϕ) $= \{0, 2, 4\}$.

   e) No. In $\mathbf{Z}_2$, $0 = 1 + 1 = 5\phi + 5\phi \neq (5 + 5)\phi = 1\phi = 1$.

13.3 2 of $\mathbf{Z}$ onto $\mathbf{Z}$; 2 of $\mathbf{Z}$ into $\mathbf{Z}_2$; 1 of $\mathbf{Z}$ onto $\mathbf{Z}_2$

13.5 0 of $\mathbf{Z}_{12}$ onto $\mathbf{Z}_5$; 6 of $\mathbf{Z}_{12}$ into $\mathbf{Z}_6$; 2 of $\mathbf{Z}_{12}$ onto $\mathbf{Z}_6$; 2 of $\mathbf{Z}_{12}$ into $\mathbf{Z}_{14}$; 4 of $\mathbf{Z}_{12}$ into $\mathbf{Z}_{16}$

13.7 T T F T F F T T T F

e) If ϕ is a homomorphism of G, then $|G\phi| = |G|/|(\text{Kernel } \phi)|$.   f) See (e).

g) A mapping of every element of a group G onto the identity of any group is always a homomorphism.

h) See (g).   i) See Exercise 13.6.   j) See Example 13.1.

13.9 A_n

13.11 *Partial answer:* (Image ϕ) $= \langle a \rangle$. (Kernel ϕ) $= n\mathbf{Z}$ for some nonnegative $n \in \mathbf{Z}$ (including $n = 0, 1$).

13.13 b) *Partial answer:* If r is not relatively prime to $|G|$, (kernel ϕ_r) $\neq \{e\}$ and ϕ_r is not a map onto G. Thus for some $a \in G$, $x^r = a$ has no solution. In the extreme case that $|G|$ divides r, $x^r = a$ has no solution for any $a \in G$, $a \neq e$.

13.15 *Partial answer:* ϕ is an isomorphism if (kernel ϕ) $= \{e\}$, that is, if the center of G is the trivial group $\{e\}$.

Chapter 14

14.1 The refinements $\{0\} < 2940\mathbf{Z} < 60\mathbf{Z} < 20\mathbf{Z} < 4\mathbf{Z} < \mathbf{Z}$ of $\{0\} < 60\mathbf{Z} < 20\mathbf{Z} < \mathbf{Z}$ and $\{0\} < 2940\mathbf{Z} < 980\mathbf{Z} < 245\mathbf{Z} < 49\mathbf{Z} < \mathbf{Z}$ of $\{0\} < 245\mathbf{Z} < 49\mathbf{Z} < \mathbf{Z}$ are isomorphic.

14.3 They are all of the form $\{(0, 0)\} < H < \mathbf{Z}_5 \times \mathbf{Z}_5$, where H may be any of the subgroups $\langle (0, 1) \rangle$, $\langle (1, 0) \rangle$, $\langle (1, 1) \rangle$, $\langle (1, 2) \rangle$, $\langle (1, 3) \rangle$, and $\langle (1, 4) \rangle$ of $\mathbf{Z}_5 \times \mathbf{Z}_5$. Thus, there are six in all.

14.11 $\{\rho_0\} \times \mathbf{Z}_4$

14.13 $\{\rho_0\} \times \mathbf{Z}_4 \leq \{\rho_0\} \times \mathbf{Z}_4 \leq \{\rho_0\} \times \mathbf{Z}_4 \leq \cdots$.

Chapter 15

15.3 b) G/L

15.5 a) $K = \{0, 3, 6, 9\}$.

b) $0 + K = \{0, 3, 6, 9\}$, $1 + K = \{1, 4, 7, 10\}$, $2 + K = \{2, 5, 8, 11\}$.

c) $(0 + K)\psi = 0$, $(1 + K)\psi = 1$, $(2 + K)\psi = 2$.

15.7 a) $HN = \{0, 2, 4, 6, 8, 10, 12, 14, 16, 18, 20, 22\}$, $H \cap N = \{0, 12\}$.

b) $0 + N = \{0, 6, 12, 18\}$, $2 + N = \{2, 8, 14, 20\}$, $4 + N = \{4, 10, 16, 22\}$.

c) $0 + (H \cap N) = \{0, 12\}$, $4 + (H \cap N) = \{4, 16\}$, $8 + (H \cap N) = \{8, 20\}$.

d) $(0 + N)\psi = 0 + (H \cap N)$, $(2 + N)\psi = 8 + (H \cap N)$, $(4 + N)\psi = 4 + (H \cap N)$.

15.9 a) $0 + H = \{0, 4, 8, 12, 16, 20\}$, $1 + H = \{1, 5, 9, 13, 17, 21\}$, $2 + H = \{2, 6, 10, 14, 18, 22\}$, $3 + H = \{3, 7, 11, 15, 19, 23\}$.

b) $0 + K = \{0, 8, 16\}$, $1 + K = \{1, 9, 17\}$, $2 + K = \{2, 10, 18\}$, $3 + K = \{3, 11, 19\}$, $4 + K = \{4, 12, 20\}$, $5 + K = \{5, 13, 21\}$, $6 + K = \{6, 14, 22\}$, $7 + K = \{7, 15, 23\}$.

c) $0 + K = \{0, 8, 16\}$, $4 + K = \{4, 12, 20\}$.

d) $(0 + K) + (H/K) = H/K = \{0 + K, 4 + K\} = \{\{0, 8, 16\}, \{4, 12, 20\}\}$
$(1 + K) + (H/K) = \{1 + K, 5 + K\} = \{\{1, 9, 17\}, \{5, 13, 21\}\}$
$(2 + K) + (H/K) = \{2 + K, 6 + K\} = \{\{2, 10, 18\}, \{6, 14, 22\}\}$
$(3 + K) + (H/K) = \{3 + K, 7 + K\} = \{\{3, 11, 19\}, \{7, 15, 23\}\}$.

e) $(0 + H)\psi = (0 + K) + (H/K), (1 + H)\psi = (1 + K) + (H/K),$
$(2 + H)\psi = (2 + K) + (H/K), (3 + H)\psi = (3 + K) + (H/K).$

15.11 *Chain (3)*   *Chain (4)*   *Isomorphisms*

$\{0\} \leq \{0\} \leq \langle 12 \rangle$ $\{0\} \leq \{0\} \leq \langle 18 \rangle \leq \langle 18 \rangle$ $\{0\}/\{0\} \simeq \{0\}/\{0\} \simeq \{0\}$

$\leq \langle 6 \rangle \leq \langle 3 \rangle$ $\leq \langle 6 \rangle \leq \langle 3 \rangle \leq \mathbf{Z}_{36}$ $\langle 12 \rangle/\{0\} \simeq \langle 6 \rangle/\langle 18 \rangle \simeq \mathbf{Z}_3$

$\leq \langle 3 \rangle \leq \mathbf{Z}_{36}$ $\langle 6 \rangle/\langle 12 \rangle \simeq \langle 18 \rangle/\{0\} \simeq \mathbf{Z}_2$

$\langle 3 \rangle/\langle 6 \rangle \simeq \langle 3 \rangle/\langle 6 \rangle \simeq \mathbf{Z}_2$

$\langle 3 \rangle/\langle 3 \rangle \simeq \langle 18 \rangle/\langle 18 \rangle \simeq \{0\}$

$\mathbf{Z}_{36}/\langle 3 \rangle \simeq \mathbf{Z}_{36}/\langle 3 \rangle \simeq \mathbf{Z}_3$

Chapter 16

16.1 $X_{\rho 0} = X, X_{\rho 1} = \{C\}, X_{\rho 2} = \{m_1, m_2, d_1, d_2, C\}, X_{\rho 3} = \{C\},$
$X_{\mu_1} = \{s_1, s_3, m_1, m_2, C, P_1, P_3\}, X_{\mu_2} = \{s_2, s_4, m_1, m_2, C, P_2, P_4\},$
$X_{\delta_1} = \{2, 4, d_1, d_2, C\}, X_{\delta_2} = \{1, 3, d_1, d_2, C\}.$

16.3 $\{1, 2, 3, 4\}, \{s_1, s_2, s_3, s_4\}, \{m_1, m_2\}, \{d_1, d_2\}, \{C\}, \{P_1, P_2, P_3, P_4\}$

16.5 Every sub-G-set of a G-set X consists of a union of orbits in X under G.

16.7 a) No

b) $\{1, 2, 3, 4\}, \{s_1, s_2, s_3, s_4\}, \{P_1, P_2, P_3, P_4\}$

16.11 b) The set of points on the circle with center at the origin and passing through P

c) The cyclic subgroup $\langle 2\pi \rangle$ of $G = \mathbf{R}$

16.15 a) $K = g_0^{-1} H g_0.$

b) *Conjecture:* H and K should be conjugate subgroups of G.

16.17

	X	Y		Z		
	a	a	b	a	b	c
0	a	a	b	a	b	c
1	a	b	a	b	c	a
2	a	a	b	c	a	b
3	a	b	a	a	b	c
4	a	a	b	b	c	a
5	a	b	a	c	a	b

There are three of them.

Chapter 17

17.1 5

17.3 2

17.5 11,712

17.7 a) 45   b) 231

17.9 a) 90   b) 6,426

Chapter 18

18.1 a) 3   b) 27   c) 1, 3   d) 1, 85, 1, 51

18.7 T T T F T F T T F F

18.9 $H = \{\iota, (1, 2, 3, 4), (1, 3)(2, 4), (1, 4, 3, 2), (1, 3), (2, 4), (1, 2)(3, 4), (1, 4)(2, 3)\}$,
$K = \{\iota, (1, 3, 2, 4), (1, 2)(3, 4), (1, 4, 2, 3), (1, 2), (3, 4), (1, 3)(2, 4), (1, 4)(2, 3)\}$,
$K = (2, 3)H(2, 3)$.

Chapter 19

19.3 T T F T T T T T F F

e) This is somewhat a matter of opinion.

i) A group of order 42 can have no subgroup of order 8, for 8 does not divide 42.

19.7 e) $p(1) = 1$. $p(2) = 2$. $p(3) = 3$. $p(4) = 5$. $p(5) = 7$. $p(6) = 11$. $p(7) = 15$.

19.9 $120 = 1 + 10 + 15 + 20 + 20 + 30 + 24$,
$720 = 1 + 15 + 45 + 40 + 15 + 120 + 90 + 40 + 90 + 144 + 120$.

Chapter 20

20.3 *Partial answer:* Yes

20.5 $|ad - bc| = 1$

20.7 T T T T T F F T T F

Chapter 21

21.1 a) $a^2b^2a^3c^3b^{-2}$, $b^2c^{-3}a^{-3}b^{-2}a^{-2}$ b) $a^{-1}b^3a^4c^6a^{-1}$, $ac^{-6}a^{-4}b^{-3}a$

21.3 a) 16   b) 36   c) 36

21.5 a) 16   b) 36   c) 18

21.9 a) *Partial answer:* $\{1\}$ is a basis for $\mathbf{Z}_4$.   c) Yes

21.11 c) A blop group on S is isomorphic to the *free group* $F[S]$ on S.

Chapter 22

22.1 $(a : a^4 = 1)$. $(a, b : a^4 = 1, b = a^2)$. $(a, b, c : a = 1, b^4 = 1, c = 1)$. (Other answers are possible.)

22.3 Octic group:

	1	a	a^2	a^3	b	ab	a^2b	a^3b
1	1	a	a^2	a^3	b	ab	a^2b	a^3b
a	a	a^2	a^3	1	ab	a^2b	a^3b	b
a^2	a^2	a^3	1	a	a^2b	a^3b	b	ab
a^3	a^3	1	a	a^2	a^3b	b	ab	a^2b
b	b	a^3b	a^2b	ab	1	a^3	a^2	a
ab	ab	b	a^3b	a^2b	a	1	a^3	a^2
a^2b	a^2b	ab	b	a^3b	a^2	a	1	a^3
a^3b	a^3b	a^2b	ab	b	a^3	a^2	a	1

Quaternion group: The same as the table for the octic group except that the sixteen entries in the lower right corner are

a^2	a	1	a^3
a^3	a^2	a	1
1	a^3	a^2	a
a	1	a^3	a^2

22.9 $\mathbf{Z}_{21}$. $(a, b : a^7 = 1, b^3 = 1, ba = a^2 b)$

22.11 $\mathbf{Z}_{12}$. $\mathbf{Z}_2 \times \mathbf{Z}_6$

$A_4 \simeq (a, b, c : a^2 = b^2 = c^3 = 1, ab = ba, ca = bc, cb = abc, cab = ac)$
$\simeq (s, t : s^3 = 1, t^2 = 1, (st)^3 = 1)$.
$D_6 \simeq (a, b : a^6 = 1, b^2 = 1, ba = a^5 b)$.
$(a, b : a^3 = 1, b^4 = 1, ba = a^2 b.)$
(See Dean [7, p. 246] for a complete solution.)

Chapter 23

23.1 a) Yes   b) No. $\mathbf{Z}^+$ has no identity for addition.   c) Yes   d) Yes
e) Yes   f) Yes   g) No. Multiplication is not closed on $\{ri \,|\, r \in \mathbf{R}\}$.

23.3 a) $1, -1$ b) $(1, 1), (1, -1), (-1, 1), (-1, -1)$ c) $1, 2, 3, 4$
d) All nonzero $q \in \mathbf{Q}$
e) $(1, q, 1), (1, q, -1), (-1, q, 1), (-1, q, -1)$ for all nonzero $q \in \mathbf{Q}$ f) $1, 3$

23.21

+	$\varnothing$	$\{a\}$	$\{b\}$	S
$\varnothing$	$\varnothing$	$\{a\}$	$\{b\}$	S
$\{a\}$	$\{a\}$	$\varnothing$	S	$\{b\}$
$\{b\}$	$\{b\}$	S	$\varnothing$	$\{a\}$
S	S	$\{b\}$	$\{a\}$	$\varnothing$

$\cdot$	$\varnothing$	S	$\{a\}$	$\{b\}$
$\varnothing$	$\varnothing$	$\varnothing$	$\varnothing$	$\varnothing$
S	$\varnothing$	S	$\{a\}$	$\{b\}$
$\{a\}$	$\varnothing$	$\{a\}$	$\{a\}$	$\varnothing$
$\{b\}$	$\varnothing$	$\{b\}$	$\varnothing$	$\{b\}$

Chapter 24

24.1 0, 3, 5, 8, 9, 11

24.3 a) 0   b) 0   c) 0   d) 3   e) 12   f) 30

24.7 There are no solutions of $x^2 + 2x + 2 = 0$. 2 is a solution of $x^2 + 2x + 4 = 0$.

24.17

	1	5	7	11
1	1	5	7	11
5	5	1	11	7
7	7	11	1	5
11	11	7	5	1

It is isomorphic to $\langle \mathbf{Z}_2 \times \mathbf{Z}_2, + \rangle$.

24.19 1

Chapter 25

25.1 $\begin{pmatrix} 7 & 13 \\ 6 & 2 \end{pmatrix}$ and $\begin{pmatrix} -8 & 63 \\ 29 & 2 \end{pmatrix}$

25.5 a) $-6 + i + 13j + 2k$ b) j c) $-\frac{1}{2}i - \frac{1}{2}j$ d) $\frac{1}{50}j - \frac{3}{50}k$

25.9 $\begin{pmatrix} 1 & 0 \\ 0 & 1 \end{pmatrix}, \begin{pmatrix} 0 & 1 \\ 1 & 0 \end{pmatrix}, \begin{pmatrix} 1 & 0 \\ 1 & 1 \end{pmatrix}, \begin{pmatrix} 0 & 1 \\ 1 & 1 \end{pmatrix}, \begin{pmatrix} 1 & 1 \\ 0 & 1 \end{pmatrix},$ and $\begin{pmatrix} 1 & 1 \\ 1 & 0 \end{pmatrix}$ are units in

$M_2(F)$.

25.13 $1\rho_0 + 0\rho_1 + 1\rho_2 + 0\mu_1 + 1\mu_2 + 1\mu_3$

25.15 $\{a_1 + a_3 j \,|\, a_1, a_3 \in \mathbf{R}\}$ and $\{a_1 + a_4 k \,|\, a_1, a_4 \in \mathbf{R}\}$

Chapter 26

26.1 $\{q_1 + q_2 i \,|\, q_1, q_2 \in \mathbf{Q}\}$

26.3 $D = \mathbf{Q}$ and $D' = \mathbf{Z}$ gives an example.

26.5 T F T F T T F T T

26.13 4, for 1 and 3 are already units in $\mathbf{Z}_4$.

26.15 $\left\{ \dfrac{m}{6^n} \,\middle|\, m \in \mathbf{Z}, n \in \mathbf{Z}^+ \right\}$

Chapter 28

28.1 $N_1 = \{0\}$, $\mathbf{Z}_{12}/N_1 \simeq \mathbf{Z}_{12}$;
$N_2 = \{0, 2, 4, 6, 8, 10\}$, $\mathbf{Z}_{12}/N_2 \simeq \mathbf{Z}_2$;
$N_3 = \{0, 3, 6, 9\}$, $\mathbf{Z}_{12}/N_3 \simeq \mathbf{Z}_3$;
$N_4 = \{0, 4, 8\}$, $\mathbf{Z}_{12}/N_4 \simeq \mathbf{Z}_4$;
$N_5 = \{0, 6\}$, $\mathbf{Z}_{12}/N_5 \simeq \mathbf{Z}_6$;
$N_6 = \mathbf{Z}_{12}$, $\mathbf{Z}_{12}/\mathbf{Z}_{12} \simeq \{0\}$.

28.3 $\{(n, n) \,|\, n \in \mathbf{Z}\}$. (Other answers are possible.)

28.5 F T F T T T T F F F

28.15 *Partial answer:* By the definition in Exercise 28.15, $\sqrt{R} = R$ for every ring R. However, according to the definition in Exercise 28.11, the radical of R is not always all of R. Thus this terminology is inconsistent if $N = R$.

28.17 If $\sqrt{N}/N$ is viewed as a subring of R/N, then it is the radical of R/N, in the sense of the definition in Exercise 28.11.

Chapter 29

29.1 ϕ_1 such that $1\phi_1 = 1$. ϕ_2 such that $1\phi_2 = 0$.

29.3 Let $\phi : \mathbf{Z} \to \mathbf{Z} \times \mathbf{Z}$ be given by $n\phi = (n, 0)$ for $n \in \mathbf{Z}$.

29.5 $2\mathbf{Z} \times \mathbf{Z}$ is a maximal ideal of $\mathbf{Z} \times \mathbf{Z}$. $\mathbf{Z} \times \{0\}$ is a prime ideal that is not maximal. $4\mathbf{Z} \times \{0\}$ is an ideal that is not prime.

29.7 T F T T T F T T F T

d) A field has no proper ideals, and a simple group has no proper normal subgroups. In both cases all quotient structures are either trivial or isomorphic to the original structure.

29.15 No. Enlarging the domain to a field of quotients, you would have to have a field containing two different prime fields $\mathbf{Z}_p$ and $\mathbf{Z}_q$, which is impossible.

Chapter 30

30.1 $f(x) + g(x) = 3x^4 + 2x^3 + 4x^2 + 1$,
$f(x)g(x) = x^7 + 2x^6 + 4x^5 + x^3 + 2x^2 + x + 3$.

30.3 a) 0 b) 2 c) 2 d) -1

30.5 $0, 4 = -1$

30.7 T T T T F F T T F T

f) In $Z_6[x]$, $(2x^3)(3x^4) = 0$.

30.9 b) F c) $F[x]$

30.11 0, 1, 2, 3

30.13 a) Let F be a subfield of a field E, let $\alpha_1, \ldots, \alpha_n$ be any elements of E, and let $x_1, \ldots, x_n$ be indeterminates. The map $\phi_{\alpha_1,\ldots,\alpha_n}: F[x_1, \ldots, x_n] \to E$ defined by

$$\left(\sum a_{m_1,\ldots,m_n} x_1^{m_1} \cdots x_n^{m_n}\right)\phi_{\alpha_1,\ldots,\alpha_n} = \sum a_{m_1,\ldots,m_n} \alpha_1^{m_1} \cdots \alpha_n^{m_n}$$

for

$$\left(\sum a_{m_1,\ldots,m_n} x_1^{m_1} \cdots x_n^{m_n}\right) \in F[x_1, \ldots, x_n]$$

is a homomorphism of $F[x_1, \ldots, x_n]$ into E. Also, $x_i\phi_{\alpha_1,\ldots,\alpha_n} = \alpha_i$, and $\phi_{\alpha_1,\ldots,\alpha_n}$ maps F isomorphically, by the identity map. $\phi_{\alpha_1,\ldots,\alpha_n}$ is an **evaluation homomorphism**.

b) 558

c) Let F be a subfield of a field E. Then $(\alpha_1, \ldots, \alpha_n) \in \underbrace{(E \times E \times \cdots \times E)}_{n \text{ factors}}$ is a **zero of** $f(x_1, \ldots, x_n) \in F[x_1, \ldots, x_n]$, if $f(x_1, \ldots, x_n)\phi_{\alpha_1,\ldots,\alpha_n} = 0$.

Chapter 31

31.1 $q(x) = x^4 + x^3 + x^2 + x - 2$, $r(x) = 4x + 3$

31.3 $(x - 1)(x + 1)(x - 2)(x + 2)$

31.5 *Partial answer:* $f(x)$ is not irreducible over **R** and is not irreducible over **C**.

31.9 T T T F T F T T T F

31.11 Yes. It is of degree 3 with no zeros in Z_5. $2x^3 + x^2 + 2x + 2$

31.15 No. $\langle x^2 - 5x + 6 \rangle$ is not a maximal ideal of $Q[x]$, since $x^2 - 5x + 6 = (x - 3)(x - 2)$ is not irreducible over Q. Yes, $\langle x^2 - 6x + 6 \rangle$ is a maximal ideal of $Q[x]$.

Chapter 32

32.1 a) Yes b) Yes c) No d) Yes e) No f) Yes g) Yes h) Yes

32.3 In $Z[x]$: $(2)(2)(x^2 - x + 2)$
In $Q[x]$: $4x^2 - 4x + 8$
In $Z_{11}[x]$: $(4x + 2)(x + 4)$

32.9 *Partial answer:* $D^* - U$ is not a group under multiplication for $1 \notin (D^* - U)$.

32.11 Not every nonunit $\neq 0$ of $Z \times Z$ has a factorization into irreducibles. For example, $(1, 0)$ is not a unit, and every factorization of $(1, 0)$ has a factor of the form $(\pm 1, 0)$, which is not irreducible, since $(\pm 1, 0) = (\pm 1, 0)(1, 50)$. The only irreducibles of $\mathbf{Z} \times \mathbf{Z}$ are $(\pm 1, p)$ and $(q, \pm 1)$, where p and q are irreducibles in $\mathbf{Z}$.

32.17 $\mathbf{Z}$

Chapter 33

33.1 a) Yes b) No. (1) is violated. c) No. (1) is violated.

d) No. (2) is violated. e) Yes

33.3 $x^3 + 2x - 1$

33.5 a) Yes. $\mathbf{Z}$ is a UFD, and Theorem 32.3 applies. c) No

d) No. By Theorem 33.1, $\mathbf{Z}[x]$ a Euclidean domain would imply $\mathbf{Z}[x]$ a PID, contradicting (c).

33.7 T F T F T T T F T T

33.9 $61 = 29(49,349) + (-92)(15,555)$.

33.17 *Partial answer:* The equation $ax = b$ has a solution in $\mathbf{Z}_n$ for nonzero $a, b \in \mathbf{Z}_n$ if and only if the positive gcd of a and n in $\mathbf{Z}$ divides b.

Chapter 34

34.1 a) $5 = (1 + 2i)(1 - 2i)$. b) 7 is irreducible in $\mathbf{Z}[i]$.

c) $4 + 3i = (1 + 2i)(2 - i)$. d) $6 - 7i = (1 - 2i)(4 + i)$.

34.3 $\sigma = 1 + i, \rho = 3i$.

34.13 c) (i) order 9, characteristic 3 (ii) order 2, characteristic 2 (iii) order 5, characteristic 5

Chapter 35

35.1 a) $x^2 - 2x - 1$ b) $x^4 - 10x^2 + 1$ c) $x^2 - 2x + 2$

d) $x^6 - 3x^4 + 3x^2 - 3$ e) $x^{12} + 3x^8 - 4x^6 + 3x^4 + 12x^2 + 5$

35.3 a) Algebraic, $\deg(\alpha, F) = 2$. b) Algebraic, $\deg(\alpha, F) = 2$. c) Transcendental

d) Algebraic, $\deg(\alpha, F) = 1$. e) Algebraic, $\deg(\alpha, F) = 2$. f) Transcendental

g) Algebraic, $\deg(\alpha, F) = 1$. h) Algebraic, $\deg(\alpha, F) = 3$.

35.7 b)

+	0	1	2	α	2α	$1 + \alpha$	$1 + 2\alpha$	$2 + \alpha$	$2 + 2\alpha$
0	0	1	2	α	2α	$1 + \alpha$	$1 + 2\alpha$	$2 + \alpha$	$2 + 2\alpha$
1	1	2	0	$1 + \alpha$	$1 + 2\alpha$	$2 + \alpha$	$2 + 2\alpha$	α	2α
2	2	0	1	$2 + \alpha$	$2 + 2\alpha$	α	2α	$1 + \alpha$	$1 + 2\alpha$
α	α	$1 + \alpha$	$2 + \alpha$	2α	0	$1 + 2\alpha$	1	$2 + 2\alpha$	2
2α	2α	$1 + 2\alpha$	$2 + 2\alpha$	0	α	1	$1 + \alpha$	2	$2 + \alpha$
$1 + \alpha$	$1 + \alpha$	$2 + \alpha$	α	$1 + 2\alpha$	1	$2 + 2\alpha$	2	2α	0
$1 + 2\alpha$	$1 + 2\alpha$	$2 + 2\alpha$	2α	1	$1 + \alpha$	2	$2 + \alpha$	0	α
$2 + \alpha$	$2 + \alpha$	α	$1 + \alpha$	$2 + 2\alpha$	2	2α	0	$1 + 2\alpha$	1
$2 + 2\alpha$	$2 + 2\alpha$	2α	$1 + 2\alpha$	2	$2 + \alpha$	0	α	1	$1 + \alpha$

	0	1	2	α	2α	$1+\alpha$	$1+2\alpha$	$2+\alpha$	$2+2\alpha$
0	0	0	0	0	0	0	0	0	0
1	0	1	2	α	2α	$1+\alpha$	$1+2\alpha$	$2+\alpha$	$2+2\alpha$
2	0	2	1	2α	α	$2+2\alpha$	$2+\alpha$	$1+2\alpha$	$1+\alpha$
α	0	α	2α	2	1	$2+\alpha$	$1+\alpha$	$2+2\alpha$	$1+2\alpha$
2α	0	2α	α	1	2	$1+2\alpha$	$2+2\alpha$	$1+\alpha$	$2+\alpha$
$1+\alpha$	0	$1+\alpha$	$2+2\alpha$	$2+\alpha$	$1+2\alpha$	2α	2	1	α
$1+2\alpha$	0	$1+2\alpha$	$2+\alpha$	$1+\alpha$	$2+2\alpha$	2	α	2α	1
$2+\alpha$	0	$2+\alpha$	$1+2\alpha$	$2+2\alpha$	$1+\alpha$	1	2α	α	2
$2+2\alpha$	0	$2+2\alpha$	$1+\alpha$	$1+2\alpha$	$2+\alpha$	α	1	2	2α

35.9 It is the monic polynomial in $F[x]$ of *minimal* degree having α as a zero.

Chapter 36

36.1 $\{(0, 1), (1, 0)\}, \{(1, 1), (-1, 1)\}, \{(2, 1), (2, 1)\}$. (Other answers are possible.)

36.3 $x^2 + x + 1$

36.7 a) A subset W of a vector space V over a field F is a **subspace of** V **over** F if the induced operations of vector addition and scalar multiplication are closed on W and W is a vector space over F under these operations.

36.9 *Partial answer:* The **direct sum** $V_1 \oplus \cdots \oplus V_n$ **of the vector spaces** V_i is the set $V_1 \times \cdots \times V_n$ of vectors, together with the operations of vector addition and scalar multiplication defined by

$$(\alpha_1, \ldots, \alpha_n) + (\beta_1, \ldots, \beta_n) = (\alpha_1 + \beta_1, \ldots, \alpha_n + \beta_n)$$

and $a(\alpha_1, \ldots, \alpha_n) = (a\alpha_1, \ldots, a\alpha_n)$ for $\alpha_i, \beta_i \in V_i$ and $a \in F$.

36.15 a) A homomorphism

b) *Partial answer:* The **kernel** (or **nullspace**) of ϕ is $\{\alpha \in V \,|\, \alpha\phi = 0\}$.

c) ϕ is an isomorphism of V with V' if (kernel ϕ) $= \{0\}$ and ϕ maps V onto V'.

Chapter 37

37.1 $\langle G, \mathcal{O}, *, \cdot \rangle$. (Other notations are possible.)

37.3 $\langle V, F, \oplus, \odot, +, \cdot, \times \rangle$. (Other notations are possible.)

37.5 $\mathbf{Z}_2 \times \{0\}$ is not a characteristic subgroup of $\langle \mathbf{Z}_2 \times \mathbf{Z}_2, + \rangle$, for there exists an automorphism of $\mathbf{Z}_2 \times \mathbf{Z}_2$ mapping $\mathbf{Z}_2 \times \{0\}$ onto $\{0\} \times \mathbf{Z}_2$.

37.7 *Partial answer:* A map $\phi: G \to G'$ is an $\mathcal{O}$-**homomorphism of the** $\mathcal{O}$-**group** G **into the** $\mathcal{O}$-**group** G' if for all α, $\beta \in G$ and $a \in \mathcal{O}$, both $(\alpha + \beta)\phi = \alpha\phi + \beta\phi$ and $(\alpha a)\phi = (\alpha\phi)a$.

37.11 **A homomorphism of a (left)** R-**module** M **into a (left)** R-**module** M' is a function $\phi: M \to M'$ such that $(\alpha + \beta)\phi = \alpha\phi + \beta\phi$ and $(r\alpha)\phi = r(\alpha\phi)$ for all α, $\beta \in M$ and $r \in R$.

Chapter 38

38.1 a) $2, \{1, \sqrt{2}\}$

b) $4, \{1, \sqrt{2}, \sqrt{3}, \sqrt{6}\}$

c) $8, \{1, \sqrt{3}, \sqrt{5}, \sqrt{15}, \sqrt{2}, \sqrt{6}, \sqrt{10}, \sqrt{30}\}$

d) $6, \{1, \sqrt[3]{2}, (\sqrt[3]{2})^2, \sqrt{3}, \sqrt[3]{2}\sqrt{3}, (\sqrt[3]{2})^2(\sqrt{3})\}$

e) $6, \{1, \sqrt{2}, \sqrt[3]{2}, \sqrt{2(\sqrt[3]{2})}, (\sqrt[3]{2})^2, \sqrt{2(\sqrt[3]{2})^2}\}$

38.2 a) 4 b) 2 c) 6 d) 9 e) 2 f) 2 g) 1 h) 2

38.3 e) $\{1, \sqrt{2}\}$ f) $\{1, \sqrt{2}\}$ g) $\{1\}$ h) $\{1, \sqrt{2}\}$

38.7 T F T F F T F F F F

38.9 *Partial answer:* Extensions of degree 2^n for $n \in \mathbf{Z}^+$ are obtained.

Chapter 39

39.5 T T T F T F T T T F

Chapter 40

40.1 a) $\sqrt{2}, -\sqrt{2}$ b) $\sqrt{2}$ c) $3 + \sqrt{2}, 3 - \sqrt{2}$

d) $\sqrt{2} - \sqrt{3}, \sqrt{2} + \sqrt{3}, -\sqrt{2} - \sqrt{3}, -\sqrt{2} + \sqrt{3}$

e) $\sqrt{2} + i, \sqrt{2} - i, -\sqrt{2} + i, -\sqrt{2} - i$ f) $\sqrt{2} + i, \sqrt{2} - i$

g) $\sqrt{1 + \sqrt{2}}, -\sqrt{1 + \sqrt{2}}, \sqrt{1 - \sqrt{2}}, -\sqrt{1 - \sqrt{2}}$

h) $\sqrt{1 + \sqrt{2}}, -\sqrt{1 + \sqrt{2}}$

40.3 a) $3 - \sqrt{2}$ b) They are the same maps.

40.5 a) $\mathbf{Q}(\sqrt{2}, \sqrt{5})$ b) $\mathbf{Q}(\sqrt{2}, \sqrt{3}, \sqrt{5})$ c) $\mathbf{Q}(\sqrt{5})$ d) $\mathbf{Q}(\sqrt{3}, \sqrt{10})$

e) $\mathbf{Q}(\sqrt{6}, \sqrt{10})$ f) $\mathbf{Q}$

40.7 F F T T F T T T T T

40.9 $0\sigma_2 = 0, 1\sigma_2 = 1, \alpha\sigma_2 = 1 + \alpha, (1 + \alpha)\sigma_2 = \alpha.$ $\mathbf{Z}_2(\alpha)_{\{\sigma_2\}} = \mathbf{Z}_2.$

40.11 Let $F = \mathbf{Z}_p(x)$, where x is an indeterminate. Then the image of F under σ_p is $\mathbf{Z}_p(x^p)$, a proper subfield of $\mathbf{Z}_p(x)$. While σ_p is an isomorphic mapping for all fields E of characteristic p, σ_p *need not be onto* E, so it need not be an automorphism.

40.17 Every automorphism of $F(x)$ maps x onto some $y = (ax + b)/(cx + d)$, where $a, b, c, d \in F$ and $ad - bc \neq 0$. Conversely, each such $y \in F(x)$ gives rise to an automorphism of $F(x)$ mapping x onto y and leaving F fixed.

Chapter 41

41.1 a) The identity map of E into $\bar{\mathbf{Q}}$.

τ given by $\sqrt{2}\tau = \sqrt{2}, \sqrt{3}\tau = -\sqrt{3}, \sqrt{5}\tau = -\sqrt{5}$

b) τ_1 given by $\sqrt{2}\tau_1 = \sqrt{2}, \sqrt{3}\tau_1 = \sqrt{3}, \sqrt{5}\tau_1 = -\sqrt{5}$;

τ_2 given by $\sqrt{2}\tau_2 = \sqrt{2}, \sqrt{3}\tau_2 = -\sqrt{3}, \sqrt{5}\tau_2 = \sqrt{5}$

c) τ_1 given by $\sqrt{2}\tau_1 = \sqrt{2}, \sqrt{3}\tau_1 = \sqrt{3}, \sqrt{5}\tau_1 = -\sqrt{5}$;

τ_2 given by $\sqrt{2}\tau_2 = \sqrt{2}, \sqrt{3}\tau_2 = -\sqrt{3}, \sqrt{5}\tau_2 = \sqrt{5}$;

τ_3 given by $\sqrt{2}\tau_3 = -\sqrt{2}, \sqrt{3}\tau_3 = \sqrt{3}, \sqrt{5}\tau_3 = \sqrt{5}$;

τ_4 given by $\sqrt{2}\tau_4 = -\sqrt{2}, \sqrt{3}\tau_4 = -\sqrt{3}, \sqrt{5}\tau_4 = -\sqrt{5}$

41.3 a) $\mathbf{Q}(\pi^2)$ b) τ_1 given by $\sqrt{\pi}\tau_1 = i\sqrt{\pi}$; τ_2 given by $\sqrt{\pi}\tau_2 = -i\sqrt{\pi}$

41.5 F T F T F T T T T F

e) $\mathbf{Q}$ and $\mathbf{Q}(\sqrt{2})$ are not isomorphic, but have isomorphic algebraic closures.

Chapter 42

42.1 a) 2 b) 2 c) 4 d) 6 e) 2 f) 12

42.3 a) 1 b) 6 c) 2

42.7 Let $F = \mathbf{Q}$ and $E = \mathbf{Q}(\sqrt{2})$. Then

$$f(x) = x^4 - 5x^2 + 6 = (x^2 - 2)(x^2 - 3)$$

has a zero in E but does not split in E.

Chapter 43

43.1 $f(x) = x^4 - 4x^2 + 4 = (x^2 - 2)^2$. Here $f(x)$ is not an irreducible polynomial. Every irreducible factor of $f(x)$ has zeros of multiplicity 1 only.

43.3 F T T F F T T T T T

43.15 Compute a gcd of $f(x)$ and $f'(x)$ using the Euclidean algorithm. Then $f(x)$ has a zero of multiplicity >1 if and only if this gcd is of degree >0.

Chapter 44

44.1 $\mathbf{Z}_3(y^3, z^9)$

44.5 $\mathbf{Z}_3(y^4, z^4)$

Chapter 45

45.1 a) 3, 5 b) 3, 10, 5, 11, 14, 7, 12, 6

c) 5, 10, 20, 17, 11, 21, 19, 15, 7, 14

45.5 T F T F T F T T F T

45.9 b) It is irreducible.

Chapter 46

46.1 a) 8 b) 8 c) 8 d) 2 e) 4 f) 4 g) 2 h) 1

46.3 The order is 3. A generator of the group is σ_9, where $\alpha\sigma_9 = \alpha^9$ for $\alpha \in GF(729)$.

46.5 a) Let $\alpha_1 = \sqrt[3]{2}$, $\alpha_2 = \sqrt[3]{2}\dfrac{-1 + i\sqrt{3}}{2}$, and $\alpha_3 = \sqrt[3]{2}\dfrac{-1 - i\sqrt{3}}{2}$.

The maps are

$$\rho_0, \text{ where } \rho_0 \text{ is the identity map;}$$
$$\rho_1, \text{ where } \alpha_1\rho_1 = \alpha_2 \text{ and } i\sqrt{3}\rho_1 = i\sqrt{3};$$
$$\rho_2, \text{ where } \alpha_1\rho_2 = \alpha_3 \text{ and } i\sqrt{3}\rho_2 = i\sqrt{3};$$
$$\mu_1, \text{ where } \alpha_1\mu_1 = \alpha_1 \text{ and } i\sqrt{3}\mu_1 = -i\sqrt{3};$$
$$\mu_2, \text{ where } \alpha_1\mu_2 = \alpha_3 \text{ and } i\sqrt{3}\mu_2 = -i\sqrt{3};$$
$$\mu_3, \text{ where } \alpha_1\mu_3 = \alpha_2 \text{ and } i\sqrt{3}\mu_3 = -i\sqrt{3}.$$

b) S_3. The notation in (a) was chosen to coincide with the notation for S_3 in Example 4.1.

c)

Group lattice diagram

Field lattice diagram

46.7　F F T T T F F T F T

46.9　a) 4　b) 1　c) 36　d) 16　e) 0　f) 0　g) 0　h) 8

46.11　The splitting field of $(x^4 - 5x^2 + 6) \in \mathbf{Q}[x]$ is $\mathbf{Q}(\sqrt{2}, \sqrt{3})$, and the group is that of Example 40.4.

46.19　$G(K/(E \cap L)) = G(K/E) \vee G(K/L)$

Chapter 47

47.3　$\mathbf{Q}(\sqrt[4]{2}, i)$: $\sqrt[4]{2} + i$, $x^8 + 4x^6 + 2x^4 + 28x^2 + 1$;
　　　$\mathbf{Q}(\sqrt[4]{2})$: $\sqrt[4]{2}$, $x^4 - 2$;
　　　$\mathbf{Q}(i(\sqrt[4]{2}))$: $i(\sqrt[4]{2})$, $x^4 - 2$;
　　　$\mathbf{Q}(\sqrt{2}, i)$: $\sqrt{2} + i$, $x^4 - 2x^2 + 9$;
　　　$\mathbf{Q}(\sqrt[4]{2} + i(\sqrt[4]{2}))$: $\sqrt[4]{2} + i(\sqrt[4]{2})$, $x^4 + 8$;
　　　$\mathbf{Q}(\sqrt[4]{2} - i(\sqrt[4]{2}))$: $\sqrt[4]{2} - i(\sqrt[4]{2})$, $x^4 + 8$;
　　　$\mathbf{Q}(\sqrt{2})$: $\sqrt{2}$, $x^2 - 2$;
　　　$\mathbf{Q}(i)$: i, $x^2 + 1$;
　　　$\mathbf{Q}(i\sqrt{2})$: $i\sqrt{2}$, $x^2 + 2$;
　　　$\mathbf{Q}$: 1, $x - 1$

47.5　The group is cyclic of order 5, and its elements are

	ι	σ_1	σ_2	σ_3	σ_4
$\sqrt[5]{2} \to$	$\sqrt[5]{2}$	$\zeta(\sqrt[5]{2})$	$\zeta^2(\sqrt[5]{2})$	$\zeta^3(\sqrt[5]{2})$	$\zeta^4(\sqrt[5]{2})$

where $\sqrt[5]{2}$ is the real 5th root of 2.

47.7 The splitting field of $x^8 - 1$ over **Q** is the same as the splitting field of $x^4 + 1$ over **Q**, so a complete description is contained in Example 47.2. (This is the easiest way to answer the problem.)

47.9 a) $s_1^2 - 2s_2$ b) $s_1^2 s_2^2 - 4s_1^3 s_3 - 4s_2^3 + 18 s_1 s_2 s_3 - 27 s_3^2$

c) $\dfrac{s_1 s_2 - 3s_3}{s_3}$

Chapter 48

48.3 a) 16   b) 400   c) 2160

48.7 T T F T T F T T F

48.9 a) 4   b) *Partial answer*: $G(K/Q) \simeq \langle \mathbf{Z}_2 \times \mathbf{Z}_2, + \rangle$

48.11 3; $\mathbf{Z}_3$ is the splitting field.

Chapter 49

49.5 T T T F T F T F F T

i) $x^3 - 2x$ over **Q** gives a counterexample.

Appendix

A7 The notion of an "interesting property" has not been made precise; it is not well defined. Also, we work in mathematics with two-valued logic; a statement is either true or false, *but not both*. The assertion that not having an interesting property would be an interesting property seems to contradict this two-valued logic. We would be saying the integer both has and does not have an interesting property.

Notations

$\in, a \in S$	membership, 3
$\varnothing$	empty set, 3
$\notin, a \notin S$	nonmembership, 3
$\{x \mid P(x)\}$	set of all x such that $P(x)$, 3
$\mathbf{Z}$	integers, 4
	additive group of integers, 60
	ring of integers, 207
$\mathbf{Z}^+$	positive integers, 4
$\mathbf{Q}$	rational numbers, 4
	field of rational numbers, 207
$\mathbf{Q}^+$	positive rational numbers, 4
$\mathbf{R}$	real numbers, 4
	field of real numbers, 207
$\mathbf{R}^+$	positive real numbers, 4
$\mathbf{C}$	complex numbers, 4
	field of complex numbers, 207
$\equiv, a \equiv b \,(\text{mod } n)$	congruence, 7, 118
$*, a * b$	binary operation, 11
$\langle G, * \rangle$	group, 20
$\mathcal{G}_1, \mathcal{G}_2, \mathcal{G}_3$	group axioms, 20
e	identity element, 20
$a^{-1}, -a$	inverse of a, 30
$|S|$	order of S, 30
$B \subseteq A$	set inclusion, 31
$B \subset A$	subset $B \neq A$, 31

468

Index

reduced, 189
word problem, 198

Zassenhaus lemma, 147
zero, divisor of, 215

multiplicity of, 388
of a polynomial, 270
zero divisor, 215
Zorn's lemma, 348, 349,
    374